Cell Cycle Control and Plant Development

Cell Cycle Control and Plant Development

Edited by

DIRK INZÉ

Blackwell
Publishing

Blackwell Publishing editorial offices:
Blackwell Publishing Ltd, 9600 Garsington Road, Oxford OX4 2DQ, UK
Tel: +44 (0)1865 776868
Blackwell Publishing Professional, 2121 State Avenue, Ames, Iowa 50014-8300, USA
Tel: +1 515 292 0140
Blackwell Publishing Asia Pty Ltd, 550 Swanston Street, Carlton, Victoria 3053, Australia
Tel: +61 (0)3 8359 1011

First published 2007 by Blackwell Publishing Ltd

ISBN: 978-1-4051-5043-9

Library of Congress Cataloging-in-Publication Data

Cell cycle control and plant development / edited by Dirk Inzé.
p. cm.
ISBN-13: 978-1-4051-5043-9 (alk. paper)
ISBN-10: 1-4051-5043-2 (alk. paper)
1. Plant cell cycle. 2. Cyclin-dependent kinases. 3. Plant cells and tisues–Growth–Regulation. I. Inzé, D. (Dirk)
QK 725.C374 2007
571.8′4929—dc22

A catalogue record for this title is available from the British Library

Set in 10/12 pt Times
by Aptara Inc., New Delhi, India
Printed and bound in Singapore
by COS Printers Pte Ltd

The publisher's policy is to use permanent paper from mills that operate a sustainable forestry policy, and which has been manufactured from pulp processed using acid-free and elementary chlorine-free practices. Furthermore, the publisher ensures that the text paper and cover board used have met acceptable environmental accreditation standards.

For further information on Blackwell Publishing, visit our website:
www.blackwellpublishing.com

Contents

Contributors

Tom Beeckman Department of Plant Systems Biology, Flanders Institute for Biotechnology, and Department of Molecular Genetics, Ghent University, Technologiepark 927, B-9052 Gent, Belgium
Matthew L. Brown Department of Biological Sciences, Louisiana State University, Baton Rouge, LA 70803, USA
Christian Chevalier Unité Mixte de Recherche 619 sur la Biologie du Fruit, Institut Fédératif de Recherche 103 en Biologie Végétale Moléculaire, Institut National de la Recherche Agronomique, Université Bordeaux 1 et Université Victor Segalen-Bordeaux 2, BP 81, F-33883 Villenave d'Ornon Cedex, France
Michelle L. Churchman Department of Biological Sciences, Louisiana State University, Baton Rouge, LA 70803, USA
Sarah Jane Cookson Laboratoire d'Ecophysiologie des Plantes sous Stress Environnementaux, INRA-AGROM, F-34060 Montpellier, France
Marie Claire Criqui Institut de Biologie Moléculaire des Plantes du CNRS, 12, rue du Général Zimmer, F-67084 Strasbourg Cedex, France
Mátyás Cserháti Institute of Plant Biology, Biological Research Center, Hungarian Academy of Sciences, Temesvári krt. 62, H-6726 Szeged, Hungary
María de la Paz Sanchez Centro de Biología Molecular 'Severo Ochoa', Consejo Superior de Investigaciones Científicas, Universidad Autónoma de Madrid, Cantoblanco, E-28049 Madrid, Spain
Juan Carlos del Pozo Instituto Nacional de Investigación y Tecnología Agraria y Alimentaria, Dept. Biotecnología (INIA), Carretera de la Coruña Km 7, E-28040 Madrid, Spain
Bénédicte Desvoyes Centro de Biología Molecular 'Severo Ochoa', Consejo Superior de Investigaciones Científicas, Universidad Autónoma de Madrid, Cantoblanco, E-28049 Madrid, Spain
Lieven De Veylder Department of Plant Systems Biology, Flanders Institute for Biotechnology, and Department of Molecular Genetics, Ghent University, Technologiepark 927, B-9052 Gent, Belgium
Dénes Dudits Institute of Plant Biology, Biological Research Center, Hungarian Academy of Sciences, Temesvári krt. 62, H-6726 Szeged, Hungary
Andrew J. Fleming Department of Animal and Plant Sciences, University of Sheffield, Western Bank, Sheffield S10 2TN, United Kingdom
Larry C. Fowke Department of Biology, University of Saskatchewan, Saskatoon, SK, Canada S7N 5E2
Pascal Genschik Institut de Biologie Moléculaire des Plantes du CNRS, 12, rue du Général Zimmer, F-67084 Strasbourg Cedex, France

Christine Granier Laboratoire d'Ecophysiologie des Plantes sous Stress Environnementaux, INRA-AGROM, F-34060 Montpellier, France
Wilhelm Gruissem ETH Zurich, Institute of Plant Sciences, Universitästrasse 2, CH-8092 Zurich, Switzerland
Crisanto Gutierrez Centro de Biología Molecular 'Severo Ochoa', Consejo Superior de Investigaciones Científicas, Universidad Autónoma de Madrid, Cantoblanco, E-28049 Madrid, Spain
Gábor V. Horváth Institute of Plant Biology, Biological Research Center, Hungarian Academy of Sciences, Temesvári krt. 62, H-6726 Szeged, Hungary
Dirk Inzé Department of Plant Systems Biology, Flanders Institute for Biotechnology, and Department of Molecular Genetics, Ghent University, Technologiepark 927, B-9052 Gent, Belgium
Peter C.L. John Plant Cell Biology Group, Research School of Biological Sciences, Australian National University, Canberra City, ACT 2600, Australia
John C. Larkin Department of Biological Sciences, Louisiana State University, Baton Rouge, LA 70803, USA
Brian A. Larkins Department of Plant Sciences, University of Arizona, 303 Forbes Hall, Tucson, AZ 85721, USA
Margit Menges Institute of Biotechnology, University of Cambridge, Tennis Court Road, Cambridge CB2 1QT, United Kingdom
Pál Miskolczi Institute of Plant Biology, Biological Research Center, Hungarian Academy of Sciences, Temesvári krt. 62, H-6726 Szeged, Hungary
Bertrand Muller Laboratoire d'Ecophysiologie des Plantes sous Stress Environnementaux, INRA-AGROM, F-34060 Montpellier, France
James A.H. Murray Institute of Biotechnology, University of Cambridge, Tennis Court Road, Cambridge CB2 1QT, United Kingdom
Hong Nguyen Department of Plant Sciences, University of Arizona, 303 Forbes Hall, Tucson, AZ 85721, USA
Jeroen Nieuwland Institute of Biotechnology, University of Cambridge, Tennis Court Road, Cambridge CB2 1QT, United Kingdom
Elena Ramirez-Parra Centro de Biología Molecular 'Severo Ochoa', Consejo Superior de Investigaciones Científicas, Universidad Autónoma de Madrid, Cantoblanco, E-28049 Madrid, Spain
Paolo A. Sabelli Department of Plant Sciences, University of Arizona, 303 Forbes Hall, Tucson, AZ 85721, USA
Akie Shimotohno Department of Molecular Cell Biology, Utrecht University, Padualaan 8, NL-3584 CH Utrecht, The Netherlands
Francois Tardieu Laboratoire d'Ecophysiologie des Plantes sous Stress Environnementaux, INRA-AGROM, F-34060 Montpellier, France
Juan Antonio Torres Acosta Department of Biology, University of Saskatchewan, Saskatoon, SK, Canada S7N 5E2
Masaaki Umeda Graduate School of Biological Sciences, Nara Institute of Science and Technology, Takayama 8916-5, Ikoma, Nara 630-0101, Japan

Steffen Vanneste Department of Plant Systems Biology, Flanders Institute for Biotechnology, and Department of Molecular Genetics, Ghent University, Technologiepark 927, B-9052 Gent, Belgium
Kobe Vlieghe Department of Plant Systems Biology, Flanders Institute for Biotechnology, and Department of Molecular Genetics, Ghent University, Technologiepark 927, B-9052 Gent, Belgium
Hong Wang Department of Biochemistry, University of Saskatchewan, Saskatoon, SK, Canada S7N 5E5
Yongming Zhou Department of Biology, University of Saskatchewan, Saskatoon, SK, Canada S7N 5E2; National Key Laboratory of Crop Genetic Improvement, Huazhong Agricultural University, Wuhan 430070, China

Preface

The cell cycle is one of the most comprehensively studied biological processes, particularly given its importance for growth and development and in many human disorders. No other field as the cell cycle has benefited from an extensive interplay between research performed on a diversity of model organisms. Studies on yeast, worms, flies, frogs, mammals and plants have contributed to a kind of universal picture on how the basic cell cycle machinery is regulated. The use of these many divergent organisms also instigated an understanding of how evolution modified the basic cell cycle machinery in order to cope with the specific developmental and environmental challenges in each organism. However, much of this picture is based on experiments performed on single cells. Indeed, it is surprising to note that the role of the cell cycle machinery during development has received relatively little attention. To understand how, in different organisms, the basic cell cycle machinery integrates with development is an important scientific challenge for the coming years. In this book, the authors hope to convince the reader that plants offer exceptional opportunities to significantly contribute to such a challenge.

The first seven chapters review our current understanding of the basic machinery that propels cells from phase to phase during the cell cycle. The plant cell cycle is governed by Ser/Thr kinases, known as cyclin-dependent kinases (CDKs; reviewed in Chapterl), which require for activity the interaction with cyclins (Chapter 2). Furthermore, the activity of these CDKs is controlled by the interaction with CDK inhibitory proteins (Chapter 3), ubiquitin-dependent proteolysis (Chapter 4) and protein phosphorylation (Chapter 5). E2F/DP transcription factors (Chapter 6) also play a key role in cell cycle progression, and their activity is controlled by retinoblastoma-related (RBR) proteins (reviewed in Chapter 7).

The second part, consisting of eight chapters, summarizes our current knowledge on how the basic cell cycle machinery is integrated with development. Chapter 8 reviews how quiescent root pericycle cells enter the cell cycle during the initiation of lateral root formation. Chapter 9 discusses the role of the cell cycle during leaf development. Many plant organs contain polyploid cells resulting from a process called endoreduplication (Chapter 10). Endoreduplication also plays an important role in trichome development (Chapter 11), tomato fruit formation (Chapter 12) and endosperm development (Chapter 13). As plant development is orchestrated by plant hormones, it is not surprising that components of the cell cycle machinery are also directly modulated by hormones (Chapter 14). Finally, environmental stresses negatively affect the cell cycle and consequently growth (Chapter 15).

I sincerely hope that this book will stimulate scientists to further explore how cell cycle control works in concert with development.

Dirk Inzé

Annual Plant Reviews

A series for researchers and postgraduates in the plant sciences. Each volume in this series focuses on a theme of topical importance and emphasis is placed on rapid publication.

Titles in the series:

1. **Arabidopsis**
   Edited by M. Anderson and J.A. Roberts
2. **Biochemistry of Plant Secondary Metabolism**
   Edited by M. Wink
3. **Functions of Plant Secondary Metabolites and their Exploitation in Biotechnology**
   Edited by M. Wink
4. **Molecular Plant Pathology**
   Edited by M. Dickinson and J. Beynon
5. **Vacuolar Compartments**
   Edited by D.G. Robinson and J.C. Rogers
6. **Plant Reproduction**
   Edited by S.D. O'Neill and J.A. Roberts
7. **Protein–Protein Interactions in Plant Biology**
   Edited by M.T. McManus, W.A. Laing and A.C. Allan
8. **The Plant Cell Wall**
   Edited by J.K.C. Rose
9. **The Golgi Apparatus and the Plant Secretory Pathway**
   Edited by D.G. Robinson
10. **The Plant Cytoskeleton in Cell Differentiation and Development**
    Edited by P.J. Hussey
11. **Plant–Pathogen Interactions**
    Edited by N.J. Talbot
12. **Polarity in Plants**
    Edited by K. Lindsey
13. **Plastids**
    Edited by S.G. Moller
14. **Plant Pigments and their Manipulation**
    Edited by K.M. Davies
15. **Membrane Transport in Plants**
    Edited by M.R. Blatt
16. **Intercellular Communication in Plants**
    Edited by A.J. Fleming
17. **Plant Architecture and Its Manipulation**
    Edited by CGN Turnbull
18. **Plasmodeomata**
    Edited by K.J. Oparka
19. **Plant Epigenetics**
    Edited by P. Meyer

20. **Flowering and Its Manipulation**
Edited by C. Ainsworth
21. **Endogenous Plant Rhythms**
Edited by A. Hall and H. McWatters
22. **Control of Primary Metabolism in Plants**
Edited by W.C. Plaxton and M.T. McManus
23. **Biology of the Plant Cuticle**
Edited by M. Riederer
24. **Plant Hormone Signaling**
Edited by P. Hadden and S.G. Thomas
25. **Plant Cell Separation and Adhesion**
Edited by J.R. Roberts and Z. Gonzalez-Carranza
26. **Senescence Processes in Plants**
Edited by S. Gan
27. **Seed Development, Dormancy and Germination**
Edited by K.J. Bradford and H. Nonogaki
28. **Plant Proteomics**
Edited by C. Finnie
29. **Regulation of Transcription in Plants**
Edited by K. Grasser
30. **Light and Plant Development**
Edited by G. Whitelam
31. **Plant Mitochondria**
Edited by David C. Logan
32. **Cell Cycle Control and Plant Development**
Edited by D. Inzé

1 The growing family of plant cyclin-dependent kinases with multiple functions in cellular and developmental regulation

Dénes Dudits, Mátyás Cserháti, Pál Miskolczi and Gábor V. Horváth

1.1 Introduction

The production of new cells by division and their subsequent elongation are key cellular events in the plant life cycle. The development of plant organs of commercial value as a result of crop cultivation is directly determined by the frequency of cell division, parameters of the cell cycle and the number and size of the cells. The cell division activity is tightly controlled by molecular machinery that regulates the cell cycle progression in coordination with nutritional, hormonal, developmental and environmental signals. The orderly progression of cells through the various phases of the cell cycle and their appropriate responses to extracellular clues are governed by multiple regulatory mechanisms, including reversible protein phosphorylation, the interactions of proteins and specific protein degradation. Phosphorylation may alter the protein activity or subcellular localization, target proteins for degradation and influence the dynamic changes in protein complexes. Since protein phosphorylation is widely recognized to be the major mechanism controlling cell cycle progression, the roles of different kinases, and particularly the family of cyclin-dependent protein kinases (CDKs), are critical for cell division control. CDKs complexed with regulatory cyclin subunits drive the cell cycle by phosphorylating key target proteins that are required for cells to progress to the next phase of the cell cycle. After the early successes in the cloning of CDK genes (homologs of yeast *cdc2/CDC28* genes) from different plants (Feiler and Jacobs, 1990), extensive studies highlighted both the phylogenetically conserved and plant-specific features of CDKs of plant origin (reviewed by Mironov *et al.*, 1999; Mészáros *et al.*, 2000; De Veylder *et al.*, 2003; Dewitte and Murray, 2003; Inzé, 2005). Joubès *et al.* (2000) provided a list of plant CDKs with their phylogenetic, structural and functional properties. In the present review, we update the previous phylogenetic tree of plant CDKs and summarize the major functional characteristics of different kinases. Special attention is devoted to the role of transcriptional control in the CDK function timing. In view of the significance of transcriptional regulation, especially as it concerns the phase specificity of the gene expression of B-type CDKs, promoters are analyzed for the presence of potential transcriptional-factor-binding sites. Finally, an overview is given of the experimental findings relating to the involvement of this kinase family in the control of plant growth or organ development.

1.2 Structural diversity in the family of plant CDKs

Joubès *et al.* (2000) published the first comprehensive list of 46 putative plant CDKs from 23 species and from considerations of amino acid sequence similarities and characteristic motifs identified five classes (CDKA to CDKE). We now extend this list to include 152 CDKs from 41 plant species (Table 1.1). The newly added CDK proteins can be categorized into eight classes: CDKA to CDKG and the CDK-like kinases (CKLs), as classified for *Arabidopsis* by Menges *et al.* (2005). The CDK subfamilies form clearly recognizable branches on the phylogenetic tree and share characteristic motifs such as putative cyclin-binding domains (Figure 1.1). The interactions between CDKs and cyclins have been demonstrated via yeast two-hybrid tests, *in vitro* pull-down assays or immunoprecipitation of kinase complexes (Mészáros *et al.*, 2000; Joubès *et al.*, 2001; Nakagami *et al.*, 2002; Lee *et al.*, 2003). Accordingly, the predicted cyclin-binding motifs can serve as criteria for the classification of CDKs. As found earlier, class CDKA with the canonical PSTAIRE motif comprises the largest class, with 48 members (Table 1.1). Figure 1.1A presents a phylogenetic tree for class CDKA, showing the clustering of kinases from the same families in relation to taxonomy, as found earlier. Thus, the monocotyledonous grasses (rice, wheat and maize) form a characteristic branch. Similarly, species from families of dicotyledons such as Solanaceae, Fabaceae and Umbelliferae (Apiaceae) exhibit closely related CDKA kinases.

This extended data collection has led to the identification of a significant number of species that possess two types of CDKA kinases positioned in separate subgroups on the phylogenetic tree (Figure 1.1A). This divergence can be clearly demonstrated for monocot grass species. Figure 1.2 shows the amino acid alignments of four CDKA kinases from maize and three CDKA kinases from rice. Orysa;CDKA1 and Orysa;CDKA2 share 83% amino acid identity; only Orysa;CDKA1 (cdc20 s-1) complemented a temperature-sensitive yeast mutant of CDC28 (Hashimoto *et al.*, 1992). In the Leguminosae family, both alfalfa and soybean have functionally different PSTAIRE kinases, as indicated by the gene expression profiles or the complementation of *cdc28* yeast mutant (Hirt *et al.*, 1993; Miao *et al.*, 1993). The alfalfa CDKA kinases also differ in cyclin-interacting partners in yeast two-hybrid tests (Mészáros *et al.*, 2000). The currently available information supports nomenclature indicating two types of plant CDKA proteins, as used in Table 1.1 and Figure 1.1A. This discrimination should be based on the significant functional differences between the PSTAIRE kinase variants in the same species. Table 1.1 also includes several species, such as *Arabidopsis*, sugar beet, rapeseed, carrot and bean, which exhibit only a single representative of the CDKA group. In tobacco, four CDKA sequences are known at present. They share more than 95% amino acid identity and could be genomic or allelic variants of a single *CDKA* gene in this amphidiploid species.

Table 1.1 Searching for CDK genes in the public databases

Species	Gene abbreviation	Accession	Protein id	Gene name/ annotation
CDKA				
Allium cepa	*Allce;CDKA1;1*	AB006033	BAA21673	*cdc2* kinase
Antirrhinum majus	*Antma;CDKA1;1*	X97637	CAA66233	*cdc2a*
Antirrhinum majus	*Antma;CDKA2;1*	X97638	CAA66234	*cdc2b*
Arabidopsis thaliana	*Arath;CDKA1;1*	M59198	AAA32831	*cdc2*
Beta vulgaris	*Betvu;CDKA1;1*	Z71702	CAA96384	*cdc2*-related protein kinase
Brassica napus	*Brana;CDKA1;1*	U18365	AAA92823	*P34cdc2* homolog
Camellia sinensis	*Camsi;CDKA1;1*	AB247281	BAE80323	*CdkA*
Chenopodium rubrum	*Cheru;CDKA1;1*	Y10160	CAA71242	*cdc2*
Coffea arabica	*Cofar;CDKA1;1*	AJ496622	CAD43177	*cdc2*
Daucus carota	*Dauca;CDKA1;1*	AJ505322	CAD43850	*cdc2*
Glycine max	*Glyma;CDKA1;1*	M93139	Translated by sixpack	None
Glycine max	*Glyma;CDKA1;2*	M93140	Translated by sixpack	None
Helianthus annuus	*Helan;CDKA1;1*	AF321361	AAL37195	*cdc2a*
Helianthus tuberosus	*Heltu;CDKA2;1*	AY063462	AAL47481	*CdkA;1*
Juglans nigra × *Juglans regia*	*Jugni;CDKA1;1*	AJ439598	CAD29319	*cdc2a*
Lycopersicon esculentum	*Lyces;CDKA1;1*	Y17225	CAA76700	*cdc2A-1*
Lycopersicon esculentum	*Lyces;CDKA2;1*	Y17226	CAA76701	*cdc2A-2*
Medicago sativa	*Medsa;CDKA1;1*	M58365	AAB41817	*CDC2MS*
Medicago sativa	*Medsa;CDKA2;1*	X70707	CAA50038	*CDC2MSB*
Nicotiana tabacum	*Nicta;CDKA1;1*	L77082	AAB02567	*cdc2*
Nicotiana tabacum	*Nicta;CDKA1;2*	L77083	AAB02568	*cdc2*
Nicotiana tabacum	*Nicta;CDKA1;3*	D50738	BAA09369	*cdc2* homolog
Nicotiana tabacum	*Nicta;CDKA1;4*	AF289467	AAG01534	*CdkA:4*
Oryza sativa	*Orysa;CDKA2;1*	X60374	CAA42922	*Rcdc2-1*
Oryza sativa	*Orysa;CDKA1;1*	X60375	CAA42923	*Rcdc2-2*
Oryza sativa	*Orysa;CDKA1;2*	AC113930	AAN62789	*OJ1384D03.15*

(*Continued*)

Table 1.1 Searching for CDK genes in the public databases (*Continued*)

Species	Gene abbreviation	Accession	Protein id	Gene name/ annotation
Petroselinum crispum	*Petcr;CDKA1;1*	L34206	AAC41680	*cdc2*
Petunia hybrida	*Pethy;CDKA1;1*	Y13646	CAA73997	*cdc2*
Phaseolus vulgaris	*Phavu;CDKA1;1*	AF126737	AAD30494	*cdc2*
Physcomitrella patens	*Phypa;CDKA1;1*	AJ515321	CAD56245	*cdk-A*
Picea abies	*Picab;CDKA1;1*	X77680	CAA54746	*cdc2Pa*
Pinus contorta	*Pinco;CDA1;1*	X80845	CAA56815	*cdc2Pnc*
Pisum sativum	*Pissa;CDKA1;1*	X53035	CAA37207	*P34* protein (148 AA)
Pisum sativum	*Pissa;CDKA2;1*	AB008187	BAA33152	*cdc2*
Populus tremula × *Populus tremuloides*	*Poptr;CDKA1;1*	AF194820	AAK16652	*CDC2* homolog
Scutellaria baicalensis	*Scuba;CDKA1;1*	AB205131	BAE06268	*cdka1*
Scutellaria baicalensis	*Scuba;CDKA2;1*	AB205132	BAE06269	*cdka2*
Sesbania rostrata	*Sesro;CDKA1;1*	Z75661	CAA99991	*cdc2* kinase homolog
Solanum tuberosum	*Soltu;CDKA1;1*	U53510	AAA98856	*P34* kinase
Triticum aestivum	*Triae;CDKA1;1*	U23409	AAD10483	*cdc2TaA*
Triticum aestivum	*Triae;CDKA2;1*	U23410	AAD10484	*cdc2TaB*
Vigna acunitifolia	*Vigac;CDKA1;1*	M99497	AAA34241	*CDC2*
Vigna radiata	*Vigra;CDKA1;1*	AF129886	AAD30506	Cell division control protein 2
Vigna unguiculata	*Vigun;CDKA1;1*	X89400	CAA61581	*CDC2*
Zea mays	*Zeama;CDKA1;1*	M60526	AAA33479	Protein cdc2 kinase
Zea mays	*Zeama;CDKA2;1*	AY104247	Translated by sixpack	*PCO133106* mRNA
Zea mays	*Zeama;CDKA1;2*	BT016935	Translated by sixpack	*E04912703H09.c* mRNA
Zea mays	*Zeama;CDKA2;2*	BT018184	Translated by sixpack	*EL01N0558C03.c* mRNA
CDKB				
Antirrhinum majus	*Antma;CDKB1;1*	X97639	CAA66235	*cdc2c*
Antirrhinum majus	*Antma;CDKB2;1*	X97640	CAA66236	*cdc2d*
Arabidopsis thaliana	*Arath;CDKB1;1*	D10851	BAA01624	*CDC2b*
Arabidopsis thaliana	*Arath;CDKB1;2*	NM_001036430	NP_001031507	*CDKB1;2*

Arabidopsis thaliana	*Arath;CDKB2;1*	AC015450	AAG51960	*F14G6.14*
Arabidopsis thaliana	*Arath;CDKB2;2*	NM_101946	NP_173517	*CDKB2;2*
Camellia sinensis	*Camsi;CDKB1;1*	AB247279	BAE80321	*cdkb*
Chenopodium rubrum	*Cheru;CDKB1;1*	AJ278885	CAC17703	*cdc2b*
Dunaliella terciolata	*Dunte;CDKB2;1*	AF038570	AAD08721	*DUNCDC2*
Glycine max	*Glyma;CDKB2;1*	AY439096	AAS13369	*CDKB*
Helianthus tuberosus	*Heltu;CDKB2;1*	AY063463	AAL47482	*CdkB1;1*
Lycopersicon esculentum	*Lyces;CDKB1;1*	AJ297916	CAC15503	*cdkB1*
Lycopersicon esculentum	*Lyces;CDKB2;1*	AJ297917	CAC15504	*cdkB2*
Medicago sativa	*Medsa;CDKB1;1*	X97315	CAA65980	*cdc2MsD*
Medicago sativa	*Medsa;CDKB2;1*	X97317	CAA65982	*cdc2MsF*
Medicago sativa	*Medsa;CDKB2;2*	DQ136188	AAZ30705	*MEDSJA; Cdc2MsF*
Mesembryanthemum crystallinum	*Mescr;CDKB2;1*	AB015182	BAA28778	*cdc2* related
Nicotiana tabacum	*Nicta;CDKB1;1*	AF289465	AAG01532	*CdkB1-1*
Nicotiana tabacum	*Nicta;CDKB1;2*	AF289466	AAG01533	*CdkB1-2*
Oryza sativa	*Orysa;CDKB1;1*	AP003349	BAD82176	*P0674H09.14*
Oryza sativa	*Orysa;CDKB1;2*	D64036	BAA19553	*SS224*
Populus tremula × *Populus tremuloides*	*Poptr;CDKB2;1*	AY307372	AAP73784	*CDKB*
Scutellaria baicalensis	*Scuba;CDKB1;1*	AB205133	BAE06270	*Cdkb*
Sorghum bicolor	*Sorbi;CDKB1;1*	AY144442	Translated by sixpack	*PPTALRE* motif
Triticum aestivum	*Triae;CDKB1;1*	BT009641	Translated by sixpack	*PPTALRE* motif
Triticum aestivum	*Triae;CDKB1;2*	BT009182	Translated by sixpack	*PPTALRE* motif
Zea mays	*Zeama;CDKB1;1*	AY106440	Translated by sixpack	*PPTALRE* motif
Zea mays	*Zeama;CDKB1;2*	AY106029	Translated by sixpack	*PPTALRE* motif
CDKC				
Arabidopsis thaliana	*Arath;CDKC;1*	AL360334	CAB96683	*cdc2*-like protein kinase
Arabidopsis thaliana	*Arath;CDKC;2*	NM_125895	NP_201301	*CDKC;2*
Brassica rapa	*Brana;CDKC;1*	AC155344	Translated by sixpack	*PITAIRE* motif
Brassica rapa	*Brana;CDKC;2*	AC166741	Translated by sixpack	*PITAIRE* motif

(*Continued*)

Table 1.1 Searching for CDK genes in the public databases (*Continued*)

Species	Gene abbreviation	Accession	Protein id	Gene name/ annotation
Lycopersicon esculentum	*Lyces;CDKC;1*	AJ294903	CAC51391	*CDKC*
Lycopersicon esculentum	*Lyces;CDKC;2*	BT014075	Translated by sixpack	*PITAIRE* motif
Medicago sativa	*Medsa;CDKC;1*	X97314	CAA65979	*cdc2MsC*
Oryza sativa	*Orysa;CDKC;1*	AC105773	AAT47442	*OJ1562_H01.5*
Oryza sativa	*Orysa;CDKC;2*	AP004326	BAD88154	*OJ1294_F06.25*
Physcomitrella patens	*Phypa;CDKC;1*	AJ428950	CAD21952	*PITAIRE* motif
Pisum sativum	*Pissa;CDKC;1*	X56554	CAA39904	*P34* kinase-related protein
Zea mays	*Zeama;CDKC;1*	AY107067	Translated by sixpack	*PITAIRE* motif
CDKD				
Arabidopsis thaliana	*Arath;CDKD1;1*	NM_106028	NP_177510	*CAK2At*
Arabidopsis thaliana	*Arath;CDKD1;2*	NM_105345	NP_176847	*CDKD1;2*
Arabidopsis thaliana	*Arath;CDKD1;3*	NM_101666	NP_173244	*CDKD1;3*
Lycopersicon esculentum	*Lyces;CDKD;1*	BT013748	Translated by sixpack	*NFTALRE* motif
Medicago sativa	*Medsa;CDKD;1*	AF302013	Translated by sixpack	*NFTALRE* motif
Oryza sativa	*Orysa;CDKD1;1*	X58194	CAA41172	*cdc2+/CDC28*-related protein kinase
Ostreococcus tauri	*Ostta;CDKD;1*	AY675096	AAV68598	*CAK*
CDKE				
Arabidopsis thaliana	*Arath;CDKE;1*	AB005234	BAB10454	*cdc2*-like protein kinase
Medicago sativa	*Medsa;CDKE;1*	X97316	CAA65981	*cdc2MsE*
Mesembryanthemum crystallinum	*Mescr;CDKE;1*	AB015181	BAA28777	*cdc2* related
Oryza sativa	*Orysa;CDKE;1*	AC018727	AAG46164	*OSJNBa0056G17.11*
Zea mays	*Zeama;CDKE;1*	BT018448	Translated by sixpack	*SPTAIRE* motif
CDKF				
Arabidopsis thaliana	*Arath;CDKF;1*	AB009399	BAA28775	*cak1At*
Euphorbia esula	*Eupes;CDKF;1*	AF230740	AAF34804	*CDK*-activating kinase
Oryza sativa	*Orysa;CDKF;1*	AP004784	BAD61885	Putative CDK-activating kinase 1At
Glycine max	*Glyma;CDKF;1*	AY439095	AAS13368	*CDK*-activating kinase

CDKG				
Arabidopsis thaliana	*Arath;CDKG;1*	NM_125732	NP_201142	*AT5G63370*
Arabidopsis thaliana	*Arath;CDKG;2*	BT000694	AAL32755	*AT1G67580*
Oryza sativa	*Orysa;CDKG;1*	XM_466592	XP_466592	Putative PITSLRE alpha 2-1
Oryza sativa	*Orysa;CDKG;2*	XM_472963	XP_472963	None
Zea mays	*Zeama;CDKG;1*	AY112336	Translated by sixpack	None
Zea mays	*Zeama;CDKG;2*	BT018272	Translated by sixpack	None
CDKL				
Arabidopsis thaliana	*Arath;CDKL;1*	NM_123304	NP_198758	*At5g39420*
Arabidopsis thaliana	*Arath;CDKL;2*	NM_106093	NP_177573	*At1g74330*
Arabidopsis thaliana	*Arath;CDKL;3*	NM_101725	NP_173302	*At1g18670*
Arabidopsis thaliana	*Arath;CDKL;4*	NM_118423	NP_194025	*At4g22940*
Arabidopsis thaliana	*Arath;CDKL;5*	NM_001036933	NP_001032010	*At5g44290*
Arabidopsis thaliana	*Arath;CDKL;6*	NM_001035873	NP_001030950	*At1g03740*
Arabidopsis thaliana	*Arath;CDKL;7*	NM_124464	NP_199899	*At5g50860*
Arabidopsis thaliana	*Arath;CDKL;8*	NM_111377	NP_187156	*At3g05050*
Arabidopsis thaliana	*Arath;CDKL;9*	NM_104338	NP_175862	*At1g54610*
Arabidopsis thaliana	*Arath;CDKL;10*	NM_104567	NP_176083	*At1g57700*
Arabidopsis thaliana	*Arath;CDKL;11*	NM_100832	NP_172431	*At1g09600*
Arabidopsis thaliana	*Arath;CDKL;12*	NM_202395	NP_974124	*At1g71530*
Arabidopsis thaliana	*Arath;CDKL;13*	NM_117069	NP_192739	*At4g10010*
Arabidopsis thaliana	*Arath;CDKL;14*	NM_103	NP_174637	*At1g33770*
Arabidopsis thaliana	*Arath;CDKL;15*	NM_104184	NP_175713	*At1g53050*
Beta vulgaris	*Betvu;CDKL;1*	Z71703	CAA96385	*cdc2*-like protein kinase
Beta vulgaris	*Betvu;CDKL;2*	AJ277243	CAB89665	*crk1*
Lotus japonicus	*Lotja;CDKL;1*	AP004904	Translated by sixpack	None
Medicago truncatula	*Medtr;CDKL;1*	AC147364	Translated by sixpack	None
Medicago truncatula	*Medtr;CDKL;2*	AC146747	ABE89393	*MtrDRAFT_AC146747g18v1*
Medicago truncatula	*Medtr;CDKL;3*	AC130801	ABE83648	*MtrDRAFT_AC130801g5v1*

(*Continued*)

Table 1.1 Searching for CDK genes in the public databases (*Continued*)

Species	Gene abbreviation	Accession	Protein id	Gene name/ annotation
Medicago truncatula	*Medtr;CDKL;4*	AC149804	ABE79143	*MtrDRAFT_AC149804g37v1*
Medicago truncatula	*Medtr;CDKL;5*	AC141113	ABE94624	*MtrDRAFT_AC141113g16v1*
Medicago truncatula	*Medtr;CDKL;6*	AC148397	ABE81196	*MtrDRAFT_AC148397g6v1*
Oryza sativa	*Orysa;CDKL;1*	AK101089	Translated by sixpack	None
Oryza sativa	*Orysa;CDKL;2*	NM_193805	NP_918694	Putative *CRK1* protein
Oryza sativa	*Orysa;CDKL;3*	AK064909	Translated by sixpack	None
Oryza sativa	*Orysa;CDKL;4*	NM_188289	NP_913178	Putative *CRK1* protein
Oryza sativa	*Orysa;CDKL;5*	NM_186098	NP_910987	Putative *CRK1* protein (*cdc2*-related kinase 1)
Oryza sativa	*Orysa;CDKL;6*	XM_479002	XP_479002	Putative *CDK CDC2C*
Oryza sativa	*Orysa;CDKL;7*	AK122094	Translated by sixpack	None
Oryza sativa	*Orysa;CDKL;8*	AK105621	Translated by sixpack	None
Oryza sativa	*Orysa;CDKL;9*	AK067238	Translated by sixpack	None
Oryza sativa	*Orysa;CDKL;10*	XM_479750	XP_479750	Putative *CRK1* protein
Oryza sativa	*Orysa;CDKL;11*	AK072696	Translated by sixpack	None
Oryza sativa	*Orysa;CDKL;12*	AK068916	Translated by sixpack	None
Oryza sativa	*Orysa;CDKL;13*	XM_466234	XP_466234	Putative *CRK1* protein
Oryza sativa	*Orysa;CDKL;14*	XM_466235	XP_466235	Putative *CRK1* protein
Oryza sativa	*Orysa;CDKL;15*	AK121206	Translated by sixpack	None
Oryza sativa	*Orysa;CDKL;16*	AK100360	Translated by sixpack	None
Oryza sativa	*Orysa;CDKL;17*	AK100088	Translated by sixpack	None
Oryza sativa	*Orysa;CDKL;18*	XM_463674	XP_463674	Putative *CRK1* protein

Nucleotide sequences were searched in the NCBI nucleic acid database with blastn ($E < 1 \times 10^{-10}$). For each sequence, the species name, nucleic acid and protein accession number were noted, as well as the gene name or protein product from the individual gene. Some of the nucleic acid sequences did not have any links to protein sequences in their annotation; therefore, the protein sequence was derived from the DNA sequence by using the *sixpack* program in Linux. The genes were given a new name following the nomenclature used in the text. The *CDKA* genes were renamed according to separation into two groups (see the text).

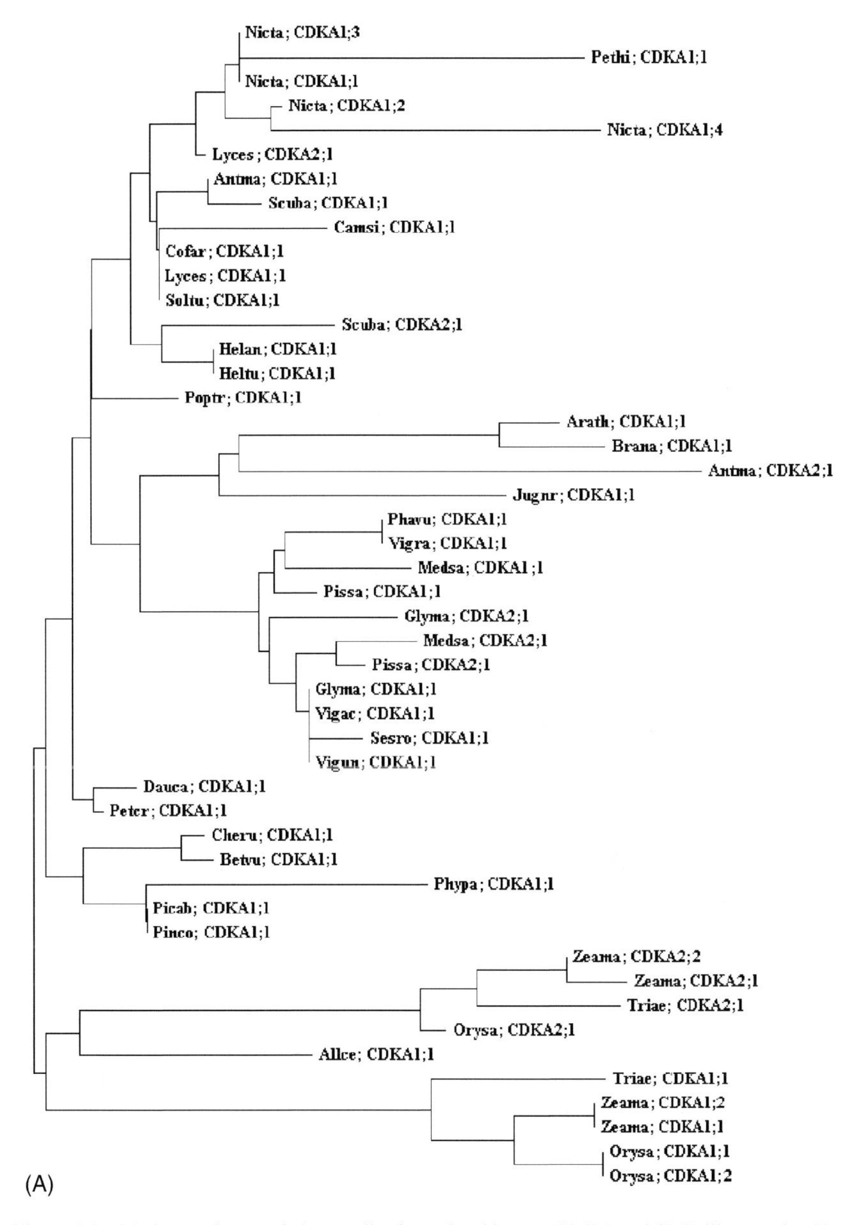

Figure 1.1 Phylogenetic tree of plant cyclin-dependent kinases (CDKs) and CDK-like proteins. Trees were drawn for the CDKA, B, C, D, E and F families. A protein multiple alignment was constructed for each of the three sequence sets, and a section of the alignment was selected which contained variable as well as conserved parts. The TreeCon program was used to draw the trees (Van de Peer and De Wachter, 1994). Bootstrap analysis was used with 1000 samples taken. The trees were ordered in such a way that the monocot species and the dicot ones would group away from each other as much as possible. (A) A-type CDKs with PSTAIRE motif; (B) B-type, plant-specific CDKs with PPTALRE or PPTTLRE motif; (C) C-type CDKs with PITAIRE motif and the closely related CKLs (CDK-LIKE); (D) D- and F-type kinases regulating CDK activators, E-type CDKs with SPTAIRE motif and G-type CDKs with PLTSLRE motif.

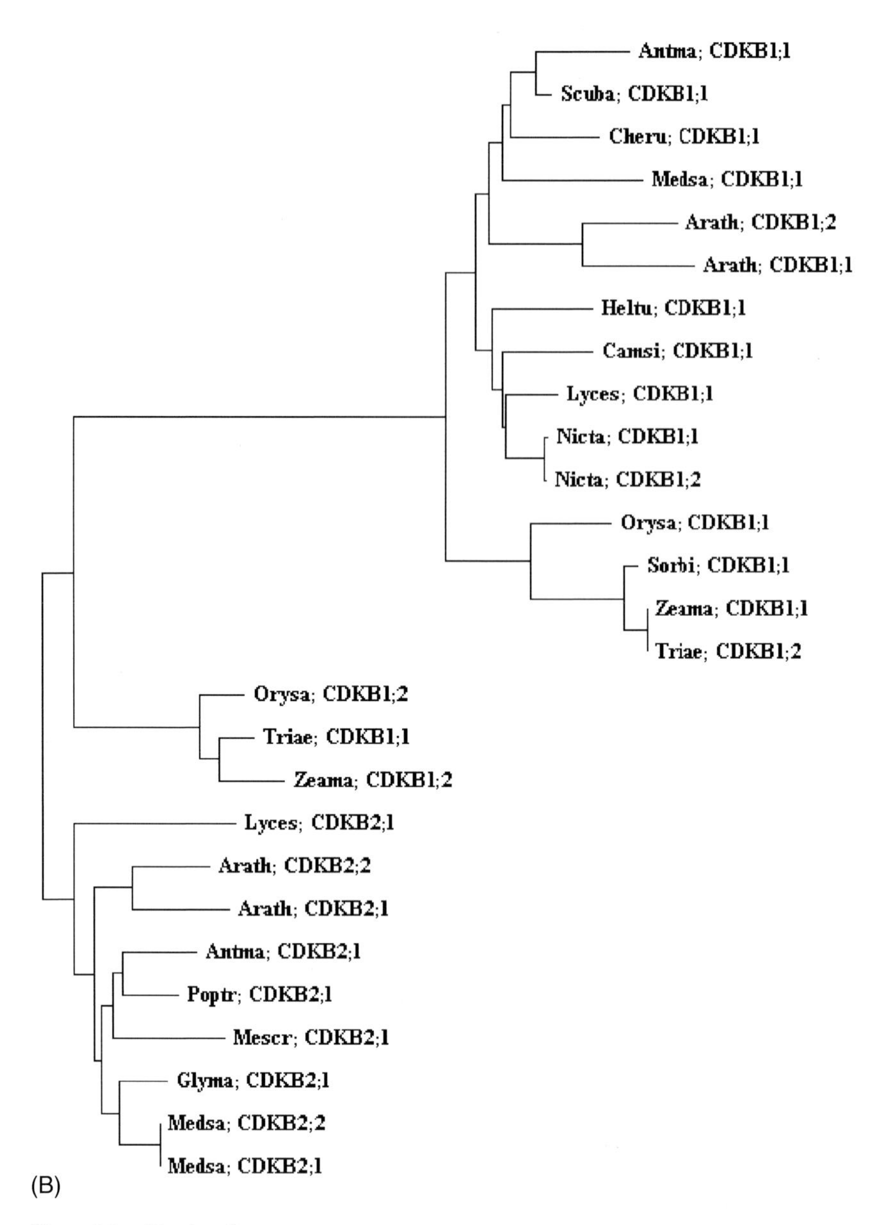

Figure 1.1 (*Continued*)

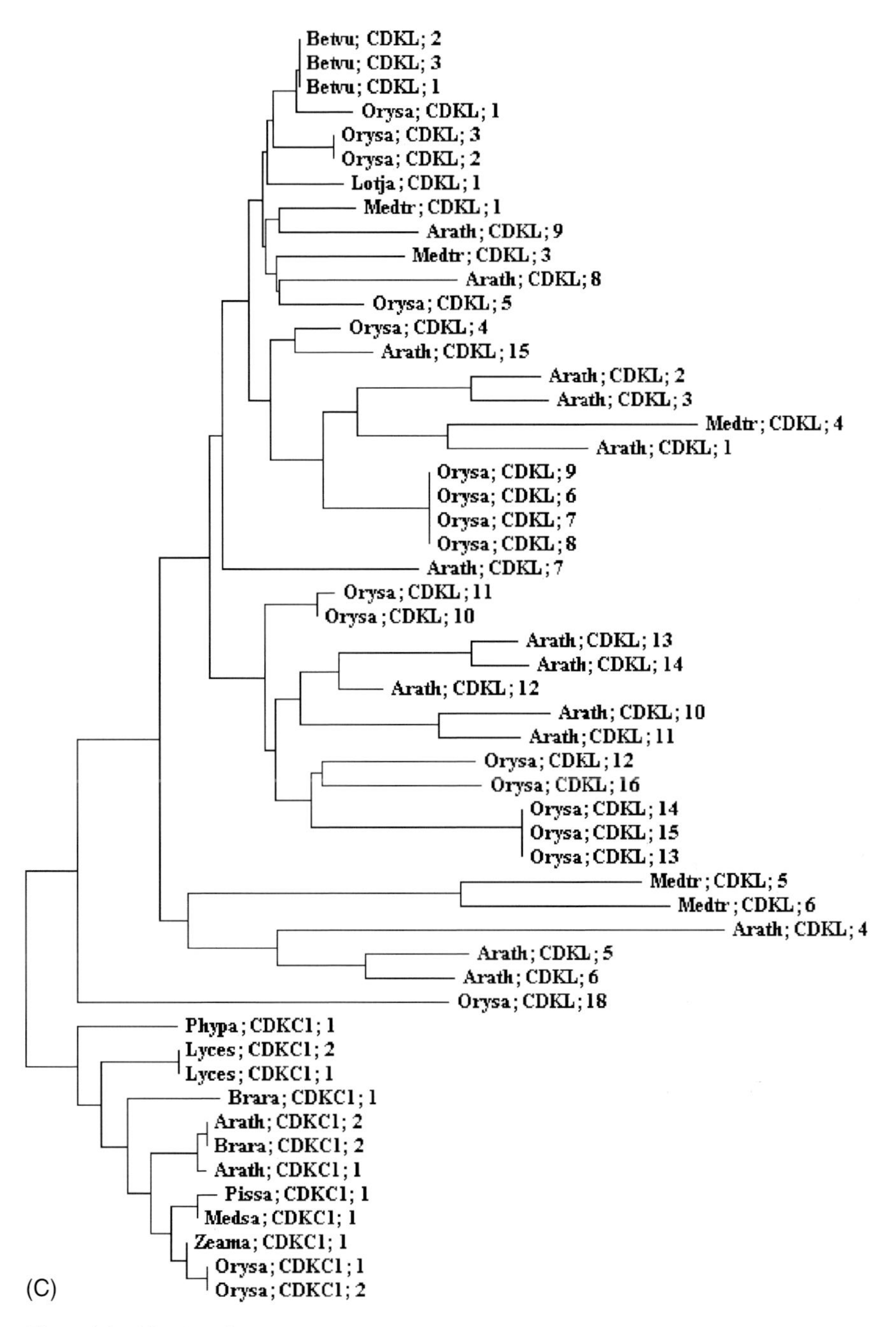

Figure 1.1 (*Continued*)

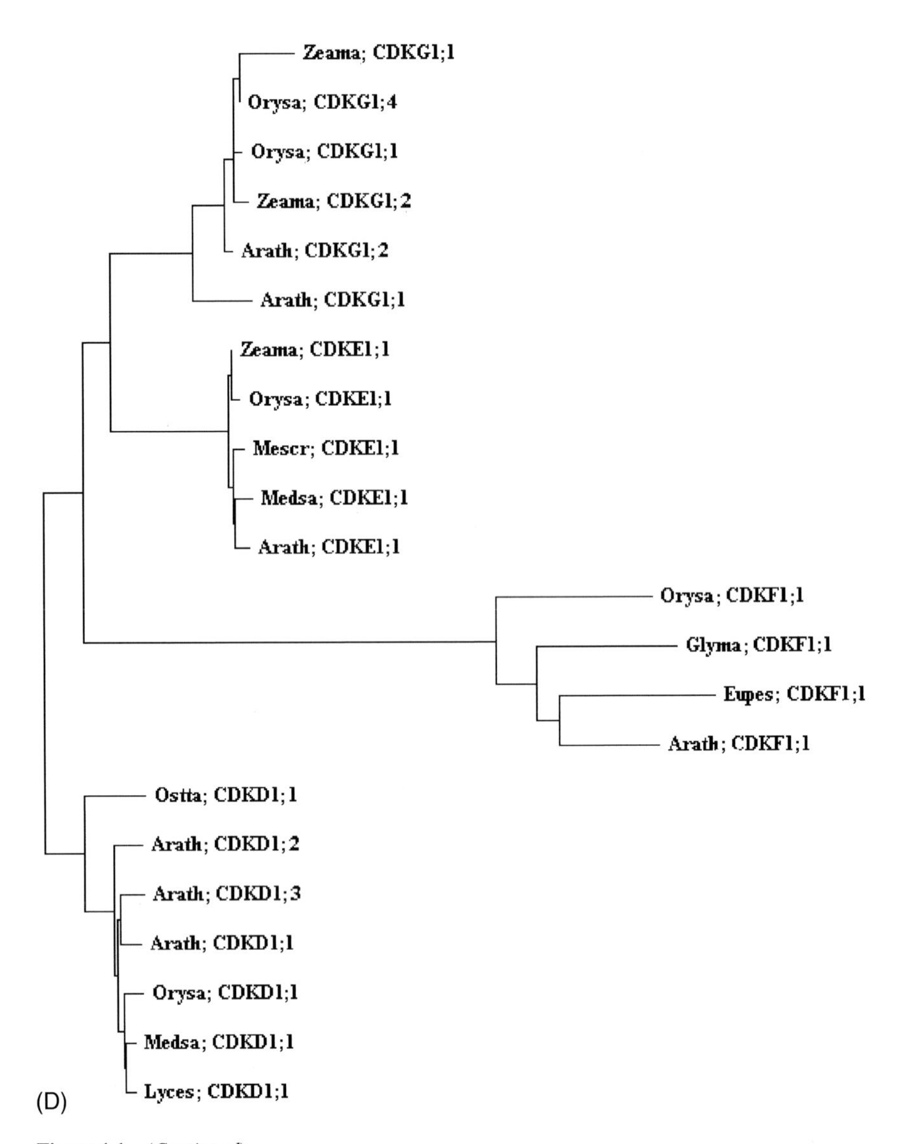

Figure 1.1 (*Continued*)

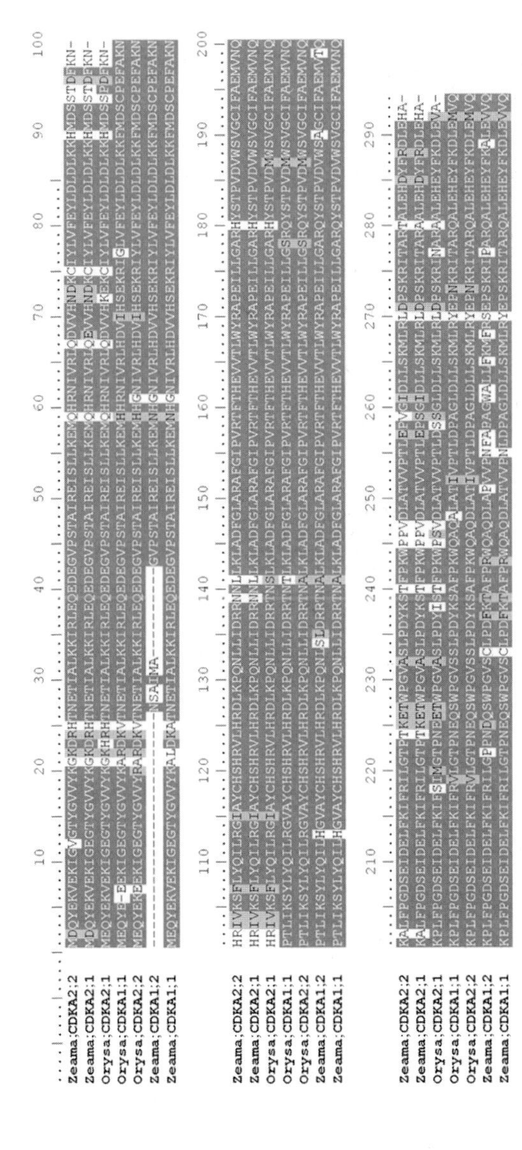

Figure 1.2 Variation of amino acid sequences in CDKA proteins from monocot grasses as maize and rice.

Class CDKB, the second largest group of plant CDKs, demands special attention: these kinases are plant specific and involved in divergent cellular functions. The discovery of the non-PSTAIRE Cdc2 kinases in *Arabidopsis* (cdc2bAt: PPTALRE; Segers *et al.*, 1996) and in *Medicago* (Cdc2MsD: PPTALRE; Cdc2MsF: PPTTLRE; Magyar *et al.*, 1997) revealed a new category of CDKs lacking counterparts in yeast or animal cells. Joubès *et al.* (2000) proposed that PPTALRE Cdc2 kinases should be placed in class CDKB1 and PPTTLRE Cdc2 kinases in class CDKB2. The present class of CDKB is definitely divided into two major groupings on the phylogenetic tree (Figure 1.1B). For dicot species such as *Arabidopsis*, tobacco and alfalfa, both class CDKB1 and class CDKB2 are represented by two members. Interestingly, class CDKB2 has not been recognized so far among monocot grass species that have CDKB kinases located on separate branches of the phylogenetic tree: as regards the PPTALRE motif, all belong in class CDKB1. However, it is quite visible in the CDKB tree that a small group of monocot *CDKB1* genes anomalously form a conspicuous cluster, which is attracted to the CDKB2 clade.

Class CDKC kinases carrying the PITAIRE cyclin-binding motifs are the closest homologs of the metazoan CDK9 proteins. Phylogenetic tree analysis shows that this group of CDKs is closely related to the CKLs, as demonstrated by Menges *et al.* (2005) for *Arabidopsis* (Figure 1.1C). In this class of CDKs, a separate subgroup of monocot grasses can be recognized on the phylogenetic tree.

Classes CDKD and CDKF are members of a kinase network that regulates CDK activity via a phosphorylation cascade. The CDK-activating kinases (CAKs), i.e. class CDKD, can be activated by CDKF kinases. Classes CDKD and CDKF are located on separate branches of the phylogenetic tree (Figure 1.1D). In contrast, classes CDKE and CDKG are less well separated, but they differ in putative cyclin-binding motifs. The first class CDKE kinases were detected in alfalfa; this unique plant sequence harbors the SPTAIRE motif (Magyar *et al.*, 1997). The CDKG kinases were identified in *Arabidopsis* as proteins containing the PLTSLRE motif (Menges *et al.*, 2005). The present search revealed CDKG kinases in rice and maize, in each case with two variants.

A novel class of CKL genes has been discovered by the Affymetrix microarray analysis (Menges *et al.*, 2005). In *Arabidopsis,* 15 CKL proteins formed a cluster well separated from the class CDKC kinases (Figure 1.1C). As indicated in Table 1.1, the present search resulted in the identification of 42 CKL proteins from a variety of plant species.

1.3 Expression profiles of CDK genes: structures and functions of promoters

The transcriptional regulation of the cell cycle control gene is an essential component in the complex regulatory machinery ensuring cell cycle progression and its coordination with morphogenetic programs and extracellular, environmental inputs. Periodic gene expression is considered to be a general feature of the eukaryotic cell cycle from yeast to man, which is dependent on multiple regulatory elements within the promoter regions and their interactions with activator or repressor proteins

(McKinney and Heitz, 1991; Koch and Nasmyth, 1994; Müller, 1995; Zhu *et al.*, 2004; Ito, 2005). Early Northern hybridization studies revealed that the accumulation of *CDKB* transcripts in G2/M cells is a unique, plant-specific characteristic of the non-PSTAIRE kinase genes. The other CDK genes in the families A, C, D and E display constitutive expression in synchronized cells and different plant organs (Martinez *et al.*, 1992; Hemerly *et al.*, 1993; Magyar *et al.*, 1993, 1997; Segers *et al.*, 1996; Sauter, 1997; Umeda *et al.*, 1999; Joubès *et al.*, 2001; Sorrell *et al.*, 2001; Freeman *et al.*, 2003; Esponosa-Ruiz *et al.*, 2004). The recent genome-wide transcript profiling of the core *Arabidopsis* cell cycle via Affymetrix microarrays confirmed that most CDK-related kinase genes were relatively constantly expressed in synchronized cells (Menges *et al.*, 2005). These studies refined previous data on the expression pattern of the *CDKB* genes, showing an early G2 peak for the *CDKB1;1/2* genes and a mitotic peak for the *CDKB2;1/2* genes. In alfalfa cells synchronized with aphidicolin, the *Medsa;CDKB1;1* (cdc2MsD) gene was expressed earlier in the G2 phase than the *Medsa,CDKB2;1* (cdc2MsF) gene in the late G2/M phase. In these cells, the *Medsa;CDKB2;1* transcripts were more abundant than those synthesized from the *Medsa;CDKB1;1* genes (Magyar *et al.*, 1997). Menges *et al.* (2005) also compared the expression profiles of the CDK genes in various *Arabidopsis* tissues and during the cell cycle reentry controlled by the sucrose level in the culture medium. Like most core cell cycle regulators, the CDK genes were expressed at a similar level in the analyzed tissues. The *CDKA* genes exhibited the highest level of expression. Tissue specificity was observed in the case of genes from the CKL3 family. Readdition of sucrose activated the CDK genes with different profiles. The *CDKB* genes were actively transcribed in the S phase, while *CDKC;1*, *CDKD;1* and *CDKG;1* responded already in the G0/G1 phase. In synchronized tobacco BY-2 cells, the C-type CDK genes demonstrated phase-specific transcript accumulation, with a peak during the late M and the early S phase (Breyne *et al.*, 2002).

Modification of the CDK gene expression by plant hormones or growth regulators plays a role in the coordination between developmental events and cell division. The treatment of leaf mesophyll protoplasts or roots of alfalfa with 2,4-dichlorophenoxyacetic acid (2,4-D) activated the *CDKA* gene (Magyar *et al.*, 1993). In that study, the accumulation of CDK transcripts correlated with an elevated level of [^{3}H]-thymidine incorporation and the expression of a replication-dependent histone H3 variant gene. When the *Arabidopsis CDKA* promoter was fused with the GUS reporter gene, it also responded to hormone treatment of the roots (Hemerly *et al.*, 1993). Cytokinins increased the GUS activity driven by the *CDKA;1* promoter in the pericycle and parenchyma cells of the vascular cylinder. This *CDKA* promoter was very active in 2,4-D-induced callus tissues. Abscisic acid (ABA) treatment completely inhibited this promoter. Wounding of the leaves resulted in the induction of *CDKA;1* around the damaged region.

The initiation of lateral root primordia from pericycle founder cells is dependent on auxin-stimulated cell cycle activation (reviewed by Casimiro *et al.*, 2003; Vanneste *et al.*, 2005). In gene expression studies, *CDKA1* transcripts were detected constitutively in the presence of auxin transport inhibitor, while the B-type CDK genes were strongly activated by auxin during lateral root initiation (Himanen *et al.*,

2002, 2004). Auxin-mediated lateral root formation requires nitric oxide (NO), as recently demonstrated in tomato (Correa-Aragunde *et al.*, 2004). The NO-mediated induction of lateral root primordia could be related to the increased expression of kinase (*CDKA;1*) and cyclin (*CYCD3:1*) and the downregulation of the CDK inhibitor (*ICK2*) genes (Correa-Aragunde *et al.*, 2006). The differential response of the CDK gene expression to hormonal signals includes the brassinolide-induced *Arath*;*CDKB1;1* gene expression in dark-grown seedlings (Yoshizumi *et al.*, 1999). In developing tomato fruits and vegetative organs, the accumulation of CDK transcripts is dependent on the cell division activity. High expressions of *Lyces;CDKB1;1* or *CDKB2;1* and *CDKC;1* genes were detected during anthesis and the subsequent 5 days following anthesis. The *Lyces;CDKA;1* gene remained active in the later stage of fruit development. The activity of the *Lyces;CDKC;1* gene was shown to be influenced by hormones and by the sugar supply (Joubès *et al.*, 2001). Interestingly, a reduction of the fruit load greatly increased the expressions of both the *CDKB2;1* and *CYCB;1* genes of tomato (Baldet *et al.*, 2006).

Several lines of experimental evidence strongly indicate a role of cell cycle phase-specific control in the transcription of the regulatory CDK genes, particularly during G2/M events. Transcripts of B-type CDKs accumulate in G2/M cells. Similar regulation is characteristic for the mitotic cyclin and kinesin genes, which have been shown to have promoters with a common *cis*-acting element, called MSA (Mitosis-Specific Activator), which serves as a binding site for Myb regulatory proteins (Tréhin *et al.*, 1999; Ito *et al.*, 2001; Araki *et al.*, 2004; Vanstraelen *et al.*, 2006). A recent analysis of the mitosis-specific promoter of the alfalfa CDK (*Medsa;CDKB2;1*) detected several putative *cis*-elements, such as MSA, ABRE (ABA-responsive element), E2F-binding site, ERE (ethylene-responsive element) and TCA-box (wound-responsive element) (Zhiponova *et al.*, 2006). Functional characterization of a short (360 bp) promoter region of this gene in transgenic lines showed that the listed putative *cis*-elements could simultaneously ensure G2/M phase-specific gene activity and responses to wounding, ethylene and 2,4-D treatments. The combination of *cis*-elements predicted to regulate cell cycle phase specificity as well as hormonal, environmental and developmental responses is a general feature of CDK promoters in plants. As listed in Table 1.2, both *Arabidopsis* and rice promoters of CDK genes share a characteristic set of *cis*-elements identified in a 2-kb-long region of their promoters by the Patch motif search program and a search of the PLACE transcription-factor-binding site databases (Higo *et al.*, 1999). The presented computer-based prediction supports the view that CDK promoters serve as integrators of a variety of signals generated by hormones (ABA, ethylene and auxin), light or developmental programs. *Cis*-elements involved in the cell cycle control, such as the E2F-binding site, MSA and specific Myb elements, have been predicted from sequence data. Interestingly, the MSA sequence motif can be recognized in all types of CDK promoters, representing either constitutive or G2/M-specific ones. Myb-like proteins have been identified as MSA-binding factors that can function as activators or repressors (Tréhin *et al.*, 1999; Ito *et al.*, 2001). The tobacco Myb transcription factor (NtmybA2) is activated by a specific cyclin–CDK complex in the G2/M phase as a consequence of the removal of the inhibitory effect of the

Table 1.2 Putative *cis*-elements in the 2-kb-long promoter region of *Arabidopsis* and rice CDK genes

	MSA	E2F-binding site	Myb	ABA response	Ethylene response	Light response	Circadian rhythm	ALFIN-1	APETALA3/ AGAMOUS
Orysa;CDKA1;1	334, 1361, 1421, 1803	89, 583		566, 1216		1368	335, 366, 1783	559	
Orysa;CDKA1;2	40, 433	141, 401, 1134, 1598	503, 587, 1213, 1258, 1289, 1498			737, 848, 955	434, 722	1915	
Orysa;CDKA2;1	1889	358, 433, 1475, 1539, 1675	498, 1238, 1415	515, 1464, 1510		1230	1947	349, 1822, 1492	1938
Orysa;CDKB1;1	148	749	959, 1196, 1684, 1782	621, 632, 669	84, 1650		641, 1024		1198
Orysa;CDKB1;2	122, 154, 242, 1921				252, 526, 633, 1341	361, 679, 751, 1285, 1725, 1958	211	1040	538, 546
Orysa;CDKC1;1	477, 1129	370	1150, 1756	690, 955	717	1916	423	1770	
Orysa;CDKC1;2	796, 1430		107, 1548		629	804, 1736, 1835	1028	1310	
Orysa;CDKD1;1			1488		1543	524, 556, 629, 1612, 1858, 1991	1156	104, 145	
Orysa;CDKE1;1	235, 587, 693, 759, 1762, 1676, 1168	330	660			250, 1668	623, 1826		
Orysa;CDKF1;1	598, 626, 1574, 1755		292, 1579	1979		1020, 1315, 1380, 1539			
Orysa;CDKG1;1	453		21, 259, 384, 1141	11, 22, 1811		201, 1178		915, 1018, 1221	
Orysa;CDKG1;2		54, 362, 1344, 1597	153, 838, 1246, 1418, 1511			704			
Arath;CDKA1;1	207, 1094, 1500, 1851, 1977		68, 207			49, 308, 890, 1180	73, 255, 109		
Arath;CDKB1;1	624, 1000, 1358, 1452, 1905	151	707, 849		169, 1666, 1997	24, 1134, 1072, 1079, 1769	532, 541, 638, 1238, 1359, 1484		

(Continued)

Table 1.2 Putative *cis*-elements in the 2-kb-long promoter region of *Arabidopsis* and rice CDK genes (*Continued*)

	MSA	E2F-binding site	Myb	ABA response	Ethylene response	Light response	Circadian rhythm	ALFIN-1	APETALA3/ AGAMOUS
Arath;CDKB1;2	125		361		18, 1921, 1955	21, 162, 243, 772, 819, 1528	1709		1598
Arath;CDKB2;1	153, 193, 571, 752, 1172, 1284, 1429			227	171	65, 890, 1072, 1570, 1575, 1611	1413	1160	
Arath;CDKB2;2	173, 955, 1239, 1294, 1877	41, 1580	987		198, 513	64, 263, 1831	1725		
Arath;CDKC1;1	83, 98, 139, 175, 273, 712, 1276, 1418, 1436, 1510, 1794	669	117			948, 1376, 1800, 1848			662
Arath;CDKC1;2	1067		831				876, 1940, 1986	331, 408, 1001	1824
Arath;CDKD1;1	113, 362, 653, 1550, 1574, 1615, 1626	1131	1080		1427		1139, 1235, 1868		
Arath;CDKD1;2	342, 461, 802, 1341, 1807		461, 1314				70, 87, 1189	201, 612, 652	
Arath;CDKD1;3	87, 487, 773, 866, 1602	132, 808	861, 1798			376	101, 1679	1009	
Arath;CDKE1;1	43, 97, 1179, 1949		82				1136, 1435		1261
Arath;CDKF1;1	257, 1090, 1769, 1804		154, 1023, 1695			299, 508, 649	286, 536, 941, 1269, 1701		1704
Arath;CDKG1;1	25, 69, 177, 241, 570, 993, 1020		141					789	
Arath;CDKG1;2	125, 274, 833, 931, 1720	50, 1475, 1907	437, 1947	327				975, 988, 1247, 1670	

We provide the number and position of potential transcription factor-binding sites. The position is indicated as base pairs from the ATG start in the 5′ direction. ABA response: ABADESI1 (RTACGTGGCR), ABREATCONSENSUS (YACGTGGC), ABREATRD22 (RYACGTGGYR), ABREAZMRAB28 (GCCACGTGGG), ABREBZMRAB28 (TCCACGTCTC), ABREMOTIFAOSOSEM (TACGTGTC), ABREOSRAB21 (ACGTSSSC), ABREZMRAB28 (CCACGTGG), ACGTABREMOTIFAOSOSEM (TACGTGTC); auxin response: AUXREPSIAA4 (KGTCCCAT), AUXRETGA2GMGH3 (TGACGTGGC); circadian rhythm: CIACADIANLELHC (CAANNNNATC), EVENINGAT (AAAATATCT); E2F-binding site: E2F1OSPCNA (GCGGGAAA), E2FANTRNR (TTTCCCGC), E2FBNTRNR (GCGGCAAA), E2FCONSENSUS (WTTSSCSS); ethylene response: ERELEE4 (AWTTCAAA); light response: INRNTPSADB (YTCANTYY); MSA element: (ACAAACGGTAA, AGACCGTTG, YCYAACGGYY); Myb element: CCA1ATLHCB1 (AAMAATCT), MYB1LEPR (GTTAGTT), MYBPLANT (MACCWAMC); developmental element: AGMOTIFNTMYB2 (AGATCCAA), ALFIN-1 (GTGGTGCTG), APETALA3/AGAMOUS (CCATTTTTGG).

C-terminal region (Araki *et al.*, 2004). In addition to MSA, other Myb elements have also been found in plant CDK promoters. In mammalian cells, the interaction of B-Myb with a CDK promoter depends on the E2F function. Since the B-Myb gene expression is under the control of E2F in the G1/S phase, E2Fs have a role in linking regulatory mechanisms determining gene activities in the G1/S and G2/M phases (Zhu *et al.*, 2004).

Recent characterizations of E2Fa-DPa-overproducing plants revealed the enhanced activation of DNA replication and subsequent stimulation of mitotic division or endoreduplication (De Veylder *et al.*, 2002; Kosugi and Ohashi, 2003). In these transformants, Boudolf *et al.* (2004b) detected the upregulation of *CDKB1;1* transcription. Mutational analysis of the E2F-binding site in the *CDKB1;1* promoter further supported the idea that expression of this mitotic kinase gene is regulated through an E2F *cis*-acting element. As shown in Table 1.2, our research on potential E2F-binding sites in CDK promoters demonstrated additional CDKs belonging to different classes of the kinase family (Arath;CDKB2;2; CDKC, D and G kinases). In rice, the promoters of several A-type CDK genes harbor a putative E2F-binding site. In tobacco cells, the overexpression of E2FB/DPA proteins caused increases in the levels of both CDKA;1 and CDKB1;1 proteins (Magyar *et al.*, 2005).

An unexpected finding from the computer-based identification of potential *cis*-elements in CDK promoters is that almost all promoters have several copies of light-responsive and circadian rhythm elements (Table 1.2). On the basis of this information we searched for expression data for *Arabidopsis* CDKs as they respond to the circadian clock. The analysis of circadian microarray data (NASCArray experiment, reference number NASCARRAY-108, Dr Kireon Edwards, University of Warwick, UK) allowed identification of the expression profiles of selected CDKs.

As shown in Figure 1.3, the reference *TOC1* gene exhibited circadian fluctuation with two expression peaks at 36 and 60 h. The *Arabidopsis CDKG1;2* gene displayed a very similar expression pattern, with moderate amplitudes. The level of expression of the *CDKA1;1* gene was high, but a pattern similar to that in the case of the *TOC1* gene could not be recognized. The *CDKB2;1* gene demonstrated a reduced activity in seedlings used for RNA isolation in this experiment. Similar expression profiles were characteristic for the majority of CDK genes. These preliminary data point to a need for the design of a more specific experimental system for studies on the potential involvement of the cell cycle genes in the circadian clock in plants.

Regulation of the expression of cell cycle genes (*cyclin B1, cdc2, wee1*) by circadian control has been demonstrated in the regenerating mouse liver (Matsuo *et al.*, 2003). The progression of colonic epithelial cells from G1 to S phase occurs between 12:00 noon and 6:00 p.m., the period during which the *cyclin D* and *E* genes are highly expressed, while the expressions of CDK inhibitor proteins (p16 and p21) are reduced (Griniatsos *et al.*, 2006). The circadian timing system adapts the cyclic physiology to geophysical time; this has been intensively studied in *Arabidopsis* (Salomé and McLung, 2005). The core oscillator is dependent on light perception and signaling, which control the clock genes (*CCA1*, *LHY* and *TOC1*). A proposal for the coupling of the circadian and cell division cycles is supported by research data pointing to the diurnal variation of plant hormones or metabolites known to

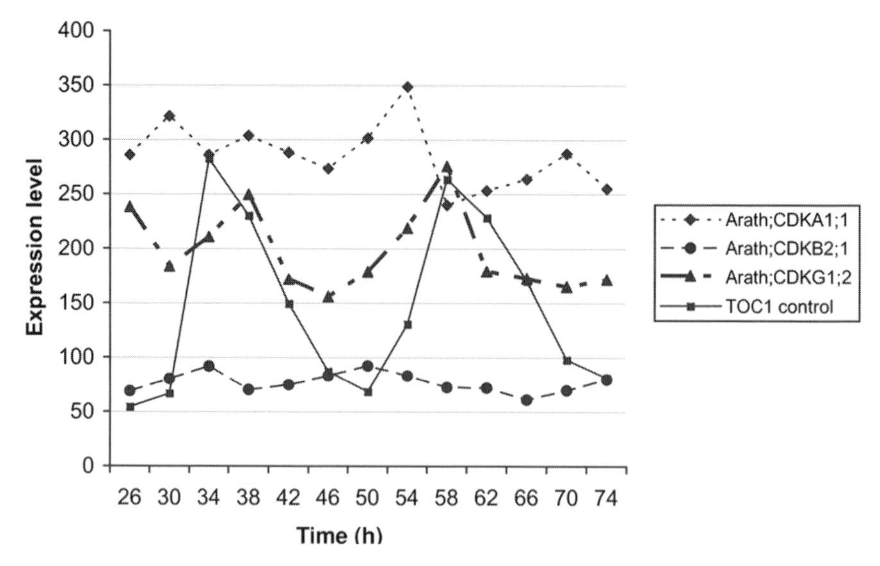

Figure 1.3 Expression of selected *Arabidopsis* CDK genes during circadian rhythm. Details are given in the text.

influence the cell cycle in plants directly. Nováková *et al.* (2005) reported diurnal changes in the levels of cytokinins, indoleacetic acid and ABA. In *Arabidopsis* rosettes, the sucrose, glucose, fructose and starch levels follow a diurnal cycle in coordination with the diurnal expression patterns of sugar-responsive genes (Bläsing *et al.*, 2005).

1.4 Diverse functions of CDK protein complexes in multiple regulatory mechanisms

The first functional data proving the existence of CDKs in plants were furnished by the complementation of yeast *Cdc2/CDC28* mutants with plant cDNAs (Colasanti *et al.*, 1991; Ferreira *et al.*, 1991; Hirt *et al.*, 1991). As indicated by the nomenclature of these cell cycle regulatory kinases, it is widely accepted that their activities are determined by specific interactions with defined cyclin partners. In contrast, only sporadic data, primarily from yeast two-hybrid tests, are available to construct a CDK–cyclin interaction map and to show experimentally that the kinase activity is dependent on the presence of a cyclin protein (Mészáros *et al.*, 2000; Stals *et al.*, 2000; Healy *et al.*, 2001; Joubès *et al.*, 2001).

In vitro pull-down assays or the coexpression of kinase and cyclin partners in insect cells have been utilized for studies on CDK–cyclin interactions. These methodologies permitted demonstration of the formation of active kinase complexes between the CYCD4;1–CDKA;1 and CYCD4;1–CDKB2;1 proteins of *Arabidopsis* (Kono *et al.*, 2003). The specific binding of rice CDKB2;1 kinase to B-type cyclins (CYCB2;1 and CYCB2;2) has been revealed by the expression of tagged proteins in insect cells (Lee *et al.*, 2003). These complexes are active in the histone H1 kinase assay. Using a baculovirus expression system, Nakagami *et al.* (2002) proved that Nicta;CYCD3;3 complexed with Nicta;CDKA;3 can phosphorylate histone H1 and retinoblastoma proteins. Formation of an active kinase complex has been demonstrated in *in vivo* tests by the coexpression of the *Medsa;CDKC;1* kinase and *Medsa;CYCT;1* cyclin genes in *Arabidopsis* protoplasts (Fülöp *et al.*, 2005). Kinases listed as cyclin dependent in Table 1.1 are identified as CDKs on the basis of protein sequence data, without experimental demonstration that the kinase function is dependent on the complex formation with a cyclin partner. Figure 1.4A presents an example to show that the *Medicago* Medsa;CDKA;1 kinase requires cyclin (Medsa;CYCD3;1) or a cell extract for the histone H1 phosphorylation function. The immunoprecipitated Medsa;CDKA1 protein complex contains the [^{35}S]-methionine-labeled Medsa;CYCD3;1 protein translated *in vitro* by using the TNT-Coupled Reticulocyte Lysate System (Figure 1.4B). Since the complex formation between selected CDKs and cyclins is a very basic regulatory component in the CDK function, there is a need for a better understanding of the nature of these interactions and the factors influencing complex formation.

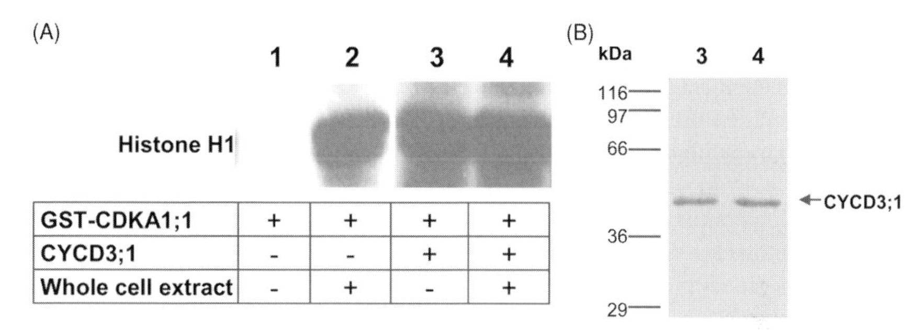

GST-CDKA1;1	+	+	+	+
CYCD3;1	-	-	+	+
Whole cell extract	-	+	-	+

Figure 1.4 Phosphorylation function of CDKA1;1 kinase is dependent on the presence of cyclin (CYCB3;1) partner. The Medsa;CDKA1;1 kinase can bind the *in vitro* translated Medsa;CYCD3;1 protein while developing a histone kinase activity: (A) Protein kinase assays of immunoprecipitates for the Medsa;CDKA1;1 complexes using histone H1 (arrow) as an added substrate. The *Escherichia coli* expressed GST–CDKA1;1 protein (1 μg) was assayed either by *in vitro* translated CYCD3;1 (5 μl) or by whole cell extract (60 μg) from alfalfa cell suspension, as indicated. The CYCD3;1 protein was translated in presence of L-[^{35}S]methionine by the rabbit TNT-Coupled Reticulocyte Lysate System (Promega, Madison, WI, USA). Plant protein extract was prepared in extraction buffer containing 25 mM Tris-HCl pH 7.6, 15 mM $MgCl_2$, 15 mM EGTA, 75 mM NaCl, 60 mM β-glycerophosphate, 1 mM DTT, 0.1% NP40, 0.1 mM Na_3VO_4, 1 mM NaF, 1 mM PMSF and protease inhibitors (Complete; Roche, Mannheim, Germany). The CDKA1;1-associated complexes were immunoprecipitated with 1 μg polyclonal antibody against its C-terminal peptide, and the details of the antibody are described in Magyar *et al.* (1997). The immunoprecipitation was allowed to proceed for 1 h at 4°C. The complexes were immobilized on protein A agarose, the washing and the kinase reactions were performed according to Magyar *et al.* (1997). (B) The Medsa;CDKA1;1 coimmunoprecipitates the *in vitro* translated Medsa;CYCD3;1. The same fractions of binding reactions (3 and 4) containing the *in vitro* translated CYCD3;1 L-[35 S]methionine-labeled products were analyzed on a 10% denaturing gel and visualized by autoradiography. The arrow indicates the position of *in vitro* translated cyclin, and molecular mass standards are indicated in kilodaltons on the left.

Plant CDKs are themselves under phosphorylation control, so they serve as substrates for kinases and phosphatases. The activation of plant CDKs requires the phosphorylation of a conserved threonine residue in the T-loop region by CDKD proteins that are functionally related to vertebrate CAKs. *Arabidopsis* CDKDs interact with At;CYCH;1 and are activated by CDKF-type kinases. The role of this phosphorylation cascade in cell cycle control and transcription has been reviewed by Umeda *et al.* (2005). CDK functions are further controlled by the phosphorylation status of the Thr14 and Tyr15 residues. Inhibitory phosphorylation of Tyr15 by the WEE1 kinases also plays a regulatory role in plant cell cycle progression. Maize, *Arabidopsis* and tomato cDNAs that encode WEE1 homologs have been cloned (Sun *et al.*, 1999; Sorrell, 2002; Gonzalez *et al.*, 2004). The overexpression of *Arabidopsis* WEE1 kinase in fission yeast led to cell cycle arrest and elongation of the cells (Sorrell, 2002). In synchronized tobacco BY-2 cells, the WEE1 transcripts accumulate in the late G1 and S phases (Gonzalez *et al.*, 2004).

The phosphorylation-based regulation of proteins is expected to be reversible, and therefore both kinases and counteracting protein phosphatases are needed to ensure the structural and functional characteristics of regulatory proteins. Since the dual-specificity phosphatase CDC25 dephosphorylates Thr15 and Tyr15 residues in yeast and mammalian cells, identification of the plant homolog of CDC25 has been a central goal of plant cell cycle research. The first functional studies were carried out on the expression of *Schizosaccharomyces pombe* CDC25 phosphatase in plant cells (Bell *et al.*, 1993; Zhang *et al.*, 1996; McKibbin *et al.*, 1998). Spcdc25-expressing BY-2 cells divided prematurely at a small cell size through shortening of the G2 phase at high CDKB1 activity, independent of the endogenous cytokinins (Orchard *et al.*, 2005). Zhang *et al.* (2005) proposed a role for cytokinins in G2/M cells in the activation of CDK by dephosphorylation of Tyr. Transgenic tobacco plants constitutively synthesizing SpCDC25 phosphatase displayed an altered cell size in both the roots and the leaves (Bell *et al.*, 1993; McKibbin *et al.*, 1998). Landrieu *et al.* (2004) recently identified an *Arabidopsis* CDC25-related protein with tyrosine phosphatase activity and stimulation of the CDK function. This protein most probably plays no role in the cell cycle but rather in arsenate detoxification (D. Inzé, personal communication). The involvement of phosphatases in the cell cycle control extends to other classes of these enzymes. The inhibition of PP2A phosphatase in cultured alfalfa cells resulted in an early increase in Medsa;CDKB2;1 (cdcMsF) kinase activity, disturbing the coordination between chromosomal and microtubule events during the G2/M transition (Ayaydin *et al.*, 2000).

The kinase functions of CDK–cyclin complexes can be modified by interplay with other structural and regulatory proteins. As components of these complexes, the SUC1/CKS1 proteins serve as docking factors. The *Arabidopsis* CKS1At protein can bind both A- and B-type CDKs (De Veylder *et al.*, 1997). Overexpression of this protein resulted in reductions in leaf size and root growth rates in transgenic *Arabidopsis* plants (De Veylder *et al.*, 2001). The CDK inhibitor proteins are currently at the focus of plant cell cycle research, and the Kip-related proteins (ICK/KRPs) of *Arabidopsis* and alfalfa are considered to mediate hormonal and stress signals toward the cell division cycle (Verkest *et al.*, 2005b; Pettkó-Szandtner *et al.*, 2006).

The *Arabidopsis* inhibitor proteins are involved in the control of endoreduplication (Verkest *et al.*, 2005a).

For a better understanding of the role of plant CDKs, further studies are needed to identify the substrates of these kinase complexes. At present, only limited *in vitro* phosphorylation data and yeast two-hybrid interaction tests highlight potential targets of the CDKs. The retinoblastoma-related proteins (RBRs) are considered to be central regulators of the G1–S phase transition through the negative regulation of E2F/DF-dependent transcription in plants (Gutiérrez, 1998; Durfee *et al.*, 2000; Inzé, 2005). The function of the RBRs depends on their phosphorylation status; CDKs can inactivate these pocket domain proteins by the production of a hyperphosphorylated form and release of the transcription factor E2F. Interactions of plant cyclins or E2F and RBRs have been demonstrated in several experiments (Ach *et al.*, 1997; Huntley *et al.*, 1998; den Boer and Murray, 2000; Roudier *et al.*, 2000; Gutiérrez *et al.*, 2002). The NtRBR1 protein from tobacco has been shown to be phosphorylated *in vitro* by the tobacco CYCD/Cdc2 complex and the CYCD3;3/CDKA complex, the NtRBR1 phosphorylation activity of which is detectable only during the G1 to S phases in synchronized tobacco BY-2 cells (Nakagami *et al.*, 1999, 2002). These results are consistent with a previous report detecting the cell-cycle-regulated kinase activity *in vivo* with ZmRBR1 as substrate, this complex effectively phosphorylating the C-terminal domain of the plant RBRs *in vitro* (Boniotti and Gutiérrez, 2001). All these studies emphasize the role of A-type CDKs in the control of the RBR/E2F pathway. Utilizing the findings of previous publications (Mészáros *et al.*, 2000; Kono *et al.*, 2003) with data indicating interactions between the B-type mitotic CDKs (Medsa;CDKB2;1, Arath;CDKB2;1) and D-type cyclins (Medsa;CYCD4, Arath;CYCD4;1, respectively), we have tested whether the alfalfa mitotic CDK can phosphorylate the C-terminal fragment of the MsRBRA;1 protein *in vitro*. As demonstrated by recent phosphorylation experiments, alfalfa RBR protein can act as substrate for both A- and B-type CDKs (P. Miskolczi, G.V. Horváth, L. Bakó, and D. Dudits, unpublished data). *In vitro* studies on the phosphorylation of the MsRBR C-terminal protein by the Medsa;CDKA1 kinase related that it is sensitive toward the inhibitory function of KRP from alfalfa (Pettkó-Szandtner *et al.*, 2006).

The plant CDK inhibitors such as the Kip-related proteins (ICK/KRPs) comprise a functionally significant class of CDK substrates. In *Arabidopsis*, ICK2 was found to be regulated posttranscriptionally through phosphorylation by CDKs and proteosome degradation (Verkest *et al.*, 2005a). In the proposed model, both CDKs can phosphorylate ICK2, and its stability is under the control of CDKB1;1 phosphorylation. Inhibition of CDKA;1 complexes by ICK2 during mitosis resulted in premature initiation of the endoreduplication cycle.

Kinesins as microtubule-associated proteins are involved in spindle formation, chromosome movement, cytokinesis and the organization of plant-specific mitotic structures (for a review, see Vanstraelen *et al.*, 2006). Kinesin-like proteins from *Arabidopsis* (KCA1 and KCA2) have two CDKA consensus phosphorylation sites. Vanstraelen *et al.* (2004) demonstrated that phosphorylation can result in conformational changes.

1.5 Developmental consequences of altered CDK functions

Transgenic approaches provide efficient methodology for the functional characterization of CDK genes and their encoded proteins. The wild-type and dominant negative (DN) mutant of *Arabidopsis CDKA* genes were overexpressed under the regulation of the constitutive CaMV 35S promoter in *Arabidopsis* and tobacco (Hemerly *et al.*, 1995). Replacement of the D147 residue with an N147 residue blocked the binding of ATP, which resulted in the loss of kinase activity without effective interaction with a cyclin. The overproduction of wild-type *CDKA;1* in *Arabidopsis* did not interfere with normal development. Regeneration of *Arabidopsis* overexpressing *DNCDKA* was unsuccessful, while a tobacco transformant with a moderate level of *DNCDKA* exhibited a lower frequency of division of enlarged cells without alteration of the morphogenic processes. When DNcdc2a kinase was expressed by an albumin promoter during embryo formation, the reduction in the rate of cell division caused a variety of abnormalities in the apical–basal embryo pattern (Hemerly *et al.*, 2000). A similar strategy has been tested in maize by overexpression of the *CDKA–WT* and the *CDKA–DN* genes in the endosperm under the control of the 27-kD γ-zein promoter (Leiva-Neto *et al.*, 2004). The increased activity of the wild-type kinase did not alter the endoreduplication. In contrast, the high levels of mutant kinase (CDKA–DN) expression reduced the endoreduplication without affecting the cell size or the starch and storage protein accumulation. These studies on tobacco and maize overexpressing DN mutant CDKs lend support to the conclusion that the CDK function in organogenesis or morphogenesis is dependent on the cell type.

The involvement of CDKB1;1 kinase in developmental regulation has been studied by different approaches. Yoshizumi *et al.* (1999) expressed the antisense *CDC2B;1* gene under the control of an inducible promoter. After dexamethasone induction, the transgenic seedlings developed short hypocotyls and open cotyledons in the dark. The short hypocotyl phenotype resulted from the reduction in cell size. Overexpression of a DN allele of *CDKB1;1* (*CDKB1;1N161*) caused reductions in size of the cotyledons and first leaves, with a decrease in their stomatal index (Boudolf *et al.*, 2004a). The lowered B-type kinase activity caused aberrant stomatal cells by blocking the cell cycle in the G2 phase (Boudolf *et al.*, 2004a). The endoreduplication was enhanced in plants that overexpressed this mutant of CDKB1;1 (Boudolf *et al.*, 2004b). Yamaguchi *et al.* (2003) made use of the overexpression of rice CDK activating kinase – CAK – gene (*Orysa;CDKD;1;1*) to achieve a transient alteration in CDK activity in tobacco leaf explants. The upregulation of CDK activities allowed auxin-induced callus formation in the absence of cytokinin.

To date, only a limited number of mutants with altered CDK functions have been identified and characterized, primarily in *Arabidopsis*. One exception is the mutation in the HUA ENHANCER 3 (*HEN3*) gene which encodes CDKE kinase (Wang and Chen, 2004). This mutation disturbed the specification of the stamen and the carpel; furthermore, staminoid petals were found in the flowers. In this mutant an increased abundance of *AG*, *AP1* and *AP2* mRNAs was detected, suggesting a negative regulatory role of HEN3 in transcription. The E-type CDKs with the SPTAIRE cyclin-binding motif are homologs of the mammalian CDK8 kinases.

The T-DNA insertional mutant of the *Arabidopsis CDKA;1;1* gene was found to be lethal in homozygosity (Iwakawa *et al.*, 2006). This loss-of-function mutation blocked the proliferation of generative cells during male gametogenesis. As shown by fluorescence microscopy of haploid cdka-1 pollen grains, pollen mitosis I (PMI) was normal, but the second division of the generative cell (PMII) was disturbed. The cdka-1 sperm-like cell could not divide, but it reached the G2 phase, and was therefore fertile, and it fused with the egg cell to initiate embryogenesis. These studies support the conclusion that CDKA;1;1 kinases can function as positive regulators of cell division and contribute to the development of the male gametophyte, embryo and endosperm.

1.6 Perspectives

CDKs are currently at the focus of plant cell cycle research, and significant progress has been achieved in the revelation of their functions, not only in the regulation of cell cycle progression but also in the control of transcription, growth and development. It is clear from the present review that the list of plant species used for studies on CDKs is growing. The genome-sequencing programs have opened up new possibilities for gene discovery and additions to CDK gene family list. Genome-wide transcript profiling data provide improved resolution in correlating CDK gene activities with organ development and responses to environmental conditions. A better understanding of the promoter function and the roles of transcription regulators will require further gene transfer experiments. We expect more work with transgenic plants as regards the testing of the functional consequences of alterations in the characteristics or activity of CDKs. Only limited information is available on specificity of the interactions between CDKs and cyclins or regulatory partners. The positioning of CDKs in phosphorylation cascades is essential, especially to establish links between cell cycle regulation and the control elements of the developmental program. Determination of the crystal structures of plant-specific CDKs and their interacting partners may be a prerequisite for the design of new molecules to alter kinase functions. The testing of new types of chemical or protein inhibitors can promote the functional characterization of plant CDKs. Extended efforts are needed to search for insertion mutants not only in *Arabidopsis* but also in rice or Fabaceae mutant collections. As yet, relatively few transgenic strategies have been tested to achieve breeding goals by the modification of CDK functions. Optimization of the plant architecture or its capability to undergo adaptation to suboptimal environmental conditions may well be a target for the genetic engineering of CDK pathways in plants.

Acknowledgments

We are grateful to Zsuzsa Keczán and Judit Szabad for their excellent help during the preparation of the manuscript, and to László Kozma Bognár for the helpful discussion on circadian rhythm. The authors express special thanks to David Durham for his excellent language correction. Gábor V. Horváth is grateful for a "János Bolyai" Research Fellowship.

References

Ach, R.A., Durfee, T., Miller, A.B., *et al.* (1997) *RRB1* and *RRB2* encode maize retinoblastoma-related proteins that interact with a plant D-type cyclin and geminivirus replication protein. *Mol Cell Biol* **17**, 5077–5086.

Araki, S., Ito, M., Soyano, T., Nishihama, R. and Machida, Y. (2004) Mitotic cyclins stimulate the activity of c-Myb-like factors for transactivation of G_2/M phase-specific genes in tobacco. *J Biol Chem* **279**, 32979–32988.

Ayaydin, F., Vissi, E., Mészáros, T., *et al.* (2000) Inhibition of serine/threonine-specific protein phosphatases causes premature activation of cdc2MsF kinase at G_2/M transition and early mitotic microtubule organization in alfalfa. *Plant J* **23**, 85–96.

Baldet, P., Hernould, M., Laporte, F., *et al.* (2006) The expression of cell-proliferation-related genes in early developing flowers is affected by a fruit load reduction in tomato plants. *J Exp Bot* **57**, 961–970.

Bell, M.H., Halford, N.G., Ormrod, J.C. and Francis, D. (1993) Tobacco plants transformed with cdc25, a mitotic inducer gene from fission yeast. *Plant Mol Biol* **23**, 445–451.

Bläsing, O.E., Gibon, Y., Günther, M., *et al.* (2005) Sugars and circadian regulation make major contributions to the global regulation of diurnal gene expression in *Arabidopsis*. *Plant Cell* **17**, 3257–3281.

Boniotti, M.B. and Gutiérrez, C. (2001) A cell cycle-regulated kinase activity phosphorylates plant retinoblastoma protein and contains, in *Arabidopsis*, a CDKA/cyclin D complex. *Plant J* **28**, 341–350.

Boudolf, V., Barôcco, R., Engler, J., *et al.* (2004a) B1 type cyclin-dependent kinases are essential for the formation of stomatal complexes in *Arabidopsis thaliana*. *Plant Cell* **16**, 945–955.

Boudolf, V., Vlieghe, K., Beemster, G.T.S., *et al.* (2004b) The plant-specific cyclin-dependent kinase CDKB1;1 and transition factor E2Fa-DPa control the balance of mitotically dividing and endoreduplicating cells in *Arabidopsis*. *Plant Cell* **16**, 2683–2692.

Breyne, P., Dreesen, R., Vandepoele, K., *et al.* (2002) Transcriptome analysis during cell division in plants. *Proc Natl Acad Sci USA* **99**, 14825–14830.

Casimiro, I., Beeckman, T., Graham, N., *et al.* (2003) Dissecting *Arabidopsis* lateral root development. *Trends Plant Sci* **8**, 165–171.

Colasanti, J., Tyers, M. and Sundaresan, V. (1991) Isolation and characterization of cDNA clones encoding a functional $p34^{cdc2}$ homologue from *Zea mays*. *Proc Natl Acad Sci USA* **88**, 3377–3381.

Correa-Aragunde, N., Graziano, M., Chevalier, C. and Lamattina, L. (2006) Nitric oxide modulates the expression of cell cycle regulatory genes during lateral root formation in tomato. *J Exp Bot* **57**, 581–588.

Correa-Aragunde, N., Graziano, M. and Lamattina, L. (2004) Nitric oxide plays a central role in determining lateral root development in tomato. *Planta* **218**, 900–905.

den Boer, B.G.W. and Murray, J.A.H. (2000) Control of plant growth and development through manipulation of cell cycle genes. *Curr Opin Plant Biol* **11**, 138–145.

De Veylder, L., Beeckman, T., Beemster, G.T.S., *et al.* (2002) Control of proliferation, endoreduplication and differentiation by the *Arabidopsis* E2Fa-Dpa transcription factor. *EMBO J* **21**, 1360–1368.

De Veylder, L., Beemster, G.T.S., Beeckman, T. and Inzé, D. (2001) *CKS1At* overexpression in *Arabidopsis thaliana* inhibits growth by reducing meristem size and inhibiting cell cycle progression. *Plant J* **25**, 617–626.

De Veylder, L., Joubès, J. and Inzé, D. (2003) Plant cell cycle transitions. *Curr Opin Plant Biol* **6**, 536–543.

De Veylder, L., Segers, G., Glab, N., *et al.* (1997) The *Arabidopsis* Cks1At protein binds the cyclin-dependent kinases Cdc2aAt and Cdc2bAt. *FEBS Lett* **412**, 446–452.

Dewitte, W. and Murray, J.A.H. (2003) Plant cell cycle. *Annu Rev Plant Biol* **54**, 235–264.

Durfee, T., Feiler, H.S. and Gruissem, W. (2000) Retinoblastoma-related proteins in plants: Homologues or orthologues of their metazoan counterparts? *Plant Mol Biol* **43**, 635–642.

Esponosa-Ruiz, A., Saxena, S., Schmidt, J., *et al.* (2004) Differential stage-specific regulation of cyclin-dependent kinases during cambial dormancy in hybrid aspen. *Plant J* **38**, 603–615.

Feiler, H.S. and Jacobs, T.W. (1990) Cell division in higher plants: A cdc2 gene, its 34-kDa product, and histone H1 kinase activity in pea. *Proc Natl Acad Sci USA* **87**, 5397–5401.

Ferreira, P.C.G., Hemerly, A.S., Villarroel, R., Van Montagu, M. and Inzé, D. (1991) The *Arabidopsis* functional homolog of the $p34^{cdc2}$ protein kinase. *Plant Cell* **3**, 531–540.

Freeman, D., Riou-Khamlichi, C., Oakenfull, E.A. and Murray, J.A.H. (2003) Isolation, characterization and expression of cyclin-dependent kinase genes in Jerusalem artichoke (*Helianthus tuberosus* L.). *J Exp Bot* **54**, 303–308.

Fülöp, K., Pettkó-Szandtner, A., Magyar, Z., *et al.* (2005) The *Medicago* CDKC;1–CYCLINT;1 kinase complex phosphorylates the carboxy-terminal domain of RNA polymerase II and promotes transcription. *Plant J* **42**, 810–820.

Gonzalez, N., Hernould, M., Delmas, F., *et al.* (2004) Molecular characterization of a WEE1 gene homologue in tomato (*Lycopersicon esculentum* Mill.). *Plant Mol Biol* **56**, 849–861.

Griniatsos, J., Michail, O.P., Theocharis, S., *et al.* (2006) Circadian variation in expression of G_1 phase cyclins D_1 and E and cyclin-dependent kinase inhibitors p16 and p21 in human bowel mucosa. *World J Gastroenterol* **12**, 2109–2114.

Gutiérrez, C. (1998) The retinoblastoma pathway in plant cell cycle and development. *Curr Opin Plant Biol* **1**, 492–497.

Gutiérrez, C., Ramirez-Parra, E., Mar Castellano, M. and del Pozo, J.C. (2002) G_1 to S transition: more than a cell cycle engine switch. *Curr Opin Plant Biol* **5**, 480–486.

Hashimoto, J., Hirabayashi, T., Hayano, Y., *et al.* (1992) Isolation and characterization of cDNA clones encoding cdc2 homologues from *Oryza sativa*: a functional homologue and cognate variants. *Mol Gen Genet* **233**, 10–16.

Healy, J.M.S., Menges, M., Doonan, J.H. and Murray, J.A.H. (2001) The *Arabidopsis* D-type cyclins CycD2 and CycD3 both interact *in vivo* with the PSTAIRE cyclin-dependent kinase Cdc2a but are differentially controlled. *J Biol Chem* **276**, 7041–7047.

Hemerly, A., de Almeida Engler, J., Bergounioux, C., *et al.* (1995) Dominant negative mutants of the CDC2 kinase uncouple cell division from iterative plant development. *EMBO J* **14**, 3925–3936.

Hemerly, A., Ferreira, P.C.G., Van Montagu, M., Engler, G. and Inzé, D. (2000) Cell division events are essential for embryo pattering and morphogenesis: studies on dominant-negative *cdc2aAt* mutants of *Arabidopsis*. *Plant J* **23**, 123–130.

Hemerly, A.S., Ferreira, P., de Almeida Engler, J., Van Montagu, M., Engler, G. and Inzé, D. (1993) *cdc2a* expression in *Arabidopsis* is linked with competence for cell division. *Plant Cell* **5**, 1711–1723.

Higo, K., Ugawa, Y., Iwamoto, M. and Korenaga, T. (1999) Plant cis-acting regulatory DNA elements (PLACE) database. *Nucleic Acids Res* **27**, 297–300.

Himanen, K., Boucheron, E., Vanneste, S., de Almeida Engler, J., Inzé, D. and Beeckman, T. (2002) Auxin-mediated cell cycle activation during early lateral root initiation. *Plant Cell* **14**, 2339–2351.

Himanen, K., Vuylsteke, M., Vanneste, S., *et al.* (2004) Transcript profiling of early lateral root initiation. *Proc Natl Acad Sci USA* **101**, 5146–5151.

Hirt, H., Páy, A., Bögre, L., Meskiene, I. and Heberle-Bors, E. (1993) *cdc2MsB* a cognate *cdc2* gene from alfalfa, complement the G1/S but not the G2/M transition of budding yeast *cdc28* mutants. *Plant J* **4**, 61–69.

Hirt, H., Páy, A., Györgyey, J., *et al.* (1991) Complementation of a yeast cell cycle mutant by an alfalfa cDNA encoding a protein kinase homologous to $p34^{cdc2}$. *Proc Natl Acad Sci USA* **88**, 1636–1640.

Huntley, R., Healy, S., Freeman, D., *et al.* (1998) The maize retinoblastoma protein homologue ZmRb-1 is regulated during leaf development and displays conserved interactions with G1/S regulators and plant cyclin D (CycD) proteins. *Plant Mol Biol* **37**, 155–169.

Inzé, D. (2005) Green light for the cell cycle. *EMBO J* **24**, 657–662.

Ito, M. (2005) Conservation and diversification of three-repeat Myb transcription factors in plants. *J Plant Res* **118**, 61–69.

Ito, M., Araki, S., Matsunaga, S., *et al.* (2001) G2/M-phase-specific transcription during plant cell cycle is mediated by c-Myb-like transcription factors. *Plant Cell* **13**, 1891–1905.

Iwakawa, H., Shinmyo, A. and Sekine, M. (2006) *Arabidopsis CDKA;1*, a *cdc2* homologue, controls proliferation of generative cells in male gametogenesis. *Plant J* **45**, 819–831.

Joubès, J., Chevalier, C., Dudits, D., *et al.* (2000) CDK-related protein kinases in plants. *Plant Mol Biol* **43**, 607–620.

Joubès, J., Lemaire-Chamley, M., Delmas, F., *et al.* (2001) A new C-type cyclin-dependent kinase from tomato expressed in dividing tissues does not interact with mitotic and G1 cyclins. *Plant Physiol* **126**, 1403–1415.

Koch, C. and Nasmyth, K. (1994) Cell cycle regulated transcription in yeast. *Curr Opin Cell Biol* **6**, 451–459.

Kono, A., Umeda-Hara, C., Lee, J., Ito, M., Uchimiya, H. and Umeda, M. (2003) *Arabidopsis* D-type cyclin CYCD4;1 is a novel cyclin partner of B2-type cyclin-dependent kinase. *Plant Physiol* **132**, 1315–1321.

Kosugi, S. and Ohashi, Y. (2003) Constitutive E2F expression in tobacco plants exhibit altered cell cycle control and morphological change in a cell type-specific manner. *Plant Physiol* **132**, 2012–2022.

Landrieu, I., da Costa, M., De Veylder, L., *et al.* (2004) A small CDC25 dual-specificity tyrosine phosphatase isoform in *Arabidopsis thaliana*. *Proc Natl Acad Sci USA* **101**, 13380–13385.

Lee, J., Das, A., Yamaguchi, M., *et al.* (2003) Cell cycle function of a rice B2-type cyclin interacting with a B-type cyclin-dependent kinase. *Plant J* **34**, 417–425.

Leiva-Neto, J. T., Grafi, G., Sabelli, P. A., *et al.* (2004) A dominant negative mutant of cyclin-dependent kinase A reduces endoreduplication but not cell size or gene expression in maize endosperm. *Plant Cell* **16**, 1854–1869.

Magyar, Z., Bakó, L., Bögre, L., Dedeoğlu, D., Kapros, T. and Dudits, D. (1993) Active cdc2 genes and cell cycle phase-specific cdc2-related kinase complexes in hormone-stimulated alfalfa cells. *Plant J* **4**, 151–161.

Magyar, Z., De Veylder, L., Atanassova, A., Bakó, L., Inzé, D. and Bögre, L. (2005) The role of the *Arabidopsis* E2FB transcription factor in regulating auxin-dependent cell division. *Plant Cell* **17**, 2527–2541.

Magyar, Z., Mészáros, T., Miskolczi, P., *et al.* (1997) Cell cycle phase specificity of putative cyclin-dependent kinase variants in synchronized alfalfa cells. *Plant Cell* **9**, 223–235.

Martinez, M.C., Jørgensen, J.-E., Lawton, M.A., Lamb, C.J. and Doerner, P.W. (1992) Spatial pattern of *cdc2* expression in relation to meristem activity and cell proliferation during plant development. *Proc Natl Acad Sci USA* **89**, 7360–7364.

Matsuo, T., Yamaguchi, S., Mitsui, S., Emi, A., Shimoda, F. and Okamura, H. (2003) Control mechanism of the circadian clock for timing of cell division *in vivo*. *Science* **302**, 255–259.

McKibbin, R.S., Halford, N.G. and Francis, D. (1998) Expression of fission yeast cdc25 alters the frequency of lateral root formation in transgenic tobacco. *Plant Mol Biol* **36**, 601–612.

McKinney, J.D. and Heitz, N. (1991) Transcriptional regulation in the eukaryotic cell cycle. *Trends Biochem Sci* **16**, 430–435.

Menges, M., de Jager, S.M., Gruissem, W. and Murray, J.A.H. (2005) Global analysis of the core cell cycle regulators of *Arabidopsis* identifies novel genes, reveals multiple and highly specific profiles of expression and provides a coherent model for plant cycle control. *Plant J* **41**, 546–566.

Mészáros, T., Miskolczi, P., Ayaydin, F., *et al.* (2000) Multiple cyclin-dependent kinase complexes and phosphatases control G_2/M progression in alfalfa cells. *Plant Mol Biol* **43**, 595–605.

Miao, G.H., Hong, Z. and Verma, D.P.S. (1993) Two functional soybean genes encoding $p34^{cdc2}$ protein kinases are regulated different plant developmental pathways. *Dev Biol* **90**, 943–947.

Mironov, V., De Veylder, L., Van Montagu, M. and Inzé, D. (1999) Cyclin-dependent kinases and cell division in plants – the nexus. *Plant Cell* **11**, 509–521.

Müller, R. (1995) Transcriptional regulation during the mammalian cell cycle. *Trends Genet* **11**, 173–178.

Nakagami, H., Kawamura, K., Sugisaka, K., Sekine, M. and Shinmyo, A. (2002) Phosphorylation of retinoblastoma-related protein by the cyclin D/cyclin-dependent kinase complex is activated at the G1/S phase transition in tobacco. *Plant Cell* **14**, 1847–1857.

Nakagami, H., Sekine, M., Murakami, H. and Shinmyo, A. (1999) Tobacco retinoblastoma-related protein phosphorylated by a distinct cyclin-dependent kinase complex with Cdc2/cyclin D *in vitro*. *Plant J* **18**, 243–252.

Nováková, M., Motyka, V., Dobrev, P.I., Malbeck, J., Gaudinová, A. and Vanková, R. (2005) Diurnal variation of cytokinin, auxin and abscisic acid levels in tobacco leaves. *J Exp Bot* **56**, 2877–2883.

Orchard, C.B., Siciliano, I., Sorrell, D.A., *et al.* (2005) Tobacco BY-2 cells expressing fission yeast *cdc25* bypass a G2/M block on the cell cycle. *Plant J* **44**, 290–299.

Pettkó-Szandtner, A., Mészáros, T., Horvath, G.V., *et al.* (2006) Activation of an alfalfa cyclin-dependent kinase inhibitor by calmodulin-like domain protein kinase. *Plant J* **46**, 111–123.

Roudier, F., Federova, H., Györgyey, J., *et al.* (2000) Cell cycle function of a *Medicago sativa* A2-type cyclin interacting with a PSTAIRE-type cyclin-dependent kinase and a retinoblastoma protein. *Plant J* **23**, 73–83.

Salomé, P.A. and McLung, C.R. (2005) What makes the *Arabidopsis* clock tick on time? A review on entrainment. *Plant Cell Environ* **28**, 21–38.

Sauter, M. (1997) Differential expression of a CAK (cdc2-activating kinase)-like protein kinase, cyclins and *cdc2* genes from rice during the cell cycle and response to gibberellin. *Plant J* **11**, 181–190.

Segers, G., Gadisseur, I., Bergounioux, C., *et al.* (1996) The *Arabidopsis* cyclin-dependent kinase gene cdc2bAt is preferentially expressed during S and G_2-phases of the cell cycle. *Plant J* **10**, 601–612.

Sorrell, D.A. and Marchbank, A. (2002) A *WEE1* homologue from *Arabidopsis thaliana*. *Planta* **215**, 518–522.

Sorrell, D.A., Menges, M., Healy, J.M.S., *et al.* (2001) Cell cycle regulation of cyclin-dependent kinases in tobacco cultivar bright yellow-2 cells. *Plant Physiol* **126**, 1214–1223.

Stals, H., Casteels, P., Van Montagu, M. and Inzé, D. (2000) Regulation of cyclin-dependent kinases in *Arabidopsis thaliana*. *Plant Mol Biol* **43**, 583–593.

Sun, Y., Dilkes, B.P., Zhang, C., *et al.* (1999) Characterization of maize (*Zea mays* L.) Wee1 and its activity in developing endosperm. *Proc Natl Acad Sci USA* **96**, 4180–4185.

Tréhin, C., Glab, N., Perennes, C., Planchais, S. and Bergounioux, C. (1999) M phase-specific activation of the *Nicotiana sylvestris Cyclin B1* promoter involves multiple regulatory elements. *Plant J* **17**, 263–273.

Umeda, M., Shimotohno, A. and Yamaguchi, M. (2005) Control of cell division and transcription by cyclin-dependent kinase-activating kinases in plants. *Plant Cell Physiol* **46**, 1437–1442.

Umeda, M., Umeda-Hara, C., Yamaguchi, M., Hashimoto, J. and Uchimiya, H. (1999) Differential expression of genes for cyclin-dependent protein kinases in rice plants. *Plant Physiol* **119**, 31–40.

Van de Peer, Y. and De Wachter, R. (1994) TREECON for Windows, a software package for the construction and drawing of evolutionary trees for the Microsoft Windows environment. *Comp Appl Biosci* **10**, 569–570.

Vanneste, S., Maes, L., De Smet, I., *et al.* (2005) Auxin regulation of cell cycle and its role during lateral root initiation. *Physiol Plant* **123**, 139–146.

Vanstraelen, M., Torres Acosta, J.A., De Veylder, L., Inzé, D. and Geelen, D. (2004) A plant-specific subclass of C-terminal kinesins contains a conserved A-type cyclin-dependent kinase site implicated in folding and dimerization. *Plant Physiol* **135**, 1417–1429.

Vanstraelen, M., Inzé, D. and Geelen, D. (2006) Mitosis-specific kinesins in *Arabidopsis*. *Trends Plant Sci* **11**, 167–175.

Verkest, A., de O. Manes, C.-L., Vercruysse, S., *et al.* (2005a) The cyclin-dependent kinase inhibitor KRP2 controls the onset of the endoreduplication cycle during *Arabidopsis* leaf development through inhibition of mitotic CDKA;1 kinase complexes. *Plant Cell* **17**, 1723–1736.

Verkest, A., Weinl, C., Inzé, D., De Veylder, L. and Schnittger, A. (2005b) Switching the cell cycle. Kip-related proteins in plant cell cycle control. *Plant Physiol* **139**, 1099–1106.

Wang, W. and Chen, X. (2004) HUA ENHANCER3 reveals a role for a cyclin-dependent protein kinase in the specification of floral organ identity in *Arabidopsis*. *Development* **131**, 3147–3156.

Yamaguchi, M., Kato, H., Yoshida, S., Yamamura, S., Uchimiya, H. and Umeda, M. (2003) Control of *in vitro* organogenesis by cyclin-dependent kinase activities in plants. *Proc Natl Acad Sci USA* **100**, 8019–8023.

Yoshizumi, T., Nagata, N., Shimada, H. and Matsui, M. (1999) An *Arabidopsis* cell cycle-dependent kinase-related gene, *CDC2b1*, plays a role in regulating seedling growth in darkness. *Plant Cell* **11**, 1883–1895.

Zhang, K., Diederich, L. and John, P.C.L. (2005) The cytokinins requirement for cell division in cultured *Nicotiana plumbaginifolia* cells can be satisfied by yeast Cdc25 protein tyrosine phosphatase. Implications for mechanisms of cytokinins response and plant development. *Plant Physiol* **137**, 308–316.

Zhang, K., Letham, D. A. and John, P.C.L. (1996) Cytokinin controls the cell cycle at mitosis by stimulating the tyrosine dephosphorylation and activation of p^{34}cdc2-like H1 histone kinase. *Planta* **200**, 2–12.

Zhiponova, M. K., Pettkó-Szandtner, A., Stelkovics, É., *et al.* (2006) Mitosis-specific promoter of the alfalfa cyclin-dependent kinase gene (Medsa;CDKB2;1) is activated by wounding and ethylene in a non-cell division-dependent manner. *Plant Physiol* **140**, 693–703.

Zhu, W., Giangrande, P. H. and Nevins J.R. (2004) E2Fs link the control of G1/S and G2/M transcription. *EMBO J* **23**, 4615–4626.

2 The plant cyclins

Jeroen Nieuwland, Margit Menges and James A.H. Murray

2.1 Introduction

Cyclins are the regulatory partner of the cyclin-dependent kinase (CDK)–cyclin heterodimeric protein kinase complex. The discovery of cyclins in sea urchin eggs by Tim Hunt and his colleagues in 1983 provided the crucial hint pointing to the biochemical mechanisms underlying the oscillator that drives the cell cycle and revealed the basic cell cycle engine conserved in all eukaryotes (Evans *et al.*, 1983; Novak *et al.*, 1998). In the original paper reporting the discovery of cyclins, only one pair of cyclins is described, corresponding to S phase (A-type) and M phase (B-type) cyclins which were later shown to associate with a single kinase subunit, turning out to be Cdk1 (Lohka *et al.*, 1988). Today, a large group of animal, yeast and plant cyclins is known, forming partners with a number of different CDKs, and cyclin association is recognized as a key type of regulation of serine–threonine protein kinases. Cyclin binding is a powerful strategy, since the CDK is inactive without cyclin association, so that regulated synthesis, destruction or localization of the cyclin subunit provides multiple levels of control for cyclin activation of the CDK subunit, as well as control on the possible association of different cyclins and CDKs. This regulatory module is largely combinatorial but by no means exclusively utilized in cell cycle control, with a number of other examples of CDK–cyclin regulation involved in transcription and response to phosphate availability (Lenburg and O'Shea, 1996; Fisher, 2005).

2.1.1 Cyclins and the cell cycle oscillator

The basic principle of the cell cycle oscillator is that the regulated synthesis and destruction of cyclins provides waves of CDK activity that drive the two crucial transitions of the cell cycle – the entry from G1 into S phase and from G2 into M phase. Theoretically, one may argue that only one cyclin would be sufficient for a working cell cycle oscillator and indeed, Fisher and Nurse proved in yeast that a single cyclin was able to promote both DNA synthesis and mitosis (Fisher and Nurse, 1996).

The single cyclin oscillator model, however, turned out to be too simplistic, as many different cyclins have subsequently been found. Moreover, somewhat different principles underlie cyclin regulation at the two main control points. G1–S cyclins are often responsive to external signals, such as hormones, or internal signals, such as growth rate or cell size, and are typically unstable proteins, rapidly turned over by the ubiquitinylation-proteasome pathway. Combined with regulated expression by such signals, this leads to their rapid disappearance on removal of the stimulus. Such behaviour is typical of D-type cyclins, which therefore report to the cell cycle on favourable conditions for division and as such are a 'bolt-on' to the core oscillator, providing the possibility for cellular, environmental and developmental oversight of the onset of a cell cycle.

In contrast, S phase (cyclin A) and mitotic cyclins (cyclin B) show regulated synthesis and destruction in a cell-cycle-dependent manner, and so their abundance depends on the position of the cell in the cycle rather than on extrinsic signals.

In animals, at least 13 cyclin classes (called cyclin A to L and T) have been identified whereas in plants more than 100 cyclins have been isolated from various plants. Genome-wide analysis of Arabidopsis yielded 50 putative cyclins which can be grouped in nine classes (Wang *et al.*, 2004a; Menges *et al.*, 2005), of which five classes have not been found in animals, although clear evidence for cell cycle roles exists only for cyclins A, B, D and H. The reader should keep in mind that a cyclin is defined as a protein containing a conserved region called the cyclin box, a region carrying the CDK-binding site (Hunt, 1991; Pines, 1995a). However, this does not necessitate that every protein with this domain has a function in the core cell cycle machinery. The plethora of cyclin reflects the fact that development of complex multicellular organisms requires spatial and temporal control of cell division. Both plants and animals need to modulate cell division in order to achieve their developmental plan and regulate this primary growth to adapt to their changing environment. Plants, in particular, need to adapt their indeterminate pattern of growth in a flexible manner, since they are unable to move to a more optimal location. Much of the current knowledge of the biochemical mechanisms regulating the cell cycle is derived from animal and yeast research, and we are just beginning to uncover the analogous and plant-specific regulatory mechanisms in the plant cell cycle. This chapter will discuss the advances in studies of cyclin proteins.

2.2 The plant cyclin family

The homology allowed the nomenclature of plant cyclins to be standardized by international agreement as described (Renaudin *et al.*, 1996), and all plant cyclins should be named and referred to according to these proposals. The naming uses, where necessary, two letters to describe the species as the first letter of the generic and specific name (e.g. At for *Arabidopsis thaliana*), followed by the three letters CYC and a further letter designating the major group of cyclins as defined by homology with other higher eukaryotes. This then is followed by a subgroup number, e.g. CYCA3 (not conserved outside plants), and a specific gene number where multiple genes are present in one plant species (e.g. CYCA3;3). This final gene number is not conserved between species.

2.2.1 Phylogenetic relationships between animal and plant cyclins

Plant cyclins share homology with the A-, B-, C-, H- and L-type classes of mammalian cyclins, although the subgroups within these classes are not conserved between plants and animals (Renaudin *et al.*, 1996; Yamaguchi *et al.*, 2000; Wang *et al.*, 2004a; La *et al.*, 2006). Arabidopsis homologues of the human E-, F-, G-, I- and UNG2-cyclins do not exist (Wang *et al.*, 2004a). In *Arabidopsis*, 50 cyclin homologues have been defined in nine classes with 23 subgroups and two additional cyclin proteins not belonging to the other classes (Figure 2.1) (Renaudin *et al.*, 1996; Yamaguchi *et al.*, 2000; Barrôco *et al.*, 2003; Wang *et al.*, 2004a; Menges *et al.*, 2005). These 'orphan' cyclins are CYCJ18 (Abrahams *et al.*, 2001), here renamed as CYCQ1;1, and a cyclin-like protein CYL;1 (Menges *et al.*, 2005).

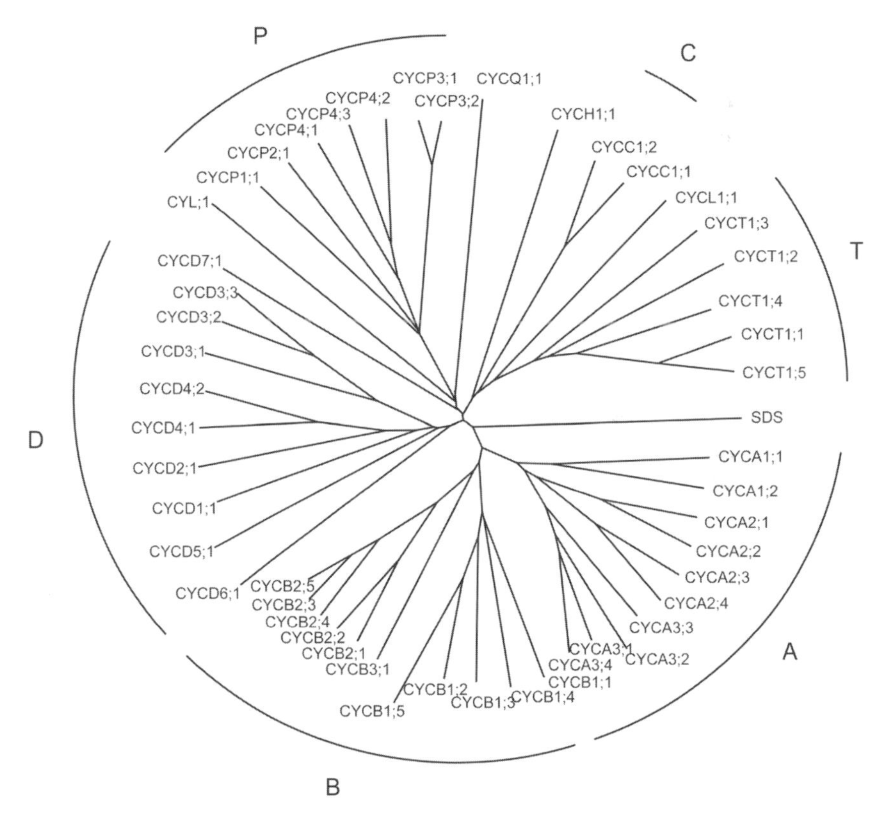

Figure 2.1 Phylogenetic tree analysis of Arabidopsis cyclin genes. For multiple sequence alignment, the full protein sequence of 50 cyclin genes was used (ClustalW, version 1.83, http://www.arabidopsis.org/cgi-bin/bulk/sequences/seqtoclustalw. pl; Thompson *et al.*, 1994). The multiple sequence alignment was used to construct an unrooted neighbour-joining tree based on sequence similarity (http://www.genebee.msu.su/services/phtree˙full.html). Corresponding gene IDs of cyclins classified here are listed in Table 1.1 (Torres Acosta *et al.*, 2004; Wang *et al.*, 2004a; Menges *et al.*, 2005). CYCQ1;1 is the new name we propose for CYCJ18 (Abrahams *et al.*, 2001).

In rice at least 50 cyclins have been found, and although rice shares ten classes of cyclins (A-, B-, C-, D-, H-, L-, SDS-, Q-, T- and P-types) with *Arabidopsis*, the F-type cyclins reported in rice have not been identified in *Arabidopsis* or any other plant species (La *et al.*, 2006). Their relationship with other cyclin groups, as well as their function, remains to be determined. It is interesting to note that the cyclin family in plants (Table 2.1) appears to be larger than those in other eukaryotes. For example, the *Caenorhabditis elegans* genome has 34 cyclins (Plowman *et al.*, 1999) while the number in humans is at least 22 (Nakamura *et al.*, 1995; Pines, 1995b).

2.2.2 *Cyclin domains*

Cyclins contain a conserved region of 250 amino acids called the cyclin core consisting of two domains: cyclin N and cyclin C (Nugent *et al.*, 1991). The cyclin N domain spans the CDK-binding region and is 100 amino acids long. This domain is also called the cyclin box and is the defining domain for cyclins. The cyclin C domain (not to be confused with the class of cyclin Cs, or CYCC proteins), on the other hand, is less conserved and is present in most but not all cyclins, suggesting a specific but perhaps not a critical function of this domain. In *Arabidopsis*, the SDS cyclin and all A-, B- and D-cyclins, except CYCB1;5 and CYCD5;1, possess the cyclin C domain whereas the C-, H-, L-, J-, T-, Q- and P-type cyclins in *Arabidopsis* lack the cyclin C domain. In rice, the SDS cyclin lacks the cyclin C domain but in general the situation is comparable to that in *Arabidopsis*, with rice A-, B-, D-cyclins containing the cyclin C domain, with the exception in both species of CYCD5;1. The rice-specific F-cyclins also contain the cyclin C domain (La *et al.*, 2006). As discussed below, some of the cyclins have a destruction box (Dbox), in addition to the cyclin core, and this Dbox is present in all A- and B-cyclins except CYCA3;3, CYCB2;5 and CYCB3;1 (Wang *et al.*, 2004a). Lastly, most of the cyclins also contain a PEST motif (single AA code) which may confer protein instability (Rechsteiner and Rogers, 1996; Wang *et al.*, 2004a).

2.2.3 *A-type cyclins*

In the early years of plant cyclin research, the A- and B-type cyclins were the first cyclins to be cloned using a PCR-based strategy which made use of the conserved CDK-binding cyclin box (Hata *et al.*, 1991; Hemerly *et al.*, 1992; Dewitte and Murray, 2003). The CYCAs are phylogenetically closest related to CYCBs, which explains the ambiguities in the original annotation of A- and B-type cyclins 10 years ago (Renaudin *et al.*, 1996; Inzé *et al.*, 1998).

The A-type cyclins are subdivided into three different subclasses, CYCA1, CYCA2 and CYCA3 (Renaudin *et al.*, 1996; Chaubet-Gigot, 2000) and this original proposal has been further confirmed by more recent phylogenetic analyses in both *Arabidopsis* (Wang *et al.*, 2004a; Menges *et al.*, 2005) and rice (La *et al.*, 2006), which revealed no new subclasses of CYCA in monocots or dicots. The

Table 2.1 Plant cyclins

Cyclin name	Locus ID or accession numbers	References
AtCYCA1;1	At1g44110	Menges *et al.*, 2005
AtCYCA1;2 TAM	At1g77390	Wang *et al.*, 2004b; Menges *et al.*, 2005
AtCYCA2;1	At5g25380, Z31589	Ferreira *et al.*, 1994a; Burssens *et al.*, 2000a, b; Beeckman *et al.*, 2001; Menges *et al.*, 2005
AtCYCA2;2	At5g11300, Z31402	Ferreira *et al.*, 1994a; Menges *et al.*, 2005
AtCYCA2;3	At1g15570	Menges *et al.*, 2005; Imai *et al.*, 2006
AtCYCA2;4	At1g80370	Menges *et al.*, 2005
AtCYCA3;1	At5g43080	Menges *et al.*, 2005
AtCYCA3;2	At1g47210	Menges *et al.*, 2005
AtCYCA3;3	At1g47220	Menges *et al.*, 2005
AtCYCA3;4	At1g47230	Menges *et al.*, 2005
AtCYCB1;1	At4g37490, X62279	Hemerly *et al.*, 1992; Ferreira *et al.*, 1994b; Doerner *et al.*, 1996; Niebel *et al.*, 1996; Ach *et al.*, 1997; Colon-Carmona *et al.*, 1999; Donnelly *et al.*, 1999; Burssens *et al.*, 2000b; Beeckman *et al.*, 2001; Boucheron *et al.*, 2002; Planchais *et al.*, 2002; Menges *et al.*, 2005
AtCYCB1;2	At5g06150, L27223	Day *et al.*, 1996; Ito *et al.*, 1998
AtCYCB1;3	At3g11520, L27224	Day and Reddy, 1998; Menges *et al.*, 2005
AtCYCB1;4	At2g26760	Day and Reddy, 1998; Menges *et al.*, 2005
AtCYCB1;5	At1g34460	Menges *et al.*, 2005
AtCYCB2;1	At2g17620, Z31400	Ferreira *et al.*, 1994a; Ito *et al.*, 1998; Menges *et al.*, 2005
AtCYCB2;2	At4g35620, Z31401	Ferreira *et al.*, 1994a; Menges *et al.*, 2005
AtCYCB2;3	At1g20610	
AtCYCB2;4	At1g76310	Menges *et al.*, 2005
AtCYCB2;5	At1g20590	Menges *et al.*, 2005
AtCYCB3;1	At1g16330	Menges *et al.*, 2005
AtCYCC1;1	At5g48640	Menges *et al.*, 2005
AtCYCC1;2	At5g48630	Menges *et al.*, 2005
AtCYCD1;1	At1g70210, X83369	Soni *et al.*, 1995; Fuerst *et al.*, 1996; Cho *et al.*, 2004; Masubelele *et al.*, 2005; Menges *et al.*, 2005
AtCYCD2;1	At2g22490, X83370	Soni *et al.*, 1995; Fuerst *et al.*, 1996; Riou-Khamlichi *et al.*, 1999, 2000; Zhou *et al.*, 1999; Cockcroft *et al.*, 2000; Boniotti and Gutierrez, 2001; Healy *et al.*, 2001; Boucheron *et al.*, 2005; Masubelele *et al.*, 2005; Menges *et al.*, 2005; Nakai *et al.*, 2006
AtCYCD3;1	At4g34160, X83371	Soni *et al.*, 1995; Fuerst *et al.*, 1996; Ach *et al.*, 1997; Riou-Khamlichi *et al.*, 1999, 2000; Zhou *et al.*, 1999; Hu *et al.*, 2000; Healy *et al.*, 2001; Jasinski *et al.*, 2002; Schnittger *et al.*, 2002b; Dewitte *et al.*, 2003; Planchais *et al.*, 2004; Boucheron *et al.*, 2005; Masubelele *et al.*, 2005; Menges *et al.*, 2005, 2006

(*Continued*)

Table 2.1 Plant cyclins (*Continued*)

Cyclin name	Locus ID or accession numbers	References
AtCYCD3;2	At5g67260	Swaminathan *et al.*, 2000; Menges *et al.*, 2005
AtCYCD3;3	At3g50070	Menges *et al.*, 2005
AtCYCD4;1	At5g65420	De Veylder *et al.*, 1999; Schnittger *et al.*, 2002b; Kono *et al.*, 2003; Masubelele *et al.*, 2005; Menges *et al.*, 2005
AtCYCD4;2	At5g10440	Menges *et al.*, 2005; Kono *et al.*, 2006
AtCYCD5;1	At4g37630	Menges *et al.*, 2005
AtCYCD6;1	At4g03270	Menges *et al.*, 2005
AtCYCD7;1	At5g02110	
AtCYCH;1	At5g27620	Menges *et al.*, 2005
AtCYCL1;1	At2g26430	Forment *et al.*, 2002; Menges *et al.*, 2005
AtCYCT1;1	At1g35440	
AtCYCT1;2	At4g19560	
AtCYCT1;3 AtCYCT	At1g27630	Barrôco *et al.*, 2003; Menges *et al.*, 2005
AtCYCT1;4 AtCYCT-like2	At4g19600	Barrôco *et al.*, 2003; Menges *et al.*, 2005
AtCYCT1;5 AtCYCT-like1	At5g45190	Barrôco *et al.*, 2003; Menges *et al.*, 2005
AtCYCP1;1	At3g63120	See references for list of other P-type cyclins – Torres Acosta *et al.*, 2004; Menges *et al.*, 2005
AtCYCP2;1	At3g21870	Torres Acosta *et al.*, 2004; Menges *et al.*, 2005
AtCYCP3;1	At2g45080	Torres Acosta *et al.*, 2004; Menges *et al.*, 2005
AtCYCP3;2	At3g60550	Torres Acosta *et al.*, 2004; Menges *et al.*, 2005
AtCYCP4;1	At2g44740	Torres Acosta *et al.*, 2004; Menges *et al.*, 2005
AtCYCP4;2	At5g61650	Torres Acosta *et al.*, 2004; Menges *et al.*, 2005
AtCYCP4;3	At5g07450	Torres Acosta *et al.*, 2004; Menges *et al.*, 2005
AtCYL;1	At4g34090	Menges *et al.*, 2005
AtSDS	At1g14750	Azumi *et al.*, 2002; Menges *et al.*, 2005
AtCYCQ1;1	At2g01905	Abrahams *et al.*, 2001
OsCYCA1;1	Os01g13260	Umeda *et al.*, 1999; Lee *et al.*, 2003, 2006
OsCYCA1;2	Os01g13220	
OsCYCA1;3	Os12g20320	
OsCYCA1;4	Os05g14730	
OsCYCA2;1	Os12g31810	
OsCYCA3;1	Os03g41100	
OsCYCA3;2	Os12g39210	
OsCYCB1;1	Os01g59120	
OsCYCB1;2	Os02g55680	
OsCYCB1;3	Os01g17400	
OsCYCB1;4	Os02g41720	
OsCYCB1;5	Os05g41390	
OsCYCB2;1	Os04g47580	Umeda *et al.*, 1999; Cooper *et al.*, 2003; Lee *et al.*, 2003, 2006
OsCYCB2;2	Os06g51110	Sauter *et al.*, 1995; Cooper *et al.*, 2003; Lee *et al.*, 2003, 2006
OsCYCC1;1	Os09g32720, D86925	Lee *et al.*, 2006

Table 2.1 (*Continued*)

Cyclin name	Locus ID or accession numbers	References
OsCYCD1;1	Os09g21450	
OsCYCD1;2	Os06g12980	
OsCYCD1;3	Os08g32540	Rohila *et al.*, 2006
OsCYCD2;1	Os07g42860	Rohila *et al.*, 2006
OsCYCD2;2	Os06g11410	
OsCYCD2;3	Os03g27420	Rohila *et al.*, 2006
OsCYCD3;1	Os09g02360	
OsCYCD4;1	Os09g29100	
OsCYCD4;2	Os08g37390	Lee *et al.*, 2006
OsCYCD5;1	Os12g39830	
OsCYCD5;2	Os03g42070	
OsCYCD5;3	Os03g10650	Rohila *et al.*, 2006
OsCYCD6;1	Os07g37010	Rohila *et al.*, 2006
OsCYCD7;1	Os11g47950	
OsCYCF1;1	Os02g39470	
OsCYCF1;2	Os02g39220	
OsCYCF1;3	Os02g39240	
OsCYCF1;4	Os02g39230	
OsCYCF2;1	Os02g39420	
OsCYCF2;2	Os02g39260	
OsCYCF2;3	Os02g38820	
OsCYCF3;1	Os03g11030	
OsCYCF3;2	Os03g11040	
OsCYCH1;1	Os03g52750	Rohila *et al.*, 2006; Lee *et al.*, 2006
OsCYCL1;1	Os01g27940	
OsCYCP1;1	Os04g53680	
OsCYCP2;1	Os04g46660	
OsCYCP3;1	Os05g33040	
OsCYCP4;1	Os10g41430	
OsCYCT1;1	Os02g04010	
OsCYCT1;2	Os02g24190	
OsCYCT1;3	Os11g05850	
OsCYCT1;4	Os12g30020	
OsSDS	Os03g12420	
OsCYCQ1;1	Os03g13480	
BnCYCA1;1	L25406	Szarka *et al.*, 1995
AcCYCA1;1	D82349	Uchida *et al.*, 1996
GmCYCA1;1	D50870	Kouchi *et al.*, 1995
NtCYCA1;1	D50735, X92966	Setiady *et al.*, 1995, 1997; Reichheld *et al.*, 1996; Uemukai *et al.*, 2005
NtCYCA1;2 (maybe allelic variant of NtCYCA1;1)	X92967	Reichheld *et al.*, 1996
ZmCYCA1;1	U10077	Renaudin *et al.*, 1994; Mews *et al.*, 1997, 2000
LeCYCA1;1	AJ243451	Joubès *et al.*, 2000
ZmCYCA1;2	U50064	Hsieh and Wolniak, 1998
BnCYCA2;1	L25405	Szarka *et al.*, 1995
GmCYCA2;1	D50869	Kouchi *et al.*, 1995

(*Continued*)

Table 2.1 Plant cyclins (*Continued*)

Cyclin name	Locus ID or accession numbers	References
MsCYCA2;1	X85783, L245816	Meskiene *et al.*, 1995; Russinova *et al.*, 1995
NtCYCA2;1	D50736	Setiady *et al.*, 1995, 1997; Nakagami *et al.*, 1999
LeCYCA2;1	AJ243452	Joubès *et al.*, 2000
MsCYCA2;2	AJ243499	Roudier *et al.*, 2000, 2003
DcCYCA3;1	S49312	Hata *et al.*, 1991
GmCYCA3;1	D50868	Kouchi *et al.*, 1995
CrCYCA3;1	D86385, D86387	Ito *et al.*, 1997
LeCYCA3;1	AJ243453	Joubès *et al.*, 2000
NtCYCA3;1	X92964, D89636	Reichheld *et al.*, 1996; Ito *et al.*, 1997; Genschik *et al.*, 1998; Criqui *et al.*, 2001
NtCYCA3;2	X92965	Reichheld *et al.*, 1996; Wyrzykowska *et al.*, 2002; Yu *et al.*, 2003
NtCYCA3;3	X93467	Reichheld *et al.*, 1996
AmCYCB1;1	X76122	Fobert *et al.*, 1994
AmCYCB1;2	X76123	Fobert *et al.*, 1994
GmCYCB1;1	X62820, Z26331	Hata *et al.*, 1991; Kouchi *et al.*, 1995
GmCYCB1;2	X62303	Hata *et al.*, 1991
GmCYCB1;3	D50871	Kouchi *et al.*, 1995; Ito *et al.*, 1998
NtCYCB1;1	Z37978	Qin *et al.*, 1995, 1996; Genschik *et al.*, 1998; Criqui *et al.*, 2000, 2001; Weingartner *et al.*, 2003, 2004; Świątek *et al.*, 2004
NsCYCB1;1	Y08992	Tréhin *et al.*, 1997
LeCYCB1;1	AJ243454	Joubès *et al.*, 2000, 2001
PhCYCB1;1	AJ250315	Porceddu *et al.*, 1999
LlB1;1	U24192	Deckert *et al.*, 1996
CrCYCB1;1	D86386	Ito *et al.*, 1998
NtCYCB1;2	D50737	Setiady *et al.*, 1995, 1997; Uemukai *et al.*, 2005
PcCYCB1;1	L34207	Logemann *et al.*, 1995
LeCYCB1;1	AJ011108	Kvarnheden *et al.*, 2000
ZmCYCB1;1	U10079	Renaudin *et al.*, 1994; Mews *et al.*, 1997, 2000
ZmCYCB1;2	U10078	Renaudin *et al.*, 1994; Mews *et al.*, 1997, 2000
ZmCYCB1;3	U66608	Sun *et al.*, 1999
MsCYCB2;1	X82039, X68740	Hirt *et al.*, 1992; Meskiene *et al.*, 1995
LeCYCB2;1	AJ243455	Joubès *et al.*, 2000, 2001
ZmCYCB2;1	U10076	Renaudin *et al.*, 1994; Mews *et al.*, 1997, 2000
MsCYCB2;2	X82040, X68741	Hirt *et al.*, 1992; Meskiene *et al.*, 1995; Weingartner *et al.*, 2003
MsCYCB2;3	X78504	Savouré *et al.*, 1995
AmCYCD1;1 (cyc D1)	AJ250396	Gaudin *et al.*, 2000; Koroleva *et al.*, 2004
HtCYCD1;1	AY063460	Freeman *et al.*, 2003
NtCYCD2;1	AJ011892	Sorrell *et al.*, 1999
TaCYCD2;1		Wang *et al.*, 2006
ZmCYCD2	AF351189	Gutierrez *et al.*, 2005
NtCYCD3;1	AJ011893	Sorrell *et al.*, 1999
HtCYCD3;1	AY063461	Freeman *et al.*, 2003
MsCYCD3;1	X88864	Dahl *et al.*, 1995; Favery *et al.*, 2002

Table 2.1 (*Continued*)

Cyclin name	Locus ID or accession numbers	References
AmCYCD3;1	AJ250397	Gaudin *et al.*, 2000
LeCYCD3;1	AJ002588, AJ245415	Joubès *et al.*, 2000, 2001; Kvarnheden *et al.*, 2000
LlCYCD3;1		Li *et al.*, 2003
AmCYCD3;2	AJ250398	Gaudin *et al.*, 2000
NtCYCD3;2	AJ011894	Sorrell *et al.*, 1999
LlCYCD3;2		Li *et al.*, 2003
LeCYCD3;2	AJ002589	Kvarnheden *et al.*, 2000
LeCYCD3;3	AJ002590	Kvarnheden *et al.*, 2000
NtCYCD3;3	AB015222	Nakagami *et al.*, 1999, 2002; Uemukai *et al.*, 2005
MtCYCT;1	AAR01224	Fülöp *et al.*, 2005

The cyclins found by whole genome sequencing of *Arabidopsis* (Wang *et al.*, 2004a; Menges *et al.*, 2005) and rice (La *et al.*, 2006) are listed with their locus ID. For other organisms only the cyclins that have been reported in journal publications are listed with their accession numbers and references. For a list of old cyclin names see Renaudin *et al.* (1996). Ac, *Adiantum capillus*; Am, *Antirrhinum majus*; At, *Arabidopsis thaliana*; Bn, *Brassica napus*; Cr, *Catharanthus roseus*; Dc, *Daucus carota*; Gm, *Glycine max*; Ht, *Helianthus tuberosus*; Le, *Lycopersicon esculentum*; Ll, *Lagenaria leucantha*; Mt, *Medicago truncatula*; Ms, *Medicago sativa*; Ns, *Nicotiana sylvestris*; Nt, *Nicotiana tabacum*; Os, *Oryza sativa*; Pc, *Petroselinum crispum*; Ph, *Petunia hybrida*; Ta, *Triticum aestivum*; Zm, *Zea mays*.

motif LVEVxEEY, proposed as the signature for all CYCAs (Renaudin *et al.*, 1996; Chaubet-Gigot, 2000), was found in most *Arabidopsis* CYCAs, although some CYCAs have functionally conserved substituted amino acid residues (Wang *et al.*, 2004a). A related LVxxTLYL motif has been found in all the A-type cyclins of rice (La *et al.*, 2006).

It is not unlikely that the three different CYCA classes have distinct functions in plants. They are differentially regulated in the cell cycle, as CYCA3 cyclins are expressed from G1/S and during S phase whereas all *CYCA1* and *CYCA2* genes show a peak in expression at G2/M (Menges *et al.*, 2005). The same expression timing was also observed for tobacco CYCA3 (from G1/S transition) and CYCA1/CYCA2 (from mid-S phase, peaking in G2/M) (Reichheld *et al.*, 1996). Based on work on the tobacco CYCA3;2, a functional analogy has been proposed between this CYCA3 cyclin and the animal-specific E-type cyclins, which governs the G1/S transition in animals (Yu *et al.*, 2003). Nevertheless, the function of NtCYCA3;2 depends on developmental context, as it was shown that although it could induce cell division in the shoot apical meristem and leaf primordia when locally overexpressed, it did not have a significant effect on overall plant growth when the overexpression was performed on the whole-plant level (Wyrzykowska *et al.*, 2002). Interestingly, these data suggest that control of cell division in localized regions is important for morphogenesis.

In a yeast two-hybrid experiment involving the A2-type cyclin CYCA2;2 from *Medicago sativa*, it was shown that this cyclin was able to interact with both MsCDKA1;1 (Cdc2MsA) and the maize retinoblastoma (RBR) protein (Roudier *et al.*, 2000). This is interesting, as human cyclin A also interacts with RB (Knudsen *et al.*, 1998). Further analysis of MsCYCA2;2 showed that it was expressed in the meristems and proliferating cells of lateral roots and nodule primordia. MsCYCA2;2 was shown to be upregulated by auxin which affected the expression pattern of MsCYCA2;2 by shifting it from the phloem to the xylem poles of the root (Roudier *et al.*, 2003).

2.2.4 B-type cyclins

The plant B-type cyclins were amongst the first cyclins to be isolated in plants (Hata *et al.*, 1991; Hemerly *et al.*, 1992), and they are closest related to the A-type cyclins in terms of their amino acid sequence. B-type cyclins have been subdivided in three classes, CYCB1, CYCB2 and CYCB3 (Renaudin *et al.*, 1996; Vandepoele *et al.*, 2002), which recently has been confirmed by global analysis of *Arabidopsis* genome sequences (Wang *et al.*, 2004a; Menges *et al.*, 2005). CYCB1 and CYCB2 groups exist in both monocotyledonous and eudicotyledonous species but the CYCB3 group has not been found in the rice genome or any other monocot species thus far (Wang *et al.*, 2004a; La *et al.*, 2006). The CYCB1 group in *Arabidopsis* and rice is by far the largest with five members for both species (Wang *et al.*, 2004a; La *et al.*, 2006). The typical cyclin B signature (H/Qx(K/R/Q)(F/L) motif has been found in all the B-type cyclins of *Arabidopsis*.

Besides the similarities in sequence between A- and B-cyclins, the cell cycle expression profiles in synchronized *Arabidopsis* cells of CYCBs are indistinguishable from those of CYCA1 and CYCA2 cyclins (Menges *et al.*, 2005), with a peak in early mitosis. This mitotic-specific expression pattern of B-type cyclins is dependent on MSA elements in their promoters (Ito *et al.*, 1997, 1998, 2001).

The biochemical activity of plant CYCBs was shown to be conserved with the animal cyclin B by microinjection of tobacco CYCB1;1 mRNA into *Xenopus* oocytes. Tobacco CYCB1;1 was sufficient to overcome the natural G2/M arrest in the oocytes (Qin *et al.*, 1996). Ectopically expressed CYCB1;1 in *Arabidopsis* was shown to stimulate cell division in root, resulting in faster growth rate (Doerner *et al.*, 1996). On the other hand, ectopic expression of CYCB1;2 was able to induce extra cell division in trichomes of *Arabidopsis*, cells that normally endoreduplicate, indicating that plant CYCB1 are both functioning in and sufficient to promote the G2/M transition. Interestingly, ectopically expressed CYCB1;1 was not able to induce extra cell division in the same cells, suggesting specialized functions for the B-type cyclins (Schnittger *et al.*, 2002a). CYCB2 appears to control progression into mitosis because ectopic expression of a CYCB2 protein from alfalfa in tobacco plants leads to earlier mitosis (Weingartner *et al.*, 2003). In rice, CYCB2 was shown to associate with the mitotic CDKB2;1 and to co-locate to metaphase chromosomes located at the centre of metaphase cells, and then to disappear after cells pass through metaphase (Lee *et al.*, 2003). The chromosome association and destruction after metaphase was previously reported for two CYCB1 proteins of tobacco (Criqui *et al.*, 2001).

2.2.5 D-type cyclins

The first main control point during the plant cell cycle is in the G1/S boundary at which the cell commits itself to DNA synthesis. D-type cyclins are proposed to be the primary sensors of external conditions and therefore mediate the control of this first checkpoint in plants by directing phosphorylation of the plant homologue of the retinoblastoma protein, RBR (Boniotti and Gutierrez, 2001; Dewitte and Murray, 2003; de Jager *et al.*, 2005; Uemukai *et al.*, 2005), which is rate limiting for S-phase entry (Menges *et al.*, 2006). Plant CYCDs were first identified functionally on their ability to rescue yeast strains lacking endogenous G1 (CLN) cyclins (Dahl *et al.*, 1995; Soni *et al.*, 1995), and indeed the very low homology with animal cyclin D of around 20% amino acid identity makes homology-based cloning impractical. Originally, Soni *et al.* (1995) cloned three D-type cyclins, now known as CYCD1;1, CYCD2;1 and CYCD3;1. Genome sequencing now shows that the complete CYCD clade in *Arabidopsis* consists of ten members in seven subclasses. Although their low similarity to animal D-type cyclins means that an orthologous relationship could not be established (Wang *et al.*, 2004a), the domain responsible for Rb binding is conserved between animals and plants. This domain, consisting of an LxCxE motif, has been demonstrated to be necessary for the interaction *in vitro* between CYCD2;1 and CYCD3;1 and RBR (Ach *et al.*, 1997; Huntley *et al.*, 1998). The *Arabidopsis* CYCD4;2 and CYCD6;1 sequences do not have this canonical LxCxE motif and CYCD5;1 has a slightly divergent motif (Vandepoele *et al.*, 2002; Dewitte and Murray, 2003).

In *Arabidopsis*, it was shown that both CYCD2;1 and CYCD3;1 are complexed in cultured cells with CDKA but not with detectable levels of CDKB1;1 (Healy *et al.*, 2001). In fact, the antibody used also cross-reacted with all other CDKBs in *Arabidopsis*, including CDKB2, showing that the CYCDs do not detectably interact with any CDKB in immunoprecipitation from cultured cells (unpublished data). In contrast, CYCD4;1 was found to bind to CDKB *in vitro*, although this has not been confirmed by immunoprecipitation from cell extracts (Kono *et al.*, 2003). The control of the G1/S transition by CYCDs has therefore been proposed to depend on CDKA activity, as CDKA remains present throughout the cell cycle, whereas CDKB activity is limited to S phase and later (Sorrell *et al.*, 2001; de Jager *et al.*, 2005).

CYCD2;1 and CYCD3;1 are the best studied D-type cyclins in *Arabidopsis*. Their expression is regulated by extrinsic signals, such as sucrose, but the protein levels of CYCD2;1 remain largely constant upon sucrose removal whereas the CYCD3;1 protein levels follow the transcript kinetics and decline more rapidly (Riou-Khamlichi *et al.*, 2000; Planchais *et al.*, 2004). The constant protein level of CYCD2;1 notwithstanding, the kinase activity associated with both CYCD2;1 and CYCD3;1 dropped after sucrose removal. Immunoprecipitation demonstrated that in stationary phase cells, CYCD2;1 is not associated with CDKA but is sequestered in an unknown complex, rendering the CYCD2;1-associated kinase inactive without affecting the protein levels (Healy *et al.*, 2001). Apart from being regulated by sucrose availability, the expression of CYCD3;1, but not that of CYCD2;1, responds strongly to plant hormones, especially cytokinin and brassinosteroids (Riou-Khamlichi *et al.*, 1999; Hu *et al.*, 2000; Oakenfull *et al.*, 2002).

Using *Arabidopsis* cell cultures, it was shown that CYCD3;1 is a highly unstable phosphoprotein, degraded by a proteasome-dependent pathway, and its presence is dependent on continuous translation (Planchais *et al.*, 2004). Furthermore, when the *Arabidopsis* cells were treated with the proteasome inhibitor MG132, the hyper-phosphorylated form of CYCD3;1 accumulated, indicating a link between CYCD3;1 phosphorylation and degradation. In fact, this connection has been often made between phosphorylation and ubiquitin-dependent proteolysis (Vierstra, 2003).

2.2.6 Other cyclins

Published data about C-, H-, L-, P-, Q- (J18)-, T- and SDS-type cyclins found in *Arabidopsis* are rather limited compared to those about the A-, B- and D-type cyclins. The members of these 'other' cyclins that are present on the *Arabidopsis* genome-wide ATH1 GeneChip do not show clear cell cycle regulation (Plates 2.1 and 2.2; Menges *et al.*, 2005).

Although La and coworkers do not report a rice C-type cyclin in their analysis of the rice genome, there is a genbank accession of a rice C-type cyclin (D86925) (Lee *et al.*, 2006), which corresponds to the locus Os09g32720 in the rice genome. Little is known about the function of C-type cyclins in plants, apart from expression profiles generated by microarray analyses (see for overview Plates 2.1–2.3; Figures 2.2–2.4). Its function may be similar to CYCH, as it was found in animals that CYCC binds and activates both CDK8, which in turn can phosphorylate the C-terminal domain (CTD) of the large subunit of RNA polymerase II, and the cell-cycle-related CDK3 (Lew *et al.*, 1991; Rickert *et al.*, 1996; Hoeppner *et al.*, 2005).

H-type cyclins have been found in *Arabidopsis*, rice and poplar (Yamaguchi *et al.*, 2000; Vandepoele *et al.*, 2002). In animal cells the CYCH homologue associates with the CDK7/$p40^{MO15}$ protein to form an active CDK-activating kinase (CAK) that phosphorylates cell cycle CDKs (CDK1, 2, 4 and 6) within their activation segment (T-loop) as well as forming part of the general transcription factor TFIIH, which phosphorylates the CTD of RNA polymerase II (Fisher and Morgan, 1994; Wallenfang and Seydoux, 2002; Fisher, 2005). In rice, CYCH was shown to interact with the CDK7 family member R2 and activates the resulting CAK (Yamaguchi *et al.*, 2000). In *Arabidopsis*, the single CYCH gene encodes a protein that interacts with the three putative CDKDs (CDK7 homologues) present in the *Arabidopsis* genome, leading to kinase activities of CDKD;2 and CDKD;3 which phosphorylate both CDKs and the CTD of RNA polymerase II (Umeda *et al.*, 1998; Shimotohno *et al.*, 2003; Shimotohno *et al.*, 2004; Umeda *et al.*, 2005). *Arabidopsis*, like yeast, also possesses a monomeric CAK called CDKF with exclusively CDK-activating phosphorylation activity.

Cyclin L is related to cyclin C and has been identified in flies worms and mammals and interacts with CDK11 (PITSLRE kinase) and phosphorylates the CTD of RNA polymerase II and splicing factor SC35 (Dickinson *et al.*, 2002). The *CYCL1;1*

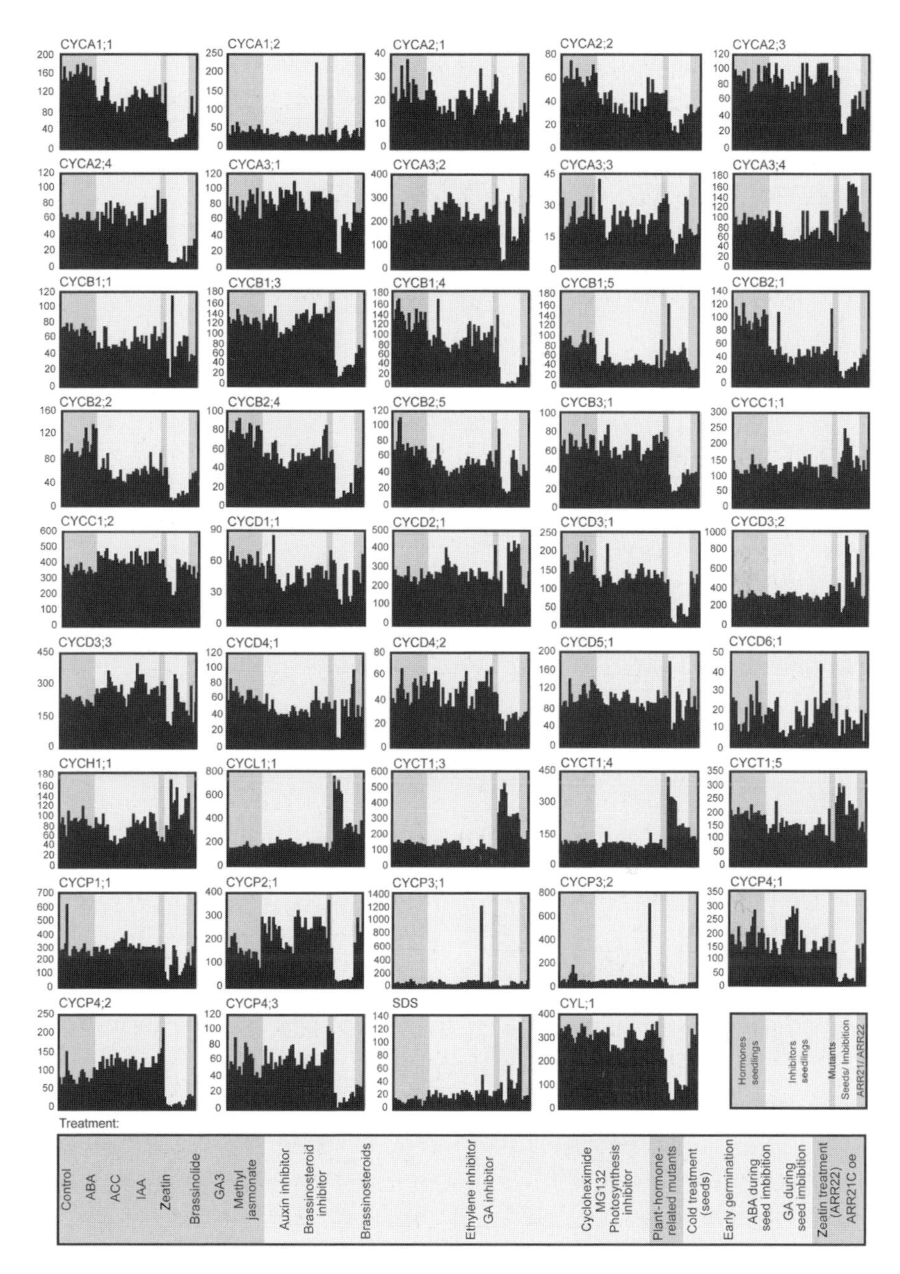

Figure 2.2 Global transcript profiling analysis of Arabidopsis cyclin genes in response to hormone treatment and during germination. Each graph represents the detected averaged absolute signal across a wide range of hormone-related experiments. The datasets used for the analysis were downloaded from TAIR. For more detailed information on the individual experiments in the graphs presented here, visit our web page (http://www.biot.cam.ac.uk/jahm/cyclins).

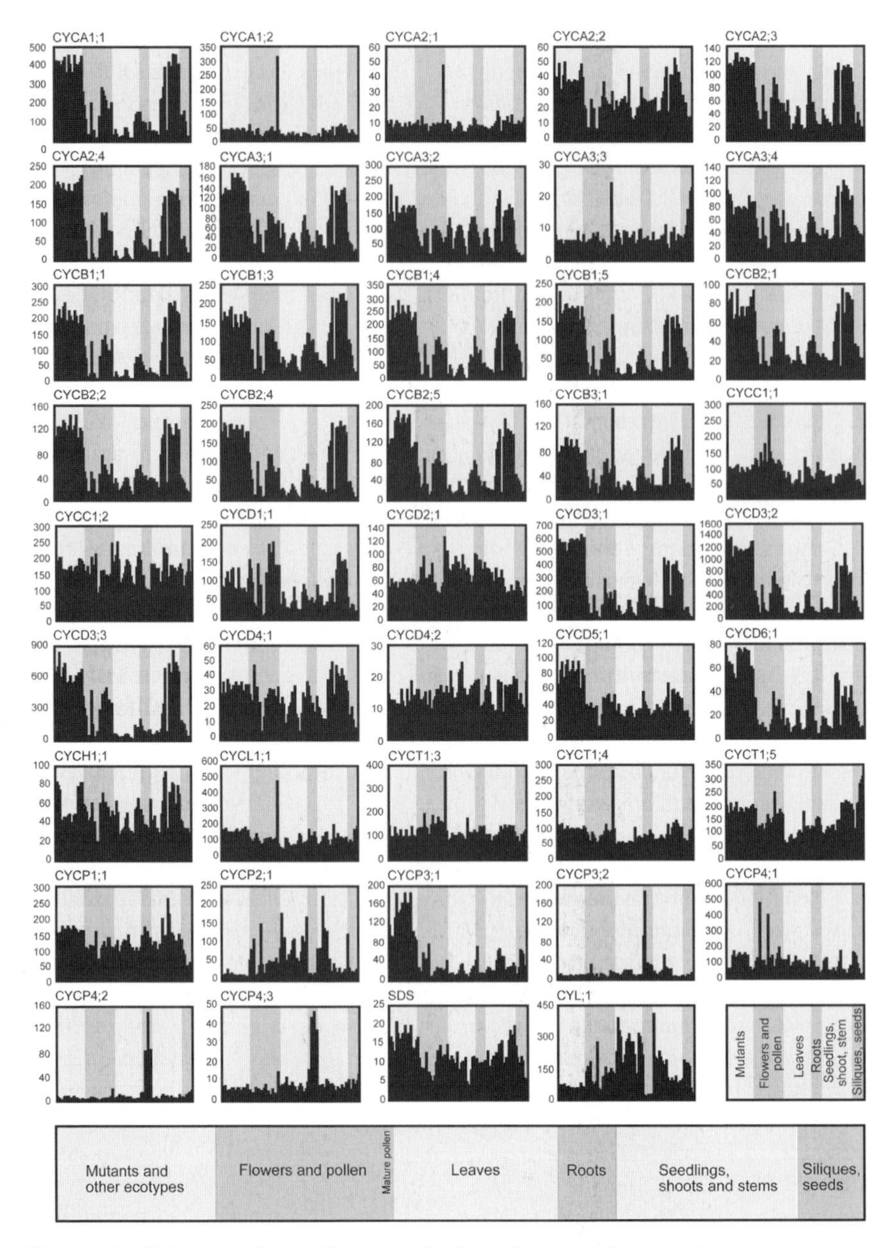

Figure 2.3 Global transcript profiling analysis of Arabidopsis cyclin genes in developmental series. Each graph represents the detected averaged absolute signal across a wide range of tissues taken at various developmental stages. For more detailed information see our web page (http://www.biot.cam.ac.uk/jahm/cyclins).

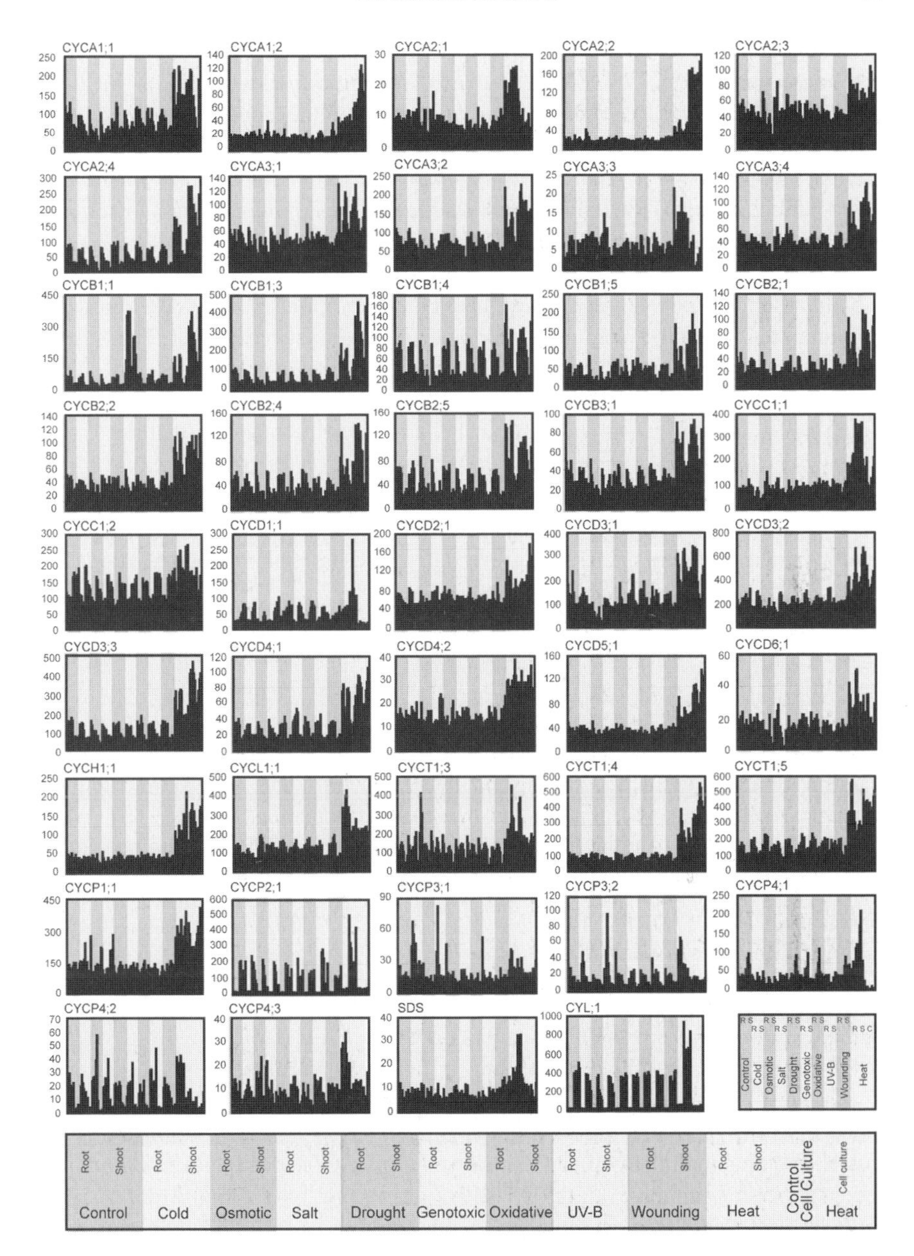

Figure 2.4 Global transcript profiling analysis of Arabidopsis cyclin genes in response to stress. Each graph represents the detected averaged absolute signal across a wide range of experiments after applying various stress treatments to roots, shoots or cell culture. All stress treatments are applied to root and shoot samples. The heat series is the most extensively studied stress treatment and expression profiles are followed after applying heat stress and also later during the recovery phase using root and shoot samples as well as cell culture. For more detailed information see http://www.biot.cam.ac.uk/jahm/cyclins.

gene of *Arabidopsis* was isolated in a screen for salt tolerance by expressing plant cDNAs in yeast. CYCL1;1, also called AtRCY1, conferred tolerance to LiCl and NaCl when expressed in yeast, but the tolerance was dependent only on the 'alternating arginine-rich' (RS) domain, characteristic of the SR family of splicing factors, independently of the nature of the full-length protein (Forment *et al.*, 2002). This finding as well as the transcription profile of CYCL1;1 upon stress treatments, shown in Figure 2.4, suggest that CYCL1;1 does not necessarily function directly in salt tolerance, although perhaps it regulates splicing of salt tolerance genes.

The plant P-type cyclins are designated as such because they share homology with the PHO80 component of the 'PHO-regulatory' system in yeast. This system regulates gene expression in response to phosphate starvation and contains a CDK–cyclin complex, PHO85/PHO80 respectively, an example of such a complex not functioning in cell cycle regulation. In *Arabidopsis*, it was shown that the P-type cyclins are differentially expressed in plant tissues. CYCPs were tested in a yeast two-hybrid system for interaction with CDK proteins and all members were able to interact with CDKA;1, whereas CYCP1;1 binds to CDKB1;1 as well. Furthermore, it was shown that only CYCP4;2 can restore the response to phosphate starvation in the *pho80* yeast mutant (Torres Acosta *et al.*, 2004).

In the same screen that we used to clone CYCDs by complementation (Soni *et al.*, 1995), a further highly divergent cyclin, originally called CYCJ18, was also identified which also complemented the same yeast cln^{-} mutant (Abrahams *et al.*, 2001). However, until late 2005, the J18-cyclin has been misannotated in the *Arabidopsis* genome, and both Wang and coworkers and La and coworkers used this incorrectly annotated sequence in their analysis, explaining why they could not detect its homologue in rice although it is in fact present as Os03g13480 (Wang *et al.*, 2004a; La *et al.*, 2006). To comply with the conventional cyclin nomenclature and to avoid future confusion with the previous misannotation in *Arabidopsis*, we propose to rename CYCJ18 as CYCQ1;1 with the gene ID At2g01905.

The plant T-type cyclins are related to the animal K/L/T-cyclins (Wang *et al.*, 2004a). Furthermore, it was shown by yeast two-hybrid studies that a CYCT homologue was able to bind *Arabidopsis* CDKC and a *Medicago* CDKC–CYCT1 complex phosphorylates the CTD of RNA polymerase II to promote transcription, suggesting that plant CYCTs have similar functions as animal CYCTs (Barrôco *et al.*, 2003; Fülöp *et al.*, 2005).

The SDS (SOLO DANCERS) cyclin in *Arabidopsis* is a highly divergent cyclin expressed specifically in male and female meiotic cells and is required for meiotic recombination (Azumi *et al.*, 2002). Although sharing only about 30% similarity with other cyclins, SDS was shown to be able to interact with CDKs in a yeast two-hybrid system, proving it is a real cyclin (Azumi *et al.*, 2002). The SDS protein does contain putative PEST elements and has both the cyclin N and cyclin C domains, and although not present in animals, it is conserved in rice (La *et al.*, 2006).

Animal cyclin F is an unusual cyclin that contains an F-box that can form SCF complexes implicated in degradation of target proteins, and is responsible for the nuclear localization of cyclin B1–CDK1 complexes by binding to the cyclin B1 subunit (Fung *et al.*, 2002). The rice F-type cyclins are evolutionarily more related to A-, B- and D-type cyclins from other species, including plants and animals, suggesting similar functions (La *et al.*, 2006). Apart from the finding that transcript of some rice CYCFs were detected in roots but not in shoots, nothing is known about the F-type cyclins at the moment (La *et al.*, 2006).

2.3 Expression of cyclins during the cell cycle

2.3.1 The G1 checkpoint

By genome-wide analysis of a synchronized *Arabidopsis* cell suspension culture, the expression timing of the core cell cycle regulators has been identified (Menges *et al.*, 2002, 2003, 2005). These cell cultures were synchronized in either re-entry of the cell cycle from G1 phase by sucrose starvation or at the G1/S boundary by a block/release with the inhibitor aphidicolin (Menges and Murray, 2002). Furthermore, by analyzing 327 publicly available microarray datasets, it was found that most of the core cell cycle genes are expressed across almost all plant tissues, although some show strong tissue specificity (Menges *et al.*, 2005).

As expected, cyclins showed distinct regulation through the cell cycle in the experiments with the *Arabidopsis* cell suspension culture (Plates 2.1 and 2.2). As cells re-enter the cell cycle through G1, the earliest cyclins that were detected are CYCD3;3 and CYCD5;1, belonging to the class of D-type cyclins (see Plates 2.1 and 2.2 and Figure 2.5). The expression levels of CYCD3;3 were fivefold higher than those of CYCD5;1, suggesting that CYCD3;3–CDKA provides the major part of CDK activity. Later in G1 phase, CYCD3;1, CYCD4;1 and CYCD4;2 levels rise to supplement the early G1 cyclins and peak at the G1/S boundary (Menges *et al.*, 2005). This G1/S expression pattern confirms early analysis of CYCD3;1 in

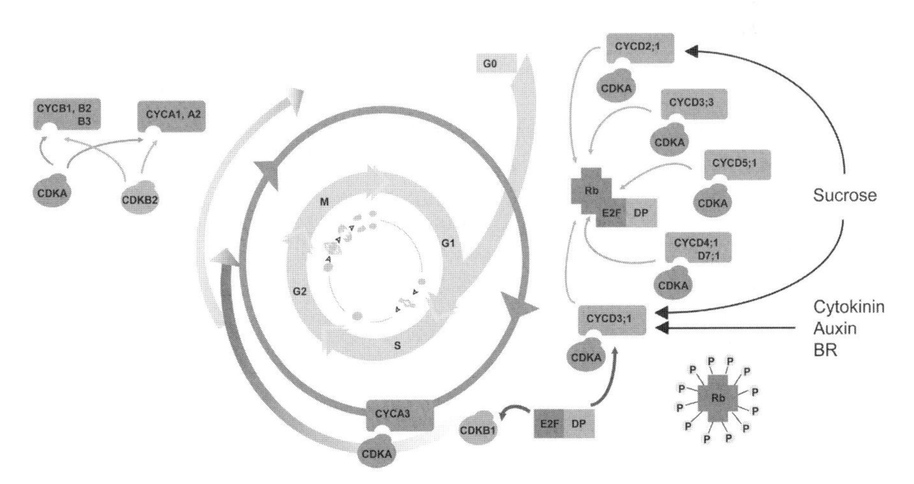

Figure 2.5 Transcriptional regulation of cell cycle control in Arabidopsis (Modified from Menges *et al.*, 2005.) For further details, see text.

Arabidopsis (Menges and Murray, 2002) and a CYCD3 orthologue in alfalfa (Dahl *et al.*, 1995). Experiments using aphidicolin synchronization ('cell cycle' panel in Plate 2.2) showed that CYCD5;1, CYCD4;1 and CYCD4;2 oscillate in the G1 phase of the subsequent cycle but that the early G1-phase- and late G1-phase regulation of CYCD3;3 and CYCD3;1, respectively, is characteristic of cell cycle re-entry and is not observed in the next cycle (Menges *et al.*, 2005). CYCD6;1 and CYCD7;1, which are not strongly regulated during re-entry, show a late G1 expression pattern, similar to that of CYCD4;1 and CYCD4;2 (Menges *et al.*, 2005). These data from the *Arabidopsis* cell suspension culture indicate that progression of the cell cycle through the G1 and G1/S phases is driven by sequential waves of different CYCD-associated kinase activity and that re-entry in the cell cycle and subsequent cycling seems differentially regulated by D-type cyclins.

The *CYCD3;1* and *CYCD3;3* genes from *Arabidopsis* are highly expressed compared to the other D-type cyclins, suggesting that the *CYCD3* genes play the lead role in G1 and G1/S control for cell cycle re-entry. This is supported by studies showing that CYCD3;1 is a rate-limiting factor for G1–S transition (Menges *et al.*, 2006), and overexpression of CYCD3;1 in *Arabidopsis* leads to more cell divisions and inhibited cells from exiting the mitotic cell cycle (Dewitte *et al.*, 2003). The question now is what caused the increase in CYCD3 transcript levels which permits cells to progress beyond the G1 checkpoint? In animals, transcription of D-type cyclins is completely dependent on the presence of serum growth factors (Sherr, 1993). In plants, it was shown that CYCD3;1 expression is dependent on cytokinins and sucrose on both the transcript (Soni *et al.*, 1995; Riou-Khamlichi *et al.*, 1999, 2000; Oakenfull *et al.*, 2002) and protein level (Planchais *et al.*, 2004). However, the regulation on transcript and protein levels differs significantly between D-type cyclins, as it was shown that although CYCD2;1 and CYCD3;1 mRNA levels are inducible by sucrose (Riou-Khamlichi *et al.*, 2000), sucrose removal caused proteasome-dependent destruction of the CYCD3;1 protein, whereas CYCD2;1 proteins remained stable (Planchais *et al.*, 2004). The degradation of cyclin proteins is essential for the cell cycle oscillator and will be further discussed below.

2.3.2 S phase

Once plant cells pass the G1–S checkpoint, they commit to complete DNA synthesis. This event is marked by major shift in the expression of core cell cycle genes (Menges *et al.*, 2005). However, in contrast to the situation in G1 and the G1/S boundary, the expression patterns of the cell cycle regulators including the cyclins are very similar in cells synchronized by aphidicolin and those re-entering the cycle after sucrose starvation. The CYCA3 group of cyclins in *Arabidopsis* is transcriptionally activated at the G1/S transition and peak at S phase (Plate 2.1 and 2.2), which has also been found for *CYCA3* genes in soybean and in tobacco BY2 cells (Kouchi *et al.*, 1995; Setiady *et al.*, 1995; Reichheld *et al.*, 1996). *Arabidopsis* CYCA3;1 and CYCA3;2 levels decline gradually during S phase until they reach their base levels at G2 whereas CYCA3;4 remains constantly expressed through S phase.

2.3.3 G2–M

After DNA synthesis, the cyclin torch passes from the CYCA3s to the CYCA1, CYCA2 and CYCB1–B3 groups which peak at the G2/M transition and decline during mitosis (Ferreira *et al.*, 1994a; Fobert *et al.*, 1994; Kouchi *et al.*, 1995; Setiady *et al.*, 1995; Ito *et al.*, 1997; Mironov *et al.*, 1999; Chaubet-Gigot, 2000; Menges *et al.*, 2005). Since these subgroups are conserved in all higher plants, it is likely that they have distinct functional differences. Most of the genes that are induced in the mitotic phase have mitosis-specific activation (MSA) sites in their promoter regions (Menges *et al.*, 2005), which have been shown to be necessary and sufficient for phase-specific regulation (Ito *et al.*, 1998). The MSA site is activated by three repeat c-Myb-like transcription factors in tobacco (Ito *et al.*, 2001; Araki *et al.*, 2004), and the AtMYB3R4 Myb protein has been proposed to regulate the expression of the MSA-dependent genes in *Arabidopsis* (Menges *et al.*, 2005).

During mitosis, cyclin activity has to be destroyed in order to progress through the metaphase–anaphase transition, exit mitosis and return to the G1 phase. In animals, proteolysis of the A- and B-type cyclins is carried out by the anaphase-promoting complex (APC/C), which is a multisubunit ubiquitin ligase (Townsley and Ruderman, 1998). This destruction is dependent on a characteristic protein domain, the destruction box or Dbox. In plants, the Dbox pathway has been first identified in two A- and B-type cyclins from tobacco, where it was shown that the Dbox caused degradation specifically at the exit of mitosis and that it is proteasome dependent (Genschik *et al.*, 1998). Tobacco CYCB1;1, CYCB1;3, CYCB2 and CYCA3;1 are localized to the nucleus as cells enter mitosis and they are destroyed at specific phases during mitosis (Criqui *et al.*, 2000, 2001; Weingartner *et al.*, 2003). Later, more detailed studies revealed that the mitotic tobacco cyclins CYCB1;1 and CYCB1;3 become associated with chromosomes during prophase and are then destroyed after metaphase (Criqui *et al.*, 2000, 2001). In addition, a study of the tobacco CYCA3;2 protein fused with GFP suggested that CYCA3 proteins are also destroyed at M phase, although the timing was not established (Yu *et al.*, 2003). Differences between Dboxes of different CYCA groups may suggest subtly different timing of destruction (Reichheld *et al.*, 1996).

2.4 Cyclins in plant development

The large diversity in expression and potential for biochemical interactions with different partners make cyclins fit for the task to regulate cell division in plant development. Unfortunately, given the large number of potentially redundant genes, this also implies a major challenge to untangle the functions of cyclins during the life cycle of plants. At the moment, there are several tools and approaches available for the plant research community to attack the problem. Here we present a global analysis of *Arabidopsis* cyclin expression in plant development, using public transcript expression databases derived from experiments using Affymetrix

ATH1 GeneChip arrays (Figures 2.2–2.4). This analysis shows absolute expression levels, but because it is derived from a large number of independent experiments in different laboratories often repeating similar experiments, any anomalous results are readily apparent. However, the data from arrays are limited in spatial and temporal resolution and not all defined cyclin genes are represented by probes on the GenChip. To investigate the transcript expression of cyclins on the cellular level, *in situ* hybridization and promoter-reporter lines can be used (for example, Kono *et al.*, 2003; Masubelele *et al.*, 2005). Nevertheless, extrapolation to protein levels and kinase activity should be done with caution, since cyclins are also heavily regulated on the protein level (see Chapter 4).

Gene misexpression is widely used as an approach to study the function of a particular gene. However, many cyclins have overlapping expression both in the cell cycle (Menges *et al.*, 2005) as well as during plant development (Figure 2.3), which will certainly cause functional redundancy and hence difficulties in detecting phenotypes in the case of downregulation or loss-of-function analyses of a single gene. Overexpression generally gives rise to results that reflect an overall increase in cell division rather than necessarily the specific function of an individual gene. Indeed, any change in the function of a cell cycle component may have an effect on cell division, leading possibly to a developmental effect. Therefore, one would ideally like to perform spatio-temporal analysis of the dividing cell pool *in planta*. The range of tools to do this is currently limited to the mitotic phase. An important mitotic marker has been made firstly by bringing the GUS reporter under control of the CYCB1;1 promoter (Hemerly *et al.*, 1993) and was perfected later in Peter Doerner's laboratory by fusing a cyclin destruction box to GUS, resulting in a reporter that is produced at the G2/M transition and rapidly degraded after anaphase (Colon-Carmona *et al.*, 1999). Hence, staining for GUS activity results in a 'spotty' pattern, which identifies individual cells caught in G2/M phase (see, for example, Plate 2.3). A fusion of *Arabidopsis* histone 2B with YFP has also been used as a marker for chromatin and enabled the observation of nuclei and mitoses in living root meristems (Boisnard-Lorig *et al.*, 2001; Wildwater *et al.*, 2005). Although reporter fusions have been made using noncyclin promoters of S-phase-specific genes, they do not mark cells in S phase exclusively, since the proteins are not degraded after S phase or mitosis.

Despite the limitations of ectopic expression, considerable insight has been obtained into the role of cell division itself as well as specific cyclin groups. In understanding the phenotypes observed in whole plants, it is important to appreciate that overall plant and organ growth depends on two types of cell cycles. The first is the conventional mitotic cycle, whereas the second involves repeated S phases without an intervening mitosis. Such endocycles lead to the production of polytene chromosomes containing multiple replicated chromatides and a doubling of nuclear DNA content for each repeated cycle. Whereas mitotic cycles are associated with meristematic cells and the early phases of growth of newly formed organs, endocycles are associated with the increase in cellular volume that drives most of plant organ growth. Hence, increased or reduced growth may result from effects on mitotic or endocycles, or both. Overexpression of the mitotic cyclin B1;1 in *Arabidopsis* resulted in an increased rate of overall growth, particularly in roots (Doerner *et al.*, 1996; Li *et al.*, 2005). This was supported by overexpression of rice CYCB2 in rice roots, which resulted in a higher rate of growth without affecting cell size (Lee *et al.*, 2003). These results suggest that the G2/M transition may be rate limiting for growth, particularly in roots.

An analogous effect was observed when *Arabidopsis* CYCD2;1 was expressed in transgenic tobacco (Cockcroft *et al.*, 2000), but in this case an increased growth rate of the whole plant was observed. However, the effect was limited to heterologous expression, since the tobacco *CYCD2* gene did not produce the same effect (unpublished data). The analysis showed that the G1 phase was shortened in faster growing transgenic plants, showing that growth rate can be enhanced by promoting the G1/S transition.

However, very different effects were found when the *CYCD3;1* gene was overexpressed in *Arabidopsis*. Expression of CYCD3;1 is hormone responsive, particularly to cytokinin, and indeed explants from leaves ectopically expressing CYCD3;1 were able to form proliferating green callus in the absence of exogenous cytokinin (Riou-Khamlichi *et al.*, 1999). Further analysis showed that CYCD3;1 promotes the G1/S transition and pushes cells from G1 into S phase, without reducing overall cell cycle length. As a result, cells spend more time in G2 phase (Dewitte *et al.*, 2003; Menges *et al.*, 2006). In the context of leaf development, CYCD3;1-overexpressing plants have small curled leaves composed of a large number of small, relatively undifferentiated, cells. Indeed, CYCD3;1 reduced the level of endoreduplication and caused the major mode of leaf growth to switch from a cell expansion-driven to a cell-division-driven process. This suggests a possible role for CYCD3;1 in the switch from mitosis to endocycles during organ growth. Overexpression of tobacco CYCA3;2 also reduced endoreduplication and cellular differentiation (Yu *et al.*, 2003). In contrast, *Medicago* CYCA2;2 overexpression did not increase protein levels and had no phenotype, whereas an antisense construct blocked somatic embryo regeneration (Roudier *et al.*, 2003).

An interesting alternative system to investigate cyclin function is overexpression in trichomes. These single cell structures normally contain a single endoreduplicated nucleus. However, ectopic expression of CYCD3;1 results in additional mitotic cycles to produce multicelled trichomes (Schnittger *et al.*, 2002a, b) as does expression of CYCB1;2 but not CYCB1;1. This shows that either ectopic CYCD3;1 or CYCB1;2 is sufficient to convert a trichome endocycle into a mitotic cycle.

Recently, it was found that there are four phases of cell cycle activation during *Arabidopsis* germination relating to morphological changes (Plate 2.3). Using genome-wide analysis of transcripts accumulated during these phases in germination, major changes in transcript levels of >2000 genes have been reported (Masubelele *et al.*, 2005). In Plate 2.3, we show the analysis of expression of the cyclins during *Arabidopsis* germination, including the mid-imbibition phase. Almost all cyclins present on the Affymetrix ATH1 GeneChip show distinct expression patterns during germination, indicative of the major changes occurring during metabolic, developmental and cell cycle reactivation.

CYCA3;3 shows a peak of expression during mid-imbibition, while others including two T-type cyclins and CYCL are upregulated during both mid-imbibition and the end of stratification (T0), likely reflecting the activation of the transcriptional machinery. At the time points before and during the activation of the root apical meristem (until 33 h), six out of the ten D-type cyclins are upregulated as well as two CYCA3s and the SDS cyclin. The highest mitotic activity was detected at later stages (phase III, 33–42 h), when cell division in the cotyledons and the root apical meristem is activated, and this coincides with the upregulation of cyclins implied in G2/M transition, such as the A1-, A2- and B-type cyclins. CYCP2;1 is the most differentially regulated cyclin and reaches high expression at the last stages (phase IV) of germination when the shoot apical meristem is fully activated and lateral organs are initiated (Masubelele *et al.*, 2005).

The D-type cyclins that are upregulated in the early phases of germination are rate limiting for the activation of cell division, as loss-of-function mutants of these CYCDs showed a delay in cell division and germination despite the absence of any gross overall phenotype (Masubelele *et al.*, 2005). As mentioned before, it has been shown that D-type cyclins can be regulated by external conditions and plant hormones. But are other cyclins also responsive to these signals? We compiled data from publicly available genome-wide microarrays of *Arabidopsis* used in experiments related to hormone signalling including hormone treatment, inhibitor and germination experiments (Figure 2.2).

As expected, several D-type cyclins show clear differential expression in hormone and inhibitor experiments with the notable exception of CYCD3;2, which shows a constant high basal level of expression but is upregulated during germination experiments, confirming the data of Plate 2.3. Besides the D-type cyclins, other classes also show differences in expression patterns in hormone-related experiments, including the A2-, A3- and P-type cyclins. Interestingly, some cyclins are found to be specifically high in germination experiments, compared to expression profiles detected in experiments after various hormone and inhibitor treatments of 7-day-old seedlings. A similar expression profile to CYCD3;2 is observed for CYCL1;1, CYCT1;3 and CYCT1;4. In contrast, other cyclins such as CYCA1;1, CYCA2;4, CYCB1;3, CYCB1;4, CYCP2;1 and CYCP4;1 show a inverse expression pattern with lower expression during early germination (Plate 2.3 and Figure 2.2). Strikingly, the expression of the CYCP3s and CYCA1;2 is greatly enhanced by treatment of seedlings with the protein synthesis inhibitor cycloheximide. CYCP1;1 and CYCP4;2 show upregulation after treatment with ABA.

Numerous studies have shown that cyclins show specific expression on organ, tissue and cellular level. The publication of an expression map of *Arabidopsis* development made it possible to have a bird's eye view of cyclin expression during development at the organ level (Schmid *et al.*, 2005), and Figure 2.3 shows that virtually all cyclins show differential expression. It also becomes clear that the maximum and basal transcript levels differ substantially between the cyclins, even within the different subclasses. However, such differences may reflect overall low expression, or differentiated expression, at the individual cell level. Several cyclins show very specific organ expression patterns, such as CYCA1;2, CYCA2;1, CYCA3;3,

CYCL1;1, CYCT1;3 and CYCT1;4, which are expressed specifically in mature pollen. Three members of the P-type cyclins, CYCP3;2, CYCP4;2 and CYCP4;3, appear to be root specific. Most of the mitotic cyclins, the CYCA3s, CYCH1;1 and some of the D-type cyclins show a common expression pattern that may possibly be a 'background' pattern reflecting the concentration of dividing cells in the samples. Notable exceptions to this are CYCA1;2, CYCA2;1 and CYCA3;3, which look like having specific roles in meiosis or male gametophyte development.

The sessile lifestyle of plants makes it necessary for them to respond and adapt to environmental changes. Light and nutrient availability, temperature and biotic stress impact on the rate and sometimes the pattern of plant growth and hence also influence the cell cycle. To have a general idea how various stress parameters may change *Arabidopsis* cyclin transcript levels, we generated a further overview using publicly available microarray data (Figure 2.4).

An example of a cyclin responding to heat stress is CYCD4;1, which shows downregulation of gene expression 1h after applying heat stress in roots and cell culture and recovers when cells are allowed to recover under normal temperature conditions for 3–6 h. The classes of A-, B-, C-, D-, P- and H-type cyclins have at least one member that shows a similar response to heat stress and change in expression similar to CYCD4;1. In contrast, CYCA1;2 and CYCT1;4 show upregulation in response to heat stress and lower expression after recovery. Out of the cyclin classes, the P-type cyclins respond the most to different stresses. CYCP1;1 and CYCP3;1 are upregulated by cold, salt and drought stress where CYCP3;1 expression is also promoted by UV-B treatment. CYCT1;3 is specifically induced by change in temperature (cold and heat), whereas CYCP3;2 is responsive to salt treatment and heat stress. Intriguingly, genotoxic stress has a specific and large effect on one cyclin, the mitotic cyclin CYCB1;1. Oxidative stress and wounding do not show a clear effect on the expression of any cyclin present in the microarray data. Since many of the stress treatments are presented alternatively for root and shoot samples, the difference in expression of a number of cyclins is very apparent from Figure 2.4, notably the shoot-specific expression of CYCP2;1, CYCC1;2 and CYL;1 and the predominance of root expression of CYCA2;4, CYCD3;3 and CYCD4;1, as well as of CYCP4;2.

2.5 Concluding remarks

The last few years have seen continuing progress in our understanding of the biochemistry of the plant cell cycle, but the most striking advances have resulted from the availability of complete genome sequence information for *Arabidopsis* and, more recently, rice. Combined with microarray technologies, this has allowed not only the definition of the large family of plant cyclins, but also a global understanding of their regulation. The challenge for the next few years is to create a developmental framework that incorporates the full gamut of cyclins, so we can understand both why plants encode such large numbers of cyclin genes and how they cooperate to integrate and control cell division and other cellular processes with development.

Acknowledgments

We thank Walter Dewitte for help on the manuscript and Klaus Herbermann for assistance on bioinformatic analysis. Jeroen Nieuwland was supported by a Marie Curie Fellowship of the European Commission (010666) and Margit Menges was funded by BBSRC grant BBS/B/13268.

References

Abrahams, S., Cavet, G., Oakenfull, E.A., *et al.* (2001) A novel and highly divergent Arabidopsis cyclin isolated by complementation in budding yeast. *Biochim Biophys Acta* **1539**, 1–6.

Ach, R.A., Durfee, T., Miller, A.B., *et al.* (1997) RRB1 and RRB2 encode maize retinoblastoma-related proteins that interact with a plant D-type cyclin and geminivirus replication protein. *Mol Cell Biol* **17**, 5077–5086.

Araki, S., Ito, M., Soyano, T., Nishihama, R. and Machida, Y. (2004) Mitotic cyclins stimulate the activity of c-Myb-like factors for transactivation of G2/M phase-specific genes in tobacco. *J Biol Chem* **279**, 32979–32988.

Azumi, Y., Liu, D., Zhao, D., *et al.* (2002) Homolog interaction during meiotic prophase I in Arabidopsis requires the SOLO DANCERS gene encoding a novel cyclin-like protein. *EMBO J* **21**, 3081–3095.

Barrôco, R.M., De Veylder, L., Magyar, Z., Engler, G., Inzé, D. and Mironov, V. (2003) Novel complexes of cyclin-dependent kinases and a cyclin-like protein from *Arabidopsis thaliana* with a function unrelated to cell division. *Cell Mol Life Sci* **60**, 401–412.

Beeckman, T., Burssens, S. and Inzé, D. (2001) The peri-cell-cycle in *Arabidopsis*. *J Exp Bot* **52**, 403–411.

Boisnard-Lorig, C., Colon-Carmona, A., Bauch, M., *et al.* (2001) Dynamic analyses of the expression of the HISTONE:YFP fusion protein in Arabidopsis show that syncytial endosperm is divided in mitotic domains. *Plant Cell* **13**, 495–509.

Boniotti, M.B. and Gutierrez, C. (2001) A cell-cycle-regulated kinase activity phosphorylates plant retinoblastoma protein and contains, in Arabidopsis, a CDKA/cyclin D complex. *Plant J* **28**, 341–350.

Boucheron, E., Guivarc'h, A., Azmi, A., Dewitte, W., Van Onckelen, H. and Chriqui, D. (2002) Competency of *Nicotiana tabacum* L. stem tissues to dedifferentiate is associated with differential levels of cell cycle gene expression and endogenous cytokinins. *Planta* **215**, 267–278.

Boucheron, E., Healy, J.H., Bajon, C., *et al.* (2005) Ectopic expression of Arabidopsis CYCD2 and CYCD3 in tobacco has distinct effects on the structural organization of the shoot apical meristem. *J Exp Bot* **56**, 123–134.

Burssens, S., de Almeida Engler, J., Beeckman, T., *et al.* (2000a) Developmental expression of the *Arabidopsis thaliana* CycA2;1 gene. *Planta* **211**, 623–631.

Burssens, S., Himanen, K., van de Cotte, B., *et al.* (2000b) Expression of cell cycle regulatory genes and morphological alterations in response to salt stress in *Arabidopsis thaliana*. *Planta* **211**, 632–640.

Chaubet-Gigot, N. (2000) Plant A-type cyclins. *Plant Mol Biol* **43**, 659–675.

Cho, J.W., Park, S.C., Shin, E.A., *et al.* (2004) Cyclin D1 and p22ack1 play opposite roles in plant growth and development. *Biochem Biophys Res Commun* **324**, 52–57.

Cockcroft, C.E., den Boer, B.G., Healy, J.M. and Murray, J.A. (2000) Cyclin D control of growth rate in plants. *Nature* **405**, 575–579.

Colon-Carmona, A., You, R., Haimovitch-Gal, T. and Doerner, P. (1999) Technical advance: spatio-temporal analysis of mitotic activity with a labile cyclin–GUS fusion protein. *Plant J* **20**, 503–508.

Cooper, B., Hutchison, D., Park, S., *et al.* (2003) Identification of rice (*Oryza sativa*) proteins linked to the cyclin-mediated regulation of the cell cycle. *Plant Mol Biol* **53**, 273–279.

Criqui, M.C., Parmentier, Y., Derevier, A., Shen, W.H., Dong, A. and Genschik, P. (2000) Cell cycle-dependent proteolysis and ectopic overexpression of cyclin B1 in tobacco BY2 cells. *Plant J* **24**, 763–773.

Criqui, M.C., Weingartner, M., Capron, A., *et al.* (2001) Sub-cellular localisation of GFP-tagged tobacco mitotic cyclins during the cell cycle and after spindle checkpoint activation. *Plant J* **28**, 569–581.

Dahl, M., Meskiene, I., Bögre, L., *et al.* (1995) The D-type alfalfa cyclin gene cycMs4 complements G1 cyclin-deficient yeast and is induced in the G1 phase of the cell cycle. *Plant Cell* **7**, 1847–1857.

Day, I.S. and Reddy, A.S. (1998) Isolation and characterization of two cyclin-like cDNAs from Arabidopsis. *Plant Mol Biol* **36**, 451–461.

Day, I.S., Reddy, A.S. and Golovkin, M. (1996) Isolation of a new mitotic-like cyclin from Arabidopsis: complementation of a yeast cyclin mutant with a plant cyclin. *Plant Mol Biol* **30**, 565–575.

Deckert, J., Jelenska, J., Gwozdz, E.A. and Legocki, A.B. (1996) The isolation of lupine cDNA clone coding for putative cyclin protein. *Biochimie* **78**, 90–94.

de Jager, S.M., Maughan, S., Dewitte, W., Scofield, S. and Murray, J.A. (2005) The developmental context of cell-cycle control in plants. *Semin Cell Dev Biol* **16**, 385–396.

De Veylder, L., de Almeida Engler, J., Burssens, S., *et al.* (1999) A new D-type cyclin of *Arabidopsis thaliana* expressed during lateral root primordia formation. *Planta* **208**, 453–462.

Dewitte, W. and Murray, J.A. (2003) The plant cell cycle. *Annu Rev Plant Biol* **54**, 235–264.

Dewitte, W., Riou-Khamlichi, C., Scofield, S., *et al.* (2003) Altered cell cycle distribution, hyperplasia, and inhibited differentiation in Arabidopsis caused by the D-type cyclin CYCD3. *Plant Cell* **15**, 79–92.

Dickinson, L.A., Edgar, A.J., Ehley, J. and Gottesfeld, J.M. (2002) Cyclin L is an RS domain protein involved in pre-mRNA splicing. *J Biol Chem* **277**, 25465–25473.

Doerner, P., Jorgensen, J.E., You, R., Steppuhn, J. and Lamb, C. (1996) Control of root growth and development by cyclin expression. *Nature* **380**, 520–523.

Donnelly, P.M., Bonetta, D., Tsukaya, H., Dengler, R.E. and Dengler, N.G. (1999) Cell cycling and cell enlargement in developing leaves of Arabidopsis. *Dev Biol* **215**, 407–419.

Evans, T., Rosenthal, E.T., Youngblom, J., Distel, D. and Hunt, T. (1983) Cyclin: a protein specified by maternal mRNA in sea urchin eggs that is destroyed at each cleavage division. *Cell* **33**, 389–396.

Favery, B., Complainville, A., Vinardell, J.M., *et al.* (2002) The endosymbiosis-induced genes ENOD40 and CCS52a are involved in endoparasitic-nematode interactions in *Medicago truncatula*. *Mol Plant Microbe Interact* **15**, 1008–1013.

Ferreira, P., Hemerly, A., de Almeida Engler, J., *et al.* (1994a) Three discrete classes of *Arabidopsis* cyclins are expressed during different intervals of the cell cycle. *Proc Natl Acad Sci USA* **91**, 11313–11317.

Ferreira, P.C., Hemerly, A.S., Engler, J.D., van Montagu, M., Engler, G. and Inzé, D. (1994b) Developmental expression of the Arabidopsis cyclin gene *cyc1At*. *Plant Cell* **6**, 1763–1774.

Fisher, D.L. and Nurse, P. (1996) A single fission yeast mitotic cyclin B p34cdc2 kinase promotes both S-phase and mitosis in the absence of G1 cyclins. *EMBO J* **15**, 850–860.

Fisher, R.P. (2005) Secrets of a double agent: CDK7 in cell-cycle control and transcription. *J Cell Sci* **118**, 5171–5180.

Fisher, R.P. and Morgan, D.O. (1994) A novel cyclin associates with MO15/CDK7 to form the CDK-activating kinase. *Cell* **78**, 713–724.

Fobert, P.R., Coen, E.S., Murphy, G.J. and Doonan, J.H. (1994) Patterns of cell division revealed by transcriptional regulation of genes during the cell cycle in plants. *EMBO J* **13**, 616–624.

Forment, J., Naranjo, M.A., Roldan, M., Serrano, R. and Vicente, O. (2002) Expression of Arabidopsis SR-like splicing proteins confers salt tolerance to yeast and transgenic plants. *Plant J* **30**, 511–519.

Freeman, D., Riou-Khamlichi, C., Oakenfull, E.A. and Murray, J.A. (2003) Isolation, characterization and expression of cyclin and cyclin-dependent kinase genes in Jerusalem artichoke (*Helianthus tuberosus* L.). *J Exp Bot* **54**, 303–308.

Fuerst, R.A., Soni, R., Murray, J.A. and Lindsey, K. (1996) Modulation of cyclin transcript levels in cultured cells of *Arabidopsis thaliana*. *Plant Physiol* **112**, 1023–1033.

Fülöp, K., Pettkó-Szandtner, A., Magyar, Z., *et al.* (2005) The *Medicago* CDKC;1–CYCLINT;1 kinase complex phosphorylates the carboxy-terminal domain of RNA polymerase II and promotes transcription. *Plant J* **42**, 810–820.

Fung, T.K., Siu, W.Y., Yam, C.H., Lau, A. and Poon, R.Y. (2002) Cyclin F is degraded during G2–M by mechanisms fundamentally different from other cyclins. *J Biol Chem* **277**, 35140–35149.

Gaudin, V., Lunness, P.A., Fobert, P.R., *et al.* (2000) The expression of D-cyclin genes defines distinct developmental zones in snapdragon apical meristems and is locally regulated by the Cycloidea gene. *Plant Physiol* **122**, 1137–1148.

Genschik, P., Criqui, M.C., Parmentier, Y., Derevier, A. and Fleck, J. (1998) Cell cycle-dependent proteolysis in plants. Identification of the destruction box pathway and metaphase arrest produced by the proteasome inhibitor mg132. *Plant Cell* **10**, 2063–2076.

Gutierrez, R., Quiroz-Figueroa, F. and Vazquez-Ramos, J.M. (2005) Maize cyclin D2 expression, associated kinase activity and effect of phytohormones during germination. *Plant Cell Physiol* **46**, 166–173.

Hata, S., Kouchi, H., Suzuka, I. and Ishii, T. (1991) Isolation and characterization of cDNA clones for plant cyclins. *EMBO J* **10**, 2681–2688.

Healy, J.M., Menges, M., Doonan, J.H. and Murray, J.A. (2001) The Arabidopsis D-type cyclins CycD2 and CycD3 both interact in vivo with the PSTAIRE cyclin-dependent kinase Cdc2a but are differentially controlled. *J Biol Chem* **276**, 7041–7047.

Hemerly, A., Bergounioux, C., Van Montagu, M., Inzé, D. and Ferreira, P. (1992) Genes regulating the plant cell cycle: isolation of a mitotic-like cyclin from *Arabidopsis thaliana*. *Proc Natl Acad Sci USA* **89**, 3295–3299.

Hemerly, A.S., Ferreira, P., de Almeida Engler, J., Van Montagu, M., Engler, G. and Inzé, D. (1993) cdc2a expression in Arabidopsis is linked with competence for cell division. *Plant Cell* **5**, 1711–1723.

Hirt, H., Mink, M., Pfosser, M., *et al.* (1992) Alfalfa cyclins: differential expression during the cell cycle and in plant organs. *Plant Cell* **4**, 1531–1538.

Hoeppner, S., Baumli, S. and Cramer, P. (2005) Structure of the mediator subunit cyclin C and its implications for CDK8 function. *J Mol Biol* **350**, 833–842.

Hsieh, W.L. and Wolniak, S.M. (1998) Isolation and characterization of a functional A-type cyclin from maize. *Plant Mol Biol* **37**, 121–129.

Hu, Y., Bao, F. and Li, J. (2000) Promotive effect of brassinosteroids on cell division involves a distinct CycD3-induction pathway in Arabidopsis. *Plant J* **24**, 693–701.

Hunt, T. (1991) Cyclins and their partners: from a simple idea to complicated reality. *Semin Cell Biol* **2**, 213–222.

Huntley, R., Healy, S., Freeman, D., *et al.* (1998) The maize retinoblastoma protein homologue ZmRb-1 is regulated during leaf development and displays conserved interactions with G1/S regulators and plant cyclin D (CycD) proteins. *Plant Mol Biol* **37**, 155–169.

Imai, K.K., Ohashi, Y., Tsuge, T., *et al.* (2006) The A-type cyclin CYCA2;3 is a key regulator of ploidy levels in Arabidopsis endoreduplication. *Plant Cell* **18**, 382–396.

Inzé, D., Dudits, D. and Francis, D. (1998) *Plant Cell Division*. Portland Press, London.

Ito, M., Araki, S., Matsunaga, S., *et al.* (2001) G2/M-phase-specific transcription during the plant cell cycle is mediated by c-Myb-like transcription factors. *Plant Cell* **13**, 1891–1905.

Ito, M., Iwase, M., Kodama, H., *et al.* (1998) A novel cis-acting element in promoters of plant B-type cyclin genes activates M phase-specific transcription. *Plant Cell* **10**, 331–341.

Ito, M., Criqui, M.-C., Sakabe, M., *et al.* (1997) Cell-cycle-regulated transcription of A- and B-type plant cyclin genes in synchronous cultures. *Plant J* **11**, 983–992.

Jasinski, S., Riou-Khamlichi, C., Roche, O., Perennes, C., Bergounioux, C. and Glab, N. (2002) The CDK inhibitor NtKIS1a is involved in plant development, endoreduplication and restores normal development of cyclin D3;1-overexpressing plants. *J Cell Sci* **115**, 973–982.

Joubès, J., Lemaire-Chamley, M., Delmas, F., *et al.* (2001) A new C-type cyclin-dependent kinase from tomato expressed in dividing tissues does not interact with mitotic and G1 cyclins. *Plant Physiol* **126**, 1403–1415.

Joubès, J., Walsh, D., Raymond, P. and Chevalier, C. (2000) Molecular characterization of the expression of distinct classes of cyclins during the early development of tomato fruit. *Planta* **211**, 430–439.

Knudsen, E.S., Buckmaster, C., Chen, T.T., Feramisco, J.R. and Wang, J.Y. (1998) Inhibition of DNA synthesis by RB: effects on G1/S transition and S-phase progression. *Genes Dev* **12**, 2278–2292.

Kono, A., Ohno, R., Umeda-Hara, C., Uchimiya, H. and Umeda, M. (2006) A distinct type of cyclin D, CYCD4;2, involved in the activation of cell division in Arabidopsis. *Plant Cell Rep* **25**, 540–545.

Kono, A., Umeda-Hara, C., Lee, J., Ito, M., Uchimiya, H. and Umeda, M. (2003) Arabidopsis D-type cyclin CYCD4;1 is a novel cyclin partner of B2-type cyclin-dependent kinase. *Plant Physiol* **132**, 1315–1321.

Koroleva, O.A., Tomlinson, M., Parinyapong, P., *et al.* (2004) CycD1, a putative G1 cyclin from *Antirrhinum majus*, accelerates the cell cycle in cultured tobacco BY-2 cells by enhancing both G1/S entry and progression through S and G2 phases. *Plant Cell* **16**, 2364–2379.

Kouchi, H., Sekine, M. and Hata, S. (1995) Distinct classes of mitotic cyclins are differentially expressed in the soybean shoot apex during the cell cycle. *Plant Cell* **7**, 1143–1155.

Kvarnheden, A., Yao, J.L., Zhan, X., O'Brien, I. and Morris, B.A. (2000) Isolation of three distinct CycD3 genes expressed during fruit development in tomato. *J Exp Bot* **51**, 1789–1797.

La, H., Li, J., Ji, Z., *et al.* (2006) Genome-wide analysis of cyclin family in rice (*Oryza sativa* L.). *Mol Genet Genomics* **275**, 374–386.

Lee, H., Auh, C.-K., Kim, D., Lee, T.-K. and Lee, S. (2006) Exogenous cytokinin treatment maintains cyclin homeostasis in rice seedlings that show changes of cyclin expression when the photoperiod is rapidly changed. *Plant Physiol Biochem* **44**, 248–252.

Lee, J., Das, A., Yamaguchi, M., *et al.* (2003) Cell cycle function of a rice B2-type cyclin interacting with a B-type cyclin-dependent kinase. *Plant J* **34**, 417–425.

Lenburg, M.E. and O'Shea, E.K. (1996) Signaling phosphate starvation. *Trends Biochem Sci* **21**, 383–387.

Lew, D.J., Dulic, V. and Reed, S.I. (1991) Isolation of three novel human cyclins by rescue of G1 cyclin (Cln) function in yeast. *Cell* **66**, 1197–1206.

Li, C., Potuschak, T., Colón-Carmona, A., Gutiérrez, R.A. and Doerner, P. (2005) Arabidopsis TCP20 links regulation of growth and cell division control pathways. *Proc Natl Acad Sci USA* **102**, 12978–12983.

Li, Y., Yu, J.Q., Ye, Q.J., Zhu, Z.J. and Guo, Z.J. (2003) Expression of CycD3 is transiently increased by pollination and N-(2-chloro-4-pyridyl)-*N′*-phenylurea in ovaries of *Lagenaria leucantha*. *J Exp Bot* **54**, 1245–1251.

Logemann, E., Wu, S.C., Schröder, J., Schmelzer, E., Somssich, I.E. and Hahlbrock, K. (1995) Gene activation by UV light, fungal elicitor or fungal infection in *Petroselinum crispum* is correlated with repression of cell cycle-related genes. *Plant J* **8**, 865–876.

Lohka, M.J., Hayes, M.K. and Maller, J.L. (1988) Purification of maturation-promoting factor, an intracellular regulator of early mitotic events. *Proc Natl Acad Sci USA* **85**, 3009–3013.

Masubelele, N.H., Dewitte, W., Menges, M., *et al.* (2005) D-type cyclins activate division in the root apex to promote seed germination in Arabidopsis. *Proc Natl Acad Sci USA* **102**, 15694–15699.

Menges, M., de Jager, S.M., Gruissem, W. and Murray, J.A. (2005) Global analysis of the core cell cycle regulators of Arabidopsis identifies novel genes, reveals multiple and highly specific profiles of expression and provides a coherent model for plant cell cycle control. *Plant J* **41**, 546–566.

Menges, M., Hennig, L., Gruissem, W. and Murray, J.A. (2002) Cell cycle-regulated gene expression in Arabidopsis. *J Biol Chem* **277**, 41987–42002.

Menges, M., Hennig, L., Gruissem, W. and Murray, J.A. (2003) Genome-wide gene expression in an Arabidopsis cell suspension. *Plant Mol Biol* **53**, 423–442.

Menges, M. and Murray, J.A. (2002) Synchronous Arabidopsis suspension cultures for analysis of cell-cycle gene activity. *Plant J* **30**, 203–212.

Menges, M., Samland, A.K., Planchais, S. and Murray, J.A. (2006) The D-type cyclin CYCD3;1 is limiting for the G1-to-S-phase transition in Arabidopsis. *Plant Cell* **18**, 893–906.

Meskiene, I., Bögre, L., Dahl, M., *et al.* (1995) cycMs3, a novel B-type alfalfa cyclin gene, is induced in the G0-to-G1 transition of the cell cycle. *Plant Cell* **7**, 759–771.

Mews, M., Sek, F.J., Moore, R., Volkmann, D., Gunning, B.E.S. and John, P.C.L. (1997) Mitotic cylin distribution during maize cell division: implications for the sequence diversity and function of cyclins in plants. *Protoplasma* **200**, 128–145.

Mews, M., Sek, F.J., Volkmann, D. and John, P.C.L. (2000) Immunodetection of four mitotic cyclins and the Cdc2a protein kinase in the maize root: their distribution in cell development and dedifferentiation. *Protoplasma* **212**, 236–249.

Mironov, V.V., De Veylder, L., Van Montagu, M. and Inzé, D. (1999) Cyclin-dependent kinases and cell division in plants – the nexus. *Plant Cell* **11**, 509–522.

Nakagami, H., Kawamura, K., Sugisaka, K., Sekine, M. and Shinmyo, A. (2002) Phosphorylation of retinoblastoma-related protein by the cyclin D/cyclin-dependent kinase complex is activated at the G1/S-phase transition in tobacco. *Plant Cell* **14**, 1847–1857.

Nakagami, H., Sekine, M., Murakami, H. and Shinmyo, A. (1999) Tobacco retinoblastoma-related protein phosphorylated by a distinct cyclin-dependent kinase complex with Cdc2/cyclin D in vitro. *Plant J* **18**, 243–252.

Nakai, T., Kato, K., Shinmyo, A. and Sekine, M. (2006) Arabidopsis KRPs have distinct inhibitory activity toward cyclin D2-associated kinases, including plant-specific B-type cyclin-dependent kinase. *FEBS Lett* **580**, 336–340.

Nakamura, T., Sanokawa, R., Sasaki, Y.F., Ayusawa, D., Oishi, M. and Mori, N. (1995) Cyclin I: a new cyclin encoded by a gene isolated from human brain. *Exp Cell Res* **221**, 534–542.

Niebel, A., de Almeida Engler, J., Hemerly, A., *et al.* (1996) Induction of *cdc2a* and *cyc1At* expression in *Arabidopsis thaliana* during early phases of nematode-induced feeding cell formation. *Plant J* **10**, 1037–1043.

Novak, B., Csikasz-Nagy, A., Gyorffy, B., Chen, K. and Tyson, J.J. (1998) Mathematical model of the fission yeast cell cycle with checkpoint controls at the G1/S, G2/M and metaphase/anaphase transitions. *Biophys Chem* **72**, 185–200.

Nugent, J.H., Alfa, C.E., Young, T. and Hyams, J.S. (1991) Conserved structural motifs in cyclins identified by sequence analysis. *J Cell Sci* **99**, 669–674.

Oakenfull, E.A., Riou-Khamlichi, C. and Murray, J.A. (2002) Plant D-type cyclins and the control of G1 progression. *Philos Trans R Soc Lond B Biol Sci* **357**, 749–760.

Pines, J. (1995a) Cyclins and cyclin-dependent kinases: a biochemical view. *Biochem J* **308**, 697–711.

Pines, J. (1995b) Cell cycle. Confirmational change. *Nature* **376**, 294–295.

Planchais, S., Perennes, C., Glab, N., Mironov, V., Inzé, D. and Bergounioux, C. (2002) Characterization of cis-acting element involved in cell cycle phase-independent activation of Arath;CycB1;1 transcription and identification of putative regulatory proteins. *Plant Mol Biol* **50**, 111–127.

Planchais, S., Samland, A.K. and Murray, J.A. (2004) Differential stability of Arabidopsis D-type cyclins: CYCD3;1 is a highly unstable protein degraded by a proteasome-dependent mechanism. *Plant J* **38**, 616–625.

Plowman, G.D., Sudarsanam, S., Bingham, J., Whyte, D. and Hunter, T. (1999) The protein kinases of *Caenorhabditis elegans*: a model for signal transduction in multicellular organisms. *Proc Natl Acad Sci USA* **96**, 13603–13610.

Porceddu, A., Reale, L., Lanfaloni, L., *et al.* (1999) Cloning and expression analysis of a *Petunia hybrida* flower specific mitotic-like cyclin. *FEBS Lett* **462**, 211–215.

Qin, L.X., Perennes, C., Richard, L., *et al.* (1996) G2-and early-M-specific expression of the NTCYC1 cyclin gene in *Nicotiana tabacum* cells. *Plant Mol Biol* **32**, 1093–1101.

Qin, L.X., Richard, L., Perennes, C., Gadal, P. and Bergounioux, C. (1995) Identification of a cell cycle-related gene, cyclin, in *Nicotiana tabacum* (L.). *Plant Physiol* **108**, 425–426.

Rechsteiner, M. and Rogers, S.W. (1996) PEST sequences and regulation by proteolysis. *Trends Biochem Sci* **21**, 267–271.

Reichheld, J.P., Chaubet, N., Shen, W.H., Renaudin, J.P. and Gigot, C. (1996) Multiple A-type cyclins express sequentially during the cell cycle in *Nicotiana tabacum* BY2 cells. *Proc Natl Acad Sci USA* **93**, 13819–13824.

Renaudin, J.P., Colasanti, J., Rime, H., Yuan, Z. and Sundaresan, V. (1994) Cloning of four cyclins from maize indicates that higher plants have three structurally distinct groups of mitotic cyclins. *Proc Natl Acad Sci USA* **91**, 7375–7379.

Renaudin, J.P., Doonan, J.H., Freeman, D., *et al.* (1996) Plant cyclins: a unified nomenclature for plant A-, B- and D-type cyclins based on sequence organization. *Plant Mol Biol* **32**, 1003–1018.

Rickert, P., Seghezzi, W., Shanahan, F., Cho, H. and Lees, E. (1996) Cyclin C/CDK8 is a novel CTD kinase associated with RNA polymerase II. *Oncogene* **12**, 2631–2640.

Riou-Khamlichi, C., Huntley, R., Jacqmard, A. and Murray, J.A. (1999) Cytokinin activation of Arabidopsis cell division through a D-type cyclin. *Science* **283**, 1541–1544.

Riou-Khamlichi, C., Menges, M., Healy, J.M. and Murray, J.A. (2000) Sugar control of the plant cell cycle: differential regulation of Arabidopsis D-type cyclin gene expression. *Mol Cell Biol* **20**, 4513–4521.

Rohila, J.S., Chen, M., Chen, S., *et al.* (2006) Protein–protein interactions of tandem affinity purification-tagged protein kinases in rice. *Plant J* **46**, 1–13.

Roudier, F., Fedorova, E., Györgyey, J., *et al.* (2000) Cell cycle function of a *Medicago sativa* A2-type cyclin interacting with a PSTAIRE-type cyclin-dependent kinase and a retinoblastoma protein. *Plant J* **23**, 73–83.

Roudier, F., Fedorova, E., Lebris, M., *et al.* (2003) The *Medicago* species A2-type cyclin is auxin regulated and involved in meristem formation but dispensable for endoreduplication-associated developmental programs. *Plant Physiol* **131**, 1091–1103.

Russinova, E., Slater, A., Atanassov, A.I. and Elliott, M.C. (1995) Cloning novel alfalfa cyclin sequences – a RACE-PCR approach. *Cell Mol Biol* **41**, 703–714.

Sauter, M., Mekhedov, S.L. and Kende, H. (1995) Gibberellin promotes histone H1 kinase activity and the expression of cdc2 and cyclin genes during the induction of rapid growth in deepwater rice internodes. *Plant J* **7**, 623–632.

Savouré, A., Fehér, A., Kaló, P., *et al.* (1995) Isolation of a full-length mitotic cyclin cDNA clone *CycIIIMs* from *Medicago sativa*: chromosomal mapping and expression. *Plant Mol Biol* **27**, 1059–1070.

Schmid, M., Davison, T.S., Henz, S.R., *et al.* (2005) A gene expression map of *Arabidopsis thaliana* development. *Nat Genet* **37**, 501–506.

Schnittger, A., Schöbinger, U., Bouyer, D., Weinl, C., Stierhof, Y.D. and Hülskamp, M. (2002a) Ectopic D-type cyclin expression induces not only DNA replication but also cell division in *Arabidopsis* trichomes. *Proc Natl Acad Sci USA* **99**, 6410–6415.

Schnittger, A., Schöbinger, U., Stierhof, Y.D. and Hulskamp, M. (2002b) Ectopic B-type cyclin expression induces mitotic cycles in endoreduplicating *Arabidopsis* trichomes. *Curr Biol* **12**, 415–420.

Setiady, Y.Y., Sekine, M., Hariguchi, N., Yamamoto, T., Kouchi, H. and Shinmyo, A. (1995) Tobacco mitotic cyclins: cloning, characterization, gene expression and functional assay. *Plant J* **8**, 949–957.

Setiady, Y.Y., Sekine, M., Yamamoto, T., Kouchi, H. and Shinmyo, A. (1997) Expression pattern of tobacco cyclin genes. *Plant Cell Rep* **16**, 368–372.

Sherr, C.J. (1993) Mammalian G1 cyclins. *Cell* **73**, 1059–1065.

Shimotohno, A., Matsubayashi, S., Yamaguchi, M., Uchimiya, H. and Umeda, M. (2003) Differential phosphorylation activities of CDK-activating kinases in *Arabidopsis thaliana. FEBS Lett.* **534**, 69–74.

Shimotohno, A., Umeda-Hara, C., Bisova, K., Uchimiya, H. and Umeda, M. (2004) The plant-specific kinase CDKF;1 is involved in activating phosphorylation of cyclin-dependent kinase-activating kinases in Arabidopsis. *Plant Cell* **16**, 2954–2966.

Soni, R., Carmichael, J.P., Shah, Z.H. and Murray, J.A. (1995) A family of cyclin D homologs from plants differentially controlled by growth regulators and containing the conserved retinoblastoma protein interaction motif. *Plant Cell* **7**, 85–103.

Sorrell, D.A., Combettes, B., Chaubet-Gigot, N., Gigot, C. and Murray, J.A. (1999) Distinct cyclin D genes show mitotic accumulation or constant levels of transcripts in tobacco bright yellow-2 cells. *Plant Physiol* **119**, 343–352.

Sorrell, D.A., Menges, M., Healy, J.M., *et al.* (2001) Cell cycle regulation of cyclin-dependent kinases in tobacco cultivar Bright Yellow-2 cells. *Plant Physiol* **126**, 1214–1223.

Sun, Y., Flannigan, B.A. and Setter, T.L. (1999) Regulation of endoreduplication in maize (*Zea mays* L.) endosperm. Isolation of a novel B1-type cyclin and its quantitative analysis. *Plant Mol Biol* **41**, 245–258.

Swaminathan, K., Yang, Y., Grotz, N., Campisi, L. and Jack, T. (2000) An enhancer trap line associated with a D-class cyclin gene in Arabidopsis. *Plant Physiol* **124**, 1658–1667.

Świątek, A., Azmi, A., Stals, H., Inzé, D. and Van Onckelen, H. (2004) Jasmonic acid prevents the accumulation of cyclin B1;1 and CDK-B in synchronized tobacco BY-2 cells. *FEBS Lett* **572**, 118–122.

Szarka, S., Fitch, M., Schaerer, S. and Moloney, M. (1995) Classification and expression of a family of cyclin gene homologues in *Brassica napus*. *Plant Mol Biol* **27**, 263–275.

Thompson, J.D., Higgins, D.G. and Gibson, T.J. (1994) CLUSTAL W: improving the sensitivity of progressive multiple sequence alignment through sequence weighting, position-specific gap penalties and weight matrix choice. *Nucleic Acids Res* **22**, 4673–4680.

Torres Acosta, J.A., de Almeida Engler, J., Raes, J., *et al.* (2004) Molecular characterization of Arabidopsis PHO80-like proteins, a novel class of CDKA;1-interacting cyclins. *Cell Mol Life Sci* **61**, 1485–1497.

Townsley, F.M. and Ruderman, J.V. (1998) Functional analysis of the *Saccharomyces cerevisiae* UBC11 gene. *Yeast* **14**, 747–757.

Tréhin, C., Ahn, I.O., Perennes, C., Couteau, F., Lalanne, E. and Bergounioux, C. (1997) Cloning of upstream sequences responsible for cell cycle regulation of the *Nicotiana sylvestris* CycB1;1 gene. *Plant Mol Biol* **35**, 667–672.

Uchida, K., Muramatsu, T., Tachibana, K., Kishimoto, T. and Furuya, M. (1996) Isolation and characterization of the cDNA for an A-like cyclin in *Adiantum capillus-veneris* L. *Plant Cell Physiol* **37**, 825–832.

Uemukai, K., Iwakawa, H., Kosugi, S., *et al.* (2005) Transcriptional activation of tobacco E2F is repressed by co-transfection with the retinoblastoma-related protein: cyclin D expression overcomes this repressor activity. *Plant Mol Biol* **57**, 83–100.

Umeda, M., Bhalerao, R.P., Schell, J., Uchimiya, H. and Koncz, C. (1998) A distinct cyclin-dependent kinase-activating kinase of *Arabidopsis thaliana*. *Proc Natl Acad Sci USA* **95**, 5021–5026.

Umeda, M., Iwamoto, N., Umeda-Hara, C., Yamaguchi, M., Hashimoto, J. and Uchimiya, H. (1999) Molecular characterization of mitotic cyclins in rice plants. *Mol Gen Genet* **262**, 230–238.

Umeda, M., Shimotohno, A. and Yamaguchi, M. (2005) Control of cell division and transcription by cyclin-dependent kinase-activating kinases in plants. *Plant Cell Physiol* **46**, 1437–1442.

Vandepoele, K., Raes, J., De Veylder, L., Rouzé, P., Rombauts, S. and Inzé, D. (2002) Genome-wide analysis of core cell cycle genes in Arabidopsis. *Plant Cell* **14**, 903–916.

Vierstra, R.D. (2003) The ubiquitin/26S proteasome pathway, the complex last chapter in the life of many plant proteins. *Trends Plant Sci* **8**, 135–142.

Wallenfang, M.R. and Seydoux, G. (2002) cdk-7 is required for mRNA transcription and cell cycle progression in *Caenorhabditis elegans* embryos. *Proc Natl Acad Sci USA* **99**, 5527–5532.

Wang, F., Huo, S.N., Guo, J. and Zhang, X.S. (2006) Wheat D-type cyclin Triae;CYCD2;1 regulate development of transgenic Arabidopsis plants. *Planta* **224**, 1129–1140.

Wang, G., Kong, H., Sun, Y., *et al.* (2004a) Genome-wide analysis of the cyclin family in Arabidopsis and comparative phylogenetic analysis of plant cyclin-like proteins. *Plant Physiol* **135**, 1084–1099.

Wang, Y., Magnard, J.L., McCormick, S. and Yang, M. (2004b) Progression through meiosis I and meiosis II in Arabidopsis anthers is regulated by an A-type cyclin predominately expressed in prophase I. *Plant Physiol* **136**, 4127–4135.

Weingartner, M., Criqui, M.C., Meszaros, T., *et al.* (2004) Expression of a nondegradable cyclin B1 affects plant development and leads to endomitosis by inhibiting the formation of a phragmoplast. *Plant Cell* **16**, 643–657.

Weingartner, M., Pelayo, H.R., Binarova, P., *et al.* (2003) A plant cyclin B2 is degraded early in mitosis and its ectopic expression shortens G2-phase and alleviates the DNA-damage checkpoint. *J Cell Sci* **116**, 487–498.

Wildwater, M., Campilho, A., Perez-Perez, J.M., *et al.* (2005) The retinoblastoma-related gene regulates stem cell maintenance in Arabidopsis roots. *Cell* **123**, 1337–1349.

Wyrzykowska, J., Pien, S., Shen, W.H. and Fleming, A.J. (2002) Manipulation of leaf shape by modulation of cell division. *Development* **129**, 957–964.

Yamaguchi, M., Fabian, T., Sauter, M., *et al.* (2000) Activation of CDK-activating kinase is dependent on interaction with H-type cyclins in plants. *Plant J* **24**, 11–20.

Yu, Y., Steinmetz, A., Meyer, D., Brown, S. and Shen, W.H. (2003) The tobacco A-type cyclin, Nicta;CYCA3;2, at the nexus of cell division and differentiation. *Plant Cell* **15**, 2763–2777.

Zhou, J.R., Gugger, E.T., Tanaka, T., Guo, Y., Blackburn, G.L. and Clinton, S.K. (1999) Soybean phytochemicals inhibit the growth of transplantable human prostate carcinoma and tumor angiogenesis in mice. *J Nutr* **129**, 1628–1635.

3 CDK inhibitors

Hong Wang, Yongming Zhou, Juan Antonio Torres Acosta and Larry C. Fowke

3.1 Introduction

The cyclin-dependent kinase (CDK), essential for cell cycle regulation in eukaryotic cells, can be inhibited by protein binding and protein phosphorylation and dephosphorylation (Morgan, 1997). The term *CDK inhibitor* (CKI) generally refers to a protein that can inhibit CDK activities through binding to the CDK complex. CDK inhibitor genes were initially identified in mammalian cells and yeast around 1993 and 1994 by many different laboratories using a variety of approaches (see reviews by Sherr and Roberts, 1995; Mendenhall, 1998). The mammalian CDK inhibitors have received considerable attention and have been investigated extensively because of their implications in human diseases, particularly cancer. Because of the large number of original publications describing the identification and functions of mammalian CDK inhibitors, the following brief historical perspective will refer to reviews only unless original papers are uniquely relevant.

The mammalian CDK inhibitors are classified into Cip/Kip and INK4 families, based on the structural similarities and the specificity of CDK inhibition (Sherr and Roberts, 1995). The Cip/Kip family consists of $p27^{Kip1}$, $p21^{Cip1}$ and $p57^{Kip2}$, while the INK4 family consists of $p16^{INK4a}$, $p15^{INK4b}$, $p18^{INK4c}$ and $p19^{INK4d}$. The Cip/Kip inhibitors have an N-terminal CDK-inhibitory domain and can bind to and inhibit cyclin D–CDK4/6 and cyclin E/A–CDK2 complexes. Crystal structure analysis shows that the human $p27^{Kip1}$ binds to the cyclin A–CDK2 complex as an extended structure interacting with both cyclin A and CDK2 (Russo *et al.*, 1996). The INK4 inhibitors consist of four or five repeating structural units known as ankyrin repeats (Sherr and Roberts, 1995). They have a narrower interacting spectrum and inhibit only cyclin D–CDK4/6 complexes. The CDK inhibitors in mammalian cells regulate the activity of G1 CDKs by physically blocking CDK activation or substrate/ATP access (Pavletich, 1999).

Mammalian CDK inhibitors have been thoroughly investigated due to their role in cell cycle regulation and potential as tumor suppressors. A large body of experimental evidence indicates that they have important roles in a variety of processes in mammalian growth and development (see reviews by Harper and Elledge, 1996; Vidal and Koff, 2000; Denicourt and Dowdy, 2004; Griffin and Shankland, 2004; Musgrove *et al.*, 2004). Mice with one or more CKI genes disrupted display a range of different phenotypes. For instance, $p27^{Kip1}$ knockout mice display a number of changes, including a gene dosage-dependent increase in body size, female infertility and deafness (Vidal and Koff, 2000). Much research on mammalian CDK inhibitors has been devoted to their possible involvement in tumorogenesis and prognosis of cancer. The $p16^{INK4a}$ gene has been recognized as a tumor suppressor because of its frequent loss of function in different human tumors (Sherr, 2000). Loss of $p27^{Kip1}$ protein has good potential to be an independent prognostic marker for a number of different tumors (Tsihlias *et al.*, 1999). Although mammalian CDK inhibitors have been traditionally associated with regulation of CDKs and thus the cell cycle, it is now clear that they are also important for several other processes such as apoptosis, cell shape regulation and cell migration (Denicourt and Dowdy, 2004; Griffin and Shankland, 2004).

At about the same time that mammalian CDK inhibitors were discovered, CDK inhibitors were also isolated from yeast, three from the budding yeast *Saccharomyces cerevisiae* (Mendenhall, 1993; Peter and Herskowitz, 1994; Schneider *et al.*, 1994) and one from the fission yeast *Schizosaccharomyces pombe* (Moreno and Nurse, 1994). Interestingly, the three CDK inhibitors in budding yeast have distinct functions. Sic1 is critical for G1 control and the timing of S phase (Mendenhall, 1993). Far1 is a CDK inhibitor responsible for cell cycle arrest during pheromone response (Peter and Herskowitz, 1994). Pho81 is an inhibitor of the CDK/cyclin complex Pho80–Pho85 that is not involved in cell cycle regulation but in the phosphate-responsive signal transduction pathway (Schneider *et al.*, 1994). Although the *Pho81* gene was well known previously from studies of the regulation of phosphate metabolism, its function as a CDK inhibitor was revealed only later (Schneider *et al.*, 1994). Sic1 and Far1 share little sequence similarity with the mammalian CDK inhibitors, while Pho81 contains ankyrin repeats similar to those of the mammalian INK4 family inhibitors. In addition, Far1 and Pho81 have functions not associated with CDK inhibition (Mendenhall, 1998).

In plants, CDK inhibitor activity was initially observed and suggested to be involved in endosperm development (Grafi and Larkins, 1995). The first plant CDK inhibitor gene was identified from Arabidopsis (Wang *et al.*, 1997), and other members of the Arabidopsis CDK inhibitor family were described later (Lui *et al.*, 2000; De Veylder *et al.*, 2001; Zhou *et al.*, 2002a). CDK inhibitors have also been reported for several plant species, and similar sequences are present in a number of diverse angiosperms and gymnosperms. All plant CDK inhibitors, or putative inhibitors, isolated so far contain a C-terminal domain that shares similarity with the mammalian Cip/Kip CDK inhibitors in a region of about 30 residues important for CDK inhibition (Wang *et al.*, 1997). However, the plant CDK inhibitors have no other sequence similarities with mammalian or yeast CDK inhibitors. Although most of the experimental characterization of plant CDK inhibitors has been performed with the Arabidopsis CDK inhibitors, reports have appeared describing two tobacco CDK inhibitors and, in the recent months, CDK inhibitors from several other plant species. In this chapter, we will review the current understanding of plant CDK inhibitors and, based on available experimental evidence, will attempt to draw conclusions and working suggestions regarding the expression, regulation and functions of plant CDK inhibitors. Two reviews on the subject have also been published recently (Verkest *et al.*, 2005a; Wang *et al.*, 2006).

3.2 Plant CDK inhibitors and sequence uniqueness

Seven CDK inhibitor genes have been identified in Arabidopsis (Wang *et al.*, 1997; Lui *et al.*, 2000; De Veylder *et al.*, 2001; Zhou *et al.*, 2002a). The initial cloning was completed by the yeast two-hybrid approach using Arath;CDKA;1 or Arath;CYCD3;1 as the bait (Wang *et al.*, 1997, 1998; De Veylder *et al.*, 2001; Zhou *et al.*, 2002a). Because of the sequence divergence of plant CDK inhibitors from animals and yeast, approaches based on sequence similarities would have been difficult. Table 3.1 lists the seven members of the Arabidopsis ICK/KRP family. In this review, the founding members of this family will be referred to as ICK1 (Wang *et al.*, 1997) and ICK2 (Lui *et al.*, 2000), while the other members will be referred to as KRP3 to KRP7 (De Veylder *et al.*, 2001), according to the names given when they were first described. ICK stands for interactor/inhibitor of CDK (Wang *et al.*, 1997; and also because the CDK inhibitory domains of plant and mammalian 7 inhibitors are at opposite ends), while KRP stands for Kip-related protein (De Veylder *et al.*, 2001).

One common feature of the ICK/KRP proteins is the conserved C-terminal domain, which is similar to a domain in the N-terminal region of the mammalian Cip/Kip CDK inhibitors (Wang *et al.*, 1997; Lui *et al.*, 2000; De Veylder *et al.*, 2001; Zhou *et al.*, 2002a). Overall, the sequence similarity is limited to this region of about 30 residues (Plate 3.1). For the mammalian $p27^{Kip1}$, the conserved residues in this region are mostly involved in the interaction with CDK2 (Russo *et al.*, 1996). This region in ICK1 is also shown to be required for the interaction with Arabidopsis CDKA;1 and CYCD3;1 (Wang *et al.*, 1998) as well as for the CDK inhibitory function in plants (Zhou *et al.*, 2003a). This region thus is considered a functional domain and is marked as region 1 (box 1) in Plate 3.1.

Genes similar to *ICK/KRP* genes have been identified from *Chenopodium rubrum* (Fountain *et al.*, 1999), tobacco (Jasinski *et al.*, 2002a, 2003), maize (Coelho

Table 3.1 *Arabidopsis* ICK/KRP CDK inhibitor gene family

Gene name	Alternative name	Genomic IDs
ICK1	*KRP1*	At2g23430
ICK2	*KRP2*	At3g50630
KRP3	*ICK6*	At5g48820
KRP4	*ICK7*	At2g32710
KRP5		At3g24810
KRP6	*ICK4*	At3g19150
KRP7	*ICK5*	At1g49620

The names by which the genes were described first are used, and the alternative names are also listed. The founding members of this gene family were named ICKs (Wang *et al.*, 1997; Lui *et al.*, 2000), and the other members were named *KRPs* (De Veylder *et al.*, 2001). For clarification, the genomic locus identifiers from the Arabidopsis genomic sequencing project (The Arabidopsis Genome Initiative, 2000) are listed.

et al., 2005), alfalfa (Pettko-Szandtner *et al.*, 2006) and tomato (Bisbis *et al.*, 2006). Some were identified by the classical yeast two-hybrid approach while others by mining sequence databases using the known ICK/KRP proteins. Sequence searches have found similar genes from a number of diverse angiosperm and gymnosperm species (H. Wang, unpublished data). The existence of multiple ICK/KRP-related proteins in the databases for different angiosperm (mostly crop) species should not come as a surprise. The conservation of the CDK inhibitory domain between ICK/KRP CDK inhibitors and the animal Cip/Kip CDK inhibitors suggests that this domain may have evolved in an ancestral species common to animals and plants. Thus, most, if not all, plants are expected to contain this type of gene. The evolution of this family of genes in plants and the phylogenetic relationship among different members will be interesting to explore.

Compared to the CDK, cyclin, E2F/DP and Rb families of plant core cell cycle regulators (see corresponding chapters in this book), the ICK/KRP family of plant CDK inhibitors is less similar to its animal counterpart in terms of primary sequence and structural layout. This very limited sequence similarity and the opposite location of the conserved region imply some fundamental differences between plant and animal CDK inhibitors. Since for the most part the sequences in ICK/KIP proteins are unique to plant CDK inhibitors, the functional and regulatory properties of the nonconserved sequence would likely be different from the animal CDK inhibitors. Therefore, this family of plant cell cycle regulators may represent a major difference between plants and animals in terms of cell cycle control. They provide a good opportunity to understand more specific differences between plants and animals, regarding the molecular machinery controlling cell proliferation. In this regard, it is also interesting to note that in the Arabidopsis genome, no gene similar to the INK4 family of animal CDK inhibitors has been identified (Vandepoele *et al.*, 2002).

Despite the good conservation in the C-terminus, Arabidopsis ICK/KRP proteins are very different from each other for most of the molecule. Sequence analysis reveals a number of conserved motifs and putative signal sequences of known functions (Plate 3.1). Of the conserved motifs, motifs 2, 3, 5 and 6 are the same as described by De Veylder *et al.* (2001). As noted, the C-terminal domain (marked by box number 1 in Plate 3.1), required for interactions with CDKA and D-type cyclins (Wang *et al.*, 1998), is the most conserved feature among all proteins of this family. There is indirect evidence that motif 2 may be involved in the interaction with cyclin as well (Wang *et al.*, 1998). Little functional information is available for other conserved motifs indicated by the boxes in Plate 3.1.

There are some putative signal sequences for potential functions. Putative monopartite nuclear localization signals (NLSs) are found in ICK1, ICK2 and KRP7, while putative bipartite NLSs are present in KRP4 and KRP5. Among them, the monopartite NLS in ICK1 has been shown to be a strong NLS (Zhou *et al.*, in press). There are putative CDK phosphorylation sites and putative PEST sequences (potential cleavage sites for proteolysis). These putative signal sequences need to be viewed with caution, since their functions remain to be determined. In addition, because of incomplete and inexact information on signal sequences in plant proteins, the bioinformatic analysis may also omit functional sequences in the proteins.

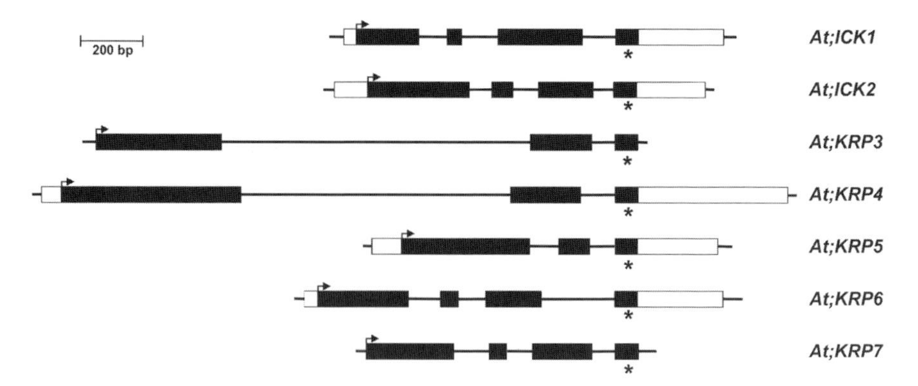

Figure 3.1 Analysis and comparison of Arabidopsis ICK/KRP proteins and the genomic organization of the genes. Exon–intron organization of Arabidopsis *ICK/KRP* genes. Closed and open boxes correspond to translational and upstream-transcribed regions, respectively. Lines represent introns. Arrows indicate the ATG translational start sites. The asterisk denotes the most conserved exon among the family, which encodes most of the conserved domain (box 1 in Plate 3.1). Scale bar is shown.

Nevertheless, the conserved motifs and putative sequences for known functions are useful as a guide for experimental investigation of the functions of this family of proteins.

Comparison of overall and conserved sequences among the seven Arabidopsis ICK/KRPs reveals that three pairs are more related to each other than to others: ICK1 and ICK2, KRP3 and KRP4 (by motifs 4, 5 and 6) and KRP 6 and KRP7 (by motif 7 and 8) (Plate 3.1). KRP5 is more related to KRP3 and KRP4 by motifs 3, 5 and 6, although it lacks motif 4. Phylogenetic analysis using full-length protein sequences showed a similar result. Furthermore, the analysis of exon–intron organization suggests a similar relationship: ICK1 and ICK2 (three introns), KRP3 and KRP4 (two introns with the first intron being very long) and KRP 6 and KRP7 (three introns) (Figure 3.1). KRP5 is more related to KRP3 and KRP4 than to others, although its first intron is much shorter than that in KRP3 and KRP4 (Figure 3.1). Interestingly, the junction between the last intron and last exon is conserved so that the last exon encodes the same segment of the C-terminal-conserved domain for all seven ICK/KRPs (Figure 3.1), as noted previously (Lui *et al.*, 2000; Jasinski *et al.*, 2002a). Thus, sequence similarity and exon–intron organization appear to distinguish three subgroups of inhibitors: (1) ICK1 and ICK2, (2) KRP3, KRP4 and KRP5 (with KRP5 being slightly more distant among the three) and (3) KRP6 and KRP7.

3.3 Expression

To ensure proper levels at the right places and right times, the expression of many cell cycle regulators is tightly controlled. Expression of *ICK/KRP* genes shows clear but generally moderate levels of tissue variability (Wang *et al.*, 1998; Lui *et al.*, 2000; De Veylder *et al.*, 2001; Jasinski *et al.*, 2002a; Pettko-Szandtner *et al.*, 2006). For Arabidopsis *ICK/KRP*s, different patterns of expression were observed among leaf, root, inflorescence and flower tissues (Wang *et al.*, 1998; Lui *et al.*, 2000; De Veylder *et al.*, 2001). Interestingly, a higher level of Arabidopsis *ICK1* expression was found in leaves of 5-week-old plants compared to other tissues (Wang *et al.*, 1998), and the tobacco *NtKIS1a* was expressed more strongly in older flower buds (Jasinski *et al.*, 2002a), indicating that the two CDK inhibitor genes are upregulated in nonproliferating and senescing tissues. The expression of Arabidopsis *ICK/KRP*s has been analyzed by *in situ* hybridization and the results show some levels of variability (Ormenese *et al.*, 2004). Although they are expressed in most tissues, *ICK1* and *ICK2* expression is absent from shoot apical meristems and vascular cells, while *KRP4* and *KRP5* are expressed mostly in proliferating cells (Ormenese *et al.*, 2004). The different patterns of expression suggest possible functional differences among the *ICK/KRP* genes during plant development.

Cell proliferation in plants is regulated by plant hormones and environmental conditions. It is thus important to understand how the cell cycle machinery is connected to the regulatory pathways of hormones. At present, knowledge is still limited. Since D-type cyclins and CDK inhibitors are mostly involved in the regulation of cell cycle start and the G1/S transition, they are good candidates for regulating the cell cycle in response to hormonal and environmental factors. It has been shown that D-type cyclins are induced and regulated by different factors (see Chapter 2).

Cell division in plants is inhibited by abscisic acid (ABA) (Himmelbach *et al.*, 1998) and abiotic stress (Bögre *et al.*, 2000). *ICK1* expression was induced by ABA and low-temperature treatments (Wang *et al.*, 1998). Induction by ABA was also observed with alfalfa *KRPMt* (Pettko-Szandtner *et al.*, 2006). *ICK1* expression is also induced by salt stress (NaCl treatment) (Ruggiero *et al.*, 2004). It was further shown that in an Arabidopsis T-DNA insertion mutant that has defective ABA biosynthesis, both ABA and *ICK1* levels did not show an increase in response to salt stress (Ruggiero *et al.*, 2004), suggesting that the induction of *ICK1* by salt treatment might be mediated by ABA. Furthermore, in transgenic plants overexpressing a transcriptional factor that binds to the ABA-responsive element, *ICK1* along with a set of ABA and stress-regulated genes showed an increase in expression (Kang *et al.*, 2002). These results together indicate that ABA can induce the expression of a CDK inhibitor such as *ICK1*, which may inhibit the cell cycle under stress. The expression of *ICK2*, on the other hand, is suppressed by auxin (Richard *et al.*, 2001; Himanen *et al.*, 2002).

The anatomical simplicity of the root makes it a good model to study cell proliferation and organ initiation and growth. Lateral roots originate from pericycle cells, and one key factor influencing lateral root initiation is auxin. When lateral root initiation was inhibited by treatment with the auxin transport inhibitor NPA (*N*-1-naphthylphthalamic acid), *ICK2* was expressed at a higher level in pericycles of Arabidopsis plants compared to control plants, but was greatly downregulated following addition of auxin, indicating that *ICK2* might be involved in lateral root initiation (Himanen *et al.*, 2002). Expression of *ICK/KRP* genes, particularly *ICK1*, *ICK2*, *KRP3* and *KRP7*, appeared to be affected in a mutant of PROPORZ1, which is essential for the developmental switch from cell proliferation to differentiation (Sieberer *et al.*, 2003). However, the biological implications for the changes of *ICK/KRP* expression are not clear, since the changes do not follow a pattern that is simply explainable at the moment. Results from these studies demonstrate that some members of the *ICK/KRP* family are regulated by intrinsic developmental and external signals.

Because of the lack of a good and easily accessible synchronized cell system, it has been traditionally more difficult to study cell cycle phase-specific genes in plants than in yeast and animals. Good efforts have been made to generate synchronized Arabidopsis cell cultures (Menges and Murray, 2002). Using those cell cultures, sequential peaking of *ICK/KRP* expression was observed with *KRP3* and *KRP4* in S phase, *KRP4* and *ICK1* in G2, and *KRP6* in G1/S phase (Menges *et al.*, 2005). Since some of the peaks were not very prominent, those patterns of expression need to be further confirmed. The data have shown more clearly that the expression of both *ICK1* and *ICK2* decreased when the cells were released from cell cycle arrest and remained relatively low during the subsequent cell cycle (Menges and Murray, 2002; Menges *et al.*, 2005), indicating that ICK1 and ICK2 may function to integrate antimitotic signals and arrest the cell cycle.

Most of the differences in transcript levels among different tissues and under different conditions described above can presumably be attributed to differences in transcription activities associated with temporary and spatial regulation. However, posttranscriptional regulation may also affect the transcript and protein levels. In tobacco, the *NtKIS1* gene could be spliced into two variant transcripts, *NtKIS1a* and *NtKIS1b* (Jasinski *et al.*, 2002a). The NtKIS1a protein is similar to ICK/KRP proteins, while NtKIS1b lacks the C-terminal domain. Surprisingly, it was observed that in an *in vitro* kinase assay, NtKIS1b was a strong competitor of NTKIS1a, suggesting that NtKIS1b suppresses the CDK inhibitory activity of NtKIS1a (Jasinski *et al.*, 2002a). Since NtKIS1b itself did not interact with NtCDKA;1 and neither did it show CDK inhibitory activity, the mechanism for this antagonistic effect against NtKIS1a remains unknown.

3.4 Interactions with cell cycle proteins and CDK inhibition

One basic biochemical property of a CDK inhibitor protein is its ability to interact with the CDK complex. The plant ICK/KRP proteins reported so far interact with both D-type cyclins and A-type CDK or with D-type cyclins alone by the yeast two-hybrid analysis. ICK1 and ICK2 were initially identified from two independent yeast two-hybrid screens using Arabidopsis CDKA;1 or CYCD3;1 as the bait, while additional ICK/KRP proteins were identified only from the screen using CYCD3;1 as the bait (Zhou *et al.*, 2002a), indicating possible differences among ICK/KRP proteins in terms of interaction with CDKA. It was further shown that in the yeast two-hybrid system, Arabidopsis ICK1, ICK2, KRP3 and KRP4 interacted with CDKA;1 while KRP6 (ICK4) and KRP7 (ICK5) showed no interaction (Zhou *et al.*, 2002a). It is interesting to note that KRP6 and KRP7 are more closely related to each other, based on conserved motifs and exon–intron organization (Plate 3.1). The results indicate that there are likely differences among ICK/KRP proteins regarding their ability to interact with the A-type CDK. In addition, it has been consistently shown that in the yeast two-hybrid system, plant CDK inhibitors do not interact with B-type as well as C-type CDKs tested (Lui *et al.*, 2000; De Veylder *et al.*, 2001; Jasinski *et al.*, 2002a; Zhou *et al.*, 2002a; Bisbis *et al.*, 2006; Pettko-Szandtner *et al.*, 2006).

All plant CDK inhibitors described so far can interact with D-type cyclins in the yeast two-hybrid system. Results from co-expression of the D-type cyclin in plants indicate that the D-type cyclin can rescue the mutant phenotype resulting from the expression of a CDK inhibitor, suggesting that CDK inhibitors and D-type cyclins also interact in plants (Jasinski *et al.*, 2002b; Zhou *et al.*, 2003b). The region in ICK1 that interacts with Arath;CYD3;1 is likely immediately before the conserved region that interacts with Arath;CDKA;1 (Wang *et al.*, 1998). This region shows a good level of similarity among the CDK inhibitors (De Veylder *et al.*, 2001; Zhou *et al.*, 2002a; Coelho *et al.*, 2005; Bisbis *et al.*, 2006). Studies also show that Arabidopsis ICK1, NtKIS1a and two tomato CDK inhibitors do not interact with mitotic cyclins tested (Jasinski *et al.*, 2002a; Zhou *et al.*, 2003b; Bisbis *et al.*, 2006). The specific interactions with particular CDKs and cyclins would likely define their inhibitory specificity, which is expected for CDK complexes containing CDKA or CYCD or both. For the mammalian CDK inhibitor $p27^{Kip1}$, both the interactions with CDK and cyclins contribute to CDK inhibition. At the moment, we do not know the relative contribution of the interaction with CDK versus the interaction with cyclin toward CDK inhibition (Wang *et al.*, 1998).

Recently, it has been reported that ICK1 lacking the C-terminal region could still interact with CYCD3;1 (Jakoby *et al.*, 2006). However, this interaction was not observed in an earlier study (Wang *et al.*, 1998). One possible reason for the discrepancy could be the different yeast two-hybrid systems used. Weak interactions may be observed in one system but not in the other. Alternatively, we have observed that CYCD3;1 itself has a low level of activation activity when fused to a DNA-binding domain vector (Zhou *et al.*, 2002a). This activity may become more apparent if a more sensitive (or less stringent) yeast two-hybrid system is used. It is important in such an analysis to include a control protein that does not interact with CYCD3;1. The specificity of the reported interaction of the N-terminal region with CYCD3;1 thus needs to be confirmed.

Strong experimental evidence shows that ICK/KRP proteins can inhibit CDK activity *in vitro* and *in vivo*. The ability to inhibit CDK activity has been shown for recombinant proteins of several ICK/KRPs from Arabidopsis (Wang *et al.*, 1997, 1998; Lui *et al.*, 2000), tobacco (Jasinski *et al.*, 2002a), maize (Coelho *et al.*, 2005) and tomato (Bisbis *et al.*, 2006). The level of inhibition is concentration dependent (Wang *et al.*, 1997; Lui *et al.*, 2000; Coelho *et al.*, 2005; Bisbis *et al.*, 2006). The CDK inhibitory activity is further confirmed in transgenic plants (Wang *et al.*, 2000; De Veylder *et al.*, 2001; Jasinski *et al.*, 2002a). Analysis of deletion constructs show that the C-terminal conserved domain is required for the interaction of ICK1 with the CDK complex and for inhibition of its activity in plants (Zhou *et al.*, 2003a).

On the other hand, removal of the N-terminal region had no effect on the ability of the mutant ICK1$^{109-191}$ protein to interact with and inhibit the CDK complex (Zhou *et al.*, 2003a). Thus, the C-terminal domain confers CDK inhibitory function to plant CDK inhibitors as in the animal Cip/Kip inhibitors.

CDKs may form complexes with different cyclins. One important aspect in understanding the functions of CDK inhibitors is to know which CDK complexes are the specific targets of the inhibitors during the cell cycle as well as in different plant organs and tissues. Addressing this question requires clear knowledge regarding the CDK complexes formed and functioning in plant cells. However, current understanding of the CDK–cyclin partner relationship existing in plant cells is poor, making it difficult to determine the specific CDK targets of ICK/KRP proteins. Since CDK inhibitors function through protein binding, the target CDK complexes are expected to contain at least one protein that can interact with ICK/KRP proteins. Since ICK/KRP proteins interact with CYCD and CDKA, the CDKA- or CYCD-containing CDK complexes would be the likely targets. It has also been shown that the Arabidopsis CDKA, not CDKB, is the *in vivo* partner of Arabidopsis CYCD2 and CYCD3 (Healy *et al.*, 2001). Thus, CDKA/CYCD complexes can be presumed to be the major targets. However, the observations that an alfalfa CDKA interacts with an A2-type cyclin (Roudier *et al.*, 2001), the Arabidopsis CYCD4;1 interacts with CDKB2;1 (Kono *et al.*, 2003) and tobacco CYCDs interact with CDKB (Kawamura *et al.*, 2006) suggest that other CDKA- or CYCD-containing complexes can also be targets.

Using an affinity column, recombinant ICK2 was found to retain the Arabidopsis CDKA;1 complex, but not the CDKB1;1 complex (Verkest *et al.*, 2005b). Similarly, Arabidopsis CDKA;1, but not CDKB1;1, rescued the mutant phenotype from ICK1 misexpression in trichomes (Schnittger *et al.*, 2003). These results are in good agreement with results from protein–protein interaction experiments showing that CDK inhibitors interact with A-type, but not with the B-type, CDKs. Results from other studies however add more complexity. Two maize inhibitors, Zeama;KRP;1 and Zeama;KRP;2, were found to inhibit Zeama;CYCA1;3 and Zeama;CYCD5;1 complexes, but not the Zeama;CYCB1;3 complex (Coelho *et al.*, 2005). When an alfalfa CDK inhibitor KRPMt was tested against CDK complexes precipitated using different antisera, KRPMt inhibited the activity of alfalfa CDKA;1, and CDKB2;1 complexes, but had little effect on CDKB1;1 and CYCA2;1 complexes (Pettko-Szandtner *et al.*, 2006). Using Arabidopsis CYCD2;1/CDKA;1 and CYCD2;1/CDKB2;1 complexes expressed in insect cells, Nakai *et al.* (2006) reported that recombinant Arabidopsis ICK/KRP proteins inhibited the CYCD2;1/CDKA;1 and CYCD2;1/CDKB2;1 complexes to a similar extent (Nakai *et al.*, 2006). Although results from different studies are not always consistent with each other regarding which specific CDK complexes are the more likely targets, results from the majority of these studies indicate that plant CDK inhibitors possess specificity towards certain CDK–cyclin complexes. The data from immunoprecipitation experiments are made more difficult to interpret and compare by several factors. First, usually only one component of the CDK–cyclin complex is detected while the other component is unknown. Second, the specificities for different antisera used for immunoprecipitation may vary. Third, it remains to be established whether the cyclin subtypes (e.g. A2-type) from different species, for instance from Arabidopsis and alfalfa, are functionally equivalent. Despite the uncertainty about other types of CDK–cyclin complexes, accumulating evidence suggests that CDKA/CYCD complexes are a major target of the ICK1/KRP proteins. CDK complexes containing either a CYCD or a CDKA may also be targets.

3.5 Protein stability and modifications

Protein destruction or proteolysis plays an important role in the cell cycle, since timely removal of certain proteins at the correct stage is crucial for the orderly progression of the cell cycle. Because of the moderate levels of gene expression observed, the levels of ICK/KRP proteins are also expected to be relatively low. Even in transgenic Arabidopsis plants expressing *GFP–ICK1*, despite high levels of transcripts, the level of GFP–ICK1 protein was low (Zhou *et al.*, 2003a). The level of ICK2 was also found to be low in transgenic plants (Verkest *et al.*, 2005b). Previously, it was observed that removal of the N-terminal region results in much stronger interaction of the mutant ICK1 with both CDKA;1 and CYCD3;1 (Wang *et al.*, 1998). Subsequent work showed that the ICK1$^{109-191}$ lacking the N-terminal region accumulated to a much higher level in transgenic plants compared to the full-length ICK1, suggesting that the N-terminal region confers instability on ICK1 (Zhou *et al.*, 2003a). As a result, ICK1$^{109-191}$ had a stronger effect than wild-type ICK1 in inhibiting growth and altering morphology of transgenic Arabidopsis plants (Zhou *et al.*, 2003a). It is not known whether N-terminal regions of other ICK/KRPs have a similar functional property.

The SCF (Skp-cullin-F-box protein) and APC (anaphase-promoting complex) are two ubiquitin ligase (E3) complexes responsible for the ubiquitin-mediated destruction of target cell cycle proteins (see Chapter 4). The SCF complex degrades the Cip/Kip family of mammalian CDK inhibitors (Cardozo and Pagano, 2004) as well as the CDK inhibitors Sic1, Far1 and Rum1 in budding and fission yeast (Toda *et al.*, 1999; Spruck and Strohmaier, 2002). The substrate specificity of the SCF complex is conferred by the F-box proteins. In mammalian cells, the F-box protein Skp2 specifies the proteolysis of Cip/Kip inhibitors (Cardozo and Pagano, 2004).

By analogy to the situation in animals and yeast, ICK/KRP proteins may also be subject to regulation by the SCF complex. This suggestion is supported by some indirect evidence. MG132 is an aldehyde peptide inhibitor of the 26S proteosome, which degrades proteins that have been polyubiquitinated by ubiquitin ligases such as the SCF complex. A recent study showed that treatment of MG132 increased the stability of ICK1 and confirmed that the N-terminal region played a major role (Jakoby *et al.*, 2006). MG132 treatment also increased the stability of recombinant ICK2 protein added to Arabidopsis cell extracts, suggesting that the proteolysis of ICK2 may be mediated by the 26S proteosome (Verkest *et al.*, 2005b).

The proteolysis of the mammalian $p27^{Kip1}$ by SCF in complex with Skp2 requires phosphorylation of $p27^{Kip1}$ at threonine 187 in the C-terminus (Spruck and Strohmaier, 2002). It would be interesting to know whether this type of regulation exists for plant CDK inhibitors. Recently, it was shown that recombinant ICK2 could be phosphorylated *in vitro* by CDKA;1 and CDKB1;1 immunoprecipitates (Verkest *et al.*, 2005b). ICK2 protein added to Arabidopsis cell extract was more stable with the addition of the specific CDK inhibitor olomoucine (Verkest *et al.*, 2005b). Furthermore, in plants expressing a dominant negative mutant of CDKB1;1, the level of ICK2 protein was higher, suggesting that phosphorylation of ICK2 regulates its level by decreasing its stability (Verkest *et al.*, 2005b). On the other hand, the alfalfa CDK inhibitor KRPMt could be phosphorylated *in vitro* by a calmodulin-like domain protein kinase (Pettkó-Szandtner *et al.*, 2006). Furthermore, the phosphorylated KRPMt showed increased CDK inhibitory activity. It remains to be seen whether this enhanced activity of KRPMt is also conserved *in vivo*. Obviously, more work is needed to determine the kinases responsible for phosphorylating plant CDK inhibitors and more importantly the effect of phosphorylation on their functions and stability.

3.6 Cellular localization

Localization to particular compartments or cellular regions is an important aspect of function and control for cell cycle regulators (Pines, 1999). A key molecular feature for the functions of ICK/KRP CDK inhibitors is their ability to interact with D-type cyclins and A-type CDKs. Studying their cellular localization and how localization is controlled would provide further understanding as to the cellular compartment in which the cell cycle regulators interact with each other. The tobacco CDK inhibitors NtKIS1a and NtKIS2 and Arabidopsis ICK1 fused to GFP are localized in the nucleus of interphase cells (Jasinski *et al.*, 2002a; Zhou *et al.*, 2003a; Weinl *et al.*, 2005). Analyses show that a sequence 'RRGTKRKL' located at residues 80–87 in the central region of ICK1 is a strong NLS (Zhou *et al.*, 2006).

When the NLS in the central region of ICK1 is mutated, the mutant ICK1 is still strongly localized in the nucleus (Zhou *et al.*, 2006). Analyses show that the nuclear localization of ICK1 is controlled by multiple sequences (Jakoby *et al.*, 2006; Zhou *et al.*, 2006). The N-terminal, central and C-terminal regions can all confer nuclear localization of the respective GFP fusion protein (Zhou *et al.*, 2006). In addition, these sequences differ in the ability and extent of nuclear localization as well as in the pattern of subnuclear distribution. The N-terminal NLS is also responsible for specifying a punctate pattern of subnuclear localization (Jakoby *et al.*, 2006; Zhou *et al.*, 2006). This pattern, however, does not occur in all cells, and interestingly the C-terminal sequence is found to suppress the punctate pattern (Zhou *et al.*, 2006). Although the punctate structures overlap with chromocenters in some cells, they do not in other cells (Jakoby *et al.*, 2006; Zhou *et al.*, 2006). The patterns and dynamic nature of the nuclear localization indicate that ICK1 protein is highly regulated in plant cells.

It is interesting to compare the nuclear localization of ICK1 and NtKIS1a to the cellular localization of CDKA and D-type cyclins, with which the CDK inhibitors interact. CDKAs appear to show both nuclear and cytoplasmic localization in interphase cells (Bögre *et al.*, 1997; Stals *et al.*, 1997; Weingartner *et al.*, 2001). One intriguing feature of CDKA cellular distribution is its co-localization with microtubule arrays, including the pre-prophase band, the spindle and the phragmoplast (Colasanti *et al.*, 1993; Stals *et al.*, 1997; Weingartner *et al.*, 2001). Arabidopsis CYCD3;1 was recently found to be predominantly localized in the nucleus (Koroleva *et al.*, 2005). Based on the cellular localization, it is tempting to speculate that ICK1 may act to inhibit CDKA activity or antagonize the function of D-type cyclins in the nucleus.

Since ICK1 and CDKA display different patterns of cellular distribution and they also interact with each other, one interesting question is whether the interaction affects their cellular distribution. If it does, which of the two has the dominant influence over the other? If the interaction affects the distribution, the distribution pattern of one interacting protein would be 'shifted' toward the pattern of the interacting protein that has a stronger influence. When ICK1 tagged with CFP (cyan fluorescent protein) and CDKA tagged with YFP (yellow fluorescent protein) were co-expressed, CFP–ICK1 remained exclusively nuclear, while YFP–CDKA became much more localized in the nucleus (Zhou *et al.*, in press). On the other hand, the exclusive nuclear localization of ICK1 was not weakened by CDKA. Furthermore, when the C-terminal region required for interacting with CDKA was deleted from ICK1, its ability to mediate the nuclear transport of CDKA was impaired (Zhou *et al.*, 2006). These observations clearly show that ICK1 could mediate the transport of CDKA into the nucleus, and this transport depends on the specific interaction between ICK1 and CDKA.

The exclusive nuclear localization and highly regulated patterns of subnuclear distribution suggest that the nuclear localization of ICK1 is critical for its role in regulating the cell cycle. Based on current understanding, the nuclear localization of ICK1 could have three functional roles: to inhibit CDKA activity in the nucleus, to counter the activity of D-type cyclins and to regulate the cellular distribution of CDKA by transporting it to the nucleus. Although further research is needed to define the biological significance of the nuclear localization by plant CDK inhibitors, the importance of cellular localization for the mammalian CDK inhibitor $p27^{Kip1}$ has been clearly established. $p27^{Kip1}$ is normally localized in the nucleus, and a classical bipartite NLS in the C-terminal region is responsible for this nuclear localization (Polyak *et al.*, 1994; Toyoshima and Hunter, 1994). When $p27^{Kip1}$ is mislocalized or relocalized to the cytoplasm, its ability to inhibit CDK2 and cell growth is impaired (Jiang *et al.*, 2000; Liu *et al.*, 2000). It was observed that increased cytoplasmic localization of $p27^{Kip1}$ instead of the normal nuclear localization occurs in certain tumors (Slingerland *et al.*, 2000; Shin *et al.*, 2002; Viglietto *et al.*, 2002) and is associated with poor prognosis in breast and ovarian cancers (Liang *et al.*, 2002; Rosen *et al.*, 2005). Interestingly, the mislocalization of $p27^{Kip1}$ is due to the phosphorylation of threonine 157 within the bipartite NLS, impairing the nuclear import of $p27^{Kip1}$ (Liang *et al.*, 2002; Shin *et al.*, 2002; Viglietto *et al.*, 2002). Thus, we can expect that the nuclear localization of ICK1 and NtKIS1a has a functional role in plants. It is also interesting to note that plant CDK inhibitors are not necessarily cell autonomous, since analysis of ICK1 expression in trichomes showed that ICK1 can move from trichomes to trichome-neighboring cells (Weinl *et al.*, 2005). This observation suggests the possibility that a plant cell expressing *ICK1* may affect adjacent cells through intercellular movement, and presumably the extent of the influence would depend on the level of *ICK1* expression in the source cell.

3.7 CDK inhibitors and plant growth and development

One important area of plant cell cycle research is the role of cell division in plant growth and development. There are several related fundamental questions. One question that has long been debated is whether cell division plays an active role in determining plant growth and the size of organs. Also, since cell division is an integral part of plant growth and development, it is important to ask how the cell cycle machinery establishes crosstalk with the pathways for regulating plant growth and development. Throughout the life of a plant, meristematic cells, sometimes referred to as 'stem' cells, are maintained in various locations, such as shoot meristems, root meristems, lateral buds and floral meristems. There are many unanswered questions related to the cell cycle and cell differentiation. For example, how are the meristematic cells maintained and, equally important, what determines if and when proliferating cells should exit the cell cycle and differentiate? Are cell division control and cell size control coupled? There are competing theories in the literature as to whether cell division is a 'master' or 'slave' of plant growth (Kaplan and Hagemann, 1991; Jacobs, 1997). Experimental results obtained over the last decade indicate that cell division is neither a 'slave' nor a 'master' of plant growth; rather cell division and plant growth are intertwined. The challenge is to define in more specific terms how these processes are connected mechanistically and to determine the extent of influence that cell division control has on plant growth as well as on morphogenesis. This topic has been reviewed extensively (e.g. Mizukami, 2001; Traas and Doonan, 2001; Beemster *et al.*, 2003; de Jager *et al.*, 2005; Gutierrez, 2005). Several chapters in this book also review the role of cell cycle in several developmental processes. In this section, we shall discuss the results concerning only the CDK inhibitors.

CDK inhibitor genes provide convenient tools to perturb cell divisions *in vivo* rather than through external means. It is interesting to observe that overexpression of *ICK1*, *ICK2*, *KRP6*, *NtKIS1a, NtKIS2* (from tobacco) and *ICDK* (from *C. rubrum*) under the control of the cauliflower mosaic virus 35S promoter showed similar effects in studies by different laboratories. The transgenic Arabidopsis plants were smaller, with altered leaf morphology (Plate 3.2A–D) (Wang *et al.*, 2000; De Veylder *et al.*, 2001; Jasinski *et al.*, 2002b; Zhou *et al.*, 2002a). These results clearly show that overexpression of a plant CDK inhibitor not only results in growth inhibition, but also leads to leaf serrations (Plate 3.2B). Thus, it appears that the expression of a CDK inhibitor not only reduces the number of cell division, but may also change the pattern of cell division during leaf development in these transgenic plants. In the case of *ICK1* overexpression, early flowering and a reduction in leaf number were also observed, although it is not known whether these changes are a direct or an indirect effect of *ICK1* expression (Wang *et al.*, 2000). Since CDKA and CYCD are the likely targets of ICK/KRP proteins, these results also support the role of the CDKA and/or CYCD complexes in regulating cell proliferation and plant growth.

The effect of overexpressing or misexpressing a CDK inhibitor can be better studied when it is expressed tissue specifically. In transgenic plants overexpressing *ICK1* driven by the 35S promoter, profound effects were observed on the growth of most organs, including leaves, roots, stems, sepals and petals, as well as on the morphology of leaves and petals (Wang *et al.*, 2000). It is not known whether some of the effects might be caused indirectly by primary effects on certain processes such as root growth. Therefore, we targeted the expression of *ICK1* to developing petals of *Brassica* plants using the Arabidopsis *AP3* promoter (Zhou *et al.*, 2002b). In addition to missing some petals, transgenic *AP3::ICK1 Brassica* plants produced smaller petals with various altered shapes, including filamentous, tubular and wrinkled petals (Plate 3.2E–J). It was also observed that various shapes of petals could be produced on the same plant (Zhou *et al.*, 2002b). This study represents the first case in which the expression of a cell cycle regulator has changed not only the shape of a plant organ, but also its basic architecture, with petals ranging from blade-like to tubular in shape.

Considering that petal shape in wild-type *Brassica* plants, as in other plants, is highly consistent, the drastic changes in petal shape shown in Plate 3.2 are very intriguing. The reduction of petal size could be attributed to inhibition of cell division by the expression of *ICK1*. However, what is the mechanism responsible for the change from the normal blade-like petals to tubular petals (Plate 3.2H–J)? One possibility is that inhibition of cell division due to *ICK1* expression in the petal primordia disrupted the normal pattern of petal development and triggered an alternative pattern of petal morphogenesis (Zhou *et al.*, 2002b). It is also possible that this switch in pattern of petal development might be due to the overall inhibition of the CDK regulatory pathway by ICK1. Currently, we do not know how this switch takes place. The results, however, do demonstrate that an alternative pattern of petal development leading to tubular petals must have been activated in the *AP3::ICK1 Brassica* plants. The existence of such a developmental route leading to tubular-shaped petals is supported by the observation that in the Arabidopsis *petal loss* (*ptl*) mutant, about 25% petals were tubular and 30% were filamentous (Griffith *et al.*, 1999). The *PTL* gene encodes a trihelix transcription factor, but its mechanism in regulating petal development is not clear (Brewer *et al.*, 2004).

ICK1 has also been expressed in specific cell types, such as pollen and trichomes (Zhou *et al.*, 2002b; Schnittger *et al.*, 2003). The expression of *ICK1* in pollen using a pollen-specific promoter resulted in a reduction in pollen viability as well as number of nuclei in pollen grains (Zhou *et al.*, 2002b). These plants also showed reduced fertility. The expression of *ICK1* in trichomes of Arabidopsis plants using the trichome-specific *GL2* promoter resulted in smaller cells and a decreased number of trichome branches (Schnittger *et al.*, 2003). Interestingly, the trichomes died early. The death of cells in trichomes in response to *ICK1* misexpression raises the question as to why no global pattern of cell death is observed in plants overexpressing *ICK1* or other CDK inhibitors. Perhaps, the cell death phenotype is determined by cell type as well as by the level of the CDK inhibitor expressed. It is likely that certain cell types are more sensitive to *ICK1*-induced cell death. However, cell death in trichomes is not observed in the plants overexpressing *ICK1* and other CDK inhibitors using the 35S promoter. This discrepancy may be due to a higher level of *ICK1* expression in trichomes as a result of using the trichome-specific pGL2 promoter rather than the 35S promoter.

The effects of overexpressing a CDK inhibitor on organ size raise interesting questions regarding the relationship between cell number and cell size. For example, does cell size remain the same when the cell number is reduced? This question is fundamentally important in the context of plant and organ size control. There exist two opposing classical theories in terms of explaining how the shape and size of an organ are controlled, the cell theory and the organismal theory (Kaplan and Hagemann, 1991; Jacobs, 1997; Day and Lawrence, 2000). According to the cell theory, the cell is the basic unit of a multicellular organism and thus is the primary determinant of the size and morphology of the organism. According to the organismal theory, the size and shape of an organism are determined by the organism as a whole, independent of individual constituent cells. Transgenic plants overexpressing a CDK inhibitor driven by the 35S promoter had fewer cells, but cells were larger (Wang *et al.*, 2000; De Veylder *et al.*, 2001; Jasinski *et al.*, 2002b), indicating that a compensatory mechanism exists at the organ(ismal) level to increase cell size in response to a reduction in cell number. However, in transgenic *Brassica AP3::ICK1* plants, cell number in smaller petals was reduced, but cell size remained unchanged (Zhou *et al.*, 2002b). Further, *ICK1* misexpression in trichomes actually reduced both the cell number and cell size (Schnittger *et al.*, 2003). Thus, there is no simple inverse relationship between cell number and cell size. The varying effects on cell size observed in different studies evaluating expression of CDK inhibitors may be due to differences in tissue type or the level and time of inhibitor expression. These results contribute to the realization that neither the cell theory nor the organismal theory is sufficient to explain the complex relationship between cell number and size in the context of growth control. They highlight the need for a more sophisticated model, which has a foundation in molecular mechanisms.

The results from the *ICK/KRP* overexpression studies show that this family of proteins can inhibit the activities of CDK complexes in plants, likely through the interactions with CDKA and/or CYCD. Co-expression experiments further show that expression of a D-type cyclin can rescue the phenotype resulting from the expression of the tobacco CDK inhibitor or *ICK1* (Jasinski *et al.*, 2002b; Schnittger *et al.*, 2003; Zhou *et al.*, 2003a). It was further observed that when *ICK1* was overexpressed, Arabidopsis *CYCD2;1* and *CYCD3;1* were upregulated (Zhou *et al.*, 2003a), indicating that increasing the level of a CDK inhibitor induces an increase in the level of positive regulators. These results indicate that interactions between CDK inhibitors and D-type cyclins observed in the yeast two-hybrid system (Wang *et al.*, 1998; Jasinski *et al.*, 2002b) are biologically relevant. More importantly, these results illustrate that the cell cycle and plant growth are affected by the relative

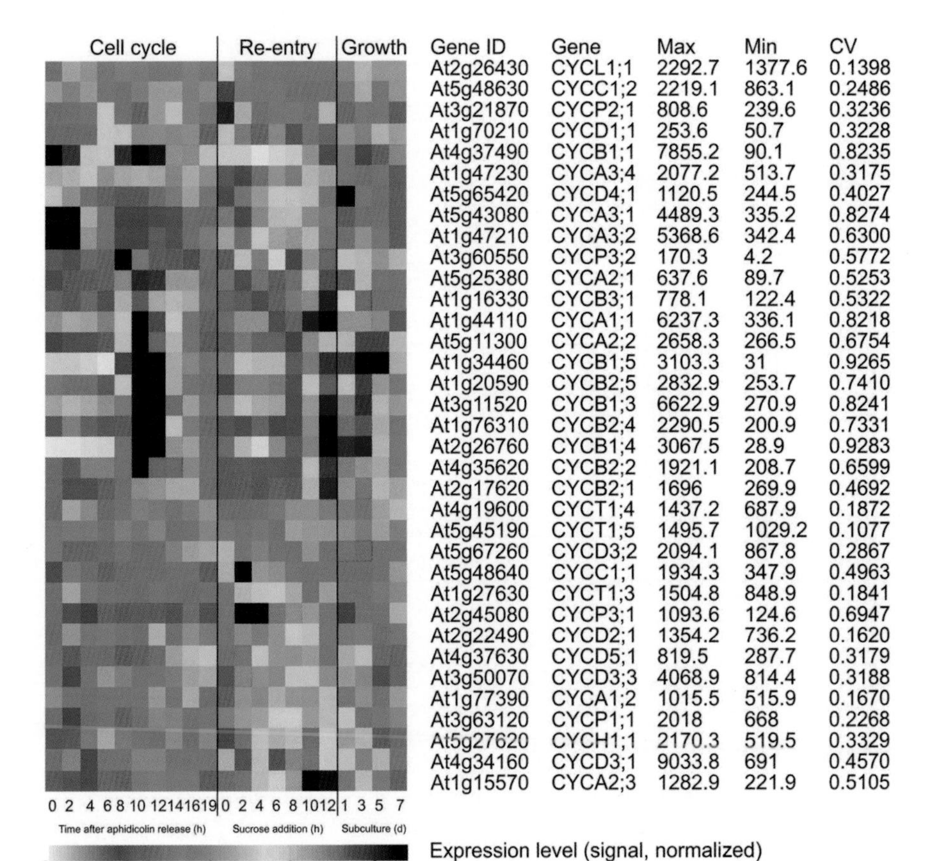

Gene ID	Gene	Max	Min	CV
At2g26430	CYCL1;1	2292.7	1377.6	0.1398
At5g48630	CYCC1;2	2219.1	863.1	0.2486
At3g21870	CYCP2;1	808.6	239.6	0.3236
At1g70210	CYCD1;1	253.6	50.7	0.3228
At4g37490	CYCB1;1	7855.2	90.1	0.8235
At1g47230	CYCA3;4	2077.2	513.7	0.3175
At5g65420	CYCD4;1	1120.5	244.5	0.4027
At5g43080	CYCA3;1	4489.3	335.2	0.8274
At1g47210	CYCA3;2	5368.6	342.4	0.6300
At3g60550	CYCP3;2	170.3	4.2	0.5772
At5g25380	CYCA2;1	637.6	89.7	0.5253
At1g16330	CYCB3;1	778.1	122.4	0.5322
At1g44110	CYCA1;1	6237.3	336.1	0.8218
At5g11300	CYCA2;2	2658.3	266.5	0.6754
At1g34460	CYCB1;5	3103.3	31	0.9265
At1g20590	CYCB2;5	2832.9	253.7	0.7410
At3g11520	CYCB1;3	6622.9	270.9	0.8241
At1g76310	CYCB2;4	2290.5	200.9	0.7331
At2g26760	CYCB1;4	3067.5	28.9	0.9283
At4g35620	CYCB2;2	1921.1	208.7	0.6599
At2g17620	CYCB2;1	1696	269.9	0.4692
At4g19600	CYCT1;4	1437.2	687.9	0.1872
At5g45190	CYCT1;5	1495.7	1029.2	0.1077
At5g67260	CYCD3;2	2094.1	867.8	0.2867
At5g48640	CYCC1;1	1934.3	347.9	0.4963
At1g27630	CYCT1;3	1504.8	848.9	0.1841
At2g45080	CYCP3;1	1093.6	124.6	0.6947
At2g22490	CYCD2;1	1354.2	736.2	0.1620
At4g37630	CYCD5;1	819.5	287.7	0.3179
At3g50070	CYCD3;3	4068.9	814.4	0.3188
At1g77390	CYCA1;2	1015.5	515.9	0.1670
At3g63120	CYCP1;1	2018	668	0.2268
At5g27620	CYCH1;1	2170.3	519.5	0.3329
At4g34160	CYCD3;1	9033.8	691	0.4570
At1g15570	CYCA2;3	1282.9	221.9	0.5105

Plate 2.1 Affymetrix ATH1 microarray analysis of the expression of cyclin genes. Three treatments/analyses are shown: *Arabidopsis* MM2d suspension culture cells (Menges and Murray, 2002) after aphidicolin block release, synchronizing cells at the G1/S boundary (labelled 'cell cycle'), sucrose removal and resupply causing re-entry into the cell cycle ('re-entry') and growth following routine subculture ('growth') (Menges *et al.*, 2003, 2005). Samples were taken every 2 h and the last time point in the growth experiment corresponds to early stationary phase when the culture would again be subcultured. The right panel shows gene ID, gene name, maximum (Max) and minimum (Min) detected signals across all three experiments of all cyclins detected as expressed. A coefficient of variation (CV) was calculated by dividing the standard deviation by the mean signal (CV = SD/mean). The absolute detected signals were normalized by dividing the signal by the mean and normalized signal are shown in the colored representation across all three experiments. A predominantly green color in a row highlights genes that are constantly expressed, blue or white represents downregulated genes and yellow, red and black upregulated genes.

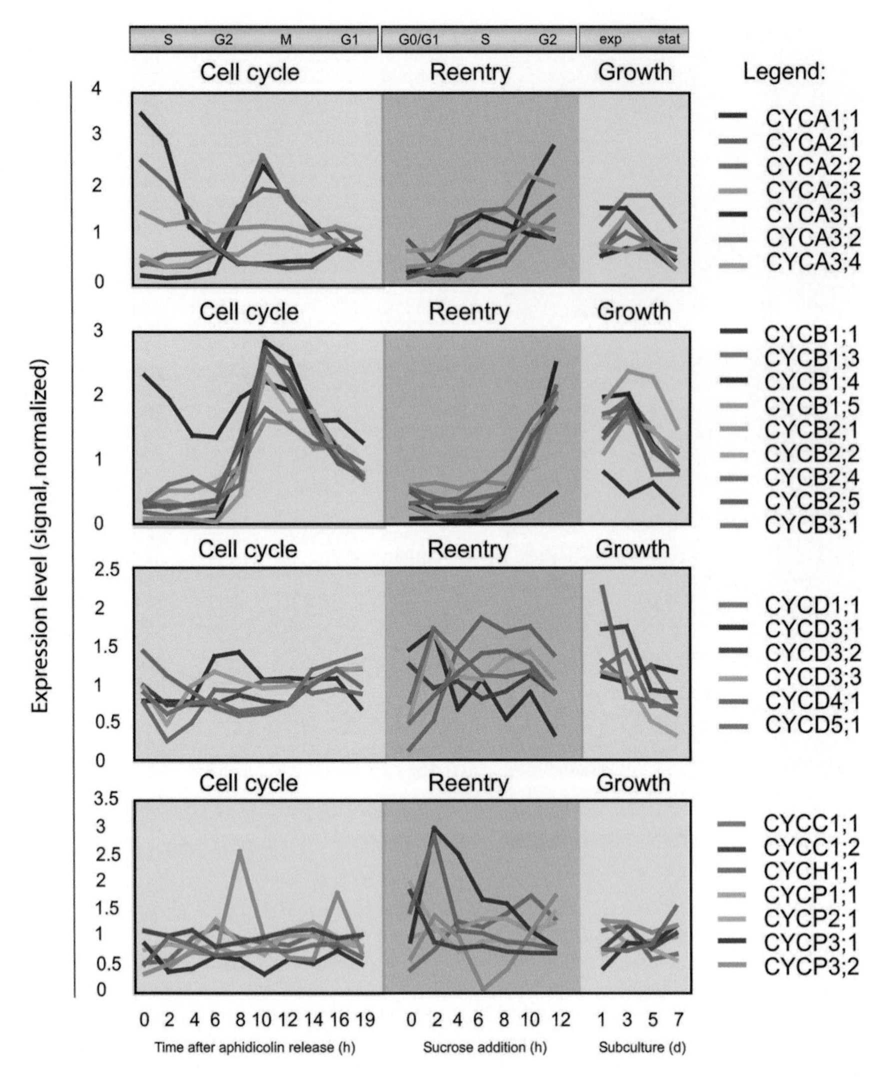

Plate 2.2 Graphical representation of expression of cyclin genes in *Arabidopsis* cell culture as shown in Plate 2.1. Normalized data from cell cycle progression (aphidicolin-induced synchrony), re-entry (starvation-induced synchrony) and unperturbed growth experiments were normalized as described in Plate 2.1. Note that only cyclin genes showing CV >0.2 in the cell cycle experiments are shown graphically.

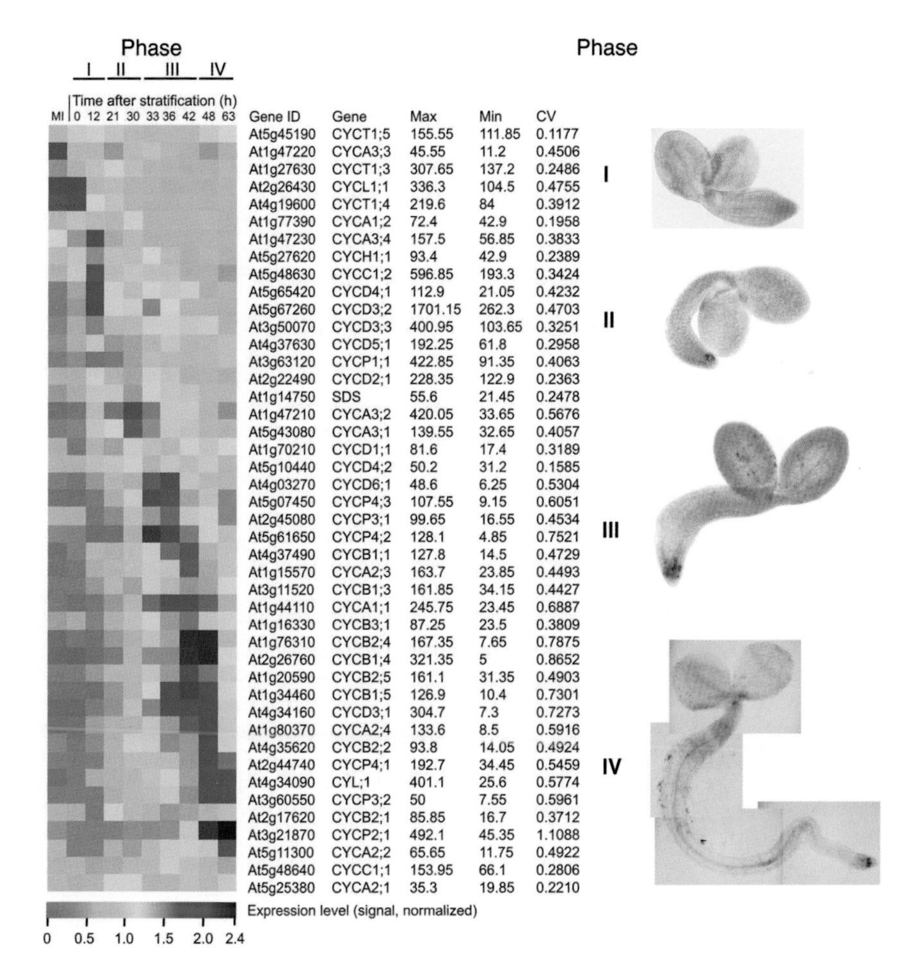

Plate 2.3 Transcript levels of all cyclin genes expressed during seed germination (Masubelele *et al.*, 2005). Seeds were sown on a double layer of pre-wetted filter paper and imbibed at 4°C for 3 days in the dark to ensure synchronous germination before moving to Conviron TC30 cabinets (Controlled Environments, Manitoba, Canada) under continuous white light (170–200 μmol m^{-2} s^{-1}) at 22°C on moist filter paper. The first sample was taken at mid-imbibition (MI, 36 h after start of water uptake at 4°C). Subsequent samples are taken after stratification at time points as indicated. Averaged absolute signals were normalized as described in Plate 2.1 and subjected to hierarchical cluster analysis (GeneMaths, version 2.01). Maximum (Max) and minimum (Min) detected signals and the CV of expression are shown for all genes called present on at least one array. All data are the average of duplicate GeneChip arrays. A predominantly cream or light orange color in a row indicates genes that are relatively constantly expressed; orange to red color represents downregulated genes and light to dark blue color represents upregulated genes. Seed germination involves four phases of cell cycle activation. In phase I, the embryo expands without mitosis, while in phase II cell division is activated in the RAM. The SAM and cotyledon cells become activated in phase III. In phase IV the cells in the cotyledons stop to proliferate and mitotic divisions are initiated in the leaf primordia and lateral root initials.

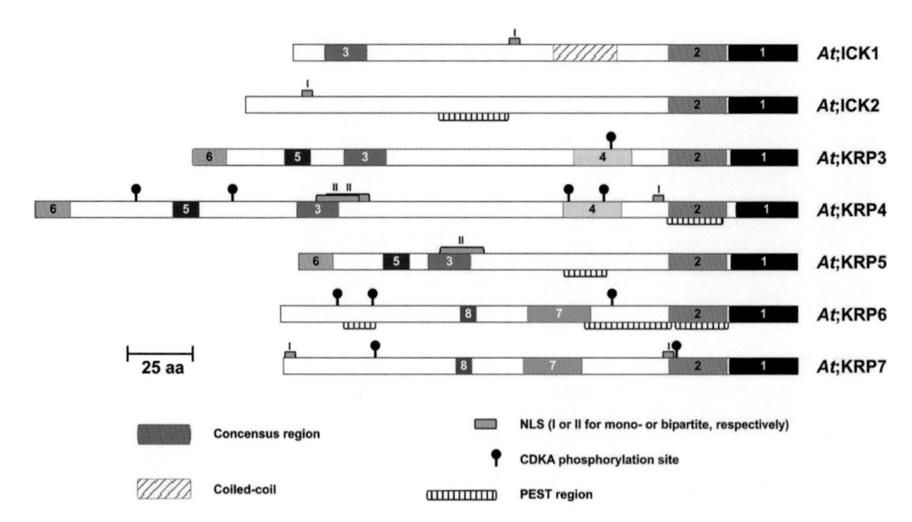

Plate 3.1 Analysis and comparison of *Arabidopsis* ICK/KRP proteins and the genomic organization of the genes. Conserved regions and sequences with putative functions. The conserved regions are marked by shaded boxes. The C-terminal conserved domain containing two connected motifs is marked as box number 1, since this domain is the hallmark for this family of proteins and is critical for CDK inhibitory function. The other motifs are numbered from 2 to 8, of which motifs 2, 3, 5 and 6 were also described by De Veylder *et al.* (2001) Sequences with putative functions are marked either above or below the bar representing the protein. Putative NLSs were identified using scan-Prosite (http://au.expasy.org/tools/scanprosite/) or PSPORTII (http://psort.hgc.jp/form.html), either as a monopartite NLS (I) or a bipartite NLS (II). Putative phosphorylation sites are motifs that match the consensus sequence [TS]-P-x-[RKQSL]. Potential PEST sequences were analyzed using PESTFind http://www.at.embnet.org/embnet/tools/bio/PESTfind/. Sequences with a score of more than +8.0 are marked. The scale bar is indicated.

Plate 3.2 Phenotypes of transgenic plants expressing *ICK1*. (A) Wild-type (left) and *35S::ICK1* (right and inset) *Arabidopsis* plants showing reduced size and changes in morphology of the *35S::ICK1* plant. (B) Arabidopsis leaves removed from the wild-type (top row), four independent *35S::ICK1* transgenic and the control *35S::ICK1* plants. Note the smaller and serrated leaves from the *35S::ICK1* plants. (C,D) Comparison of a flower from the wild-type (C) and *35S::ICK1* plant (D). (E–J) Petals from control (E) and transformed *AP3::ICK1 Brassica* (F–J) plants showing absence of petals or changes in petal morphology. (F,G) Petals are missing in the flowers. (H–J) Tubular or filamentous petals are produced. (A–D) are from Wang *et al.* (2000) in *Plant Journal* **24**, 616–617, reproduced with permission of Blackwell Publishing. (E–J) are from Zhou *et al.* (2002b) in *Planta* **215**, 251, reproduced with permission of Springer Science and Business Media.

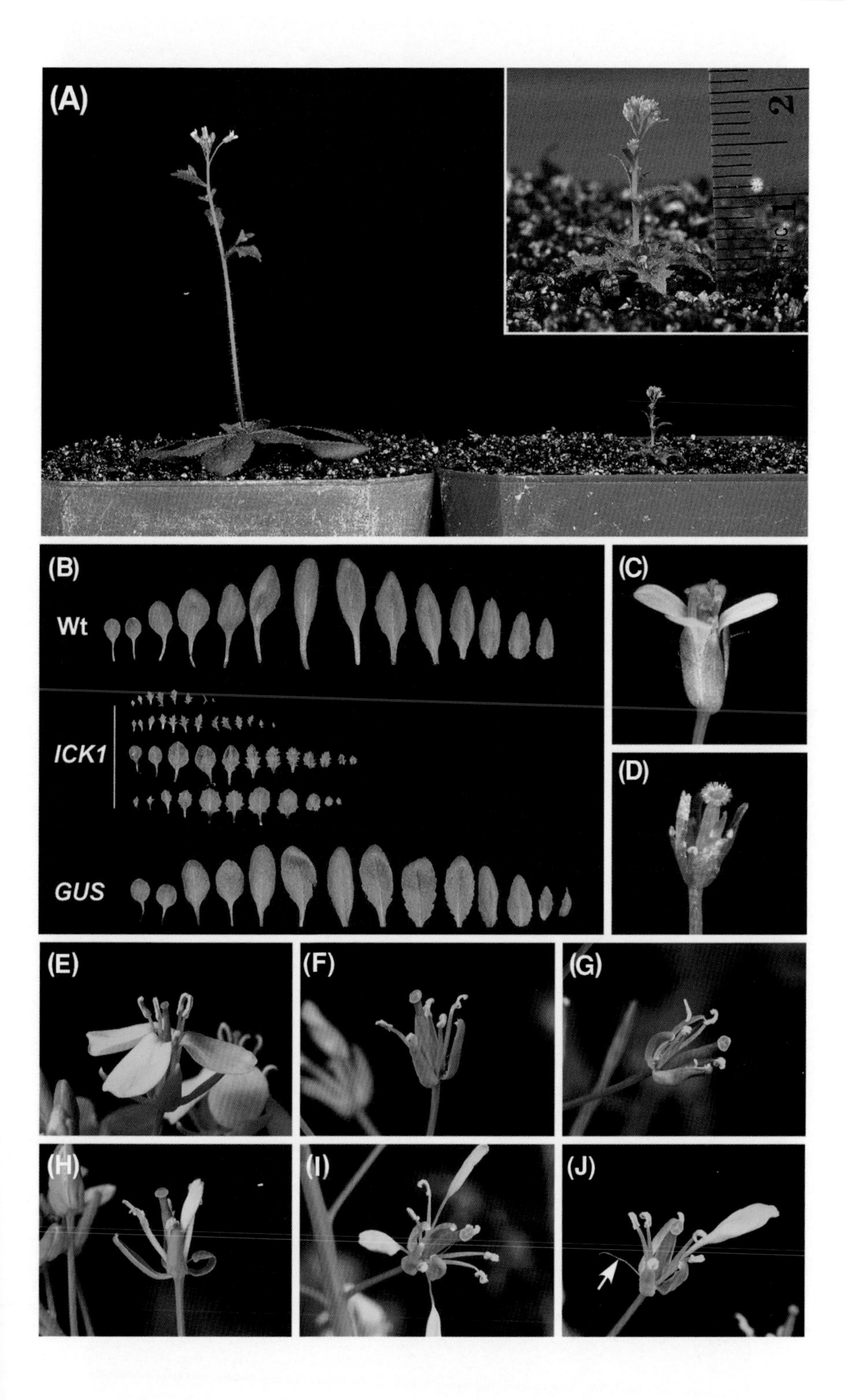
(A)
2
1
(B)
Wt
ICK1
GUS
(C)
(D)
(E)
(F)
(G)
(H)
(I)
(J)

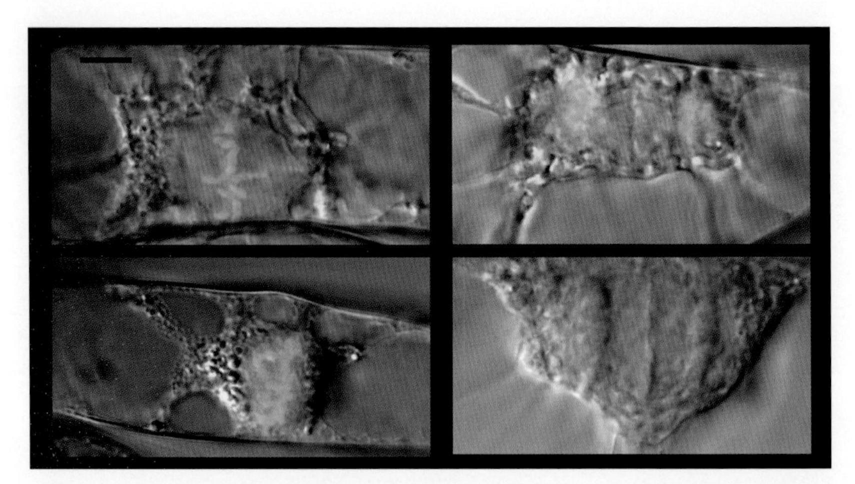

Plate 4.1 B1-type cyclin stability in plant BY2 cells. The two lower pictures illustrate GFP-tagged cyclin B1;1 subcellular localization during mitosis: metaphase (right) and telophase (left) in transgenic tobacco BY2 cells. The high fluorescent-lighted chromosomes indicate that the B1-type cyclins may be involved in the regulation of chromosome structure organization. In telophase no cyclin B1;1–GFP is detectable anymore because of its rapid turnover. The two upper pictures illustrate non-degradable cyclin B1;1 (mutDBox)–GFP fusion protein stability during mitosis: metaphase (right) and telophase (left). Here the cyclin–GFP fusion protein, in which the Dbox has been mutated, remains stable in telophase. The photographs correspond to the fluorescent confocal plane superimposed to the differential interference contrast (DIC) image. Bar = 10 μm.

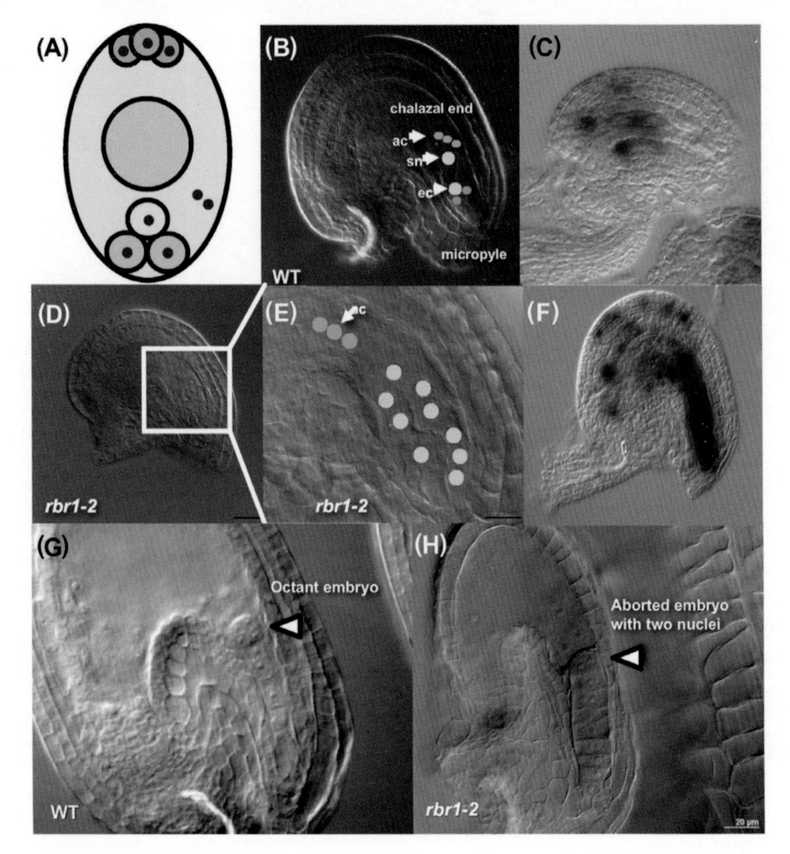

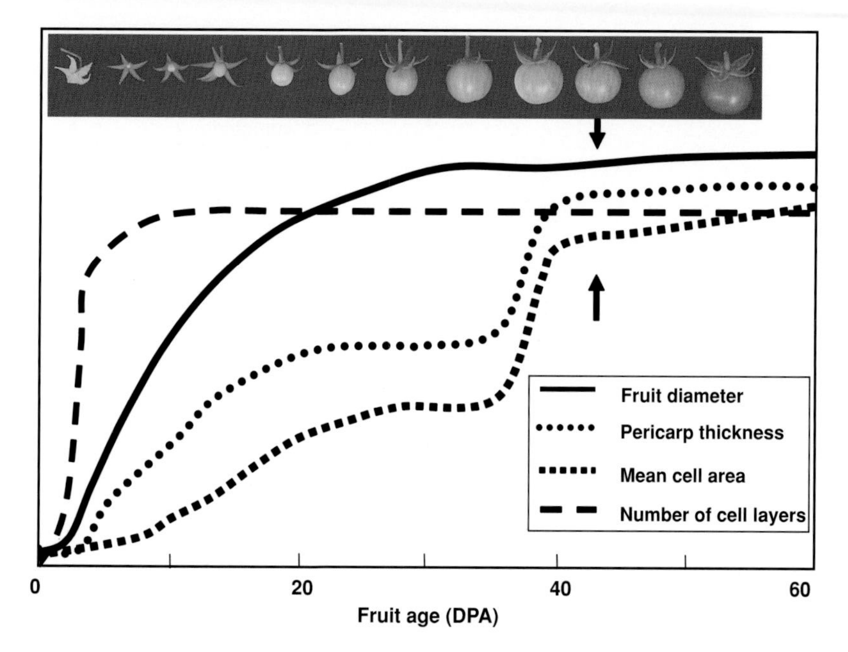

Plate 12.1 Developmental characteristics of tomato fruit growth (adapted from Cheniclet *et al.*, 2005). Tomato fruit growth as witnessed by the increase in fruit diameter results first from cell division activities which amplify very rapidly the number of cell layers inside the pericarp. Concomitantly cell enlargement starts as early as 3 to 4 days post-anthesis (DPA) and contribute to increase the pericarp thickness. In less than 8–10 days, the number of cell layers is fixed within the pericarp: fruit growth only proceeds from cell expansion afterwards. Note the sudden burst of pericarp thickness and mean cell area at the onset of ripening, namely the Breaker stage (indicated with vertical arrows) which corresponds to the first visible change of colour due to carotenoid accumulation.

Plate 7.1 Plant RBR function is required for correct gametophytic, endosperm and early embryo development (Ebel *et al.*, 2004). Arabidopsis mutants heterozygous for *RBR1* loss-of-function alleles develop normally but show maternally controlled gametophytic lethality that results in abortion of approximately 50% ovules (Ebel *et al.*, 2004). Analysis of unfertilized *RBR1*$^-$ gametophytes revealed an overproliferation of nuclei in the position of the egg cell apparatus and the diploid nucleus of the central cell. (A) Schematic drawing of the female gametophyte showing the egg cell apparatus (bottom: yellow, egg cell; orange, two synergids), the three antipodal cells (top, green cells), and the diploid endosperm nucleus in the central cells (black dots). (B) Cleared unfertilized wild-type ovule showing the position of the female gametophyte. For better visualization the nuclei have been colored according to the schematic drawing in A (ec, egg cell; sn, endosperm nucleus; ac, antipodal cells). (C) Expression of a *CYCB1-promoter::GUS* construct in ovules. The GUS staining reflects cell cycle activity in the sporophytic ovule wall. Note that GUS staining is clearly absent from the micropylar end of the ovule, indicating that the egg apparatus has withdrawn from the mitotic cycle prior to fertilization. (D) Cleared unfertilized ovule in *RBR1/rbr1-2* plants containing a *RBR1*$^-$ gametophyte in which nuclei in the egg cell apparatus are overproliferating. (E) Enlarged view of the cleared ovule in (D) showing the micropylar end containing the unfertilized *RBR1*$^-$ female gametophyte. The nuclei of the hyperproliferating egg cell apparatus are highlighted (amber) for easier visibility. Note that the antipodal cells (green) are positioned normally and do not overproliferate. (F) To demonstrate the overproliferation phenotype of the *RBR1*$^-$ female gametophyte, the *RBR1/rbr1-2* mutant was crossed with a *CYCB1-promoter::GUS* line, which marks nuclei entering M phase. The strong GUS activity in the micropylar region clearly demonstrates that the cells of the unfertilized egg cell apparatus cannot be maintained in their differentiated state in the absence of functional RBR. (G) Cleared wild-type ovule after fertilization showing an octant-stage embryo. (H) Embryo development in Arabidopsis requires a functional *RBR1* gene. Female *RBR*$^-$ gametophytes become fertilized at a very low frequency. The embryo is unable to develop, however, and becomes arrested at a very early stage with abnormal cell divisions and cells with supernumerary nuclei.

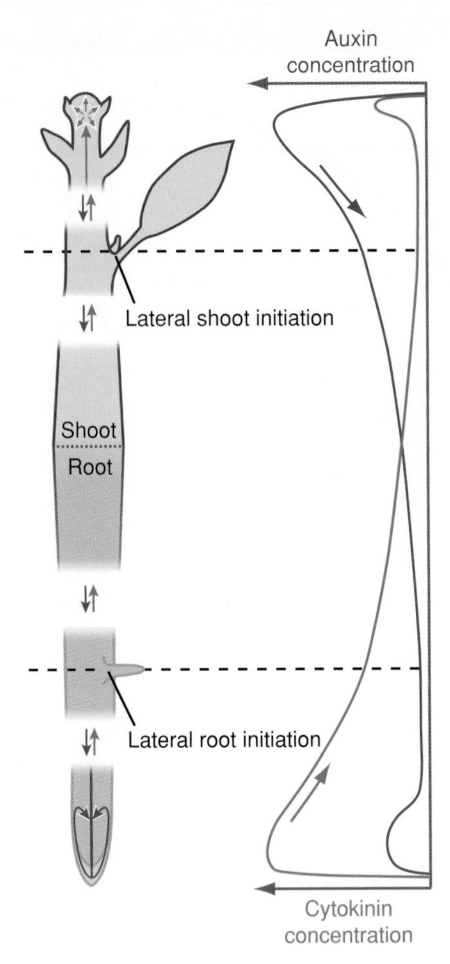

Plate 14.2 Trends in tissue concentrations of auxin and cytokinin through the plant. Auxin synthesized in the shoot is frequently the key limiting factor for cell division in the root, whereas cytokinin, largely synthesized in the root, has this role in the shoot. Coloured arrows indicate hormone movement. Auxin is concentrated in the cell division zone at the root tip through recycling of the hormone via auxin efflux regulator PIN proteins. At the shoot apex, cytokinin concentration is locally amplified by its induction of the homeobox transcription factor STM that expresses the cytokinin-synthesizing enzyme IPT7. Gradients are consistent with extractable hormone levels, the expression zones of cytokinin biosynthetic enzymes and activity of auxin- or cytokinin-responsive promoters. It is considered likely that cell division in lateral meristems begins where ratios of auxin and cytokinin become optimal owing to increased distance from the dominant terminal meristem. Shoot lateral meristems are activated as shoot tissue matures when auxin level has declined relative to cytokinin, which occurs more extensively when transgenes or mutations raise cytokinin or reduce auxin. In the root, lateral meristems are initiated in the zone where cytokinin has declined and auxin has risen, which is more extensive and closer to the tip when a transgene or mutation increases auxin or reduces cytokinin. This diagram is elaborated from earlier concepts of control by hormone ratio (Skoog and Miller, 1957; Torrey, 1962; Klee and Romano, 1994; Werner *et al.*, 2003). Other references are given in the text.

levels of CDK inhibitors and interacting D-type cyclins. This view can likely be extended to other cell cycle regulators, since the final CDK activity depends on the integrated input from various components of the cell cycle machinery as well as protein phosphorylation and proteolysis. Thus, an improved understanding of the role of cell cycle regulation in growth control will also require a more complete understanding of the interplay among different factors.

So far, all results regarding the effects of ICK/KRP proteins have been obtained by transgenic overexpression or misexpression. Although results from these studies show that cell division is important for plant growth as well as morphogenesis, the results are less informative about the specific roles for each of the CDK inhibitors in plants. Currently, we know little about the *in vivo* functions for each of the seven *ICK/KRP* genes in wild-type Arabidopsis plants. For this type of understanding, we need to resort to approaches that can downregulate the expression and activity of *ICK/KRP*s. One approach is to look for possible phenotypes associated with the T-DNA knockout mutants. The likely functional redundancies among the ICK/KRP proteins make this part of the work more challenging. Thus far, there have been no reports of phenotypic effects with T-DNA insertion mutants of the Arabidopsis *ICK/KRP* genes.

3.8 Cell cycle phase transitions

In mammalian cells, the Cip/Kip family of inhibitors act on CDK4/6 and CDK2 complexes while INK4 inhibitors act on only cyclin D–CDK4/6 complexes (Morgan, 1997). Therefore, they are mainly involved in G1 activation and the G1 to S transition during the cell cycle. The plant CDK inhibitors interact with D-type cyclins and most of them interact also with A-type CDKs (Wang *et al.*, 1998; Lui *et al.*, 2000; De Veylder *et al.*, 2001; Zhou *et al.*, 2002a; Bisbis *et al.*, 2006; Pettko-Szandtner *et al.*, 2006). The primary functions of the CDK inhibitors are likely a result of interactions with the CYCD and CDKA complexes. D-type cyclins are believed to be mainly involved in regulating the G1/S transition (Dewitte and Murray, 2003; Chapter 2). Results from several studies indicate that D-type cyclins may also be important for the G2/M transition (Sorrell *et al.*, 1999; Schnittger *et al.*, 2002; Kono *et al.*, 2003; Koroleva *et al.*, 2004). CDKA is likely involved in both the G1/S and G2/M transitions (Stals and Inzé, 2001; Dewitte and Murray, 2003). Thus, the CDK inhibitors are most likely involved in the G1/S transition and may also play a role in G2/M transition.

Several lines of indirect evidence are consistent with a role for CDK inhibitors in the activation of the cell cycle and the G1/S transition. First, it has consistently been shown that transgenic overexpression of a plant CDK inhibitor can block the cell cycle (Wang *et al.*, 2000; De Veylder *et al.*, 2001; Jasinski *et al.*, 2002b; Zhou *et al.*, 2002a). The observed decreases rather than increases in ploidy level with strong expression of CDK inhibitors imply a major effect at the G1/S transition rather than the G2/M transition. Second, the levels of *ICK1* and *ICK2* transcripts decreased during reentry into the cell cycle in synchronized Arabidopsis cell cultures

(Menges and Murray, 2002; Menges *et al.*, 2005), suggesting that downregulation of inhibitors is required for activation of the cell cycle.

Evidence has also been obtained for possible roles of CDK inhibitors in the G2 and M phases. For example, ICK1 microinjected into *Tradescantia* stamen hair cells slowed mitosis significantly by increasing the metaphase transit time (Cleary *et al.*, 2002). This effect is likely due to the inhibition of CDKA by ICK1. Second, in synchronized Arabidopsis cells, *KRP4* transcript peaked in early G2, and *ICK1* in late G2 (Menges *et al.*, 2005), although the peaks were not very prominent. Furthermore, in transgenic Arabidopsis plants weakly overexpressing *ICK1* or *ICK2*, leaf cells entered the endocycle earlier and ploidy levels were increased (Verkest *et al.*, 2005b; Weinl *et al.*, 2005). These data indicate that weak expression of *ICK1* or *ICK2* inhibits mitotic CDKA activity but has less effect on CDKA activity in the G1 phase. However, more direct evidence regarding the specific roles of CDK inhibitors is needed. In addition, a better understanding of the roles for different CDKA–cyclin complexes would provide further insight into the specific roles of ICK/KRP proteins during the cell cycle.

3.9 Cell cycle exit and endoreduplication

Plant cells may exit from the proliferative mitotic cell cycle and start to differentiate. During differentiation, certain cells may undergo endoreduplication also called the endocycle, leading to polyploidy. Endoreduplication can be considered a modified cell cycle consisting of only G1 and S phases. Endoreduplication is very common in plants, particularly in angiosperms. For instance, most differentiated cells in Arabidopsis plants are polyploids (Galbraith *et al.*, 1991). During the normal mitotic cell cycle, cells are licensed to undergo only one round of DNA replication, followed by mitosis. Based on cell cycle regulation, the switch from the mitotic cycle to the endocycle requires inhibition of mitosis and, at the same time, activation of or permission for multiple rounds of DNA replication (S phase). Different mechanisms, primarily involving cell cycle regulators, may be utilized by plants to accomplish the switch from the mitotic cycle to the endocycle (Larkins *et al.*, 2001; Kondorosi and Kondorosi, 2004). One suggested mechanism for endoreduplication is through inhibition of M-phase CDKs and increased activity of S-phase CDKs (Grafi and Larkin, 1995). Alternatively, inhibition of mitosis can be achieved by the activation of the APC, which degrades mitotic cyclins (Kondorosi and Kondorosi, 2004).

A detailed examination of endoreduplication is beyond the scope of this chapter (see Chapter 10). Here we will focus on results concerning the CDK inhibitors. In wild-type Arabidopsis plants, the level of *ICK1* expression is highest in leaves of mature 5-week-old plants (Wang *et al.*, 1998). Similarly, the data from *in situ* hybridization indicate that levels of *ICK1* and *ICK2* are much higher in nonmitotic cells than in mitotic cells (Ormenese *et al.*, 2004). Taken together, these results imply that ICK1 and ICK2 may be involved in regulating exit from the cell cycle. Results indicating the involvement of ICK/KRP protein in endoreduplication have been obtained from transgenic studies.

Results from several studies have shown that strong overexpression of a CDK inhibitor blocks endoreduplication and reduces ploidy levels (De Veylder *et al.*, 2001; Jasinski *et al.*, 2002b; Zhou *et al.*, 2002a; Schnittger *et al.*, 2003). Since CDKA complexes are the likely targets of ICK/KRP proteins, these results suggest that CDKA is important for endoreduplication. This suggestion is supported by the finding that expression of a dominant negative mutant of maize CDKA significantly reduced endoreduplication in maize endosperm (Leiva-Neto *et al.*, 2004). However, more recently it was found that while strong expression of *ICK1* and *ICK2* inhibited endoreduplication, their weak expression actually promoted it (Verkest *et al.*, 2005b; Weinl *et al.*, 2005). Similarly, overexpression of a maize CDK inhibitor resulted in one additional round of endoreduplication in some maize embryonic calli (Coelho *et al.*, 2005). These results indicate that the effect of CDK inhibitors is dosage dependent.

There are two likely reasons to explain the apparently contradictory effects of CDK inhibitors on endoreduplication (Figure 3.2). According to one suggestion, ICK/KRP proteins may exhibit preferential binding to CDK complexes that control the G2/M transition (Verkest *et al.*, 2005). At a low level of CDK inhibitor expression, the cell cycle proceeds normally. When ICK/KRP expression increases to a moderate level (in weakly expressing transgenic plants), the ICK/KRP-regulated CDK complex for G2/M transition is preferentially inhibited, while the CDK complexes for DNA replication are not affected. Thus, the cell enters the endocycle (Figure 3.2A). When the ICK/KRP level is high (in strongly expressing transgenic plants), both the mitotic and endocycle are arrested. Although this suggestion can explain the dual effects, the proposed mechanism for preferential binding of ICK/KRP proteins to A-type mitotic CDKs is not consistent with the experimental results regarding the targets of ICK/KRP proteins. First, as discussed above, the major target of ICK/KRP proteins is likely CDKA–CYCD complexes, which are believed to have a major role in regulating the G1/S transition. Second, in several studies it has also been shown that ICK/KRP proteins do not interact with B-type CDKs or mitotic cyclins.

It is also possible that the dual effects of ICK/KRP expression on endoreduplication may be due to the differences in the level of CDK activity required for entering mitosis and S phase (Verkest *et al.*, 2005b). According to this suggestion, the mitotic cycle requires a higher level of CDKs than the S phase. When ICK/KRP expression increases to a moderate level (in weakly expressing transgenic plants), activity of CDKs decreases below the threshold for maintaining the mitotic cycle and the cell enters the endocycle (Figure 3.2B). The available evidence regarding the biochemical properties of ICK/KRP proteins is more consistent with this second model. However, there are problems. For instance, this model implies that similar CDK complexes (likely CDKA-containing complexes) are critical for both the G1/S and G2/M transitions. This raises the question as to how B-type CDKs and mitotic cyclins, which are believed to regulate the G2/M transition (Porceddu *et al.*, 2001; Stals and Inzé, 2001; Dewitte and Murray, 2003), fit into this model. Future experiments may determine which model is functional or provide data to modify these existing models.

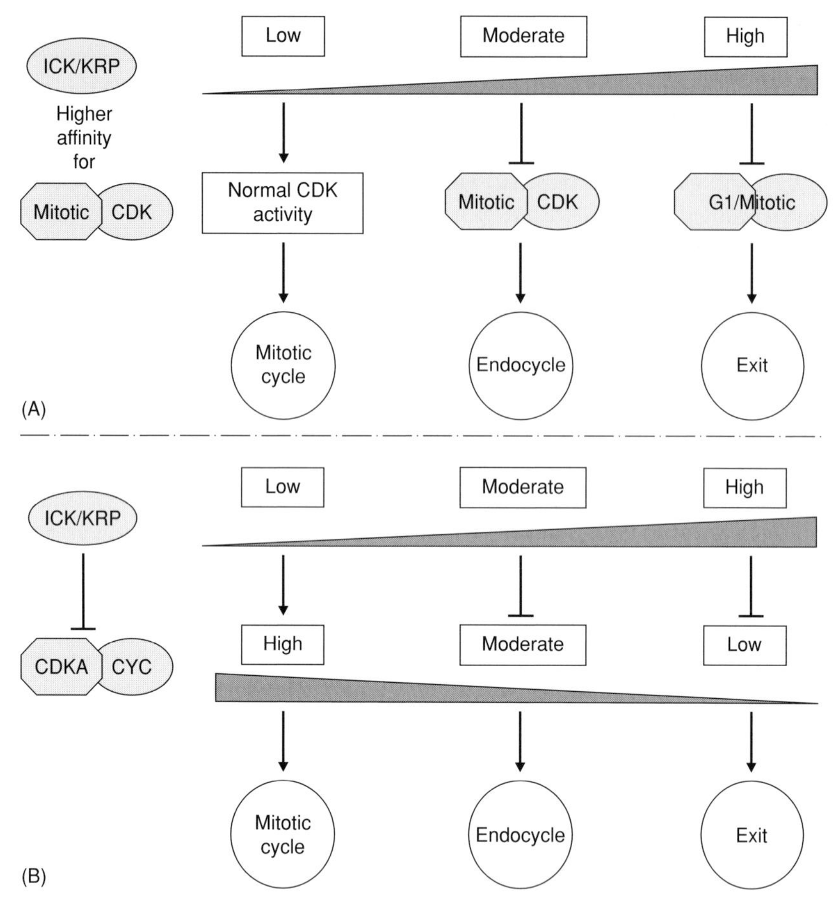

Figure 3.2 Two models to explain the effect of CDK inhibitor expression and CDK activity on endoreduplication and cell cycle exit. Existing results show that strong expression of a CDK inhibitor inhibits both cell division and endoreduplication, while weak expression inhibits cell division but promotes endoreduplication. (A) A model that is based on the assumption that ICK/KRP proteins have stronger binding to A-type CDK complexes which regulate the G2/M transition. A moderate increase of ICK/KRP expression inhibits mitosis but not S phase. At a higher level of expression, both mitosis and endoreduplication are inhibited. (B) A model based on the assumption that the level of ICK/KRP-regulated CDK activity required for entering mitosis is higher than that for entering S phase. Thus, proliferative cells have a high level of ICK/KRP-regulated CDK activity. With a moderate increase of ICK/KRP expression, the ICK/KRP-regulated CDK activity decreases to a level that no longer supports the mitotic cycle, while G1 and S phases continue.

3.10 Concluding remarks

The ICK/KRP CDK inhibitors represent a family of cell cycle regulators that diverge significantly from the CDK inhibitors in animals and yeast. Therefore, plant CDK inhibitor genes provide a good opportunity to understand how plants differ from animals in terms of cell cycle machinery. *In vitro* and *in vivo* results have demonstrated CDK inhibitory function for some members of the ICK/KRP family.

Because of good sequence conservation in the CDK inhibition region (Plate 3.1), generic CDK inhibitory function can be assumed for other members of this family. It will be interesting to determine the molecular and cellular functions conferred by the sequences outside this conserved region. Some initial understanding has been obtained with Arabidopsis ICK1 and ICK2. The N-terminal region of ICK1 confers protein instability and further studies are necessary to determine the exact mechanism for this instability. Furthermore, multiple sequences in ICK1 are involved in the regulation of the nuclear localization and subnuclear distribution.

Why are there seven CDK inhibitors in Arabidopsis? The two major reasons are either the need for temporal and spatial regulation of expression or the need for different specificities toward different CDK complexes. Some evidence has been obtained to support these suggestions. Tissue-specific variation in expression level has been described. The seven Arabidopsis *ICK/KRP* genes show some interesting patterns during activation of and progression through the cell cycle, indicating that they may have different roles at different stages of the cell cycle. For instance, decreased expression of *ICK1* and *ICK2* occurs upon the entry into the cell cycle, while *KRP4* and *ICK1* appear to show a peak of expression in the G2 phase. However, despite recent progress, we still know little about the specific role of each member of the ICK/KRP family in different tissues and during the cell cycle. Sequence analysis reveals potentially interesting differences among the seven Arabidopsis CDK inhibitors. ICK1, for example, is the only one possessing a strong coiled-coil region (Plate 3.1). The sequence, expression and yeast two-hybrid interaction analyses all indicate certain differences among the seven members. Future work will no doubt clarify the differences among the inhibitors by providing more specific information regarding the regulation and function for each *ICK/KRP* gene. The conservation of the CDK inhibitory domain between plant and animal CDK inhibitors indicates that CDK inhibitor genes will likely be present in most, if not all, plants. However, apart from Arabidopsis, only a few CDK inhibitors have been partially characterized from other species. Studies of different plant CDK inhibitors, particularly from distant plant species, should provide insight into the evolution and functions of this important component of plant cell cycle machinery.

From a cell biology perspective, further information is required to dissect the role of these CDK inhibitors during the cell cycle as well as upon entry into and exit from the cell cycle. The assignment of cell cycle phase-specific functions is made difficult partly by the lack of clear roles for different CDKs and cyclins. We know from protein–protein interaction data that the targets of ICK/KRP proteins are likely D-type cyclins and A-type CDKAs. D-type cyclins, in general, are believed to play a major role in the activation of the cell cycle and the G1/S transition. A-type CDKs are also important for the G1/S transition. Thus, it is perhaps safe to conclude that ICK/KRP CDK inhibitors have a major role in the G1/S transition. On the other hand, A-type CDKs and D-type cyclins are also implicated in the G2/M transition. Further, A-type CDKs can also form a complex with B-type cyclins. Inhibition of CDKA complexes may affect the G2/M transition. It has been observed that weak expression of *ICK1* and *ICK2* actually promotes early entry into the endocycle and increases the ploidy level in Arabidopsis leaves. Thus, plant CDK inhibitors may also regulate the G2/M transition.

There have been no reports describing the effect of downregulating plant CDK inhibitors. As with many other multigene families, functional redundancies may make this type of analysis more difficult. However, since the CDK inhibitors are an important component of the cell cycle machinery, and transgenic expression of CDK inhibitors has shown profound effects on cell division, endoreduplication, plant growth and plant morphology, we can expect that plant CDK inhibitors have important roles in plants.

Notes added at proofing stage

The following studies that are of particular relevance to plant CDK inhibitors have been published after the submission of the manuscript. A family of five CDK inhibitors from rice have been reported by Barrôco *et al.* (2006, *Plant Physiol* **142**, 1053–1064). The effects of overexpressing one rice inhibitor gene, *Orysa;KRP1*, include reduced cell number in leaves, larger cell size, reduced seed filling and reduced endoreduplication in endosperm. Churchman *et al.* (2006, *Plant Cell* **18**, 3145–3157) describes a different inhibitor gene *SIAMESE* (*SIM*), initially known for its mutant phenotype of reduced ploidy levels in trichomes. SIM as well as the related proteins share a conserved motif in the CYCD-interacting region of ICK/KRP CDK inhibitors and interacts with D-type cyclins and CDKA;1, although direct evidence of CDK inhibition remains to be established. Overexpression of *SIM* resulted in much smaller plants, enlarged cells and interestingly increased ploidy levels in Arabidopsis plants (Churchman *et al.*, 2006). Bird *et al.* (2007, *Plant Cell Rep*, doi: 10.1007/s00299-006-0294-3) show that all seven Arabidopsis CDK inhibitors are localized in the nucleus. In terms of sequences that confer nuclear localization, excluding the C-terminal CDK inhibitory region, ICK2 has one additional region that can confer nuclear localization of GFP, while all other six members have two or more regions (sequences) able to confer nuclear localization (Bird *et al.*, 2007).

Acknowledgments

We (L.C.F. and H.W.) gratefully acknowledge financial support from the Natural Sciences and Engineering Research Council of Canada.

References

The Arabidopsis Genome Initiative (2000) Analysis of the genome sequence of the flowering plant *Arabidopsis thaliana*. *Nature* **408**, 796–815.

Beemster, G.T.S., Fiorani, F. and Inzé, D. (2003) Cell cycle: the key to plant growth control? *Trends Plant Sci* **8**, 154–158.

Bisbis, B., Delmas, F., Joubes, J., *et al.* (2006) Cyclin-dependent kinase (CDK) inhibitors regulate the CDK–cyclin complex activities in endoreduplicating cells of developing tomato fruit. *J Biol Chem* **281**, 7374–7383.

Bögre, L., Meskiene, I., Heberle-Bors, E. and Hirt, H. (2000) Stressing the role of MAP kinases in mitogenic stimulation. *Plant Mol Biol* **43**, 705–718.

Bögre, L., Zwerger, K., Meskiene, I., *et al.* (1997) The cdc2Ms kinase is differently regulated in the cytoplasm and in the nucleus. *Plant Physiol* **113**, 841–852.

Brewer, P.B., Howles, P.A., Dorian, K., *et al.* (2004) *PETAL LOSS*, a trihelix transcription factor gene, regulates perianth architecture in the Arabidopsis flower. *Development* **131**, 4035–4045.

Cardozo, T. and Pagano, M. (2004) The SCF ubiquitin ligase: insights into a molecular machine. *Nat Rev Mol Cell Biol* **5**, 739–751.

Cleary, A.L., Fowke, L.C., Wang, H. and John, P.C.L. (2002) The effect of ICK1, a plant cyclin-dependent kinase inhibitor, on mitosis in living plant cells. *Plant Cell Rep* **20**, 814–820.

Coelho, C.M., Dante, R.A., Sabelli, P.A., *et al.* (2005) Cyclin-dependent kinase inhibitors in maize endosperm and their potential role in endoreduplication. *Plant Physiol* **138**, 2323–2336.

Colasanti, J., Cho, S.O., Wick, S. and Sundaresan, V. (1993) Localization of the functional $p34^{cdc2}$ homolog of maize in root tip and stomatal complex cells: association with predicted division sites. *Plant Cell* **5**, 1101–1111.

Day, S.J. and Lawrence, P.A. (2000) Measuring dimensions: the regulation of size and shape. *Development* **127**, 2977–2987.

de Jager, S.M., Maughan, S., Dewitte, W., Scofield, S. and Murray, J.A. (2005) The developmental context of cell-cycle control in plants. *Semin Cell Dev Biol* **16**, 385–396.

De Veylder, L., Beeckman, T., Beemster, G.T.S., *et al.* (2001) Functional analysis of cyclin-dependent kinase inhibitors of *Arabidopsis*. *Plant Cell* **13**, 1653–1668.

Denicourt, C. and Dowdy, S.F. (2004) Cip/Kip proteins: more than just CDKs inhibitors. *Genes Dev* **18**, 851–855.

Dewitte, W. and Murray, J.A. (2003) The plant cell cycle. *Annu Rev Plant Biol* **54**, 235–264.

Fountain, M.D., Renz, A. and Beck, E. (1999) Isolation of a cDNA encoding a G1-cyclin-dependent kinase inhibitor (ICDK) from suspension cultured photoautotrophic *Chenopodium rubrum* L. cells. *Plant Physiol* **120**, 339.

Galbraith, D.W., Harkins, K.R. and Knapp, S. (1991) Systemic endopolyploidy in *Arabidopsis thaliana*. *Plant Physiol* **96**, 985–989.

Grafi, G. and Larkins, B.A. (1995) Endoreduplication in maize endosperm: involvement of M-phase-promoting factor inhibition and induction of S-phase-related kinases. *Science* **269**, 1262–1264.

Griffin, S.V. and Shankland, S.J. (2004) Not just an inhibitor: a role for p21 beyond the cell cycle – 'The truth is rarely pure and never simple.' *J Am Soc Nephrol* **15**, 825–826.

Griffith, M.E., da Silva Conceicao, A. and Smyth, D.R. (1999) *PETAL LOSS* gene regulates initiation and orientation of second whorl organs in the *Arabidopsis* flower. *Development* **126**, 5635–5644.

Gutierrez, C. (2005) Coupling cell proliferation and development in plants. *Nat Cell Biol* **7**, 535–541.

Harper, J.W. and Elledge, S.J. (1996) Cdk inhibitors in development and cancer. *Curr Opin Genet Dev* **6**, 56–64.

Healy, J.M., Menges, M., Doonan, J.H. and Murray, J.A. (2001) The *Arabidopsis* D-type cyclins CYCD2 and CYCD3 both interact *in vivo* with the PSTAIRE cyclin-dependent kinase Cdc2a but are differentially controlled. *J Biol Chem* **276**, 7041–7047.

Himanen, K., Boucheron, E., Vanneste, S., De Almeida Engler, J., Inzé, D. and Beeckman, T. (2002) Auxin-mediated cell cycle activation during early lateral root initiation. *Plant Cell* **14**, 2339–2351.

Himmelbach, A., Iten, M. and Grill, E. (1998) Signalling of abscisic acid to regulate plant growth. *Philos Trans R Soc Lond B Biol Sci* **353**, 1439–1444.

Jacobs, T. (1997) Why do plant cells divide? *Plant Cell* **9**, 1021–1029.

Jakoby, M.J., Weinl, C., Pusch, S., *et al.* (2006) Analysis of the subcellular localization, function and proteolytic control of the Arabidopsis CDK inhibitor ICK1/KRP1. *Plant Physiol* **141**, 1293–1305.

Jasinski, S., Leite, C.S., Domenichini, S., *et al.* (2003) NtKIS2, a novel tobacco cyclin-dependent kinase inhibitor is differentially expressed during the cell cycle and plant development. *Plant Physiol Biochem* **41**, 667–676.

Jasinski, S., Perennes, C., Bergounioux, C. and Glab, N. (2002a) Comparative molecular and functional analyses of the tobacco cyclin-dependent kinase inhibitor NtKIS1a and its spliced variant NtKIS1b1. *Plant Physiol* **130**, 1871–1882.

Jasinski, S., Riou-Khamlichi, C., Roche, O., Perennes, C., Bergounioux, C. and Glab, N. (2002b) The CDK inhibitor NtKIS1a is involved in plant development, endoreduplication and restores normal development of cyclin D3;1-overexpressing plants. *J Cell Sci* **115**, 973–982.

Jiang, Y., Zhao, R.C. and Verfaillie, C.M. (2000) Abnormal integrin-mediated regulation of chronic myelogenous leukemia CD34+ cell proliferation: BCR/ABL up-regulates the cyclin-dependent kinase inhibitor, $p27^{Kip}$, which is relocated to the cell cytoplasm and incapable of regulating cdk2 activity. *Proc Natl Acad Sci* USA **97**, 10538–10543.

Kang, J.Y., Choi, H.I., Im, M.Y. and Kim, S.Y. (2002) Arabidopsis basic leucine zipper proteins that mediate stress-responsive abscisic acid signaling. *Plant Cell* **14**, 343–357.

Kaplan, D.R. and Hagemann, W. (1991) The relationship of cell and organism in vascular plants. *BioScience* **41**, 693–703.

Kawamura, K., Murray, J.A., Shinmyo, A. and Sekine, M. (2006) Cell cycle regulated D3-type cyclins form active complexes with plant-specific B-type cyclin-dependent kinase *in vitro*. *Plant Mol Biol* **61**, 311–327.

Kondorosi, E. and Kondorosi, A. (2004) Endoreduplication and activation of the anaphase-promoting complex during symbiotic cell development. *FEBS Lett* **567**, 152–157.

Kono, A., Umeda-Hara, C., Lee, J., Ito, M., Uchimiya, H. and Umeda, M. (2003) Arabidopsis D-type cyclin CYCD4;1 is a novel cyclin partner of B2-type cyclin-dependent kinase. *Plant Physiol* **132**, 1315–1321.

Koroleva, O.A., Tomlinson, M.L., Leader, D., Shaw, P. and Doonan, J.H. (2005) High-throughput protein localization in Arabidopsis using *Agrobacterium*-mediated transient expression of GFP–ORF fusions. *Plant J* **41**, 162–174.

Koroleva, O.A., Tomlinson, M., Parinyapong, P., *et al.* (2004) CycD1, a putative G1 cyclin from *Antirrhinum majus*, accelerates the cell cycle in cultured tobacco BY-2 cells by enhancing both G1/S entry and progression through S and G2 phases. *Plant Cell* **16**, 2364–2379.

Larkins, B.A., Dilkes, B.P., Dante, R.A., Coelho, C.M., Woo, Y.M. and Liu, Y. (2001) Investigating the hows and whys of DNA endoreduplication. *J Exp Bot* **52**, 183–192.

Leiva-Neto, J.T., Grafi, G., Sabelli, P.A., *et al.* (2004) A dominant negative mutant of cyclin-dependent kinase A reduces endoreduplication but not cell size or gene expression in maize endosperm. *Plant Cell* **16**, 1854–1869.

Liang, J., Zubovitz, J., Petrocelli, T., *et al.* (2002) PKB/Akt phosphorylates p27, impairs nuclear import of p27 and opposes p27-mediated G1 arrest. *Nat Med* **8**, 1153–1160.

Liu, X., Sun, Y., Ehrlich, M., *et al.* (2000) Disruption of TGF-beta growth inhibition by oncogenic ras is linked to $p27^{Kip1}$ mislocalization. *Oncogene* **19**, 5926–5935.

Lui, H., Wang, H., DeLong, C., Fowke, L.C., Crosby, W.L. and Fobert, P.R. (2000) The *Arabidopsis* Cdc2a-interacting protein ICK2 is structurally related to ICK1 and is a potent inhibitor of cyclin-dependent kinase activity *in vitro*. *Plant J* **21**, 379–385.

Mendenhall, M.D. (1993) An inhibitor of $p34^{CDC28}$ protein kinase activity from *Saccharomyces cerevisiae*. *Science* **259**, 216–219.

Mendenhall, M.D. (1998) Cyclin-dependent kinase inhibitors of *Saccharomyces cerevisiae* and *Schizosaccharomyces pombe*. *Curr Top Microbiol Immunol* **227**, 1–24.

Menges, M., de Jager, S.M., Gruissem, W. and Murray, J.A. (2005) Global analysis of the core cell cycle regulators of *Arabidopsis* identifies novel genes, reveals multiple and highly specific profiles of expression and provides a coherent model for plant cell cycle control. *Plant J* **41**, 546–566.

Menges, M. and Murray, J.A. (2002) Synchronous *Arabidopsis* suspension cultures for analysis of cell-cycle gene activity. *Plant J* **30**, 203–212.

Mizukami, Y. (2001) A matter of size: developmental control of organ size in plants. *Curr Opin Plant Biol* **4**, 533–539.

Moreno, S. and Nurse, P. (1994) Regulation of progression through the G1 phase of the cell cycle by the rum1+ gene. *Nature* **367**, 236–242.

Morgan, D.O. (1997) Cyclin-dependent kinases: engines, clocks, and microprocessors. *Annu Rev Cell Dev Biol* **13**, 261–291.

Musgrove, E.A., Davison, E.A. and Ormandy, C.J. (2004) Role of the CDK inhibitor p27 (Kip1) in mammary development and carcinogenesis: insights from knockout mice. *J Mammary Gland Biol Neoplasia* **9**, 55–66.

Nakai, T., Kato, K., Shinmyo, A. and Sekine, M. (2006) Arabidopsis KRPs have distinct inhibitory activity toward cyclin D2-associated kinases, including plant-specific B-type cyclin-dependent kinase. *FEBS Lett* **580**, 336–340.

Ormenese, S., de Almeida Engler, J., De Groodt, R., De Veylder, L., Inzé, D. and Jacqmard, A. (2004) Analysis of the spatial expression pattern of seven Kip related proteins (KRPs) in the shoot apex of *Arabidopsis thaliana*. *Ann Bot* **93**, 575–580.

Pavletich, N.P. (1999) Mechanisms of cyclin-dependent kinase regulation: structures of Cdks, their cyclin activators, and Cip and INK4 inhibitors. *J Mol Biol* **287**, 821–828.

Peter, M. and Herskowitz, I. (1994) Direct inhibition of the yeast cyclin-dependent kinase Cdc28–Cln by Far1. *Science* **265**, 1228–1231.

Pettkó-Szandtner, A., Mészáros, T., Horváth, G.V., *et al.* (2006) Activation of an alfalfa cyclin-dependent kinase inhibitor by calmodulin-like domain protein kinase. *Plant J* **46**, 111–123.

Pines, J. (1999) Four-dimensional control of the cell cycle. *Nature Cell Biol* **1**, E73–E79.

Polyak, K., Lee, M.H., Erdjument-Bromage, H., *et al.* (1994) Cloning of $p27^{Kip1}$, a cyclin-dependent kinase inhibitor and a potential mediator of extracellular antimitogenic signals. *Cell* **78**, 59–66.

Porceddu, A., Stals, H., Reichheld, J.P., *et al.* (2001) A plant-specific cyclin-dependent kinase is involved in the control of G2/M progression in plants. *J Biol Chem* **276**, 36354–36360.

Richard, C., Granier, C., Inzé, D. and De Veylder, L. (2001) Analysis of cell division parameters and cell cycle gene expression during the cultivation of *Arabidopsis thaliana* cell suspensions. *J Exp Bot* **52**, 1625–1633.

Rosen, D.G., Yang, G., Cai, K.Q., *et al.* (2005) Subcellular localization of $p27^{kip1}$ expression predicts poor prognosis in human ovarian cancer. *Clin Cancer Res* **11**, 632–637.

Roudier, F., Fedorova, E., Györgyey, J., *et al.* (2001) Cell cycle function of a *Medicago sativa* A2-type cyclin interacting with a PSTAIREtype cyclin-dependent kinase and a retinoblastoma protein. *Plant J* **23**, 73–83.

Ruggiero, B., Koiwa, H., Manabe, Y., *et al.* (2004) Uncoupling the effects of abscisic acid on plant growth and water relations. Analysis of sto1/nced3, an abscisic acid-deficient but salt stress-tolerant mutant in *Arabidopsis*. *Plant Physiol* **136**, 3134–3147.

Russo, A.A., Jeffrey, P.D., Patten, A.K., Massague, J. and Pavletich, N.P. (1996) Crystal structure of the $p27^{Kip1}$ cyclin-dependent-kinase inhibitor bound to the cyclin A–Cdk2 complex. *Nature* **382**, 325–331.

Schneider, K.R., Smith, R.L. and O'Shea, E.K. (1994) Phosphate-regulated inactivation of the kinase PHO80–PHO85 by the CDK inhibitor PHO81. *Science* **266**, 122–126.

Schnittger, A., Schöbinger, U., Bouyer, D., Weinl, C., Stierhof, Y.D. and Hülskamp, M. (2002) Ectopic D-type cyclin expression induces not only DNA replication but also cell division in Arabidopsis trichomes. *Proc Natl Acad Sci USA* **99**, 6410–6415.

Schnittger, A., Weinl, C., Bouyer, D., Schöbinger, U. and Hülskamp, M. (2003) Misexpression of the cyclin-dependent kinase inhibitor ICK1/KRP1 in single-celled *Arabidopsis* trichomes reduces endoreduplication and cell size and induces cell death. *Plant Cell* **15**, 303–315.

Sherr, C.J. (2000) The Pezcoller lecture: cancer cell cycles revisited. *Cancer Res* **60**, 3689–3695.

Sherr, C.J. and Roberts, J.M. (1995) Inhibitors of mammalian G1 cyclin-dependent kinases. *Genes Dev* **9**, 1149–1163.

Shin, I., Yakes, F.M., Rojo, F., *et al.* (2002) PKB/Akt mediates cell-cycle progression by phosphorylation of $p27^{Kip1}$ at threonine 157 and modulation of its cellular localization. *Nat Med* **8**, 1145–1152.

Sieberer, T., Hauser, M.T., Seifert, G.J. and Luschnig, C. (2003) PROPORZ1, a putative Arabidopsis transcriptional adaptor protein, mediates auxin and cytokinin signals in the control of cell proliferation. *Curr Biol* **13**, 837–842.

Slingerland, J. and Pagano, M. (2000) Regulation of the cdk inhibitor p27 and its deregulation in cancer. *J Cell Physiol* **183**, 10–17.

Sorrell, D.A., Combettes, B., Chaubet-Gigot, N., Gigot, C. and Murray, J.A. (1999) Distinct cyclin D genes show mitotic accumulation or constant levels of transcripts in tobacco bright yellow-2 cells. *Plant Physiol* **119**, 343–352.

Spruck, C.H. and Strohmaier, H.M. (2002) Seek and destroy: SCF ubiquitin ligases in mammalian cell cycle control. *Cell Cycle* **1**, 250–254.

Stals, H., Bauwens, S., Traas, J., Van Montagu, M., Engler, G. and Inzé, D. (1997) Plant CDC2 is not only targeted to the pre-prophase band, but also co-localizes with the spindle, phragmoplast, and chromosomes. *FEBS Lett* **418**, 229–234.

Stals, H. and Inzé, D. (2001) When plant cells decide to divide. *Trends Plant Sci* **6**, 359–364.

Toda, T., Ochotorena, I. and Kominami, K. (1999) Two distinct ubiquitin-proteolysis pathways in the fission yeast cell cycle. *Philos Trans R Soc Lond B Biol Sci* **354**, 1551–1557.

Toyoshima, H. and Hunter, T. (1994) p27, a novel inhibitor of G1 cyclin–Cdk protein kinase activity, is related to p21. *Cell* **78**, 67–74.

Traas, J. and Doonan, J.H. (2001) Cellular basis of shoot apical meristem development. *Int Rev Cytol* **208**, 161–206.

Tsihlias, J., Kapusta, L. and Slingerland, J. (1999) The prognostic significance of altered cyclin-dependent kinase inhibitors in human cancer. *Annu Rev Med* **50**, 401–423.

Vandepoele, K., Raes, J., De Veylder, L., Rouzé, P., Rombauts, S. and Inzé, D. (2002) Genome-wide analysis of core cell cycle genes in Arabidopsis. *Plant Cell* **14**, 903–916.

Verkest, A., Weinl, C., Inzé, D., De Veylder, L. and Schnittger, A. (2005a) Switching the cell cycle. Kip-related proteins in plant cell cycle control. *Plant Physiol* **139**, 1099–1106.

Verkest, A., Manes, C.L., Vercruysse, S., *et al.* (2005b) The cyclin-dependent kinase inhibitor KRP2 controls the onset of the endoreduplication cycle during *Arabidopsis* leaf development through inhibition of mitotic CDKA;1 kinase complexes. *Plant Cell* **17**, 1723–1736.

Vidal, A. and Koff, A. (2000) Cell-cycle inhibitors: three families united by a common cause. *Gene* **247**, 1–15.

Viglietto, G., Motti, M.L., Bruni, P., *et al.* (2002) Cytoplasmic relocalization and inhibition of the cyclin-dependent kinase inhibitor $p27^{Kip1}$ by PKB/Akt-mediated phosphorylation in breast cancer. *Nat Med* **8**, 1136–1144.

Wang, H., Fowke, L.C. and Crosby, W.L. (1997) A plant cyclin-dependent kinase inhibitor gene. *Nature* **386**, 451–452.

Wang, H., Qi, Q., Schorr, P., Cutler, A.J., Crosby, W.L. and Fowke, L.C. (1998) ICK1, a cyclin-dependent protein kinase inhibitor from *Arabidopsis thaliana* interacts with both Cdc2a and CYCD3 and its expression is induced by abscisic acid. *Plant J* **15**, 501–510.

Wang, H., Zhou, Y. and Fowke, L.C. (2006) The emerging importance of CDK inhibitors in the regulation of the plant cell cycle and related processes. *Can J Bot* **84**, 640–650.

Wang, H., Zhou, Y., Gilmer, S., Whitwill, S. and Fowke, L.C. (2000) Expression of the plant cyclin-dependent kinase inhibitor ICK1 affects cell division, plant growth and morphology. *Plant J* **24**, 613–623.

Weingartner, M., Binarova, P., Drykova, D., *et al.* (2001) Dynamic recruitment of Cdc2 to specific microtubule structures during mitosis. *Plant Cell* **13**, 1929–1943.

Weinl, C., Marquardt, S., Kuijt, S.J., *et al.* (2005) Novel functions of plant cyclin-dependent kinase inhibitors, ICK1/KRP1, can act non-cell-autonomously and inhibit entry into mitosis. *Plant Cell* **17**, 1704–1722.

Zhou, Y., Fowke, L.C. and Wang, H. (2002a) Plant CDK inhibitors: studies of interactions with cell cycle regulators in the yeast two-hybrid system and functional comparisons in transgenic *Arabidopsis* plants. *Plant Cell Rep* **20**, 967–975.

Zhou, Y., Li, G., Brandizzi, F., Fowke, L.C. and Wang, H. (2003a) The plant cyclin-dependent kinase inhibitor ICK1 has distinct functional domains for *in vivo* kinase inhibition, protein instability and nuclear localization. *Plant J* **35**, 476–489.

Zhou, Y., Niu, H., Brandizzi, F., Fowke, L.C. and Wang, H. (2006) Molecular control of nuclear and subnuclear targeting of the plant CDK inhibitor ICK1 and ICK1-mediated nuclear transport of CDKA. *Plant Mol Biol* **62**, 261–278.

Zhou, Y., Wang, H., Gilmer, S., Whitwill, S. and Fowke, L.C. (2003b) Effects of co-expressing the plant CDK inhibitor ICK1 and D-type cyclin genes on plant growth, cell size and ploidy in *Arabidopsis thaliana*. *Planta* **216**, 604–613.

Zhou, Y., Wang, H., Gilmer, S., Whitwill, S., Keller, W. and Fowke, L.C. (2002b) Control of petal and pollen development by the plant cyclin-dependent kinase inhibitor ICK1 in transgenic *Brassica* plants. *Planta* **215**, 248–257.

4 The UPS: an engine that drives the cell cycle

Pascal Genschik and Marie Claire Criqui

4.1 The molecular machinery mediating ubiquitin-dependent proteolysis

Regulation of protein stability through the ubiquitin proteasome system (UPS) is now considered as a major mechanism underlying many cellular and organismal processes, such as cell division, DNA repair, quality control of newly produced proteins, developmental pathways, important parts of immune defense and, in plants, light and phytohormone signal transduction (Ciechanover *et al.*, 2000; Pickart, 2001; Smalle and Vierstra, 2004).

4.1.1 Ubiquitylation reaction

Degradation via the UPS is a two-step process: the protein is first tagged by covalent attachment of ubiquitin and subsequently degraded by a multicatalytic protease complex called the 26S proteasome (Figure 4.1). Conjugation of ubiquitin to the protein involves a cascade of three enzymes: E1, E2 and E3. Ubiquitin-activating enzyme (E1) forms a high-energy thioester intermediate, E1-S$\sim$Ubi, which is then transesterified to one of the several ubiquitin-conjugating enzyme (E2). The transfer of ubiquitin from the E2-S$\sim$Ubi to an ε-NH2 group of an internal lysine residue in the target protein substrate requires an ubiquitin protein ligase (E3). By successively adding activated ubiquitin moieties to internal lysine residues on the previously conjugated ubiquitin molecule, a polyubiquitin chain is synthesized. However, the fate of a ubiquitin-conjugated protein depends on the length of the ubiquitin chain and on the type of ubiquitin–ubiquitin linkage in the chain (reviewed in Pickart and Fushman, 2004). Ubiquitin chains of at least four ubiquitin monomers, in which the C-terminus of one ubiquitin molecule is attached to the Lysine 48 of the next ubiquitin molecule in the chain, is recognized by the 26S proteasome. However, conjugated proteins can still be deubiquitinated by deubiquitylating enzymes (DUBs) prior to proteolysis. It is even considered that this is an important step of regulation, as it is the last step to reverse protein degradation (Amerik and Hochstrasser, 2004). Nevertheless, once ubiquitin-conjugated proteins finally reach their last destination, the 26S proteasome, they will be unfolded and threaded into the cylindrical central part of the 26S proteasome, where they are cleaved to peptides and the ubiquitin monomers are recycled (Hendil and Hartmann-Petersen, 2004).

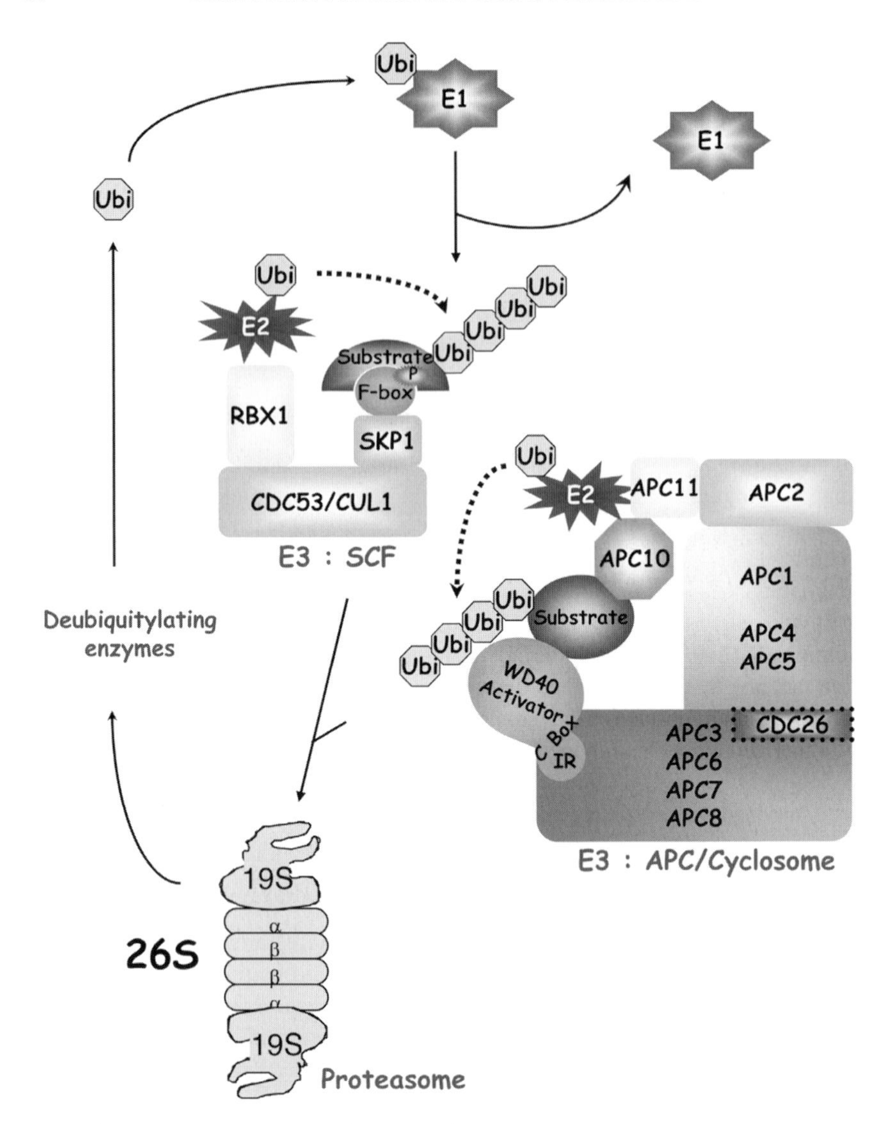

Figure 4.1 Schematic drawing of SCF- and APC/C-dependent ubiquitylation and subsequent degradation reactions. The E1 enzyme activates ubiquitin in a thioester linkage, which is trans-esterified to a ubiquitin-conjugating E2 enzyme. Transfer of ubiquitin to a substrate lysine requires its recognition by an E3 ligase, here the SCF and the APC/C. Both E3s request a RING-H2 finger subunit that interacts with the E2 enzyme to stimulate the ubiquitylation reaction. In addition, these E3s bear substrate adaptor proteins that recruit specifically the substrates to the core complex. An F-box protein assumes this role for the SCF, whereas the APC/C requests a WD40 activator protein (CDC20/FZ or CDH1/FZR) and most likely also the APC10 subunit. It is noteworthy that the neddylation pathway regulating the SCF is not represented and that the modular structure of the APC/C complexes is still poorly understood, thus its representation here is rather speculative. Substrate recognition by the SCF generally requests phosphorylation of the target proteins. Finally, multiubiquitylated proteins are recognized by the 26S proteasome and proteolyzed into peptides, and ubiquitin is recycled through the action of deubiquitylating enzymes.

4.1.2 *Ubiquitin protein ligases*

As the E3 enzymes specify the substrates, they play the most important role in the ubiquitylation reaction. Several hundred different E3s have been identified in sequenced metazoan and plant genomes, based on specific, commonly shared structural motifs. These E3 enzymes have been subdivided into two major classes: those that contain a HECT (homologous to the E6–AP C-terminus) domain and those that contain a RING (really interesting new gene) domain (Pickart, 2001). Proteins that share the RING fold form the biggest class of E3s and can be subdivided into different subclasses including RING-H2, RING-HC, RING-v, RING-D, RING-G, RING-S/T, RING-C2, the PHD finger (Coscoy and Ganem, 2003; Stone *et al.*, 2005) and the U-box (Ohi *et al.*, 2003). For the HECT domain E3s, ubiquitin is transferred to an active cystein residue on the E3 enzyme prior to its transfer to the ligase-bound substrate. In contrast to the HECT domain enzymes, RING E3s do not form a covalent intermediate with ubiquitin (Seol *et al.*, 1999).

Many RING E3s are believed to work as monomers or as homo- or heterodimers with other RING proteins, combining the substrate and E2-binding sites on the same molecule. However, the most intensively studied subclass of the RING E3s is that of the cullin-RING ligase (CRL) superfamily, the members of which form multiprotein complexes (Petroski and Deshaies, 2005; Thomann *et al.*, 2005). These E3 enzymes can be viewed as two functional modules brought together by the CULLIN proteins, acting as molecular scaffolds. The first module forms the catalytic centre and is composed of a RING-finger domain protein and an E2 enzyme. The second module can be considered as the substrate recognition module, in which a specific protein physically interacts with the target substrate. Members of the CULLIN protein family have been identified in all eukaryotes and based on phylogenetic studies fall into different subfamilies (Shen *et al.*, 2002).

Each CULLIN (CUL) protein forms a different class of E3 enzymes; among them, the best characterized complexes are the SCF (SKP1–CUL1–F-box) (Cardozo and Pagano, 2004), the ECS (ElonginC–CUL2–SOCS-box) (Ivan and Kaelin, 2001) and the CUL3–BTB (Pintard *et al.*, 2004) complexes, as well as the APC/C (anaphase-promoting complex or cyclosome) (Peters, 2002), which contains a more distant CULLIN member, called APC2. Other CULLINs (at least CUL4, CUL5 and CUL7) also form protein complexes with E3 activities (Kamura *et al.*, 2001; Dias *et al.*, 2002; Wertz *et al.*, 2004, respectively). CUL2 and CUL5 are found only in metazoans, whereas the CUL1, CUL3, CUL4 and APC2 members are conserved in metazoans and plants.

4.2 The SCF and APC/C: the two master E3s regulating the cell cycle

Successful progression through the cell cycle requires the coordinate destruction of essential regulatory proteins by the UPS (Vodermaier, 2004). Two families of ubiquitin protein ligases dominate DNA duplication and cell division: the SCF and the APC/cyclosome.

4.2.1 The SCF: an E3 regulating the G1/S transition

CULLIN1 (CUL1) forms presently the best characterized CRL enzyme (reviewed in Cardozo and Pagano, 2004). CUL1 interacts at its carboxy terminus with the RING domain protein RBX1 (also called ROC1 or HRT1) and an E2 enzyme and at its amino terminus with the adaptor protein SKP1 (Figure 4.1). F-box proteins contain a rather variable interaction domain known as the F-box that binds to SKP1. Additionally, F-box proteins carry a great variety of typical protein–protein interaction domains that confer substrate specificity for ubiquitylation. CUL1 is also regulated by the covalent linkage of an ubiquitin-like protein, RUB1/NEDD8, through the neddylation activating and conjugating enzymes (Hochstrasser, 1998). NEDD8-modification of CUL1 dissociates CAND1, an inhibitor of the SCF, and consequently promotes the binding of SKP1 and the F-box protein to CUL1 and thus the assembly of the SCF E3 ligase complex (Liu *et al.*, 2002; Zheng *et al.*, 2002). The neddylation of CUL1 is removed by the peptidase activity of the COP9-signalosome (CSN) (reviewed in Cope and Deshaies, 2003).

In plants, the so-called CUL1 (e.g. *Arabidopsis* AtCUL1) is phylogenetically distant from the yeast or metazoan CUL1 members and fall into a separate phylogenetic clade (Shen *et al.*, 2002). However, AtCUL1 forms clearly SCF-type protein complexes with the *Arabidopsis* SKP1-like proteins (named ASK1 and ASK2), the RING domain protein RBX1 and several F-box proteins (Gray *et al.*, 2001; Xu *et al.*, 2002). *Arabidopsis* counts about 700 F-box proteins (Gagne *et al.*, 2002). This number seems to be significantly higher in plants than in other eukaryotes for which the full genome sequence is available.

In fungi and also in animal cells, SCF-dependent ubiquitylation plays a critical role in the control of the cell cycle by promoting the degradation of several regulatory proteins (Cardozo and Pagano, 2004). In particular, the budding yeast SCF^{CDC4} and the mammalian SCF^{SKP2} (the name of the F-box protein being indicated in uppercase) are required to destroy the cell-cycle-dependent kinase inhibitors SIC1 and p27, respectively (reviewed in Deshaies and Ferrell, 2001; Pagano, 2004 and see below), thus promoting the entry into S phase. In plants, the SCF might also play an important role in regulating the cell cycle, as *Arabidopsis cul1* loss-of-function mutants arrest early during embryogenesis at the zygote stage (Shen *et al.*, 2002).

4.2.2 The APC/C: the E3 coordinating cell cycle progression through mitosis and G1

The APC/C is the largest and most complex E3 known so far (Figure 4.1). In vertebrates, 11 subunits have been described (Peters, 2002), while in yeast the APC/C is composed of minimum 13 subunits (Yoon *et al.*, 2002). The minimal ubiquitin ligase module of the APC/C comprises APC2, the distant member of the cullin family, and the RING-finger protein APC11. These two subunits interact with each other and with E2 ubiquitin-conjugating enzymes and have been shown to be sufficient for catalytic activity *in vitro* but without substrate specificity (Gmachl *et al.*, 2000; Leverson *et al.*, 2000; Tang *et al.*, 2001). The budding yeast APC/C contains three essential subunits with tetratricopeptide repeat (TPRs) protein–protein interaction domains: APC3/CDC27, APC8/CDC23 and APC6/CDC16, one of which (APC3/CDC27) has been implicated in binding the activating subunit CDH1 (Vodermaier *et al.*, 2003; Kraft *et al.*, 2005). Phosphorylation of these subunits during mitosis is required to activate the APC/C (Rudner and Murray, 2000; Kraft *et al.*, 2003). APC10/DOC1 characterized by the presence of a DOC domain is important for substrate recognition and/or extending the polyubiquitin chain on a substrate (Carroll and Morgan, 2002; Passmore *et al.*, 2003; Carroll *et al.*, 2005). The functions of the other subunits remain unknown.

In addition to its core components, the APC/C requires a member of the WD40 family for activity. The coactivator CDC20/FIZZY or CDH1/FIZZY-RELATED activates APC/C to ubiquitylate substrates containing characteristic destruction motifs: the Dbox, the KEN-box, the A-box and the GxEN-box in a cell-cycle-dependent manner (Zachariae *et al.*, 1998; Jaspersen *et al.*, 1999; Kramer *et al.*, 2000; Pfleger and Kirchner, 2000; Rudner and Murray, 2000). Two domains are thought to play roles in binding of these activating subunits to the APC/C: a short internal motif called the C-box (Schwab *et al.*, 2001) and a C-terminal IR dipeptide (Vodermaier *et al.*, 2003). CDC20/FZ and CDH1/FZR also contain WD40 repeats that bind directly to APC/C targets, and thus may serve as a bridge between enzyme and substrate (Burton and Solomon, 2001; Pfleger *et al.*, 2001; Burton *et al.*, 2005; Kraft *et al.*, 2005). CDC20/FZ activates APC/C during metaphase, whereas CDH1/FZR promotes APC/C activity for the exit of mitosis and during G1.

Unlike the SCF, the APC/C is mainly required to induce progression and exit from mitosis by inducing proteolysis of different cell cycle regulators including PDS1/SECURIN and CYCLIN B. The proteolytic events triggered by APC/C are required to release sister chromatid cohesion during anaphase, specify the exit from mitosis and prevent premature entry into S phase (see below).

In plants, counterparts of all known vertebrate APC/C subunits could be identified (Capron *et al.*, 2003a). In Arabidopsis, single-copy genes encode all subunits, except for APC3/CDC27, where two *CDC27*-related genes have been identified, called *CDC27A* and *CDC27B*. Interestingly, the latter gene turned out to be the *Hobbit* gene (Willemsen *et al.*, 1998; Blilou *et al.*, 2002) that is required for cell division and cell differentiation in meristems. In contrast to the *CDC27A* gene, *CDC27B/HOBBIT* transcripts mainly accumulate around G2/M in post-embryonic meristems (Blilou *et al.*, 2002). Although it remains to be established whether CDC27B/HOBBIT is a component of the plant APC/C, one interesting possibility is that plants form two types of APC/C complexes, APC/C^{CDC27A} and APC/C$^{CDC27B/HOBBIT}$, each having specific cell cycle functions. Plants have also multiple APC/C activators. The first CDH1/FZR-type APC/C activator was identified in *Medicago sativa* as a cell cycle switch, named CCS52 (Cebolla *et al.*, 1999). In Arabidopsis there are nine APC/C activators: six CDC20 isoforms and three CCS52/CDH1 isoforms. These data suggest the existence of numerous APC/C$^{CDC20/CCS52/CDC27}$ forms in Arabidopsis, which in conjunction with different E2 enzymes might have distinct as well as complementary functions in regulating the cell cycle. Furthermore, *Arabidopsis apc2* (Capron *et al.*, 2003b) and *apc6/NOMEGA* (Kwee and Sundaresan, 2003) loss-of-function mutants are impaired primarily in megagametogenesis after the first mitotic division, and the arrested female gametophytes are unable to degrade Dbox-containing reporter proteins. Thus, APC/C plays an important role in regulating the cell cycle in plants as well.

4.3 Cell cycle targets of the proteolytic machinery

It is noteworthy that the degradation of key cell cycle regulators by the UPS is only one facet of their regulation and that many other regulations operate for most of them, such as transcriptional and translational controls, sequestration and cellular localization, among others.

4.3.1 The transition from G1 to S phase

4.3.1.1 CKIs and the transition from G1 to S phase

The G1/S cell cycle transition depends on the precise timed destruction of cyclin-dependent kinase (CDK) inhibitors, also called CKIs. In yeast, the degradation of the CKI SIC1 is absolutely required for the onset of DNA replication (Schwob *et al.*, 1994). Elimination of Sic1 is achieved by the SCF^{CDC4} E3 and requires phosphorylation of SIC1 by the G1 cyclin CDK activity at multiple residues in order to be recognized by the F-box protein CDC4 (Nash *et al.*, 2001). Thus, stable forms of SIC1 that lack phosphorylation sites cause a G1 phase arrest (Verma *et al.*, 1997).

In animal cells, two classes of CKIs have been identified, the INK4 class (including p15, p16, p18 and p19) and the CIP/KIP family (including p21, p27 and p57) (reviewed in Sherr and Roberts, 1995). Interestingly, the mammalian p27 CKI also becomes unstable when cells approach the S phase and its degradation requires phosphorylation by cyclin E–CDK2 on a conserved threonine residue (Thr187) (Sheaff *et al.*, 1997). The degradation of phosphorylated p27 implies the SCF^{SKP2} E3 (SKP2 being an LRR-containing F-box protein) (Carrano *et al.*, 1999; Montagnoli *et al.*, 1999; Sutterluty *et al.*, 1999; Tsvetkov *et al.*, 1999). In addition, p27 degradation also requires CKS1, a component of the CDK complexes (Ganoth *et al.*, 2001; Spruck *et al.*, 2001). The exact role of p27 in gating the G1/S transition is more complex in mammalian cells than in yeast, as the two other CIP/KIP proteins, p21 and p57, might act redundantly with p27. As for p27, it was found that p21 is also degraded in a phosphorylation-dependent manner by SCF^{SKP2} (Bornstein *et al.*, 2003). It is now well established that the function of p27 to repress cell cycle progression is dosage dependent and p27 was found haplo-insufficient for tumour suppression (Fero *et al.*, 1998). Thus, low p27 protein level in tumour cells correlates with aggressiveness of the disease and patient mortality (Catzavelos *et al.*, 1997; Loda *et al.*, 1997). Strikingly, SKP2 is overexpressed in a broad spectrum of human cancers and its expression level correlates with tumour malignancy (Gstaiger *et al.*, 2001). Although SCF^{SKP2} targets many cell cycle regulatory proteins for degradation (reviewed in Nakayama and Nakayama, 2005), p27 seems the most important effector as p27 deficiency almost completely rescues the phenotypic aberrations seen in $SKP2^{-/-}$ mice

(Nakayama *et al.*, 2004). Moreover, p27 degradation by the SCF^{SKP2} E3 not only allows DNA replication to proceed on schedule, but also seems to regulate progression in mitosis (Nakayama *et al.*, 2004).

Consistently, p27 is stabilized in $SKP2^{-/-}$ mice, which exhibit multiple cell cycle defects but remain viable (Nakayama *et al.*, 2000). However, by replacing the murine *p27* gene with one carrying an alanine instead of threonine at position 187 (p27T187A), the protein was found stabilized in S and G2 phases of the cell cycle, but surprisingly, still remains unstable in G1 (Malek *et al.*, 2001). This finding indicated the existence of another proteolytic pathway, which is Thr187 phosphorylation- and SKP2 independent and acts in the cytosol of cells at the G0 to G1 transition (Hara *et al.*, 2001). Indeed, the likely corresponding E3 enzyme, called KPC (Kip1 ubiquitination-promoting complex), was recently purified (Kamura *et al.*, 2004). The KPC E3 consists of two proteins, KPC1 containing a RING-finger domain and KPC2 sharing a ubiquitin-like domain (UBL) and two ubiquitin-associated domains (UBA). Consistently, overexpression of KPC promoted the degradation of p27, and depletion of KPC1, by RNA interference, inhibited p27 degradation in G1 but not in S phase. Thus, p27 is regulated by two different RING-finger E3s (Figure 4.2):

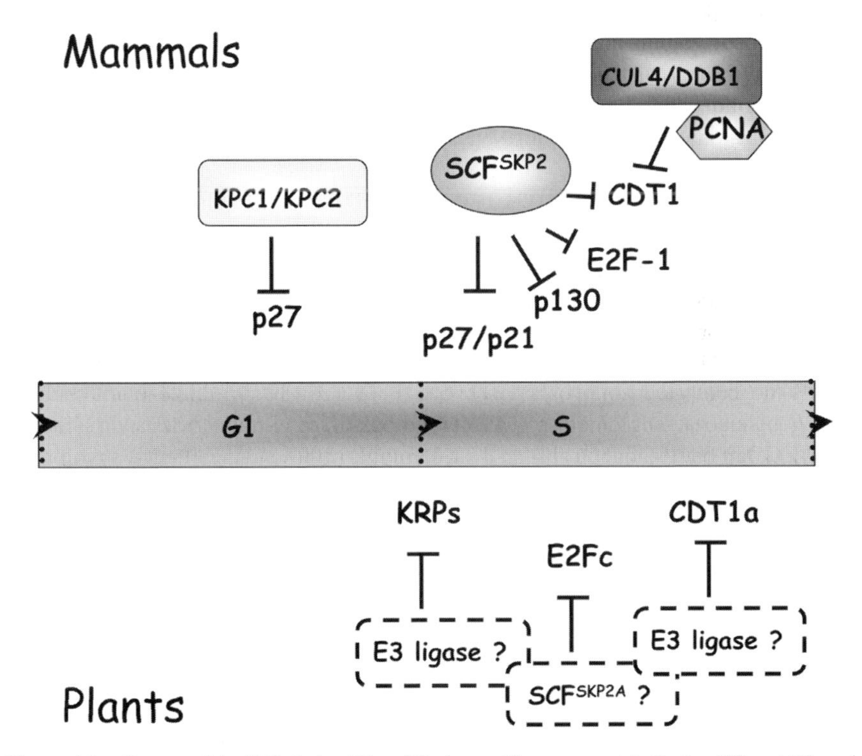

Figure 4.2 Targets of the UPS during G1 and S phases. The upper part indicates different E3s and their targets from mammals. This list is nevertheless not exhaustive; for example, cyclin E, which is degraded via at least two different E3s, is not indicated. For plants, this figure is highly speculative, as it is unknown when exactly during the cell cycle these proteins are degraded; moreover, the E3 enzymes involved remain unknown.

the KPC that mediates polyubiquitination of p27 in the cytoplasm during G1 phase and the SCF^{SKP2}, which is involved in p27 degradation in the nucleus during S and G2 phases (Hengst, 2004). However, this might not be the full picture, as it was recently found that inactivation of human cullin CUL4A (mammals encode two *CUL4* paralogues, *CUL4A* and *CUL4B*) also produces p27 stabilization and G1 cell cycle arrest, although it is not yet demonstrated that the CUL4A E3 ligase directly ubiquitinates p27 protein (Higa *et al.*, 2006).

Whereas no homologues of the INK4 class of CKIs were identified in plants, several proteins sharing a short amino acid motif with the mammalian KIP/CIP CKIs were reported (Wang *et al.*, 1997; De Veylder *et al.*, 2001; Vandepoele *et al.*, 2002). The Arabidopsis genome encodes seven such proteins called KRP1/ICK1, KRP2/ICK2, KRP3, KRP4, KRP5, KRP6 and KRP7. Despite poor homology conservation with the metazoan CIP/KIP proteins, the plant KRPs bind and inhibit several CDK complexes (most likely those formed by A-type CDKs associated to A- or D-type cyclins) and seem to play a key function in cell cycle regulation, particularly at the cell cycle exit and also at the onset of endoreduplication (reviewed in Verkest *et al.*, 2005a). However, still very little is known about the regulation of KRPs by proteolysis.

In maize, the Zeama;KRP;2 protein is subjected to proteolysis at late stages of endosperm polyploidization (Coelho *et al.*, 2005). Also, the Arabidopsis KRP2/ICK2 is a highly unstable protein, whose degradation depends on the proteasome (Verkest *et al.*, 2005b). Interestingly, the KRP2/ICK2 protein abundance is regulated by CDK-dependent phosphorylation, as the protein is not degraded in plant extracts pre-treated with olomoucine (CDK inhibitor) and because a higher KRP2/ICK2 protein level is detected in Arabidopsis mutant plants with reduced CDKB1;1 activity (Verkest *et al.*, 2005b). In addition, it was found that the deletion of the N-terminal region of Arabidopsis KRP1/ICK1 increased the protein stability (Zhou *et al.*, 2003; Weinl *et al.*, 2005). However, neither the E3 ligase nor the degradation motif that must be located in the N-terminal region of KRP1/ICK1 are known. Interestingly, two F-box proteins, similar to the metazoan SKP2, have been identified in *Arabidopsis* (del Pozo *et al.*, 2002 and our unpublished data). An obvious issue to test is whether a plant SCF^{SKP2}-like E3 is involved in KRP degradation, and if yes, when and where does the degradation occur during the cell cycle and does it request specific phosphorylation event(s).

4.3.1.2 The RBR–E2F pathway

The model for G1/S transition in animals indicates that the retinoblastoma (pRB) family of proteins, also called the pocket proteins, binds and negatively regulates the transcriptional activity of E2F in G0 and early G1. Upon stimulation by growth factors in animal cells, Cyclin–CDK complexes (CDK4/6–Cyclin D and CDK2–Cyclin E) phosphorylate pRB, resulting in the release of E2F and the transactivation of target genes (Weinberg, 1995). Thus, the E2F transcription factors are downstream effectors of the RB protein and are required for the timely regulation of numerous genes essential for DNA replication and cell cycle progression (reviewed in Bracken *et al.*, 2004). Both pRB and E2F proteins are targets of the ubiquitin pathway. Thus, the stability of the pRB protein is under the control of the RING-finger-protein MDM2 (Sdek *et al.*, 2005), whereas the stability of its cognate, the pRB-related protein p130, is regulated by the SCF^{SKP2} in a phosphorylation-dependent manner (Tedesco *et al.*, 2002). The SCF^{SKP2} E3 is also involved in the turnover of the E2F-1 transcription factor (Marti *et al.*, 1999) (Figure 4.2). However, E2F-1 protein turnover seems, in addition, under the control of other E3 ligases (Ohta and Xiong, 2001). Thus, deregulation of any E3 enzyme controlling the stability of members of the pRB or the E2F protein families might have dramatic consequences for the control of the cell cycle and might lead to human tumourigenesis.

Arabidopsis genome encodes a single retinoblastoma-related protein, called RBR, and a complex family of E2F/DP proteins (Shen, 2002; Vandepoele *et al.*, 2002). Increasing evidence suggests that the E2F–RB pathway plays a similar important role in controlling the G1-to-S phase transition in plant cells (reviewed in Gutierrez *et al.*, 2002; Inzé, 2005). Thus, overproduction of E2Fa–DPa, cyclin D3 or disruption of RBR function by virus-induced silencing all induce ectopic cell division (de Veylder *et al.*, 2002; Dewitte *et al.*, 2003; Park *et al.*, 2005). At the current stage, not much is known of the function of the UPS in the regulation of the E2F–RB pathway. Interestingly, one of the two Arabidopsis SKP2 paralogues, SKP2A, seems involved in the turnover of E2Fc, supposed to act as a repressor of transcription, and this degradation seems to depend on CDK phosphorylation (del Pozo *et al.*, 2002) (Figure 4.2). However, the demonstration that a plant SCF^{SKP2}-like E3 controls the cell cycle, is still missing. Nevertheless, the overexpression of a truncated but stable form of the E2Fc protein in Arabidopsis plants represses cell division and affects cell morphogenesis (del Pozo *et al.*, 2002). Most interestingly, E2Fc degradation was found light dependent, suggesting that the UPS could act as a molecular mechanism that links environmental signals (here light) to the cell cycle machinery.

4.3.2 Regulators that control DNA replication licensing

Cells limit their DNA replication activity to once per cell division cycle, to maintain their genomic integrity. The molecular mechanism that guarantees that many origins of replication fire only once per cell cycle has been unravelled (reviewed in Diffley, 2004; Machida *et al.*, 2005). Thus, the origin recognition complex (ORC) marks the position of replication origins in the genome for the assembly of a multiprotein pre-replicative complex (pre-RC), consisting of ORC, cell division cycle 6 (CDC6), CDC10-dependent transcript (CDT1) and mini-chromosome maintenance (MCM) proteins. Different and redundant mechanisms prevent the re-firing of origins before completion of the cell cycle; among them the UPS plays a crucial role.

Thus, the mammalian ORC1 protein is selectively released from chromatin and ubiquitylated during the progression through S phase (Li and DePamphilis, 2002), whereas in Drosophila, the ORC1 homologue is degraded at the end of M phase and throughout G1 by the APC/C E3 ligase activated by CDH1/FZR (Araki *et al.*, 2003).

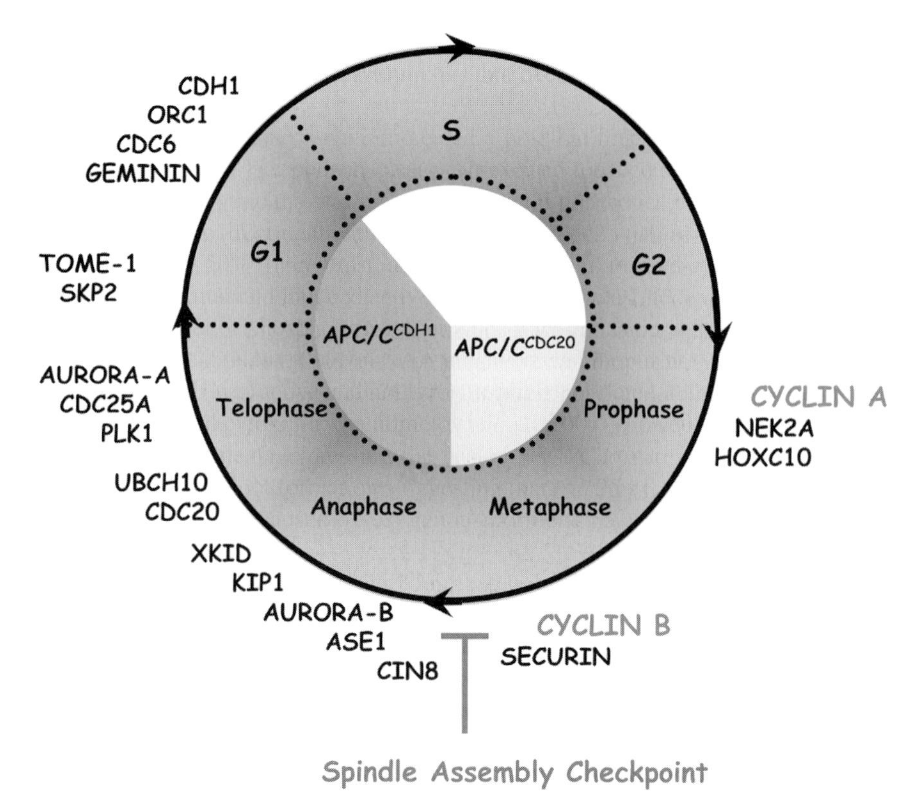

Figure 4.3 Regulation of the cell cycle by the APC/C. Targets from various organisms of the APC/C during the cell cycle. In early mitosis, the APC/C is first activated by the CDC20/FZ and later by the CDH1/FZR, allowing the E3 to extend its repertoire of substrates. Evidences for APC/C-dependent degradation of A- and B-type cyclins have also been reported in plants. If the spindle checkpoint is not satisfied, the degradation of securin via the APC/C^{CDC20} is repressed and thus cells are unable to proceed through anaphase.

Similarly, the human CDC6 is also targeted for ubiquitin-mediated proteolysis by the APC/C^{CDH1} E3 enzyme in G1 (Petersen *et al.*, 2000) (Figure 4.3). However, overexpression of a stable mutant of CDC6 is not sufficient to induce multiple rounds of DNA replication in the same cell cycle. In fission yeast, CDC18, the orthologue of CDC6, is targeted by another E3 enzyme, the SCF$^{POP1/POP2}$ E3 ligase, at the initiation of S phase (Kominami and Toda, 1997). Here, the degradation of CDC18 seems to be a critical aspect of DNA replication, as expression of a stable mutant form of CDC18, lacking all CDK phosphorylation sites, promotes high levels of overreplication (Jallepalli *et al.*, 1997).

Recent work has also highlighted the important function of the UPS in maintaining genome integrity by regulating the stability of the key replication initiator protein CDT1 and, in metazoans, its inhibitor called geminin (reviewed in Diffley, 2004; Machida *et al.*, 2005). CDT1 is evolutionarily conserved and is present in cells in G1 phase, where it is required for initiation of replication, but once origins have fired, CDT1 has to be inactivated, thereby preventing another round of replication. Depending on the organism, the mechanism of CDT1 inactivation diverges. In budding yeast, CDT1 protein levels are approximately constant throughout the cell cycle, but the protein is exported from the nucleus after G1 phase (Tanaka and Diffley, 2002). In contrast, in fission yeast and higher eukaryotes, the UPS is the way to inactivate CDT1. Indeed, the expression of a truncated form of the CDT1 protein, which is resistant to proteolysis, is able to induce re-replication in human cell lines (Nishitani *et al.*, 2004). The mechanism of CDT1 ubiquitylation and degradation is complex and requires more than one E3 enzyme. Human CDT1 protein degradation was first found to be mediated by the SCF^{SKP2} E3 ubiquitin ligase, and this pathway is dependent on CDK activity (Li *et al.*, 2003; Liu *et al.*, 2004). Nevertheless, it was later found that CDT1 degradation during S phase is most likely SKP2 independent in both mammalian cells and *Xenopus* extracts (Arias and Walter, 2005; Takeda *et al.*, 2005). However, recent findings support a model in which two independent pathways act on CDT1 degradation during S phase (Figure 4.2): (1) an SCF^{SKP2}-dependent pathway involving CDK2-dependent phosphorylation of CDT1 and (2) another pathway involving the CUL4/DDB1 E3 enzyme and PCNA, a processivity factor for DNA polymerases that interacts with an N-terminal motif of CDT1 (Arias and Walter, 2006; Senga *et al.*, 2006). The CUL4/DDB1 ubiquitin ligase was already known to mediate CDT1 ubiquitylation and subsequently degradation in response to DNA damages (Higa *et al.*, 2003; Hu *et al.*, 2004). Also in *Caenorhabditis elegans*, CUL4 seems to play a prominent role in preventing aberrant reinitiation of DNA replication, as *cul4* loss-of-function results in massive DNA re-replication producing cells with up to 100C DNA content, and this phenotype can be suppressed by removal of the *CDT1* gene (Zhong *et al.*, 2003).

In addition to the degradation pathways by the UPS, metazoans have evolved another mechanism to inhibit CDT1: a protein called geminin. Geminin inhibits licensing by binding to and inactivating CDT1 and thus preventing the loading of MCM proteins (Wohlschlegel *et al.*, 2000; Tada *et al.*, 2001). As CDT1, geminin is also regulated by the UPS. The protein accumulates in S and G2 phase, but is absent in G1 when the pre-RC are formed. Initially, geminin was discovered in a screen to identify protein substrates of the APC/C ubiquitin ligase in *Xenopus* (McGarry and Kirschner, 1998). Like mitotic cyclins (see below), geminin has a destruction box which is required for its degradation by the APC/C at the metaphase to anaphase transition. Strikingly, at least in *Xenopus*, a non-proteolytic function of APC/C-dependent ubiquitylation seems also to regulate geminin activity at the end of mitosis (Li and Blow, 2004).

In plants, little is known about the role of the UPS in regulating replication licensing. Whereas ORC, CDC6, CDT1 and MCM proteins are conserved in plants (reviewed in Gutierrez, 2005), no geminin homologue has yet been identified. Misexpression of both CDC6 and CDT1 in Arabidopsis leads to extra endoreplication cycles (Castellano *et al.*, 2001, 2004). The Arabidopsis CDC6 protein is degraded in plant extracts probably by the UPS, as the proteasome inhibitor MG132 can prevent its turnover (Castellano *et al.*, 2001). However, the E3 enzyme involved in this mechanism remains unknown. Similarly, the AtCDT1a (one of the two Arabidopsis CDT1 homologues) protein is stabilized *in planta* by blocking the proteasome activity (Castellano *et al.*, 2004). Moreover, evidences have also been provided that AtCDT1a requires CDK-dependent phosphorylation to trigger its degradation, but here again, the E3 ubiquitin ligase remains unknown (Figure 4.2).

4.3.3 Metaphase to anaphase transition

The APC/C is activated in early mitosis through Cyclin B/CDK1-dependent phosphorylation and the binding of its activator CDC20. As reflected by its name, APC/C is a major player in the metaphase to anaphase transition. A requirement for the APC/C during sister chromatid separation was deduced from experiments in *Xenopus laevis* egg extracts and in yeast (Holloway *et al.*, 1993; Irniger *et al.*, 1995). Sister chromatids are held together by the multiprotein complex called cohesin. This complex is cleaved by separase that, in turn, is inhibited by securin (Pellman and Christman, 2001). At the anaphase entry, the APC/C-dependent degradation of securin enables separase activation and as a consequence cleavage of the cohesin complex thereby allowing sister chromatid separation (Uhlmann *et al.*, 2000; Yanagida, 2000). Proteolysis of securin is ensured by APC/C^{CDC20} before anaphase onset, but degradation is maintained until the end of G1 by APC/C^{CDH1} (Nasmyth, 2001; Zur and Brandeis, 2001). Although plant orthologues of securin proteins have not yet been reported (these proteins are poorly conserved) inhibition of the proteasome during prophase blocks the cells in metaphase, indicating that securin orthologues do exist in plants (Genschik *et al.*, 1998).

The degradation of securins is especially significant in a mechanism called the spindle assembly checkpoint (Hoyt *et al.*, 1991) (Figure 4.3). To ensure balanced chromosome segregation and to avoid aneuploidy, this mechanism delays anaphase and exit from mitosis until all the chromosomes are properly attached to the spindle by means of a specialized complex called the kinetochore (Gorbsky, 2001). An unattached kinetochore becomes a source of a diffusible 'inhibitor' of the APC/C. Although the precise nature of the inhibitor is still incompletely understood, two checkpoint proteins with key functions are MAD2 (Mitotic arrest deficient 2) and BUBR1 (Budding uninhibited by benzimidazole R1), both of which can bind to CDC20, thereby sequestering it from activating the APC/C complex (Fang, 2002). Inhibition of APC/C *in vivo* may involve the transient formation of a mitotic checkpoint complex (MCC) containing BUBR1, BUB3, MAD2 and CDC20 (Yu, 2002). Unattached kinetochores serve as catalytic sites for the formation of the checkpoint complexes containing MAD2 and CDC20 (Yu, 2002). These checkpoint complexes may then diffuse away from the kinetochores to inhibit APC/C throughout the cell. This view is consistent with structural studies on MAD1, MAD2 and CDC20 (Luo *et al.*, 2000, 2002, 2004; Sironi *et al.*, 2002; de Antoni *et al.*, 2005). MAD2 interacts with MAD1, which recruits it to unattached kinetochores, and with the spindle checkpoint effector CDC20 (reviewed in Nasmyth, 2005).

In plants, colchicine (an anti-tubulin drug) has previously been used to accumulate metaphase cells in root tip meristem, indicating the existence of a similar checkpoint mechanism during plant mitosis (Levan, 1938). Moreover, in BY2 cells arrested in metaphase by anti-microtubule drugs (such as propyzamide and oryzalin), the Dbox pathway is inhibited and consequently leads to the stabilization of cyclin B1 and most probably a securin-like protein (Criqui *et al.*, 2001). The spindle checkpoint mechanism is still unknown in plants, but Yu *et al.* (1999) reported the isolation of a *MAD2* orthologue from maize and demonstrated that the protein is located at the centromeres at prophase and frequently localizes at the kinetochores at prometaphase. In the hexaploid wheat, three additional *MAD2* orthologues have been identified (Kimbara *et al.*, 2004). Their cell-cycle-dependent localization pattern was analogous to that in animals (Li and Benezra, 1996). The MAD2 protein shifts from the cytoplasm to the nuclei before cells enter metaphase, and its amount decreases towards the anaphase. Under microtubule-depolymerizing conditions, the wheat MAD2 protein accumulates at the kinetochores. Therefore, the proteins involved in spindle checkpoint regulation may also be conserved in plants. In Arabidopsis, orthologues of the inhibitory proteins likely exist, such as the spindle checkpoint components BUBR1 (At2g33560) and MAD2 (At3g25980, At1g16590), but their function has not yet been investigated. Functional analysis of the CDC20s is also lacking in plants.

4.3.4 Mitotic cyclin destruction: the essential step to exit mitosis

The mitotic cyclins were the first APC/C substrates to be characterized and were named based on their dramatic instability during mitosis. The instability of B-type cyclins in animals and yeast is conferred by a small degenerate but highly conserved motif of nine amino acids RxxLxxIxN located in the N-terminal region of these proteins and known as the destruction box (Dbox; Glotzer *et al.*, 1991). Deletion or point mutation of the Dbox inhibits cyclin proteolysis (Brandeis and Hunt, 1996; Yamano *et al.*, 1998). Destruction of cyclin B1 begins at metaphase and continues throughout mitosis and G1 phases (Clute and Pines, 1999) (Figure 4.3). Both complexes APC/C^{CDC20} and APC/C^{CDH1} mediate its degradation. Cyclin B1 proteolysis is required to inhibit CDK1 activity and, as a consequence, to induce different cell processes such as sister chromatid separation, disassembly of the mitotic spindle, chromosome decondensation, cytokinesis and reformation of the nuclear envelope (Murray and Kirschner, 1989; Luca *et al.*, 1991; Gallant and Nigg, 1992; Holloway *et al.*, 1993; Surana *et al.*, 1993).

First insights into mitotic cyclin stability in plants came from subcellular immunolocalization experiments performed in maize root tip cells (Mews *et al.*, 1997). The maize B1 cyclin (Zeama;CYCB1;2) behaves like animal B1 cyclins (Clute and Pines, 1999): it relocates to the nucleus during prophase and disappears at anaphase. Surprisingly, another B1 cyclin (Zeama;CYCB1;1) is predominantly nuclear during the entire cell cycle and does not seem to be degraded at the exit of mitosis, although it carries a Dbox motif. However, in synchronized tobacco BY2 cells, immunoblot assays showed that the endogenous cyclin B1;1 undergoes cell-cycle-dependent proteolysis and is stabilized after proteasome inhibitor treatment (Criqui *et al.*, 2000). Furthermore, by recording time-lapse images of BY2 cells expressing tobacco cyclin B1;1–GFP or cyclin B1;3–GFP fusions, cyclin degradation was found to start at the onset of anaphase and probably close to the chromosomes (Criqui *et al.*, 2001). A conserved Dbox is present in most plant mitotic cyclins (Renaudin *et al.*, 1998), with few exceptions (Vandepoele *et al.*, 2002). The importance of a functional Dbox has been investigated *in planta*. Thus, it was found that translational fusions of the N-terminal domain of Nicta;CYCB1;1 tobacco cyclin specify the degradation of the chloramphenicol acetyltransferase (CAT) reporter protein at the exit of mitosis and that this degradation is proteasome dependent (Genschik *et al.*, 1998), as the proteasome inhibitor MG132 stabilized the chimaeric protein. Mutation inside the Dbox clearly abolished the cell-cycle-specific oscillations of the fusion protein, demonstrating the functionality of this degron in plants. Moreover, a full-length non-degradable cyclin B1;1–GFP fusion protein, in contrast to the native cyclin B1;1–GFP form, was not degraded after metaphase and remained associated with chromosomes during anaphase and on subsequently decondensing chromosomes (Criqui *et al.*, 2001) (Plate 4.1).

The importance of cyclin destruction for exit of mitosis was demonstrated in *Xenopus* eggs by constructing a truncated version cyclin B. When the N-terminal region of sea urchin cyclin B was deleted, these cyclins could not be degraded and CDKs remained active. As a consequence, cells were arrested in mitosis (Murray *et al.*, 1989; Holloway *et al.*, 1993). Non-degradable mitotic cyclins also produced a mitotic arrest in *Drosophila melanogaster* (Rimmington *et al.*, 1994; Sigrist *et al.*, 1995) and HeLa cells (Gallant and Nigg, 1992). Subsequent experiments in *Saccharomyces cerevisiae* revealed that high amounts of a stabilized version of the mitotic cyclin CLB2 specifically blocked very late events in mitosis and thus mitotic exit (Surana *et al.*, 1993). These cells arrested with chromosomes segregated to opposite poles and with elongated mitotic spindles. Indestructible cyclin CDC13 arrests *Schizosaccharomyces pombe* cells in anaphase with separated and condensed chromosomes and no septa (Yamano *et al.*, 1996).

In BY2 cell suspension culture, high expression of non-degradable cyclin B1 impairs mitosis after metaphase (Weingartner *et al.*, 2004). Time-lapse analyses showed that cells initiated anaphase but lacked nuclear division and cytokinesis, leading often to endomitosis and sometimes to polynucleated cells. Moreover, it was shown that strong expression of non-degradable cyclin B1 interferes with microtubule organization and dynamics: the mid-zone spindle during anaphase, the phragmoplast during telelophase and cortical microtubules in the next G1 after fusion of the two sets of chromatids to a single nucleus. Transgenic tobacco plants expressing high levels of the non-degradable cyclin B1 showed severe developmental defects, misshaped cells in leaves, stems and roots and retarded growth of seedlings leading to post-germination death in the most extreme cases. Microtubules in cells from such seedlings are disorganized and often missing along the cell cortex (Weingartner *et al.*, 2004). Thus, strong expression of non-degradable cyclin B1 in plants like in other eukaryotic cells leads to severe cell cycle defects.

The instability of A-type cyclins, like for the B-type cyclins, is conferred by the Dbox motif in their N-terminal regions. However, cyclin A is destroyed slightly ahead of cyclin B in a variety of organisms (reviewed in Pines, 2006). In human and Drosophila cells, cyclin A starts to degrade during prometaphase and is not subjected to the spindle assembly checkpoint (Whitfield *et al.*, 1990; den Elzen and Pines, 2001; Geley *et al.*, 2001). In HeLa cells, degradation of NEK2A kinase that regulates centrosomes separation and the HOXC10 transcription factor parallels that of cyclin A (Hames *et al.*, 2001; Gabellini *et al.*, 2003) (Figure 4.3). In plant BY2 cells, the NtCYCA3;1–GFP fusion protein was found exclusively in the nucleus during G2 phase and was never detected during mitosis, suggesting that the fusion had already disappeared in late G2 phase or early prophase (Criqui *et al.*, 2001). Similar results have been obtained for an A2-type cyclin in alfalfa (Roudier *et al.*, 2000). Immunolocalization experiments showed that the Medsa;CYCA2 protein is localized in the nucleus of prophase cells but is undetectable in metaphase, indicating proteolysis in early mitosis. The different degradation pattern of the A-type and B-type cyclins may be attributed partly to the different specificities of APC/C complexes associated with different activator subunits. Nevertheless, it remains puzzling how APC/C^{CDC20} can recognize proteins in the presence of the spindle assembly checkpoint.

4.3.5 *APC^{CDC20} versus $APC^{CDH1/CCS52}$*

The APC/C activators FZ/CDC20 and FZR/CDH1 act consecutively in the cell cycle (Figure 4.3) and determine stage-specific activation and substrate selection of the APC/C (reviewed in Peters, 2002; Castro *et al.*, 2005). The switch from CDC20 to CDH1 is thought to allow degradation of many additional substrates because APC/C^{CDH1} has been shown to have broader substrate specificity than APC/C^{CDC20} (Zur and Brandeis, 2002). In addition to cyclin and securin, many other important cell cycle proteins have been proved to be targets of APC/C in yeast and mammalian cells. These include several factors controlling anaphase: the kinesin-related protein XKID, which is involved in chromosome alignment during metaphase and chromosome movements to the spindle poles throughout anaphase (Funabiki and Murray, 2000; Levesque and Compton, 2001; Castro *et al.*, 2003), two motor proteins, the kinesins KIP1 and CIN8, required to allow progression through anaphase (Gordon and Roof, 2001; Hildebrandt and Hoyt, 2001) and the mitotic spindle-associated protein ASE1 (anaphase spindle elongation 1), a component of the *S. cerevisiae* spindle and member of a conserved family of microtubule-bundling proteins implicated in central spindle formation and cytokinesis (Juang *et al.*, 1997; Yamashita *et al.*, 2005). The proteins targeted by APC/C^{CDH1} also include regulatory proteins such as the AURORA-A and B kinases that control many cell cycle events ranging from centrosomes maturation to mitotic entry (Crane *et al.*, 2004) and the phosphatase CDC25A that removes inhibitory phosphorylation from S-phase CDKs (Donzelli *et al.*, 2002). Recent reports highlight the synergy between the destructive action by APC/C and SCF, which were originally viewed as two independent pathways (reviewed in Vodermaier, 2004). TOME-1 is the first F-box protein identified as a target for degradation by APC/C^{CDH1} in G1 (Ayad *et al.*, 2003). TOME-1 mediates the destruction of mitosis-inhibitory kinase WEE1 via SCF^{TOME-1}. The reinstated WEE1 in G1 with the help of APC/C^{CDH1}, which mediates inactivation of both

CDK1/Cyclin B and SCF^{TOME-1}, prevents premature entry into mitosis during S phase. SKP2 is the second example of F-box protein targeted for degradation by the APC/C^{CDH1}, thus limiting SCF^{SKPP2} activity during the G1 phase of the cell cycle (Bashir *et al.*, 2004; Wei *et al.*, 2004).

Among the substrates degraded during mitotic exit are regulators of the APC/C: the Polo kinase (PLK1), which mediates activation of APC/C by phosphorylation (Charles *et al.*, 1998; Brassac *et al.*, 2000; Lindon and Pines, 2004), the ubiquitin-conjugating enzyme E2-C/UBCH10 (Yamanaka *et al.*, 2000) and the coactivators CDC20/FZ (Weinstein, 1997; Shirayama *et al.*, 1998) and CDH1/FZR (Listovsky *et al.*, 2004). In somatic cells, once cyclin degradation starts in anaphase, the decline in Cyclin B/CDK1 activity allows APC/C to bind CDH1. In budding yeast, the CDC14 phosphatase is responsible for dephosphorylating CDH1, but it is unclear whether this holds true in animal cells. CDC20 proteolysis by APC^{CDH1} induces APC^{CDC20} inactivation and allows the switch from APC/C^{CDC20} to APC/C^{CDH1}. At this stage of the cell cycle, APC/C^{CDH1} takes over the degradation of mitotic cyclins, preventing the premature accumulation of these proteins and a premature entry into S phase (Zachariae and Nasmyth, 1999).

CDH1-type activators named as cell cycle switch CCS52 proteins have been identified in *Medicago* species, *M. sativa* and the model legume *Medicago truncatulata* (Cebolla *et al.*, 1999; Tarayre *et al.*, 2004). *Ms/MtCCS52A* is an orthologue of the fission yeast and animal CDH1-type proteins. In fission yeast, overexpression of the plant gene triggers mitotic cyclin degradation resulting in cell division arrest, cell elongation and in a switch from mitotic cycles to endoreduplication cycles, which consists of one or several rounds of DNA synthesis in the absence of mitosis. *In planta*, expression of *CCS52* is linked to cell differentiation and endoreduplication (Vinardell *et al.*, 2003). Higher accumulation of *CCS52* mRNAs was found in the endoreduplicating and expanding *M. truncatulata* nodule cells, and reduction of the transcript level by antisense suppression of the cell cycle switch 52A was correlated with a reduction in ploidy levels in these cells. *MtCCS52B* is a homologue in *M. truncatulata* that exhibits strikingly different cell cycle and developmental regulation, when compared to *MtCCS52A* (Tarayre *et al.*, 2004).

The *Arabidopsis thaliana* genome contains three *CCS52* genes: *AtCCS52A1*, *AtCCS52A2* and *AtCCS52B*. Functional analysis of the *Medicago* CCS52A proteins led to the discovery of a conserved motif called CDH1-specific motif (CSM) that is also essential for APC/C interaction, in addition to the consensus binding motifs already described; the C-box at the N-terminus and the C-terminal IR tail that interacts with the tetratricopeptide repeat (TPR) subunits, APC3 and APC7 in the human APC/C (Vodermaier *et al.*, 2003). Consistently, the AtCCS52 proteins were able to bind to both the yeast and the *Arabidopsis* APC/C. In synchronized *Arabidopsis* cell cultures the *AtCCS52B* transcripts were detected from G2/M to M phase, while *AtCCS52A1* and *AtCCS52A2* were present from late M phase until G2, suggesting sequential action of these APC/C activators in the plant cell cycle. However, the activity of CDH1 in mammals and yeast is under tight control by post-transcriptional mechanisms including phosphorylation, interaction with inhibitory proteins and degradation mediated by APC/C^{CDH1} or the SCF (Benmaamar and Pagano, 2005).

The activity of AtCCS52 could also be regulated by phosphorylation, as potential phosphorylation sites are critical in the interaction of the Mt/MsCCS52A proteins with the core APC/C. Expression of all three *AtCCS52* genes in the heterologous system *S. pombe* led to the inhibition of cell proliferation but with distinctive phenotypic effects (Fülöp *et al.*, 2005). Overexpression of *AtCCS52B* affects neither the size nor the shape of yeast cells, in contrast to *AtCCS52A1* or *AtCCS52A2*, which trigger both elongation of yeast cells and enlargement of nuclei. Moreover, yeast cells overexpressing *AtCCS52A1* display a branched phenotype, which is reminiscent of ploidy-dependent branching of the unicellular trichomes in Arabidopsis (Hülskamp *et al.*, 1999).

4.3.6 Regulation of endoreduplication by the APC/C

Polyploidy is particularly widespread in many plant cell types, in which endoreduplication occurs during differentiation of cells (Joubès and Chevalier, 2000; Sugimoto-Shirasu and Roberts, 2003). Genetic studies suggest that plant endoreduplication is actively regulated in both entry and the number of endocycles (Hülskamp, 2004). Initiation of the endocycle programme requires prior exit from the cell cycle. Various regulatory proteins of the mitotic CDK activity have been suggested as key regulators for the mitosis to endocycle transition (Edgar and Orr-Weaver, 2001; Larkins *et al.*, 2001). Since mitosis does not take place during endoreduplication, mitotic cyclin B and CDKB are not required. Indeed, the expression of these genes is downregulated as is histone H1 kinase activity specific of M-phase CDKs (Jacqmard *et al.*, 1999; Joubès *et al.*, 1999; Boudolf *et al.*, 2004). CCS52 proteins might act as a major regulator of the endoreduplication cycles, most likely by favouring degradation of mitotic cyclins (reviewed in Kondorosi and Kondorosi, 2004). Consistent with this, overexpression of the mitotic cyclin CYCB1;2 in trichomes results in multicellular trichomes and is thus sufficient to switch from endoreduplication to mitosis (Schnittger *et al.*, 2002).

Recently, an *Arabidopsis thaliana* A-type cyclin, CYCA2;3, has been identified as a key regulator of ploidy levels in this plant (Imai *et al.*, 2006). Null mutations of *cyc2;3* semi-dominantly promote endocycles and increase the ploidy levels achieved in mature organ. Conversely, strong expression of the protein results in termination of endocycle succession. Stabilization of the protein through mutation in the Dbox sequence further enhances this reduction. Thus, the CYCA2;3 protein acts as a negative regulator of endocycles and is itself most likely negatively regulated by APC/C. One interesting possibility is that APC/C^{CCS52} regulates ploidy levels through the negative regulation of A2-type cyclins. Indeed, yeast two-hybrid screens and immunoprecipitation assays indicate that both AtCCS52A and B proteins are able to bind A- and B-type mitotic cyclins (Fülöp *et al.*, 2005). In Arabidopsis, it has been proposed that CDKA;1 activity is downregulated at the entry of endoreduplication and remains at a low level for progression of endoreduplication (Verkest *et al.*, 2005a). In this context, the pivotal role of CYCA2;3 and its turnover in the control of endocycles remains to be elucidated.

4.4 Conclusion

Compared to fungi and metazoans, the function of the UPS in the control of the plant cell cycle is still very poorly understood. In particular, we know near to nothing about the role of selective protein degradation at G1/S transition. Is this transition under the control of an SCF E3 ligase in plants as it was found in yeast and metazoans? KRPs are possible targets of such a mechanism, as at least some of them are degraded by the proteasome. The identification of the E3 enzyme(s) involved as well as the molecular mechanism(s) signalling its degradation is a clear challenge for the future. In contrast to the SCF, the function of the APC/C in plants is better documented and its role in mitosis and endocycles has been established. Nevertheless, the next step will be to identify substrates of this E3 ligase, and more than hundred might be expected. This will be a technical challenge, as the proteins cannot be simply predicted, based on the degradation motifs, which are too degenerated. It is also noteworthy that the APC/C has been involved in the degradation of non-cell-cycle proteins in metazoans. This issue might certainly be worth investigating in plants as well. In addition, an interesting issue to explore is the regulation(s) occurring on the plant APC/C itself. The large number of activators and the presence of two distinct CDC27 subunits suggest the existence of multiple APC/C complexes, which would permit the destruction of a large repertoire of substrates. Beside the SCF and the APC/C several other classes of E3 ligases have recently been involved in cell cycle control in animal models, such as the RING-finger KPC or the CUL4/DDB1 E3s among others. Are these enzymes conserved in plants and do they play similar roles in the control of the cell cycle? Moreover, it is presently unknown how all these E3s are interconnected in plants, whereas in yeast and metazoan it has already been found that certain substrates are degraded by two or even three different E3s and that some E3 components are themselves targets of other E3 classes. Last but not least, an interesting field of research will be to identify the regulators (kinases, phosphatases, other enzymes or even docking proteins) that modify and/or act on the target proteins to signal their degradation. Thus, the puzzle is far from complete and this topic will keep more than one laboratory busy in the coming years.

References

Amerik, A.Y. and Hochstrasser, M. (2004) Mechanism and function of deubiquitinating enzymes. *Biochim Biophys Acta* **1695**, 189–207.

Araki, M., Wharton, R.P., Tang, Z., Yu, H. and Asano, M. (2003) Degradation of origin recognition complex large subunit by the anaphase-promoting complex in Drosophila. *EMBO J* **22**, 6115–6126.

Arias, E.E. and Walter, J.C. (2005) Replication-dependent destruction of Cdt1 limits DNA replication to a single round per cell cycle in Xenopus egg extracts. *Genes Dev* **19**, 114–126.

Arias, E.E. and Walter, J.C. (2006) PCNA functions as a molecular platform to trigger Cdt1 destruction and prevent re-replication. *Nat Cell Biol* **8**, 84–90.

Ayad, N.G., Rankin, S., Murakami, M., Jebanathirajah, J., Gygi, S. and Kirschner, M.W. (2003) Tome-1, a trigger of mitotic entry, is degraded during G1 via the APC. *Cell* **113**, 101–113.

Bashir, T., Dorrello, N.V., Amador, V., Guardavaccaro, D. and Pagano, M. (2004) Control of the SCF(Skp2-Cks1) ubiquitin ligase by the APC/C(Cdh1) ubiquitin ligase. *Nature* **428**, 190–193.

Benmaamar, R. and Pagano, M. (2005) Involvement of the SCF complex in the control of Cdh1 degradation in S-phase. *Cell Cycle* **4**, 1230–1232.

Blilou, I., Frugier, F., Folmer, S., *et al.* (2002) The *Arabidopsis HOBBIT* gene encodes a CDC27 homolog that links the plant cell cycle to progression of cell differentiation. *Genes Dev* **16**, 2566–2575.

Bornstein, G., Bloom, J., Sitry-Shevah, D., Nakayama, K., Pagano, M. and Hershko, A. (2003) Role of the SCFSkp2 ubiquitin ligase in the degradation of p21Cip1 in S phase. *J Biol Chem* **278**, 25752–25757.

Boudolf, V., Vlieghe, K., Beemster, G.T.S., *et al.* (2004) The plant-specific cyclin-dependent kinase CDKB1;1 and transcription factor E2Fa–DPa control the balance of mitotically dividing and endoreduplicating cells in Arabidopsis. *Plant Cell* **16**, 2683–2692.

Bracken, A.P., Ciro, M., Cocito, A. and Helin, K. (2004) E2F target genes: unraveling the biology. *Trends Biochem Sci* **29**, 409–417.

Brandeis, M. and Hunt, T. (1996) The proteolysis of mitotic cyclins in mammalian cells persists from the end of mitosis until the onset of S phase. *EMBO J* **15**, 5280–5289.

Brassac, T., Castro, A., Lorca, T., *et al.* (2000) The polo-like kinase Plx1 prevents premature inactivation of the APC(Fizzy)-dependent pathway in the early Xenopus cell cycle. *Oncogene* **19**, 3782–3790.

Burton, J.L. and Solomon, M.J. (2001) D box and KEN box motifs in budding yeast Hsl1p are required for APC-mediated degradation and direct binding to Cdc20p and Cdh1p. *Genes Dev* **15**, 2381–2395.

Burton, J.L., Tsakraklides, V. and Solomon, M.J. (2005) Assembly of an APC-Cdh1-substrate complex is stimulated by engagement of a destruction box. *Mol Cell* **18**, 533–542.

Capron, A., Okresz, L. and Genschik, P. (2003a) First glance at the plant APC/C, a highly conserved ubiquitin-protein ligase. *Trends Plant Sci* **8**, 83–89.

Capron, A., Serralbo, O., Fülöp, K., *et al.* (2003b) The Arabidopsis anaphase-promoting complex or cyclosome: molecular and genetic characterization of the APC2 subunit. *Plant Cell* **15**, 2370–2382.

Cardozo, T. and Pagano, M. (2004) The SCF ubiquitin ligase: insights into a molecular machine. *Nat Rev Mol Cell Biol* **5**, 739–751.

Carrano, A.C., Eytan, E., Hershko, A. and Pagano, M. (1999) SKP2 is required for ubiquitin-mediated degradation of the CDK inhibitor p27. *Nat Cell Biol* **1**, 193–199.

Carroll, C.W., Enquist-Newman, M. and Morgan, D.O. (2005) The APC subunit Doc1 promotes recognition of the substrate destruction box. *Curr Biol* **15**, 11–18.

Carroll, C.W. and Morgan, D.O. (2002) The Doc1 subunit is a processivity factor for the anaphase-promoting complex. *Nat Cell Biol* **4**, 880–887.

Castellano, M.M., Boniotti, M.B., Caro, E., Schnittger, A. and Gutierrez, C. (2004) DNA replication licensing affects cell proliferation or endoreplication in a cell type-specific manner. *Plant Cell* **16**, 2380–2393.

Castellano, M.M., del Pozo, J.C., Ramirez-Parra, E., Brown, S. and Gutierrez, C. (2001) Expression and stability of *Arabidopsis* CDC6 are associated with endoreplication. *Plant Cell* **13**, 2671–2686.

Castro, A., Bernis, C., Vigneron, S., Labbe, J.C. and Lorca, T. (2005) The anaphase-promoting complex: a key factor in the regulation of cell cycle. *Oncogene* **24**, 314–325.

Castro, A., Vigneron, S., Bernis, C., Labbe, J.C. and Lorca, T. (2003) Xkid is degraded in a D-box, KEN-box, and A-box-independent pathway. *Mol Cell Biol* **23**, 4126–4138.

Catzavelos, C., Bhattacharya, N., Ung, Y.C., *et al.* (1997) Decreased levels of the cell-cycle inhibitor p27Kip1 protein: prognostic implications in primary breast cancer. *Nat Med* **3**, 227–230.

Cebolla, A., Vinardell, J.M., Kiss, E., *et al.* (1999) The mitotic inhibitor ccs52 is required for endoreduplication and ploidy-dependent cell enlargement in plants. *EMBO J* **18**, 4476–4484.

Charles, J.F., Jaspersen, S.L., Tinker-Kulberg, R.L., Hwang, L., Szidon, A. and Morgan, D.O. (1998) The Polo-related kinase Cdc5 activates and is destroyed by the mitotic cyclin destruction machinery in *S. cerevisiae*. *Curr Biol* **8**, 497–507.

Ciechanover, A., Orian, A. and Schwartz, A.L. (2000) Ubiquitin-mediated proteolysis: biological regulation via destruction. *BioEssays* **22**, 442–451.

Clute, P. and Pines, J. (1999) Temporal and spatial control of cyclin B1 destruction in metaphase. *Nat Cell Biol* **1**, 82–87.

Coelho, C.M., Dante, R.A., Sabelli, P.A., *et al.* (2005) Cyclin-dependent kinase inhibitors in maize endosperm and their potential role in endoreduplication. *Plant Physiol* **138**, 2323–2336.

Cope, G.A. and Deshaies, R.J. (2003) COP9 signalosome: a multifunctional regulator of SCF and other cullin-based ubiquitin ligases. *Cell* **114**, 663–671.

Coscoy, L. and Ganem, D. (2003) PHD domains and E3 ubiquitin ligases: viruses make the connection. *Trends Cell Biol* **13**, 7–12.

Crane, R., Gadea, B., Littlepage, L., Wu, H. and Ruderman, J.V. (2004) Aurora A, meiosis and mitosis. *Biol Cell* **96**, 215–229.

Criqui, M.C., Parmentier, Y., Derevier, A., Shen, W.H., Dong, A. and Genschik, P. (2000) Cell cycle-dependent proteolysis and ectopic overexpression of cyclin B1 in tobacco BY2 cells. *Plant J* **24**, 763–773.

Criqui, M.C., Weingartner, M., Capron, A., *et al.* (2001) Sub-cellular localisation of GFP-tagged tobacco mitotic cyclins during the cell cycle and after spindle checkpoint activation. *Plant J* **28**, 569–581.

de Antoni, A., Pearson, C.G., Cimini, D., *et al.* (2005) The Mad1/Mad2 complex as a template for Mad2 activation in the spindle assembly checkpoint. *Curr Biol* **15**, 214–225.

del Pozo, J.C., Boniotti, M.B. and Gutierrez, C. (2002) Arabidopsis E2Fc functions in cell division and is degraded by the ubiquitin-SCF^{AtSKP2} pathway in response to light. *Plant Cell* **14**, 3057–3071.

den Elzen, N. and Pines, J. (2001) Cyclin A is destroyed in prometaphase and can delay chromosome alignment and anaphase. *J Cell Biol* **153**, 121–136.

Deshaies, R.J. and Ferrell, J.E., Jr. (2001) Multisite phosphorylation and the countdown to S phase. *Cell* **107**, 819–822.

De Veylder, L., Beeckman, T., Beemster, G.T.S., *et al.* (2001) Functional analysis of cyclin-dependent kinase inhibitors of *Arabidopsis*. *Plant Cell* **13**, 1653–1668.

De Veylder, L., Beeckman, T., Beemster, G.T.S., *et al.* (2002) Control of proliferation, endoreduplication and differentiation by the *Arabidopsis* E2Fa-DPa transcription factor. *EMBO J* **21**, 1360–1368.

Dewitte, W., Riou-Khamlichi, C., Scofield, S., *et al.* (2003) Altered cell cycle distribution, hyperplasia, and inhibited differentiation in Arabidopsis caused by the D-type cyclin CYCD3. *Plant Cell* **15**, 79–92.

Dias, D.C., Dolios, G., Wang, R. and Pan, Z.Q. (2002) CUL7: A DOC domain-containing cullin selectively binds Skp1.Fbx29 to form an SCF-like complex. *Proc Natl Acad Sci USA* **99**, 16601–16606.

Diffley, J.F. (2004) Regulation of early events in chromosome replication. *Curr Biol* **14**, R778–R786.

Donzelli, M., Squatrito, M., Ganoth, D., Hershko, A., Pagano, M. and Draetta, G.F. (2002) Dual mode of degradation of Cdc25 A phosphatase. *EMBO J* **21**, 4875–4884.

Edgar, B.A. and Orr-Weaver, T.L. (2001) Endoreplication cell cycles: more for less. *Cell* **105**, 297–306.

Fang, G. (2002) Checkpoint protein BubR1 acts synergistically with Mad2 to inhibit anaphase-promoting complex. *Mol Biol Cell* **13**, 755–766.

Fero, M.L., Randel, E., Gurley, K.E., Roberts, J.M. and Kemp, C.J. (1998) The murine gene p27Kip1 is haplo-insufficient for tumour suppression. *Nature* **396**, 177–180.

Fülöp, K., Tarayre, S., Kelemen, Z., *et al.* (2005) Arabidopsis anaphase-promoting complexes: multiple activators and wide range of substrates might keep APC perpetually busy. *Cell Cycle* **4**, 1084–1092.

Funabiki, H. and Murray, A.W. (2000) The Xenopus chromokinesin Xkid is essential for metaphase chromosome alignment and must be degraded to allow anaphase chromosome movement. *Cell* **102**, 411–424.

Gabellini, D., Colaluca, I.N., Vodermaier, H.C., *et al.* (2003) Early mitotic degradation of the homeoprotein HOXC10 is potentially linked to cell cycle progression. *EMBO J* **22**, 3715–3724.

Gagne, J.M., Downes, B.P., Shiu, S.H., Durski, A.M. and Vierstra, R.D. (2002) The F-box subunit of the SCF E3 complex is encoded by a diverse superfamily of genes in Arabidopsis. *Proc Natl Acad Sci USA* **99**, 11519–11524.

Gallant, P. and Nigg, E.A. (1992) Cyclin B2 undergoes cell cycle-dependent nuclear translocation and, when expressed as a non-destructible mutant, causes mitotic arrest in HeLa cells. *J Cell Biol* **117**, 213–224.

Ganoth, D., Bornstein, G., Ko, T.K., *et al.* (2001) The cell-cycle regulatory protein Cks1 is required for SCF(Skp2)-mediated ubiquitinylation of p27. *Nat Cell Biol* **3**, 321–324.

Geley, S., Kramer, E., Gieffers, C., Gannon, J., Peters, J.M. and Hunt, T. (2001) Anaphase-promoting complex/cyclosome-dependent proteolysis of human cyclin A starts at the beginning of mitosis and is not subject to the spindle assembly checkpoint. *J Cell Biol* **153**, 137–148.

Genschik, P., Criqui, M.C., Parmentier, Y., Derevier, A. and Fleck, J. (1998) Cell cycle-dependent proteolysis in plants. Identification of the destruction box pathway and metaphase arrest produced by the proteasome inhibitor MG132. *Plant Cell* **10**, 2063–2076.

Glotzer, M., Murray, A.W. and Kirschner, M.W. (1991) Cyclin is degraded by the ubiquitin pathway . *Nature* **349**, 132–138.

Gmachl, M., Gieffers, C., Podtelejnikov, A.V., Mann, M. and Peters, J.M. (2000) The RING-H2 finger protein APC11 and the E2 enzyme UBC4 are sufficient to ubiquitinate substrates of the anaphase-promoting complex. *Proc Natl Acad Sci USA* **97**, 8973–8978.

Gorbsky, G.J. (2001) The mitotic spindle checkpoint. *Curr Biol* **12**, R1001–R1004.

Gordon, D.M. and Roof, D.M. (2001) Degradation of the kinesin Kip1p at anaphase onset is mediated by the anaphase-promoting complex and Cdc20p. *Proc Natl Acad Sci USA* **98**, 12515–12520.

Gray, W.M., Kepinski, S., Rouse, D., Leyser, O. and Estelle, M. (2001) Auxin regulates SCF^{TIR1}-dependent degradation of AUX/IAA proteins. *Nature* **414**, 271–276.

Gstaiger, M., Jordan, R., Lim, M., *et al.* (2001) Skp2 is oncogenic and overexpressed in human cancers. *Proc Natl Acad Sci USA* **98**, 5043–5048.

Gutierrez, C. (2005) Coupling cell proliferation and development in plants. *Nat Cell Biol* **7**, 535–541.

Gutierrez, C., Ramirez-Parra, E., Castellano, M.M. and del Pozo, J.C. (2002) G_1 to S transition: more than a cell cycle engine switch. *Curr Opin Plant Biol* **5**, 480–486.

Hames, R.S., Wattam, S.L., Yamano, H., Bacchieri, R. and Fry, A.M. (2001) APC/C-mediated destruction of the centrosomal kinase Nek2A occurs in early mitosis and depends upon a cyclin A-type D-box. *EMBO J* **20**, 7117–7127.

Hara, T., Kamura, T., Nakayama, K., Oshikawa, K., Hatakeyama, S. and Nakayama, K. (2001) Degradation of p27(Kip1) at the G(0)–G(1) transition mediated by a Skp2-independent ubiquitination pathway. *J Biol Chem* **276**, 48937–48943.

Hendil, K.B. and Hartmann-Petersen, R. (2004) Proteasomes: a complex story. *Curr Protein Pept Sci* **5**, 135–151.

Hengst, L. (2004) A second RING to destroy $p27^{Kip1}$. *Nat Cell Biol* **6**, 1153–1155.

Higa, L.A., Mihaylov, I.S., Banks, D.P., Zheng, J. and Zhang, H. (2003) Radiation-mediated proteolysis of CDT1 by CUL4-ROC1 and CSN complexes constitutes a new checkpoint. *Nat Cell Biol* **5**, 1008–1015.

Higa, L.A., Yang, X., Zheng, J., *et al.* (2006) Involvement of CUL4 ubiquitin E3 ligases in regulating CDK inhibitors Dacapo/$p27^{Kip1}$ and Cyclin E degradation. *Cell Cycle* **5**, 71–77.

Hildebrandt, E.R. and Hoyt, M.A. (2001) Cell cycle-dependent degradation of the *Saccharomyces cerevisiae* spindle motor Cin8p requires APC^{Cdh1} and a bipartite destruction sequence. *Mol Biol Cell* **12**, 3402–3416.

Hochstrasser, M. (1998) There's the rub: a novel ubiquitin-like modification linked to cell cycle regulation. *Genes Dev* **12**, 901–907.

Holloway, S.L., Glotzer, M., King, R.W. and Murray, A.W. (1993) Anaphase is initiated by proteolysis rather than by the inactivation of maturation-promoting factor. *Cell* **73**, 1393–1402.

Hoyt, M.A., Totis, L. and Roberts, B.T. (1991) *S.* cerevisiae genes required for cell cycle arrest in response to loss of microtubule function. *Cell* **66**, 507–517.

Hu, J., McCall, C.M., Ohta, T. and Xiong, Y. (2004) Targeted ubiquitination of CDT1 by the DDB1-CUL4A-ROC1 ligase in response to DNA damage. *Nat Cell Biol* **6**, 1003–1009.

Hülskamp, M. (2004) Plant trichomes: a model for cell differentiation. *Nat Rev Mol Cell Biol* **5**, 471–480.

Hülskamp, M., Schnittger, A. and Folkers, U. (1999) Pattern formation and cell differentiation: trichomes in Arabidopsis as a genetic model system. *Int Rev Cytol* **186**, 147–178.

Imai, K., Ohashi, Y., Tsuge, T., *et al.* (2006) The A-Type Cyclin CYCA2;3 is a key regulator of ploidy levels in *Arabidopsis* endoreduplication. *Plant Cell* **18**, 382–396.

Inzé, D. (2005) Green light for the cell cycle. *EMBO J* **24**, 657–662.

Irniger, S., Piatti, S., Michaelis, C. and Nasmyth, K. (1995) Genes involved in sister chromatid separation are needed for B-type cyclin proteolysis in budding yeast. *Cell* **81**, 269–278. (Erratum: *Cell* (1998) **93**, 487).

Ivan, M. and Kaelin, W.G, Jr. (2001) The von Hippel-Lindau tumor suppressor protein. *Curr Opin Genet* **11**, 27–34.

Jacqmard, A., De Veylder, L., Segers, G., *et al.* (1999) Expression of CKS1At in *Arabidopsis thaliana* indicates a role for the protein in both the mitotic and the endoreduplication cycle. *Planta* **207**, 496–504.

Jallepalli, P.V., Brown, G.W., Muzi-Falconi, M., Tien, D. and Kelly, T.J. (1997) Regulation of the replication initiator protein p65cdc18 by CDK phosphorylation. *Genes Dev* **11**, 2767–2779.

Jaspersen, S.L., Charles, J.F. and Morgan, D.O. (1999) Inhibitory phosphorylation of the APC regulator Hct1 is controlled by the kinase Cdc28 and the phosphatase Cdc14. *Curr Biol* **9**, 227–236.

Joubès, J. and Chevalier, C. (2000) Endoreduplication in higher plants. *Plant Mol Biol* **43**, 735–745.

Joubès, J., Phan, T.H., Just, D., *et al.* (1999) Molecular and biochemical characterization of the involvement of cyclin-dependent kinase A during the early development of tomato fruit. *Plant Physiol* **121**, 857–869.

Juang, Y.L., Huang, J., Peters, J.M., McLaughlin, M.E., Tai, C.Y. and Pellman, D. (1997) APC-mediated proteolysis of Ase1 and the morphogenesis of the mitotic spindle. *Science* **275**, 1311–1314.

Kamura, T., Burian, D., Yan, Q., *et al.* (2001) Muf1, a novel Elongin BC-interacting leucine-rich repeat protein that can assemble with Cul5 and Rbx1 to reconstitute a ubiquitin ligase. *J Biol Chem* **276**, 29748–29753.

Kamura, T., Hara, T., Matsumoto, M., *et al.* (2004) Cytoplasmic ubiquitin ligase KPC regulates proteolysis of $p27^{Kip1}$ at G1 phase. *Nat Cell Biol* **6**, 1229–1235.

Kimbara, J., Endo, T.R. and Nasuda, S. (2004) Characterization of the genes encoding for MAD2 homologues in wheat. *Chromosome Res* **12**, 703–714.

Kominami, K. and Toda, T. (1997) Fission yeast WD-repeat protein pop1 regulates genome ploidy through ubiquitin-proteasome-mediated degradation of the CDK inhibitor Rum1 and the S-phase initiator Cdc18. *Genes Dev* **11**, 1548–1560.

Kondorosi, E. and Kondorosi, A. (2004) Endoreduplication and activation of the anaphase-promoting complex during symbiotic cell development. *FEBS Lett* **567**, 152–157.

Kraft, C., Herzog, F., Gieffers, C., *et al.* (2003) Mitotic regulation of the human anaphase-promoting complex by phosphorylation. *EMBO J* **22**, 6598–6609.

Kraft, C., Vodermaier, H.C., Maurer-Stroh, S., Eisenhaber, F. and Peters, J.M. (2005) The WD40 propeller domain of Cdh1 functions as a destruction box receptor for APC/C substrates. *Mol Cell* **18**, 543–553.

Kramer, E.R., Scheuringer, N., Podtelejnikov, A.V., Mann, M. and Peters, J.M. (2000) Mitotic regulation of the APC activator proteins CDC20 and CDH1. *Mol Biol Cell* **11**, 1555–1569.

Kwee, H.S. and Sundaresan, V. (2003) The NOMEGA gene required for female gametophyte development encodes the putative APC6/CDC16 component of the anaphase promoting complex in Arabidopsis. *Plant J* **36**, 853–866.

Larkins, B.A., Dilkes, B.P., Dante, R.A., Coelho, C.M., Woo, Y.M. and Liu, Y. (2001) Investigating the hows and whys of DNA endoreduplication. *J Exp Bot* **52**, 183–192.

Levan, A. (1938) The effect of colchicines on root mitosis in Allium. *Physiol Plant (Lund)* **24**, 471–486.

Leverson, J.D., Joazeiro, C.A., Page, A.M., Huang, H., Hieter, P. and Hunter, T. (2000) The APC11 RING-H2 finger mediates E2-dependent ubiquitination. *Mol Biol Cell* **11**, 2315–2325.

Levesque, A.A. and Compton, D.A. (2001) The chromokinesin kid is necessary for chromosome arm orientation and oscillation, but not congression, on mitotic spindles. *J Cell Biol* **154**, 1135–1146.

Li, A. and Blow, J.J. (2004) Non-proteolytic inactivation of geminin requires CDK-dependent ubiquitination. *Nat Cell Biol* **6**, 260–267.

Li, C.J. and DePamphilis, M.L. (2002) Mammalian Orc1 protein is selectively released from chromatin and ubiquitinated during the S-to-M transition in the cell division cycle. *Mol Cell Biol* **22**, 105–116.

Li, X., Zhao, Q., Liao, R., Sun, P. and Wu, X. (2003) The SCF^{Skp2} ubiquitin ligase complex interacts with the human replication licensing factor Cdt1 and regulates Cdt1 degradation. *J Biol Chem* **278**, 30854–30858.

Li, Y. and Benezra, R. (1996) Identification of a human mitotic checkpoint gene: hsMAD2. *Science* **274**, 246–248.

Lindon, C. and Pines, J. (2004) Ordered proteolysis in anaphase inactivates Plk1 to contribute to proper mitotic exit in human cells. *J Cell Biol* **164**, 233–241.

Listovsky, T., Oren, Y.S., Yudkovsky, Y., *et al.* (2004) Mammalian Cdh1/Fzr mediates its own degradation. *EMBO J* **23**, 1619–1626.

Liu, E., Li, X., Yan, F., Zhao, Q. and Wu, X. (2004) Cyclin-dependent kinases phosphorylate human Cdt1 and induce its degradation. *J Biol Chem* **279**, 17283–17288.

Liu, J., Furukawa, M., Matsumoto, T. and Xiong, Y. (2002) NEDD8 modification of CUL1 dissociates $p120^{CAND1}$, an inhibitor of CUL1-SKP1 binding and SCF ligases. *Mol Cell* **10**, 1511–1518.

Loda, M., Cukor, B., Tam, S.W., *et al.* (1997) Increased proteasome-dependent degradation of the cyclin-dependent kinase inhibitor p27 in aggressive colorectal carcinomas. *Nat Med* **3**, 231–234.

Luca, F.C., Shibuya, E.K., Dohrmann, C.E. and Ruderman, J.V. (1991) Both cyclin A delta 60 and B delta 97 are stable and arrest cells in M-phase, but only cyclin B delta 97 turns on cyclin destruction. *EMBO J* **10**, 4311–4320.

Luo, X., Fang, G., Coldiron, M., *et al.* (2000) Structure of the Mad2 spindle assembly checkpoint protein and its interaction with Cdc20. *Nat Struct Biol* **7**, 224–249.

Luo, X., Tang, Z., Rizo, J. and Yu, H. (2002) The Mad2 spindle checkpoint protein undergoes similar major conformational changes upon binding to either Mad1 or Cdc20. *Mol Cell* **9**, 59–71.

Luo, X., Tang, Z., Xia, G., *et al.* (2004) The Mad2 spindle checkpoint protein has two distinct natively folded states. *Nat Struct Mol Biol* **11**, 338–345.

Machida, Y.J., Hamlin, J.L. and Dutta, A. (2005) Right place, right time, and only once: replication initiation in metazoans. *Cell* **123**, 13–24.

Malek, N.P., Sundberg, H., McGrew, S., Nakayama, K., Kyriakides, T.R. and Roberts, J.M. (2001) A mouse knock-in model exposes sequential proteolytic pathways that regulate p27Kip1 in G1 and S phase. *Nature* **413**, 323–327.

Marti, A., Wirbelauer, C., Scheffner, M. and Krek, W. (1999) Interaction between ubiquitin-protein ligase SCFSKP2 and E2F-1 underlies the regulation of E2F-1 degradation. *Nat Cell Biol* **1**, 14–19.

McGarry, T.J. and Kirschner, M.W. (1998) Geminin, an inhibitor of DNA replication, is degraded during mitosis. *Cell* **93**, 1043–1053.

Mews, M., Sek, F.J., Moore, R., Gunning, B.E.S. and John, P.C.L. (1997) Mitotic cyclin distribution during maize cell division: implications for the sequence diversity and function of cyclins in plants. *Protoplasma* **200**, 128–145.

Montagnoli, A., Fiore, F., Eytan, E., *et al.* (1999) Ubiquitination of p27 is regulated by Cdk-dependent phosphorylation and trimeric complex formation. *Genes Dev* **13**, 1181–1189.

Murray, A.W. and Kirschner, M.W. (1989) Cyclin synthesis drives the early embryonic cell cycle. *Nature* **339**, 275–280.

Murray, A.W., Solomon, M.J. and Kirschner, M.W. (1989) The role of cyclin synthesis and degradation in the control of maturation promoting factor activity. *Nature* **339**, 280–286.

Nakayama, K., Nagahama, H., Minamishima, Y.A., *et al.* (2000) Targeted disruption of Skp2 results in accumulation of cyclin E and p27(Kip1), polyploidy and centrosome overduplication. *EMBO J* **19**, 2069–2081.

Nakayama, K., Nagahama, H., Minamishima, Y.A., *et al.* (2004) Skp2-mediated degradation of p27 regulates progression into mitosis. *Dev Cell* **6**, 661–672.

Nakayama, K.I. and Nakayama, K. (2005) Regulation of the cell cycle by SCF-type ubiquitin ligases. *Semin Cell Dev Biol* **16**, 323–333.

Nash, P., Tang, X., Orlicky, S., *et al.* (2001) Multisite phosphorylation of a CDK inhibitor sets a threshold for the onset of DNA replication. *Nature* **414**, 514–521.

Nasmyth, K. (2001) Disseminating the genome: joining, resolving, and separating sister chromatids during mitosis and meiosis. *Annu Rev Genet* **35**, 673–745.

Nasmyth, K. (2005) How do so few control so many? *Cell* **120**, 739–746.

Nishitani, H., Lygerou, Z. and Nishimoto, T. (2004) Proteolysis of DNA replication licensing factor Cdt1 in S-phase is performed independently of geminin through its N-terminal region. *J Biol Chem* **279**, 30807–30816.

Ohi, M.D., Vander Kooi, C.W., Rosenberg, J.A., Chazin, W.J. and Gould, K.L. (2003) Structural insights into the U-box, a domain associated with multi-ubiquitination. *Nat Struct Biol* **10**, 250–255.

Ohta, T. and Xiong, Y. (2001) Phosphorylation- and Skp1-independent *in vitro* ubiquitination of E2F1 by multiple ROC-cullin ligases. *Cancer Res* **61**, 1347–1353.

Pagano, M. (2004) Control of DNA synthesis and mitosis by the Skp2-p27-Cdk1/2 axis. *Mol Cell* **14**, 414–416.

Park, J.A., Ahn, J.W., Kim, Y.K., *et al.* (2005) Retinoblastoma protein regulates cell proliferation, differentiation and endoreduplication in plants. *Plant J* **42**, 153–163.

Passmore, L.A., McCormack, E.A., Au, S.W., *et al.* (2003) Doc1 mediates the activity of the anaphase-promoting complex by contributing to substrate recognition. *EMBO J* **22**, 786–796.

Pellman, D. and Christman, M.F. (2001) Separase anxiety: dissolving the sister bond and more. *Nat Cell Biol* **3**, E207–E209.

Peters, J.M. (2002) The anaphase-promoting complex: proteolysis in mitosis and beyond. *Mol Cell* **9**, 931–943.

Petersen, B.O., Wagener, C., Marinoni, F., *et al.* (2000) Cell cycle- and cell growth-regulated proteolysis of mammalian CDC6 is dependent on APC-CDH1. *Genes Dev* **14**, 2330–2343.

Petroski, M.D. and Deshaies, R.J. (2005) Function and regulation of cullin-RING ubiquitin ligases. *Nat Rev Mol Cell Biol* **6**, 9–20.

Pfleger, C.M. and Kirschner, M.W. (2000) The KEN box: an APC recognition signal distinct from the D box targeted by Cdh1. *Genes Dev* **14**, 655–665.

Pfleger, C.M., Lee, E. and Kirschner, M.W. (2001) Substrate recognition by the Cdc20 and Cdh1 components of the anaphase-promoting complex. *Genes Dev* **15**, 2396–2407.

Pickart, C.M. (2001) Mechanisms underlying ubiquitination. *Annu Rev Biochem* **70**, 503–533.

Pickart, C.M. and Fushman, D. (2004) Polyubiquitin chains: polymeric protein signals. *Curr Opin Chem Biol* **8**, 610–616.

Pines, J. (2006) Mitosis: a matter of getting rid of the right protein at the right time. *Trends Cell Biol* **16**, 55–63.

Pintard, L., Willems, A. and Peter, M. (2004) Cullin-based ubiquitin ligases: Cul3-BTB complexes join the family. *EMBO J* **23**, 1681–1687.

Renaudin, J.-P., Savouré, A., Philippe, H., Van Montagu, M., Inzé, D. and Rouzé, P. (1998) Characterization and classification of plant cyclin sequences related to A- and B-type cyclins. In: *Plant Cell Division* (eds Francis, D., Dudits, D. and Inzé, D.). Portland Press, London, pp. 67–98.

Rimmington, G., Dalby, B. and Glover, D.M. (1994) Expression of N-terminally truncated cyclin B in the Drosophila larval brain leads to mitotic delay at late anaphase. *J Cell Sci* **107**, 2729–2738.

Roudier, F., Fedorova, E., Györgyey, J., *et al.* (2000) Cell cycle function of a *Medicago sativa* A2-type cyclin interacting with a PSTAIRE-type cyclin-dependent kinase and a retinoblastoma protein. *Plant J* **23**, 73–83.

Rudner, A.D. and Murray, A.W. (2000) Phosphorylation by Cdc28 activates the Cdc20-dependent activity of the anaphase-promoting complex. *J Cell Biol* **149**, 1377–1390.

Schnittger, A., Schöbinger, U., Stierhof, Y.D. and Hülskamp, M. (2002) Ectopic B-type cyclin expression induces mitotic cycles in endoreduplicating Arabidopsis trichomes. *Curr Biol* **12**, 415–420.

Schwab, M., Neutzner, M., Mocker, D. and Seufert, W. (2001) Yeast Hct1 recognizes the mitotic cyclin Clb2 and other substrates of the ubiquitin ligase APC. *EMBO J* **20**, 5165–5175.

Schwob, E., Bohm, T., Mendenhall, M.D. and Nasmyth, K. (1994) The B-type cyclin kinase inhibitor p40SIC1 controls the G1 to S transition in *S. cerevisiae*. *Cell* **79**, 233–244.

Sdek, P., Ying, H., Chang, D.L., *et al.* (2005) MDM2 promotes proteasome-dependent ubiquitin-independent degradation of retinoblastoma protein. *Mol Cell* **20**, 699–708.

Senga, T., Sivaprasad, U., Zhu, W., *et al.* (2006) PCNA is a co-factor for Cdt1 degradation by CUL4/DDB1 mediated N-terminal ubiquitination. *J Biol Chem* **281**, 6246–6252.

Seol, J.H., Feldman, R.M., Zachariae, W., *et al.* (1999) Cdc53/cullin and the essential Hrt1 RING-H2 subunit of SCF define a ubiquitin ligase module that activates the E2 enzyme Cdc34. *Genes Dev* **13**, 1614–1626.

Sheaff, R.J., Groudine, M., Gordon, M., Roberts, J.M. and Clurman, B.E. (1997) Cyclin E-CDK2 is a regulator of p27Kip1. *Genes Dev* **11**, 1464–1478.

Shen, W.H. (2002) The plant E2F-Rb pathway and epigenetic control. *Trends Plant Sci* **7**, 505–511.

Shen, W.H., Parmentier, Y., Hellmann, H., *et al.* (2002) Null mutation of AtCul1 causes arrest in early embryogenesis in Arabidopsis. *Mol Biol Cell* **13**, 1916–1928.

Sherr, C.J and Roberts, J.M. (1995) Inhibitors of mammalian G1 cyclin-dependent kinases. *Genes Dev* **9**, 1149–1163.

Shirayama, M., Zachariae, W., Ciosk, R. and Nasmyth, K. (1998) The Polo-like kinase Cdc5p and the WD-repeat protein Cdc20p/fizzy are regulators and substrates of the anaphase promoting complex in *Saccharomyces cerevisiae*. *EMBO J* **17**, 1336–1349.

Sigrist, S., Jacobs, H., Stratmann, R. and Lehner, C.F. (1995) Exit from mitosis is regulated by Drosophila fizzy and the sequential destruction of cyclins A, B and B3. *EMBO J* **14**, 4827–4838.

Sironi, L., Mapelli, M., Knapp, S., De Antoni, A., Jeang, K.T. and Musacchio, A. (2002) Crystal structure of the tetrameric Mad1-Mad2 core complex: implications of a 'safety belt' binding mechanism for the spindle checkpoint. *EMBO J* **21**, 2496–2506.

Smalle, J. and Vierstra, R.D. (2004) The ubiquitin 26S proteasome proteolytic pathway. *Annu Rev Plant Biol* **55**, 555–590.

Spruck, C., Strohmaier, H., Watson, M., *et al.* (2001) A CDK-independent function of mammalian Cks1: targeting of SCF(Skp2) to the CDK inhibitor p27Kip1. *Mol Cell* **7**, 639–650.

Stone, S.L., Hauksdottir, H., Troy, A., Herschleb, J., Kraft, E. and Callis, J. (2005) Functional analysis of the RING-type ubiquitin ligase family of Arabidopsis. *Plant Physiol* **137**, 13–30.

Sugimoto-Shirasu, K. and Roberts, K. (2003) 'Big it up': endoreduplication and cell-size control in plants. *Curr Opin Plant Biol* **6**, 544–553.

Surana, U., Amon, A., Dowzer, C., McGrew, J., Byers, B. and Nasmyth, K. (1993) Destruction of the CDC28/CLB mitotic kinase is not required for the metaphase to anaphase transition in budding yeast. *EMBO J* **12**, 1969–1978.

Sutterluty, H., Chatelain, E., Marti, A., *et al.* (1999) p45SKP2 promotes p27Kip1 degradation and induces S phase in quiescent cells. *Nat Cell Biol* **1**, 207–214.

Tada, S., Li, A., Maiorano, D., Mechali, M. and Blow, J.J. (2001) Repression of origin assembly in metaphase depends on inhibition of RLF-B/Cdt1 by geminin. *Nat Cell Biol* **3**, 107–113.

Takeda, D.Y., Parvin, J.D. and Dutta, A. (2005) Degradation of Cdt1 during S phase is Skp2-independent and is required for efficient progression of mammalian cells through S phase. *J Biol Chem* **280**, 23416–23423.

Tanaka, S. and Diffley, J.F. (2002) Interdependent nuclear accumulation of budding yeast Cdt1 and Mcm2–7 during G1 phase. *Nat Cell Biol* **4**, 198–207.

Tang, Z., Li, B., Bharadwaj, R., *et al.* (2001) APC2 Cullin protein and APC11 RING protein comprise the minimal ubiquitin ligase module of the anaphase-promoting complex. *Mol Biol Cell* **12**, 3839–3851.

Tarayre, S., Vinardell, J.M., Cebolla, A., Kondorosi, A. and Kondorosi, E. (2004) Two classes of the CDh1-type activators of the anaphase-promoting complex in plants: novel functional domains and distinct regulation. *Plant Cell* **16**, 422–434.

Tedesco, D., Lukas, J. and Reed, S.I. (2002) The pRb-related protein p130 is regulated by phosphorylation-dependent proteolysis via the protein-ubiquitin ligase SCF^{Skp2}. *Genes Dev* **16**, 2946–2957.

Thomann, A., Dieterle, M. and Genschik, P. (2005) Plant CULLIN-based E3s: phytohormones come first. *FEBS Lett* **579**, 3239–3245.

Tsvetkov, L.M., Yeh, K.H., Lee, S.J., Sun, H. and Zhang, H. (1999) p27^{Kip1} ubiquitination and degradation is regulated by the SCF^{Skp2} complex through phosphorylated Thr187 in p27. *Curr Biol* **9**, 661–664.

Uhlmann, F., Wernic, D., Poupart, M.A., Koonin, E.V. and Nasmyth, K. (2000) Cleavage of cohesin by the CD clan protease separin triggers anaphase in yeast. *Cell* **103**, 375–386.

Vandepoele, K., Raes, J., De Veylder, L., Rouzé, P., Rombauts, S. and Inzé, D. (2002) Genome-wide analysis of core cell cycle genes in Arabidopsis. *Plant Cell* **14**, 903–916.

Verkest, A., de O. Manes, C.-L., Vercruysse, S., *et al.* (2005a) The cyclin-dependent kinase inhibitor KRP2 controls the onset of the endoreduplication cycle during Arabidopsis leaf development through inhibition of mitotic CDKA;1 kinase complexes. *Plant Cell* **17**, 1723–1736.

Verkest, A., Weinl, C., Inzé, D., De Veylder, L. and Schnittger, A. (2005b) Switching the cell cycle. Kip-related proteins in plant cell cycle control. *Plant Physiol* **139**, 1099–1106.

Verma, R., Annan, R.S., Huddleston, M.J., Carr, S.A., Reynard, G. and Deshaies, R.J. (1997) Phosphorylation of Sic1p by G1 Cdk required for its degradation and entry into S phase. *Science* **278**, 455–460.

Vinardell, J.M., Fedorova, E., Cebolla, A., *et al.* (2003) Endoreduplication mediated by the anaphase-promoting complex activator CCS52A is required for symbiotic cell differentiation in *Medicago truncatula* nodules. *Plant Cell* **15**, 2093–2105.

Vodermaier, H.C. (2004) APC/C and SCF: controlling each other and the cell cycle. *Curr Biol* **14**, R787–R796.

Vodermaier, H.C., Gieffers, C., Maurer-Stroh, S., Eisenhaber, F. and Peters, J.M. (2003) TPR subunits of the anaphase-promoting complex mediate binding to the activator protein CDH1. *Curr Biol* **13**, 1459–1468.

Wang, H., Fowke, L.C. and Crosby, W.L. (1997) A plant cyclin-dependent kinase inhibitor gene. *Nature* **386**, 451–452.

Wei, W., Ayad, N.G., Wan, Y., Zhang, G.J., Kirschner, M.W. and Kaelin, W.G., Jr. (2004) Degradation of the SCF component Skp2 in cell-cycle phase G1 by the anaphase-promoting complex. *Nature* **428**, 194–198.

Weinberg, R.A. (1995) The retinoblastoma protein and cell cycle control. *Cell* **81**, 323–330.

Weingartner, M., Criqui, M.C., Meszaros, T., *et al.* (2004) Expression of a nondegradable cyclin B1 affects plant development and leads to endomitosis by inhibiting the formation of a phragmoplast. *Plant Cell* **16**, 643–657.

Weinl, C., Marquardt, S., Kuijt, S.J., *et al.* (2005) Novel functions of plant cyclin-dependent kinase inhibitors, ICK1/KRP1, can act non-cell-autonomously and inhibit entry into mitosis. *Plant Cell* **17**, 1704–1722.

Weinstein, J. (1997) Cell cycle-regulated expression, phosphorylation, and degradation of p55Cdc. A mammalian homolog of CDC20/Fizzy/slp1. *J Biol Chem* **272**, 28501–28511.

Wertz, I.E., O'Rourke, K.M., Zhang, Z., *et al.* (2004) Human De-etiolated-1 regulates c-Jun by assembling a CUL4A ubiquitin ligase. *Science* **303**, 1371–1374.

Whitfield, W.G., Gonzalez, C., Maldonado-Codina, G. and Glover, D.M. (1990) The A- and B-type cyclins of Drosophila are accumulated and destroyed in temporally distinct events that define separable phases of the G2–M transition. *EMBO J* **9**, 2563–2572.

Willemsen, V., Wolkenfelt, H., de Vrieze, G., Weisbeek, P. and Scheres, B. (1998) The HOBBIT gene is required for formation of the root meristem in the Arabidopsis embryo. *Development* **125**, 521–531.

Wohlschlegel, J.A., Dwyer, B.T., Dhar, S.K., Cvetic, C., Walter, J.C. and Dutta, A. (2000) Inhibition of eukaryotic DNA replication by geminin binding to Cdt1. *Science* **290**, 2309–2312.

Xu, L., Liu, F., Lechner, E., *et al.* (2002) The SCF(COI1) ubiquitin-ligase complexes are required for jasmonate response in Arabidopsis. *Plant Cell* **14**, 1919–1935.

Yamanaka, A., Hatakeyama, S., Kominami, K., Kitagawa, M., Matsumoto, M. and Nakayama, K. (2000) Cell cycle-dependent expression of mammalian E2-C regulated by the anaphase-promoting complex/cyclosome. *Mol Biol Cell* **11**, 2821–2831.

Yamano, H., Gannon, J. and Hunt, T. (1996) The role of proteolysis in cell cycle progression in *Schizosaccharomyces pombe*. *EMBO J* **15**, 5268–5279.

Yamano, H., Tsurumi, C., Gannon, J. and Hunt, T. (1998) The role of the destruction box and its neighbouring lysine residues in cyclin B for anaphase ubiquitin-dependent proteolysis in fission yeast: defining the D-box receptor. *EMBO J* **17**, 5670–5678.

Yamashita, A., Sato, M., Fujita, A., Yamamoto, M. and Toda, T. (2005) The roles of fission yeast ase1 in mitotic cell division, meiotic nuclear oscillation, and cytokinesis checkpoint signaling. *Mol Biol Cell* **16**, 1378–1395.

Yanagida, M. (2000) Cell cycle mechanisms of sister chromatid separation; roles of Cut1/separin and Cut2/securin. *Genes Cells* **5**, 1–8.

Yoon, H.J., Feoktistova, A., Wolfe, B.A., Jennings, J.L., Link, A.J. and Gould, K.L. (2002) Proteomics analysis identifies new components of the fission and budding yeast anaphase-promoting complexes. *Curr Biol* **12**, 2048–2054.

Yu, H. (2002) Regulation of APC-Cdc20 by the spindle checkpoint. *Curr Opin Cell Biol* **14**, 706–714.

Yu, H.G., Muszynski, M.G. and Kelly Dawe, R.I. (1999) The maize homologue of the cell cycle checkpoint protein MAD2 reveals kinetochore substructure and contrasting mitotic and meiotic localization patterns. *J Cell Biol* **145**, 425–345.

Zachariae, W. and Nasmyth, K. (1999) Whose end is destruction: cell division and the anaphase-promoting complex. *Genes Dev* **13**, 2039–2058.

Zachariae, W., Shevchenko, A., Andrews, P.D., *et al.* (1998) Mass spectrometric analysis of the anaphase-promoting complex from yeast: identification of a subunit related to cullins. *Science* **279**, 1216–1619.

Zheng, J., Yang, X., Harrell, J.M., *et al.* (2002) CAND1 binds to unneddylated CUL1 and regulates the formation of SCF ubiquitin E3 ligase complex. *Mol Cell* **10**, 1519–1526.

Zhong, W., Feng, H., Santiago, F.E. and Kipreos, E.T. (2003) CUL-4 ubiquitin ligase maintains genome stability by restraining DNA-replication licensing. *Nature* **423**, 885–889.

Zhou, Y., Li, G., Brandizzi, F., Fowke, L.C. and Wang, H. (2003) The plant cyclin-dependent kinase inhibitor ICK1 has distinct functional domains for *in vivo* kinase inhibition, protein instability and nuclear localization. *Plant J* **35**, 476–489.

Zur, A. and Brandeis, M. (2001) Securin degradation is mediated by fzy and fzr, and is required for complete chromatid separation but not for cytokinesis. *EMBO J* **20**, 792–801.

Zur, A. and Brandeis, M. (2002) Timing of APC/C substrate degradation is determined by fzy/fzr specificity of destruction boxes. *EMBO J* **21**, 4500–4510.

5 CDK phosphorylation

Akie Shimotohno and Masaaki Umeda

5.1 Introduction

In eukaryotes, the regulation of cell division is one of the important mechanisms that underlie several essential processes, including cell fate determination, differentiation, organ development, growth, cell death and carcinogenesis. Recently, plant cell cycle has become the focus of attention because almost all aspects of post-embryonic development in plants are primarily controlled by cell division in the meristems. Meristems are tissues including stem cells that are involved in both self-maintenance and cell production for growth. Both these processes are localized to distinct meristem domains that have different proliferation rates. Contrary to somatic cell division in animals, meristematic cell division activity continuously contributes to the formation of the plant body after embryogenesis. Another interesting reason to study the plant cell cycle is the understanding of the crosstalk between cell division and differentiation that in turn governs totipotency in plant cells.

Studies on yeasts, *Drosophila* and mammals have demonstrated that signalling pathways that regulate cell cycle progression eventually converge on the control of cyclin-dependent protein kinase (CDK) activity. The kinase activity of CDKs depends on its binding to cyclins. Similar to vertebrates, plants express several types of CDKs and cyclins (Mironov *et al.*, 1999); thus, different sets of CDK/cyclin pairs might regulate plant cell division at each stage of the cell cycle. Recently, Vandepoele *et al.* (2002) identified 61 core cell cycle genes in the *Arabidopsis* genome, and additional cyclin-like genes were updated later (Wang *et al.*, 2004). Based on the primary structure, plant CDKs have been classified into six types: CDKA to CDKF (Joubès *et al.*, 2000; Vandepoele *et al.*, 2002). Among these, CDKA and CDKB are directly involved in cell cycle progression. CDKA contains a conserved PSTAIRE motif, which is an important domain for cyclin binding. In addition, it apparently plays a role in both G1-to-S and G2-to-M phase progression. The expression of *CDKB*, which contains altered PSTAIRE sequences, is restricted from the late S to M phases. In contrast, CDKC and CDKE are assumed to play critical roles in transcriptional control. CDKC is closely related to vertebrate CDK9 and forms a complex with cyclin T to function as a positive regulator of transcription (Barrôco *et al.*, 2003; Fülöp *et al.*, 2005). In humans, cyclin T–CDK9 was identified as the catalytic subunit of positive transcription elongation factor b (P-TEFb) (Marshall and Price, 1995; Zhou *et al.*, 1998). CDKE is a homologue of mammalian CDK8, which interacts with cyclin C and exerts a negative effect on transcription as a component of the RNA polymerase II holoenzyme (Leclerc *et al.*, 1996; Maldonado *et al.*, 1996;

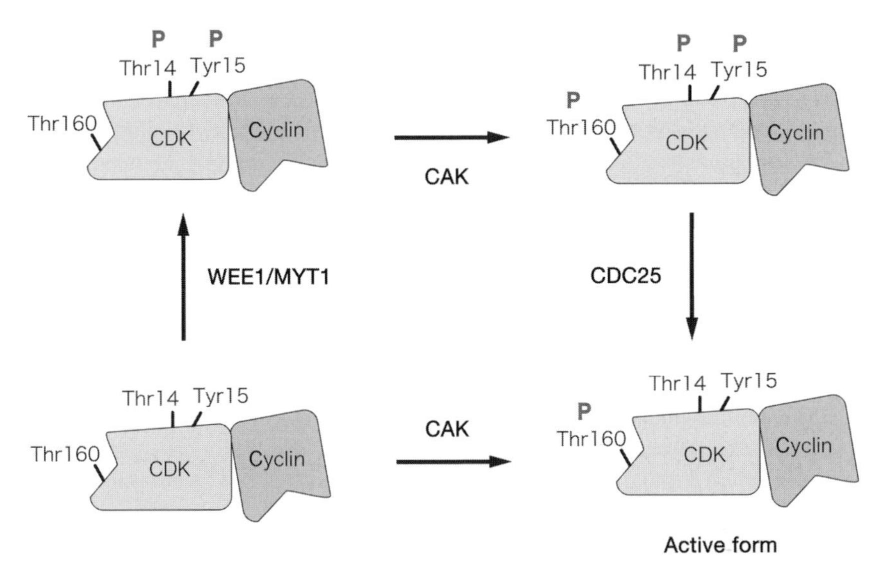

Figure 5.1 CDK regulation by activating and inhibitory phosphorylations. Phosphorylation of human CDK2 is shown as an example. WEE1 and MYT1 phosphorylate the threonine and tyrosine residues in the ATP-binding site and inhibit the CDK activity. These phosphorylations are counteracted by the dual-specificity phosphatase CDC25. CAK phosphorylates the threonine residue within the T-loop and activates the CDK activity.

Rickert *et al.*, 1996). Recently, Wang and Chen (2004) observed that Arath;CDKE;1 is encoded by *HUA ENHANCER3* (*HEN3*), whose mutation affected cell expansion in leaves and cell fate specification in floral meristems. CDKD and CDKF are CDK-activating kinases (CAKs) as described below in detail. A distinct kinase named CDKG has been recently defined based on the conserved PLTSLRE-motif and its homology to the human galactosyltransferase-associated protein kinase p58/GTA – a member of a p34 (Cdc2)-related kinase subfamily (Menges *et al.*, 2005). In mammals, it is known that p58/GTA associates specifically with cyclin D3 and that it coordinates the G2-to-M phase progression (Zhang *et al.*, 2002).

In addition to the binding of cyclin subunits, CDKs are regulated by protein phosphorylations (Figure 5.1). There are two to three inhibitory phosphorylation sites and one activating phosphorylation site. The inhibitory phosphorylations (Thr14 and Tyr15 in human CDC2 and CDK2) are carried out by the WEE1 and MYT1 kinases, and they are removed by the CDC25 phosphatase family. Activating phosphorylation occurs within the so-called T-loop on a conserved threonine residue (Thr160 and Thr161 in human CDK2 and CDC2, respectively). This event is catalyzed by CAKs. Full activation of CDKs requires binding of a cyclin, removal of inhibitory phosphorylations and the presence of an activating phosphorylation. In this chapter, we focus on recent advances in the field of CDK phosphorylation in plants and discuss the implications of these results based on comparisons with other organisms.

5.2 Overview of CAKs in yeasts and vertebrates

A CDK-activating kinase (CAK) phosphorylates the threonine residue on the T-loop region. The T-loop blocks the entry of the substrate into the catalytic cleft when the threonine residue is unphosphorylated (for review, see Nigg, 1996; Draetta, 1997). The phosphate on the threonine is inserted into a cationic pocket beneath the T-loop, and it acts as the central node for a network of hydrogen bonds spreading outwards to stabilize neighbouring interactions in both CDK and cyclin. Thus, the principal function of T-loop phosphorylation is probably stabilizing the protein–substrate interaction. This phosphoregulatory system is particularly important in the activation of a subset of CDK/cyclin complexes such as CDC2–cyclin B, where cyclin binding alone barely affects the activity.

The catalytic subunit of CAK belongs to the CDK family and has a preference for the phospho-acceptor site of TXXVVTL (in which the first threonine is phosphorylated and where X indicates lack of conservation); it is termed 'CDK7/p40$^{\mathrm{MO15}}$' in vertebrates, and its regulatory subunit is named cyclin H (Fisher and Morgan, 1994; Labbé *et al.*, 1994; Mäkelä *et al.*, 1994) (Figure 5.2). When alone, CDK7 has a low CAK activity; however, in the presence of cyclin H, CDK7 activity is significantly stimulated. In addition to cyclin H, MAT1 (*m*énage *à* *t*rois *1*) – a RING-finger protein – also interacts with CDK7 to assemble the CDK7–cyclin H complex (Devault *et al.*, 1995; Tassan *et al.*, 1995) (Figure 5.2). Unlike other CDKs, CAK exhibits another function in controlling basal transcription as well as CDK activation. That is, it phosphorylates the carboxy-terminal domain (CTD) of the largest subunit of

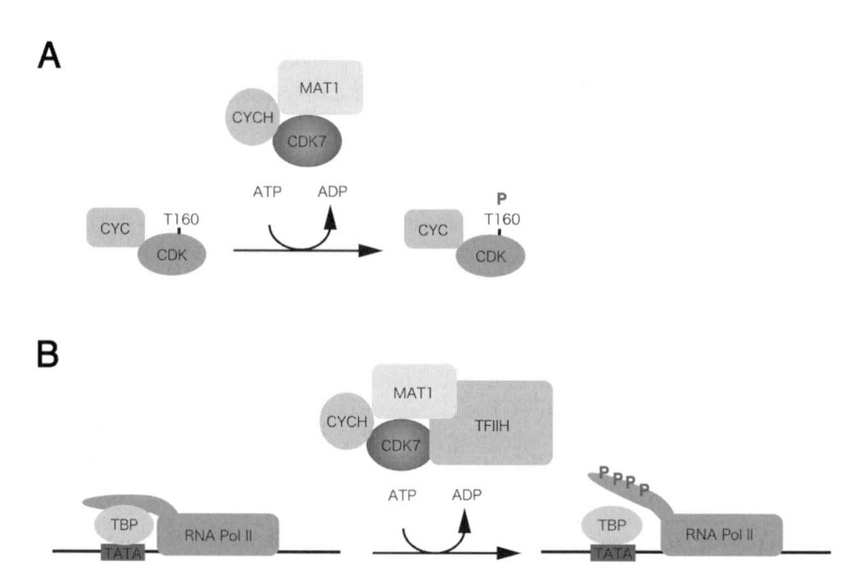

Figure 5.2 CAK controls cell cycle progression and basal transcription in mammals. (A) The trimeric CAK complex consisting of CDK7, cyclin H (CYCH) and MAT1 phosphorylates and activates CDK/cyclin complexes. (B) The CAK complex identified in TFIIH complexes phosphorylates the CTD of the largest subunit of RNA polymerase II (RNA Pol II) and controls basal transcription. (TBP, TATA-box-binding protein).

RNA polymerase II (Figure 5.2). In mammalian cells, the CTD contains up to 52 repeats of the consensus heptapeptide YSPTSPS. Null mutation in CDK7, either *in vivo* or *in vitro*, led to a drastic inhibition of CTD kinase activity (Mäkelä *et al.*, 1995; Tirode *et al.*, 1999). It is known that the amino (N)-terminal RING-finger domain of MAT1 is a requisite for CTD phosphorylation (Busso *et al.*, 2000), but not for the CAK activity *in vitro* (Tassan *et al.*, 1995). Rather, its carboxy (C)-terminal region, which interacts with the CDK7–cyclin H complex, is responsible for the stimulation of the CAK activity (Tassan *et al.*, 1995).

The heterotrimeric CAK complex also exists as an intrinsic component of the general transcription factor TFIIH, which is involved in the initiation and elongation of transcription (Roy *et al.*, 1994; Serizawa *et al.*, 1995; Shiekhattar *et al.*, 1995), and DNA repair (Frit *et al.*, 1999) (Figure 5.2). Mammalian TFIIH is a multiprotein complex of nine subunits consisting of two major subcomplexes: the core TFIIH (XPB, p34, p44, p52 and p62) and the CAK trimeric complex (Roy *et al.*, 1994; Adamczewski *et al.*, 1996). The remaining subunit, namely XPD helicase, is associated with either the core or the kinase complex and is assumed to anchor the CAK to the core TFIIH (Drapkin *et al.*, 1996; Reardon *et al.*, 1996; Coin *et al.*, 1999). The trimeric CAK complex is mainly involved in cell cycle regulation. In contrast, when associated with the core TFIIH, it is involved in transcription (Rossignol *et al.*, 1997; Yankulov *et al.*, 1997).

In the fission yeast *Schizosaccharomyces pombe*, CAK consists of Mcs6/Crk1/Mop1, Mcs2 and Pmh1, which are closely related to CDK7, cyclin H and MAT1, respectively, and it phosphorylates Cdc2 (Damagnez *et al.*, 1995). The Mcs6–Mcs2–Phm1 trimeric complex is a part of TFIIH and phosphorylates the CTD of RNA polymerase II (Buck *et al.*, 1995; Damagnez *et al.*, 1995; Lee *et al.*, 1999). Similarly, in the budding yeast *Saccharomyces cerevisiae*, Kin28p is the closest relative of CDK7; it is associated with the cyclin H homologue Ccl1p and the MAT1 homologue Rig2p/Tfb3p for the phosphorylation of the CTD as components of TFIIH (Feaver *et al.*, 1994; Cismowski *et al.*, 1995; Svejstrup *et al.*, 1996a; Faye *et al.*, 1997; Feaver *et al.*, 1997). However, this trimeric complex does not exhibit Cdc28p kinase activity. Another 44-kDa monomeric kinase, designated Cak1p/Civ1p, is involved in the activation of Cdc28p through T-loop phosphorylation *in vivo* (Espinoza *et al.*, 1996; Kaldis *et al.*, 1996; Thuret *et al.*, 1996). It shows a very low sequence similarity to other CDKs and does not possess the CTD kinase activity, indicating that CDK and CTD phosphorylations are controlled by distinct kinases in the budding yeast (Cismowski *et al.*, 1995). There are indications that Cak1p is also required for the activation of Kin28p, suggesting that Cak1p influences transcription as well as cell division (Kimmelman *et al.*, 1999).

5.3 Vertebrate-type CAK in plants

5.3.1 CDKD, cyclin H and MAT1

The threonine residues within the T-loop are also conserved in plant CDKs (except CDKC). Joubès *et al.* (2000) classified plant CDK7 homologues into the CDKD group. The first plant CAK orthologue was identified in rice; it was a Cdc2-related protein kinase named R2 (Hata, 1991), thereafter renamed Orysa;CDKD;1. The amino acid sequence is closely related to those of animal and yeast CDK7 homologues with 50–55% identity. It contains an extended C-terminal region of 92 amino acids, which is not present in animal and yeast CAKs. Fabian-Marwedel *et al.* (2002) showed that in a tobacco protoplast system, this extended region contains a nuclear localization signal (NLS). When *Orysa;CDKD;1* was overexpressed in a *cak1/civ1*-deficient mutant of budding yeast, it partially suppressed the temperature sensitivity. In addition, immunoprecipitates of rice proteins with the anti-Orysa;CDKD;1 antibody phosphorylated the threonine residue (Thr161) within the T-loop of rice CDKA;1 (Orysa;CDKA;1) and the *Arabidopsis* CTD (Yamaguchi *et al.*, 1998). These data suggest that Orysa;CDKD;1 is a functional homologue of vertebrate CAKs.

There are three *CDKD* genes in the *Arabidopsis* genome: *Arath;CDKD;1*, *Arath;CDKD;2* and *Arath;CDKD;3*, which were originally named *CAK3At*, *CAK4At* and *CAK2At*, respectively (Umeda, 2002; Shimotohno *et al.*, 2003). With the exception of Arath;CDKD;2, both Arath;CDKD;1 and Arath;CDKD;3 have the C-terminal extension along with a significant sequence similarity to that of Orysa;CDKD;1. The green fluorescent protein (GFP) fused to Arath;CDKDs showed that Arath;CDKD;1 and Arath;CDKD;3 displayed almost exclusive nuclear localization, whereas Arath;CDKD;2 localized in the cytoplasm and nuclei in *Arabidopsis* protoplasts and onion epidermal cells (Shimotohno *et al.*, 2004). This supports the notion that the C-terminal region contains the NLS, as described earlier. Enzyme-activity analysis demonstrated that Arath;CDKD;2 and Arath;CDKD;3 phosphorylated both human CDK2 and *Arabidopsis* CTD, whereas Arath;CDKD;1 lacks these kinase activities *in vitro* (Shimotohno *et al.*, 2003, 2004). Therefore, Arath;CDKD;1 may be an inactive CAK variant, or it may have a different substrate specificity. Interestingly, Arath;CDKD;2 and Arath;CDKD;3 differed in their preference for substrates. CDK2 kinase activity of Arath;CDKD;3 was higher than that of Arath;CDKD;2, whereas Arath;CDKD;2 had a higher CTD kinase activity than Arath;CDKD;3 (Shimotohno *et al.*, 2004) (Figure 5.3).

In higher plants, cyclin H homologues have been isolated from poplar, rice and *Arabidopsis* (named *Poptr;CYCH;1*, *Orysa;CYCH;1* and *Arath;CYCH;1*, respectively) (Yamaguchi *et al.*, 2000; Shimotohno *et al.*, 2004). In the cyclin box region, plant cyclin H shows approximately 60% similarity to human cyclin H. Although all *Arabidopsis* CDKDs interact with Arath;CYCH;1 in the yeast two-hybrid system, Arath;CDKD;1 showed a significantly lower interaction when compared with Arath;CDKD;2 and Arath;CDKD;3. The kinase activities of Arath;CDKD;2 and Arath;CDKD;3 on CDKs and CTD were markedly elevated on binding with cyclin H in insect cells (Shimotohno *et al.*, 2004). These results suggest that plant cyclin H is a regulatory subunit of CAK, which positively controls the CDK and CTD kinase activities of CDKD, similar to that in vertebrates and fission yeasts. Note that Orysa;CDKD;1 binds not only to cyclin H but also to cyclin C in the *in vitro* pull-down assay (Yamaguchi *et al.*, 2000), whereas human cyclin C interacts with CDK8 but not with CDK7 in the yeast two-hybrid system (Mäkelä *et al.*, 1994).

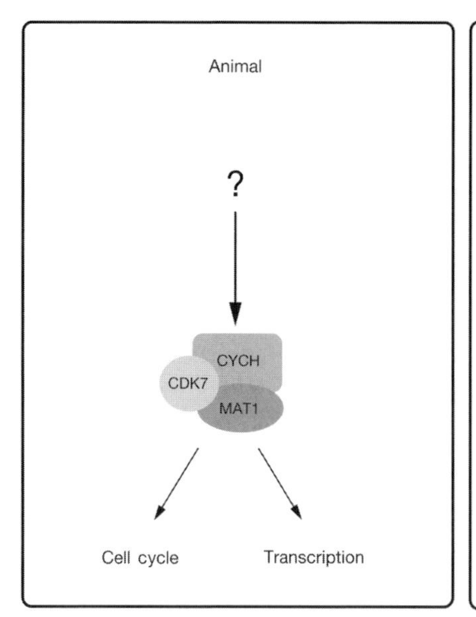

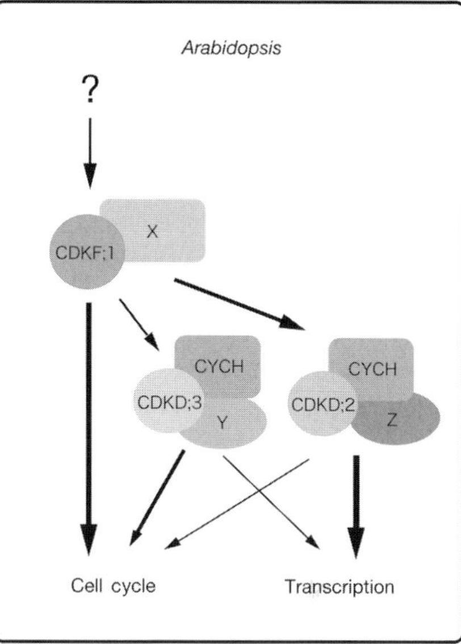

Figure 5.3 Comparison between the CAK-mediated pathways of animals and *Arabidopsis*. In animals, a functional homologue of CAK-activating kinase has not been identified, while Arath;CDKF;1 is responsible for phosphorylation and activation of *Arabidopsis* CAKs, namely Arath;CDKD;2 and Arath;CDKD;3. Arath;CDKF;1 forms a complex with unknown subunit(s) (X) other than cyclin H. Arath;CDKD;2 and Arath;CDKD;3 interact with cyclin H (CYCH) and other subunits (Y or Z) to form different protein complexes.

MAT1 homologues in vertebrates and yeasts have two motifs: an N-terminal RING-finger domain that plays a crucial role in basal transcription and the CTD phosphorylation process and a central coiled-coil domain that is associated with XPD and XPB helicases, components of TFIIH complexes (Busso *et al.*, 2000). The carboxyl terminus of MAT1 is also important for the association and activity of the vertebrate-type CAK complex (Busso *et al.*, 2000). *Arabidopsis* and rice genomes encode putative MAT1 homologues (At4g30820 in *Arabidopsis*). However, their amino acid sequences are conserved only in the coiled-coil domain, and they lack the RING-finger motif. To determine whether these genes encode for functional MAT1 homologues, further biochemical characterization is essential. CDKDs form multiple CAK complexes with distinct enzyme activities in plant cells (Yamaguchi *et al.*, 1998; Shimotohno *et al.*, 2004). However, the identity of the complex that contains the MAT1 homologues remains unknown.

5.3.2 CDKD protein complexes

In cultured *Arabidopsis* cells, Arath;CDKD;3 forms two distinct complexes with molecular masses >700 and 130 kD, respectively. The larger complex phosphorylates the CTD but not human CDK2, whereas the smaller complex showed a higher

CDK2 kinase activity but low CTD kinase activity. Such biochemical features are similar to those of CDK7-like kinases from metazoa, which form a TFIIH complex >700 kD and a 180-kD complex consisting of the three CAK subunits (Devault *et al.*, 1995; Schultz *et al.*, 2000). In contrast, a majority of the Arath;CDKD;2 kinases form a complex with a molecular mass of approximately 200 kD; this complex phosphorylates the CTD substrate. A minor Arath;CDKD;2 complex >700 kD also displayed CTD kinase activity, while both complexes showed only trace levels of CDK2 kinase activities (Shimotohno *et al.*, 2004).

The anti-Arath;CYCH;1 antibody recognized multiple Arath;CYCH;1 variants in suspension cells. However, immunoprecipitation studies showed that a 37-kD protein was mainly associated with Arath;CDKD;2 and Arath;CDKD;3, and that Arath;CYCH;1 formed a stable complex with Arath;CDKD;2 in suspension cells (Shimotohno *et al.*, 2006). Recently, several studies reported on the post-translational modifications of cyclin H in mammalian cells. The cyclin C–CDK8 complex phosphorylates the N-terminal and C-terminal α-helical domains of cyclin H and represses the CTD kinase activity of CAK, downregulating basal transcription (Akoulitchev *et al.*, 2000). The second kinase CK2 phosphorylates cyclin H at position 315, which is necessary to obtain complete CAK activity (Schneider *et al.*, 2002) and also targets the C-terminus-encoding NLS to specifically interact with the nuclear import receptors known as importins (Krempler *et al.*, 2005). The limited data available on plant cyclin H do not facilitate the speculation of whether similar regulatory mechanisms by phosphorylation may also function in plants. Further studies will enable greater understanding of the involvement of cyclin H in CAK function.

5.3.3 CDKD in cell cycle regulation and transcriptional control

Transcripts of *Orysa;CDKD;1* and *Orysa;CYCH;1* were accumulated in the S phase in partially synchronized rice suspension cells; the CTD kinase activity of Orysa;CDKD;1 was elevated in the G1 and S phases. The pattern of induction was similar to that seen for the S-phase marker, namely histone H3 (Sauter, 1997; Yamaguchi *et al.*, 2000; Fabian-Marwedel *et al.*, 2002). In the intercalary meristem of deepwater rice internodes, the transcript level and kinase activity of Orysa;CDKD;1 were significantly increased by submergence or by gibberellin treatment for the acceleration of cell division. This upregulation showed a good correlation with an increase in the number of S-phase cells at early time points after submergence (Sauter *et al.*, 1995; Sauter, 1997). Therefore, Orysa;CDKD;1 may control S-phase entry and/or progression through DNA replication and activation of downstream CDKs. To support this notion, the rice suspension cells overexpressing *Orysa;CDKD;1* accelerated S-phase progression and increased the ratio of the G2-phase cells (Fabian-Marwedel *et al.*, 2002). In contrast, microarray analysis of *Arabidopsis* cell cycle regulators with suspension-cultured cells showed that *Arath;CDKD;2* and *Arath;CDKD;3* are constantly expressed throughout the cell cycle. Rather, *Arath;CDKD;3* was upregulated after sucrose starvation and resupply in suspension cells (Menges *et al.*, 2005). This indicates that Arath;CDKD;3 may be involved in the activation of CDK activity during cell cycle re-entry.

The bacterial virulence protein VirD2 plays an important role in the nuclear import and chromosomal integration of *Agrobacterium*-transferred DNA (T-DNA) in eukaryotic host cells. Recently, Bakó *et al.* (2003) observed that VirD2 interacted with and was phosphorylated by an alfalfa CDKD – Medsa;CDKD;1 (originally designated as CAK2Ms) – *in vitro* and *in vivo*. They also observed that VirD2 was tightly associated with TATA-box-binding proteins (TBPs) in the nuclei of alfalfa cells. TBPs are key regulators in transcription/repair systems due to their interaction with the basic transcription machinery, including RNA polymerase II. Similar to vertebrate-type CAKs, Medsa;CDKD;1 interacted with and phosphorylated CTD, which could recruit TBP (Bakó *et al.*, 2003). VirD2 interactions with TBP and CDKD suggest that T-DNA integration may be mediated by widely conserved nuclear factors in eukaryotes. Mammalian TFIIH is involved in nucleotide excision repair (NER) as well as in the initiation and elongation of transcription (Svejstrup *et al.*, 1996b). Therefore, it is likely that nuclear VirD2-associated factors provide a link between T-DNA integration and transcription-coupled repair by associating with the components of the TFIIH. It will be interesting to determine whether phosphorylation of VirD2 by CDKD regulates their interactions and affect nuclear import and integration of T-DNA in *Agrobacterium*-transformed cells.

5.4 Plant-specific CAK

5.4.1 Unique features of CDKF

Arath;CDKF;1 (originally designated as *CAK1At*) has been isolated as a suppressor of *cak1* mutation in budding yeasts (Umeda *et al.*, 1998). Its amino acid sequence is related to those of vertebrate-type CAKs, but the similarities are restricted to the conserved kinase domains. Using database searches, homologues of *Arath;CDKF;1* can be identified only in plant species, namely *Euphorbia* (AF230740), rice (AK120969) and soybean (AY439095) with identities of 49.8, 35.9 and 39.3%, respectively, but not in other kingdoms (Umeda *et al.*, 2005). A unique feature of Arath;CDKF;1 among the other CAKs is that it carries an unusual insertion of 111 amino acids (amino acid position 178–288) between its kinase active site and phosphoregulatory site corresponding to the T-loop in CDKs. To determine whether this unique stretch is required for Arath;CDKF;1 activity, Shimotohno *et al.* (2004) expressed several truncated versions of *Arath;CDKF;1* in the *cak1*^{ts} mutant budding yeast strain GF2351. Interestingly, the cells expressing the variant in which this unique region was deleted were observed to suppress the temperature sensitivity of the mutant. This is consistent with the result that the same variant, which was transiently expressed in *Arabidopsis* root protoplasts, did not show a significant reduction in the CDK kinase activity in contrast to wild-type Arath;CDKF;1 (Shimotohno and Umeda, unpublished result). These results suggest that this unusual stretch is not essential for Arath;CDKF;1 activity. However, all the four plant species possess

CDKF-related kinases carrying the unique insertion with a significant amino acid similarity. Thus, this region may be involved in the control of the interactions of CDKF with specific regulatory proteins or substrates acting in plant-specific signalling pathways.

Immunoprecipitates of an *Arabidopsis* protein extract with the anti-Arath;CDKF;1 antibody phosphorylated human CDK2 at the threonine residue within the T-loop and activated its histone H1 kinase activity. However, Arath;CDKF;1 did not phosphorylate the *Arabidopsis* CTD *in vitro* and was unable to interact with Arath;CYCH;1 (Umeda *et al.*, 1998; Shimotohno *et al.*, 2004) (Figure 5.3). CTD kinase activities of *Arabidopsis* were separated from the total protein extract into the flow-through fraction by DEAE Sepharose and precipitated by $p13^{suc1}$-agarose (Umeda *et al.*, 1998), whereas Arath;CDKF;1 was immunologically detected only in the DEAE-Sepharose-bound fraction and was not associated with $p13^{suc1}$. These results indicate that Arath;CDKF;1 has a CAK activity but is distinct from vertebrate-type CAKs, including plant CDKDs, which exhibit both CDK and CTD kinase activities. The Arath;CDKF;1 protein, on the other hand, shows an apparent molecular mass of 62 kD on SDS–PAGE analysis, and it occurs in a 130-kDa protein complex with a high CAK activity in suspension cells (Shimotohno *et al.*, 2004). As Arath;CYCH;1 could not interact with Arath;CDKF;1 in yeast and plant cells (Shimotohno *et al.*, 2004, in press), it is possible that Arath;CDKF;1 shows *in vivo* association with other regulatory protein(s) that might control its activity in response to external or internal stimuli.

Despite the low sequence similarity, Arath;CDKF;1 is functionally related to the budding yeast Cak1p. This is supported by recent biochemical studies on CAKs (Tsakraklides and Solomon, 2002). These studies showed that (1) Cak1p and Arath;CDKF;1 were insensitive to the protein kinase inhibitor 5′-fluorosulfonylbenzoyladenosine (FSBA), which covalently modifies the invariant lysine in protein kinases, including CDK2 and CDK7 (Solomon *et al.*, 1993; Enke *et al.*, 1999); (2) they exhibited a preference for cyclin-free CDK2 as the substrate; (3) they lack CTD kinase activities; and (4) they did not require a highly conserved lysine, which is located in the nucleotide-binding pocket of CDK family proteins. On the other hand, sequence alignment together with a comparison of kinetic parameters suggested that Arath;CDKF;1 is included in the same family as vertebrate CDK7 rather than the budding yeast Cak1p. This classification is supported by the following features: (1) Cak1p lacks the glycine-rich motif (GXGXXG), which stabilizes ATP in the binding pocket (Kaldis *et al.*, 1996), whereas CDK7 and Arath;CDKF;1 contain this motif; (2) The threonine residue within the T-loop is conserved in CDK7 and Arath;CDKF;1, but not in Cak1p. These discrepancies may reflect evolutional diversification of CAKs and the uniqueness of Arath;CDKF;1.

5.4.2 CAK-activating kinase activity of CDKF

Vertebrate CDK7 and fission yeast Mcs6 are also phosphorylated at serine and threonine residues within the T-loops, similar to other CDKs. In fission yeasts, Csk1 has been identified as the second kinase with CAK activity, and it was observed to phosphorylate Mcs6 on the T-loop activation site (S165) and activate the

Mcs6–Mcs2 complex *in vivo* (Hermand *et al.*, 1998, 2001; Lee *et al.*, 1999). Thus, Csk1 was defined as a CAK-activating kinase (CAKAK), i.e. an upstream activating kinase of Mcs6. Kimmelman *et al.* (1999) demonstrated that in budding yeasts, Cak1p phosphorylates the T-loop of Kin28p and thereby stimulates its CTD kinase activity. This suggests that despite their low sequence similarity, budding yeast Cak1p and fission yeast Csk1 perform similar functions by the phosphorylation of Kin28p and Mcs6, respectively. In fact, detailed phylogenetic analyses revealed that Cak1p and Csk1 were included in the same family with significant bootstrap support (Liu and Kipreos, 2002).

Arath;CDKF;1 is also closely related to the fission yeast Csk1 in terms of enzyme activity. When *Arabidopsis* CAK cDNAs were introduced into a fission yeast strain that carried a disrupted *csk1*Δ gene, only transformants expressing *Arath;CDKF;1*, but not *Arath;CDKD*, were able to suppress the temperature sensitivity. Moreover, an *in vitro* kinase assay demonstrated that Arath;CDKF;1 phosphorylated the conserved serine and threonine residues within the T-loops of Arath;CDKD;2 and Arath;CDKD;3, but not of Arath;CDKD;1. In *Arabidopsis* root protoplasts, the CTD kinase activity of Arath;CDKD;2 was elevated, depending on its T-loop phosphorylation by Arath;CDKF;1 (Shimotohno *et al.*, 2004). These results suggested that Arath;CDKF;1 is a CAKAK that modulates the activity of Arath;CDKD;2 and Arath;CDKD;3, thereby controlling CDK activities and basal transcription in *Arabidopsis* (Figure 5.3). Unlike Cak1p and Csk1, the plant CDKF kinases identified thus far have a conserved phosphoregulatory site within the T-loop (Umeda *et al.*, 2005). When Arath;CDKF;1 carrying a T290A mutation was transiently expressed in *Arabidopsis* root protoplasts, the mutant protein displayed very low activity, suggesting that T-loop phosphorylation may be essential for CDKF kinase activities in plant cells (Shimotohno *et al.*, in press).

Schaber *et al.* (2002) reported that the budding yeast Cak1p functions in the SMK1 pathway, which is involved in spore formation and further meiotic development process through Cdc28p-independent mechanisms. This result indicates that Cak1p is involved not only in the cell cycle or transcriptional regulation but also in the control of early meiotic development prior to or at pre-meiotic DNA synthesis. A close relationship between Arath;CDKF;1 and Cak1p suggests that Arath;CDKF;1 may also be involved in gamete formation, but this assumption requires verification by further studies on *Arabidopsis* mutants. Recently, it has been demonstrated that the fission yeast Csk1 activates CDK9/Pch1 complexes to coordinate the transcriptional elongation process (Pei *et al.*, 2006). Although the activation of CDK9–cyclin T by CAK has not been demonstrated in metazoans, T-loop mutation in human CDK9 abolished the kinase activity (Chen *et al.*, 2004). This suggests that the phosphoregulatory mechanisms of CDK9 may also exist in higher eukaryotes. It will be interesting to determine whether Arath;CDKF;1 is associated with transcriptional regulation through the phosphorylation of a CDK9 homologue CDKC in *Arabidopsis*.

Recently, the existence of a 'second class' CAK has been reported in mammalian cells and *Drosophila* embryos (Larochelle *et al.*, 1998; Nagahara *et al.*, 1999; Kaldis and Solomon, 2000; Leclerc *et al.*, 2000). Liu *et al.* (2004) identified a novel monomeric CAK p42 in human cells; it shows sequence homology to both

Cak1p and CDK7. Although the enzyme activity *in vitro* was rather weak, it was responsible for the CDK2-activating kinase activity *in vivo*. However, p42 did not exhibit any CDK7-activating kinase activity; thus, the CAKAK in vertebrates remains unidentified (Figure 5.3). Database searches suggested that p42-related kinases do not exist in plants. In fact, CDKF-related genes have been identified in several plant species but not in other kingdoms. This indicates that the phosphorylation cascade mediated by CDKD and CDKF may receive environmental and/or hormonal signals specific to plants in order to facilitate proper development of organs.

5.5 Manipulation of *in vivo* CDK activities by CAK

In animals, CAK is involved in the activation of almost all CDKs. However, in plants, the phosphorylation and activation of endogenous CDKs have not been reported thus far. It is possible that plant CAKs may recognize cyclin–CDK complexes but not CDK monomers, and an *in vitro* kinase assay using recombinant substrates of plant CDKs might not produce successful results. Nevertheless, we showed that Arath;CDKF;1 activates Arath;CDKA;1 in *Arabidopsis* root protoplasts (Shimotohno *et al.*, in press), suggesting that plant CAKs may also be involved in the activation of overall CDK activity in tissues. This indicates that CAK genes can be a tool for manipulating the CDK activity in plant cells.

In this context, Umeda *et al.* (2000) expressed the sense or antisense gene of *Arath;CDKF;1* in *Arabidopsis* in an inducible manner. In transgenic plants, the kinase activity of Arath;CDKA;1 was reduced, and root growth was inhibited. Co-expression of *Arath;CDKA;1* suppressed the phenotype of antisense plants but not sense plants (Umeda and Umeda-Hara, unpublished results), suggesting that, in sense plants, *Arath;CDKF;1* overexpression might cause hyperphosphorylation of unknown substrates that resulted in a reduction of the Arath:CDKA;1 activity. In the root meristem of transgenic plants, columellar and cortical initial cells were differentiated into daughter cells prior to the cessation of cell division (Umeda *et al.*, 2000). This indicates that the indeterminate state of the initial cells might be maintained by CDK activities, independent of cell division.

Another example is the overexpression of *Orysa;CDKD;1* in tobacco leaf explants (Yamaguchi *et al.*, 2003). In root-inducing media, root regeneration was markedly inhibited in transgenic leaf sections. Moreover, *Orysa;CDKD;1*-expressing explants produced calli in the presence of high concentration of auxins. This phenotype was enhanced by higher expression of *Orysa;CDKD;1* or co-expression of *Orysa;CYCH;1*. This result indicates that endogenous CDK activities were elevated by *Orysa;CDKD;1* overexpression, and root regeneration converted to disorganized cellular proliferation; this resulted in the production of calli. Similar observations have been described in transgenic plants overexpressing cyclin genes. Weingartner *et al.* (2003) produced transgenic tobacco plants that ectopically expressed an alfalfa cyclin B2 gene. Leaf disc assays showed that root regeneration from transgenic sections was blocked in media having a high auxin-to-cytokinin ratio. In plants, *Arath;CYCD3;1* overexpression in *Arabidopsis* inhibited the differentiation of leaves with small polygon-shaped pavement cells and retarded the formation of lignified secondary xylem elements (Dewitte *et al.*, 2003). These observations suggest that cell differentiation is guaranteed within a particular range of CDK activity. Therefore, molecular mechanisms suppressing CDK activity under a threshold level may play an essential role in differentiation programmes during organ development.

5.6 Inhibitory phosphorylation of yeast and vertebrate CDKs

Characterization of genetic interactions of *S. pombe* mutants that have defects in cell division cycle control helped in establishing epistatic relationships among the genes. These included the *wee1*, *cdc25* and *cdc2* mutants. Early genetic experiments revealed that *wee1* and *cdc25* represent opposing forces regulating *cdc2* because the *wee1-50* phenotype was suppressed when combined with *cdc25-22* (Foe, 1989). Additionally, overexpression of the wild-type *Cdc25* in a *wee1-50* background caused the cells to undergo lethal premature mitosis – a phenomenon known as mitotic catastrophe (Russell and Nurse, 1986). This convincing genetic data that linked *wee1*, *cdc25* and *cdc2* were corroborated using biochemical studies, leading to the identification of inhibitory phosphorylations (that occur on Thr14 and Tyr15 in human CDK2) mediated by the Wee1-like protein kinases and dephosphorylated by the Cdc25 phosphatase family (for reviews, see Morgan, 1997; Solomon and Kaldis, 1998). The budding yeast homologues of Wee1 and Cdc25 are Swe1p and Mih1p, respectively (Russell *et al.*, 1989; Booher *et al.*, 1993).

WEE1 phosphorylates the tyrosine residue in the conserved glycine-rich motif (residues 11–16 in CDK2) that forms a part of the ATP-binding site cleft (Hanks and Hunter, 1995) (Figure 5.1). With the exception of CDK7, all amino acid sequences from CDK1 to CDK8 contain a tyrosine residue equivalent to Tyr15 in CDK2. CDKs 1, 2, 4 and 6 are phosphorylated on this tyrosine *in vivo* (for a review, see Solomon and Kaldis, 1998). CDKs 1 and 2 are also phosphorylated on Thr14, and this phosphorylation is catalyzed by MYT1 – a membrane-associated WEE1 homologue (Mueller *et al.*, 1995) (Figure 5.1). Thr14 is not conserved in CDK4 and CDK6 (where it is an alanine), or in CDK7 (where it is replaced by a glutamine) (Booher *et al.*, 1997). In fission yeasts, there exists a Wee1-related kinase named Mik1 that shows functional redundancy with Wee1 (Igarashi *et al.*, 1991; Lundgren *et al.*, 1991).

The kinase activity of the WEE1 and MYT1 kinases is counteracted by the dual-specificity phosphatase CDC25; this enzyme dephosphorylates both the Thr14 and Tyr15 residues (Figure 5.1). Unlike other dual-specificity phosphatases, which show limited substrate preference in general, CDC25 dephosphorylates only CDKs within the active site. Humans have three CDC25 isoforms (CDC25A, CDC25B and CDC25C); among these, CDC25A regulates G1–S transition by dephosphorylating the cyclin E–CDK2 complex. CDC25B and CDC25C dephosphorylate the cyclin A–CDK2 and the cyclin B–CDK1 complexes, respectively. Both CDC25B and CDC25C are required for the G2–M transition (Nilsson and Hoffmann, 2000).

5.7 Inhibitory phosphorylation of plant CDKs

5.7.1 Plant WEE1 kinases

Plant cDNAs encoding WEE1 homologues have been isolated from maize (Sun *et al.*, 1999), *Arabidopsis* (Sorrell *et al.*, 2002) and tomato (Gonzalez *et al.*, 2004). Maize WEE1 (Zeama;WEE1) showed maximum identity with human WEE1 (50%) and a slightly lesser identity with Wee1 from *S. cerevisiae* (40.5%) and *S. pombe* (43.1%). The kinase domain exhibited 65% identity with that of *Arabidopsis* WEE1 (Arath;WEE1).

The functionality of the isolated homologues was tested in fission yeast cells. The results revealed that the overexpression of *Zeama;WEE1* or *Arath;WEE1* in *S. pombe* significantly inhibited cell division and instigated elongation of the cells, i.e. a 3.9-fold increase in the length of the *Arath;WEE1*-expressing cells as compared with the control cells (Sun *et al.*, 1999; Sorrell *et al.*, 2002). This phenotype is observed in cases when the *S. pombe* or human *WEE1* was overexpressed in these cells (Igarashi *et al.*, 1991). The opposite phenotype would be the *S. pombe* mutant with a small cell or a 'wee' phenotype. An *in vitro* kinase assay showed that a GST-fused Zeama;WEE1 that was produced in *Escherichia coli* inhibited histone H1 kinase activity of the p13^{suc1}-absorbed CDK (Sun *et al.*, 1999). These results indicated that plant WEE1 homologues have functional properties that are characteristic of yeast and animal WEE1 kinases. However, specific phosphorylation of the tyrosine residue within the ATP-binding site of CDKs has not been described in plants. *Arath;WEE1* could not complement a *wee1 mik1* double mutant of *S. pombe*, which lacks all Wee1 kinase activity (Sorrell *et al.*, 2002). This result is surprising, particularly considering the successful complementation of similar mutants with human and *Drosophila WEE1* genes (Igarashi *et al.*, 1991; Campbell *et al.*, 1995).

The expression of *Arath;WEE1* was highest in tissues that exhibited high frequencies of cell division, such as seedlings and flowers. However, it was virtually undetectable in the stem tissue or fully expanded leaves that essentially comprised non-cycling cells (Sorrell *et al.*, 2002). In addition, tomato *WEE1* (*Lyces;WEE1*) showed enhanced expression in the areas where cell divisions occur predominantly (Gonzalez *et al.*, 2004). These results indicate a tight association between *WEE1* expression and cell division. In tobacco BY2 cells, the *Nicta;WEE1* expression is cell cycle regulated with the maximum transcript accumulation in the late G1 and S phases (Gonzalez *et al.*, 2004). This is consistent with the data of Menges *et al.* (2003), in which *Arath;WEE1* was strongly expressed in the S phase in an *Arabidopsis* cell suspension.

During fruit development in tomato, *Lyces;WEE1* was expressed in the pericarp and in the jelly-like locular tissue concomitant with endoreduplication (Gonzalez *et al.*, 2004). In maize endosperm, *Zeama;WEE1* RNA accumulated when endosperm nuclei underwent maximum endoreduplication (Sun *et al.*, 1999), implying that WEE1 plays a role in endocycle as well as in the classical cell cycle. This assumption was supported by a previously established result in maize (Grafi and Larkins, 1995). The kinase activity bound to p13^{suc1} beads increased during the early stages of endosperm development, and it then declined as the mitotic activity was reduced and endoreduplication was initiated. The amount of $p34^{cdc2}$ remained unchanged as long as endoreduplication was detected. Biochemical analyses showed that an active inhibitor of the M-phase-promoting factor (MPF) was produced in endoreduplicated cells. In contrast, the amount and activity of S-phase-related kinases were observed to be increased. These results demonstrated that in maize endosperm, endoreduplication proceeds as a result of two events: the inactivation of the mitotic $p34^{cdc2}$/cyclin B kinase and induction of S-phase-related protein kinases. It is possible that Zeama;WEE1 is responsible for the reduction in the M-phase kinase to promote the endocycle. A similar observation was reported in wheat; it was reported that water stress caused a reduction in CDK activity, and this was correlated with an increased proportion of $p34^{cdc2}$ deactivated by tyrosine phosphorylation (Schuppler *et al.*, 1998). Therefore, inhibitory phosphorylation of CDKs may be regulated by developmental programmes and stress signals.

5.7.2 Requirement for tyrosine dephosphorylation in plant cell division

No homologues of CDC25 were reported in plants, except in *Arabidopsis* where a candidate gene has been recently identified as described below. However, a few reports demonstrated a negative correlation between tyrosine phosphorylation on CDKs and cell division, indicating that some phosphatase may function in dephosphorylating and activating CDKs in plants also. Sorrell *et al.* (2003) showed that the fission yeast Cdc25 bound three members of the 14-3-3 family proteins (G-box factor-like 14 κ, λ, ω) of *Arabidopsis* in the yeast two-hybrid system and an immunoprecipitation assay. In response to DNA damage in vertebrate cells, CDC25 is phosphorylated on Ser216 and moves to the cytoplasm where it binds a 14-3-3 protein to provoke DNA damage checkpoint (Peng *et al.*, 1997). Therefore, plant 14-3-3 proteins may also inactivate Cdc25-like phosphatase(s) that generally localizes in the nuclei and enhances the CDK activity.

Overexpression studies of the fission yeast *Cdc25* in plants indicated a possible involvement of similar phosphatase(s) in cell division. When tobacco plants were transformed with the fission yeast *Cdc25* under the control of a cauliflower mosaic virus 35S gene promoter, eight of the nine transgenic plants produced leaves in which the lamina was lengthened and twisted and the internal region was ‘pocketed’ in places (Bell *et al.*, 1993). They showed precocious flowering due to flowering at an earlier stage of development and sometimes produced abnormal ‘petal-less flowers’. In the root meristem, the cells of the transformants were smaller than those of the wild type throughout the cell cycle, including at the mitotic prophase. Similar observations were also reported in tobacco leaves where an inducible expression of the fission yeast *Cdc25* resulted in variable cell size and cell proliferation in the lamina margins followed by alterations in leaf shape (Wyrzykowska *et al.*, 2002). Induction of *Cdc25* expression in cultured primary roots of tobacco resulted in a substantial increase in the number of lateral root primordia, which were significantly smaller and comprised small cells at mitosis (McKibbin *et al.*, 1998). However, it did not perturb normal development of the lateral roots. These results suggest that

Cdc25 expression created a new threshold size for division due to premature entry into mitosis, because fission yeast mutants that overexpress *Cdc25* show premature division.

To date, several reports have indicated a possible link between cytokinin signalling and CDC25-like phosphatase(s). In excised tobacco stem pith, the $p34^{cdc2}$ protein induced by auxins was inactive and could not induce cell division (Zhang *et al.*, 1996). After treatment with the fission yeast Cdc25 *in vitro*, the $p34^{cdc2}$ activity increased by 50%. In contrast, a combination of auxins and cytokinins resulted in cell proliferation and induced active $p34^{cdc2}$, which was moderately activated by Cdc25. The level of phosphotyrosine in $p34^{cdc2}$ was considerably higher on naphthaleneacetic acid (NAA) than on NAA plus benzylaminopurine (BAP), suggesting that most $p34^{cdc2}$ was active and not tyrosine phosphorylated in the combined presence of auxin and cytokinin. Suchomelová *et al.* (2004) showed that the culture of tobacco stem segments expressing the fission yeast *Cdc25* resulted in the phenotype of enhanced shoot formation, suggesting that *Cdc25* expression mimicked the developmental effect caused by exogenous hormone balance that shifted towards cytokinin.

Several data suggested that cytokinin is required at the control point in the G2 phase when cells begin to enter the stage of mitosis. In *Nicotiana tabacum* L. cv. Xanthi, a three-fold peak in the cytokinin concentration was reported in the late G2 phase (Nishinari and Syono, 1986). In tobacco BY2 cells, zeatin- and dihydrozeatin-type cytokinins showed sharp peaks at the end of the S phase and during mitosis (Redig *et al.*, 1996). Also, in BY2 cells, the cell cycle was blocked at early or late G2 by application of lovastatin – an inhibitor of the isoprenoid pathway of cytokinin biosynthesis – and zeatin was capable of overriding the effects of lovastatin inhibition of mitosis (Laureys *et al.*, 1998). Therefore, cytokinin had been assumed to play a key role in the G2 to M phase transition. A suspension culture of *Nicotiana plumbaginifolia* that lacked cytokinin, as it was incubated with auxin alone, was almost completely arrested in the G2 phase. Cytokinin stimulated the activation of cell division along with lowered levels of phosphotyrosine in $p34^{cdc2}$ (Zhang *et al.*, 1996) and an increase in CDC25-related phosphatase activity that coincided with the time of activation of tobacco CDKA activity (Zhang *et al.*, 2005). *In vivo*-activated CDKA could not be additionally activated by phosphatase, indicating that the previous *in vivo* activation was induced by phosphatase. Inducible expression of the fission yeast *Cdc25* also suppressed the G2 arrest caused by deprivation of cytokinin, and tobacco CDKA showed enhanced activity after the induction of *Cdc25*, which correlated with the period during which phosphotyrosine in CDKA declined to a basal level (Zhang *et al.*, 2005). These results suggested that Cdc25 activity reduced phosphotyrosine in CDKA and resulted in increased CDKA activity for overcoming the G2 block caused by cytokinin depletion.

Overexpression of the fission yeast *Cdc25* in tobacco BY2 cells resulted in the shortening of the G2 phase (Orchard *et al.*, 2005). Mitotic cell size was significantly smaller compared with the control line, and the cells exhibited a tendency to form double filaments of near-isodiametric cells as opposed to the single-cell filaments that are characteristic of wild-type cells. In *Cdc25*-expressing cells, Nicta;CDKB;1 activity was transiently high in the early S phase, S/G2 phase and early M phase.

This indicates that premature and sustained Nicta;CDKB;1 activity might result in premature cell division at a small mitotic size. Although *Cdc25*-expressing cells contained remarkably low levels of endogenous cytokinin regardless of lovastatin treatment, they retained mitotic competence, indicating that *Cdc25* expression bypasses a block on G2/M in cytokinin-depleted cells (Orchard *et al.*, 2005). This result and the data on *N. plumbaginifolia* cells suggest that one major function of cytokinin in the cell cycle is to cause the activation of CDC25-like phosphatase that activates CDKA and/or CDKB;1 and promotes G2-to-M phase progression.

5.7.3 A CDC25-like phosphatase and an antiphosphatase in Arabidopsis

An *Arabidopsis* gene that may code for a CDK25-like phosphatase has been recently identified and designated *Arath;CDC25* (Landrieu *et al.*, 2004). It was obtained by searching for the catalytic domain harbouring the conserved $HC(X)_5R$ motif of the active site loop of the tyrosine phosphatase proteins. It encodes a 146-amino acid protein and shows some homology with the catalytic domain of the three human CDC25 isoforms – CDC25A (26% identity/100 amino acid), CDC25B (28% identity/104 amino acid) and CDC25C (32% identity/65 amino acid), and the yeast CDC25s (Mih1 from *S. cerevisiae* and Cdc25 from *S. pombe*). While human and yeast CDC25s have a long non-conserved N-terminal extension followed by the catalytic domain, it is absent in the Arath;CDC25 protein (Bordo and Bork, 2002). The catalytic domain of Arath;CDC25 is preceded only by a putative NLS. The rice homologues Orysa;CDC25;1 and Orysa;CDC25;2 also lack the large N-terminal regulatory region. The recombinant protein of Arath;CDC25 hydrolyzed the phosphotyrosine analogue *para*-nitrophenyl phosphate. The histone H1-kinase activity of *Arabidopsis* CDKs purified with $p10^{CKS1At}$ beads increased when incubated with the Arath;CDC25 protein. Addition of the specific CDC25 inhibitor NSC95397 abolished the CDK activity in the presence of the Arath;CDC25 protein. However, tyrosine dephosphorylation on CDKs has not been demonstrated; thus, it remains unknown whether Arath;CDC25 directly targets the ATP-binding site of CDKs.

The Arath;CDC25.S72 protein, in which the Cys72 of $HC^{72}(X)_5R^{78}$ had mutated to serine, failed to hydrolyze the substrate; however, it showed a better yield than the wild-type protein in *E. coli*. The tertiary structure of the Arath;CDC25.S72 protein was then obtained by nuclear magnetic resonance (NMR) spectroscopy. The backbone of the main secondary structures of Arath;CDC25.S72 was superimposed on those of human CDC25A and CDC25B. However, Arath;CDC25.S72 has an additional zinc-binding loop in the C-terminal portion, which does not exist in the catalytic domain of human CDC25A or CDC25B. This domain is probably conserved in the rice homologues, considering the conservation of the residues involved in metal coordination. NMR mapping studies revealed the molecular interaction of Arath;CDC25.S72 with synthetic peptides of Arath;CDKA;1, corresponding to the glycine-rich loop, including Thr14 and Tyr15. Phosphorylation of Tyr15 is a necessary condition, but the equivalent addition of a phosphate to Thr14 further strengthens this physical interaction, emphasizing the dual specificity of CDC25 phosphatase.

When *Arath;CDC25* was expressed in a fission yeast, a highly significant reduction in mitotic cell length was observed as compared with wild-type cells (Sorrell *et al.*, 2005). This suggests that Arath;CDC25 can function as a mitotic accelerator in fission yeasts. However, it was unable to complement the temperature-sensitive *cdc25-22* mutant (Landrieu *et al.*, 2004). RT-PCR analysis showed that *Arath;CDC25* was expressed at low levels in all tissues examined (Sorrell *et al.*, 2005), implying that expression of *Arath;CDC25* was not enhanced in rapidly dividing cells, as compared to non-proliferative tissues in *Arabidopsis*. This is in contrast to the expression patterns of *Arath;WEE1* and a gene for the 14-3-3 protein, which binds the fission yeast Cdc25. They were upregulated in proliferative tissues (Sorrell *et al.*, 2002, 2003). Bleeker *et al.* (2006) have recently reported that Arath;CDC25 and its homologue in *Holcus lanatus* possess arsenate reductase activity. Overexpression of *Arath;CDC25* in *Arabidopsis* improved tolerance to mildly toxic levels of arsenate exposure, probably due to enhanced rates of phytochelatin accumulation that increases arsenate sequestration. Conversely, *Arabidopsis* T-DNA insertion mutants showed increased sensitivity at lower exposure levels. At present, it remains unknown whether the primary function of Arath;CDC25 lies in CDK dephosphorylation or in arsenate reduction, but it is noticeable that the T-DNA insertion mutant did not show any macroscopic phenotype for growth rate or growth habit in the absence of arsenate exposure (Bleeker *et al.*, 2006).

Recently, an antiphosphatase that negatively controls the CDK activity has been identified in *Arabidopsis*. Protein tyrosine phosphatase (PTP)-like proteins are a highly conserved protein family in eukaryotes characterized by a mutated catalytic site. Da Costa *et al.* (2006) investigated the *Arabidopsis* PTP-like protein PASTICCINO2 (PAS2) and showed that it interacts with CDKA;1 that is phosphorylated on Tyr15 and not with its unphosphorylated isoform. Loss of PAS2 function in *Arabidopsis pas2-1* mutant resulted in dephosphorylation of CDKA;1 and upregulated the kinase activity. This suggests that PAS2, like other PTP-like proteins, functions as an antiphosphatase that binds CDKA;1 and prevents its dephosphorylation by CDC25-like phosphatases. Overexpression of *PAS2* in tobacco BY2 cells slowed down cell division at the G2 to M transition and early mitosis (Da Costa *et al.*, 2006), whereas the *pas2* mutation had been characterized by ectopic cell proliferation in the apical part, which is enhanced by cytokinin (Faure *et al.*, 1998; Bellec *et al.*, 2002). These results support the idea that PAS2 would maintain CDKA;1 in a phosphorylated and inactive state, preventing the premature action of CDC25-like phosphatases. PAS2 was localized in the cytoplasm of dividing cells but moved into the nucleus upon cell differentiation (Da Costa *et al.*, 2006), suggesting that the balance between cell division and differentiation may depend on the interaction between CDKA;1 and the antiphosphatase PAS2.

5.8 Conclusion and perspectives

In plants, CDKD is related to vertebrate-type CAKs in terms of the primary sequence and enzyme activities. CDKD requires a regulatory partner cyclin H for exhibiting

CDK and CTD kinase activities. In *Arabidopsis*, multiple CDKDs display distinct substrate preferences and form different protein complexes, suggesting that they may play different roles in the cell cycle, transcription, DNA repair and chromosomal integration of *Agrobacterium* T-DNA (Figure 5.3). CDKF is a plant-specific CAK that has a high CAK activity, but does not possess CTD kinase activity. It functions without cyclin H and forms an active 130-kDa complex in *Arabidopsis* cells. CDKF is also involved in phosphorylation and activation of CDKDs; thus, it may be a CAKAK regulating basal transcription as well as the cell cycle (Figure 5.3). The phosphoregulatory system mediated by multiple CAKs implies that CDKF and/or its partner subunit(s) may perceive hormonal and developmental signals that control the CDKF activity and affect cell proliferation and transcriptional activities. Therefore, understanding CDKF regulation will be particularly important in revealing the manner in which the CAKAK—CAK–CDK phosphorylation cascade participates in response to internal or external stimuli in plant cells.

In plants, genes for WEE1 kinase and CDC25-like phosphatase have been identified, but phosphorylation or dephosphorylation of specific threonine and tyrosine residues on CDKs was not demonstrated thus far. Identification of target site(s) will facilitate our understanding of the manner in which the inhibitory phosphorylation occurs on different types of plant CDKs and regulates their activities *in vivo*. It is possible that cytokinin controls cell division through the dephosphorylation of CDKs; thus, in G2 to M phase transition, the cytokinin signal and CDK phosphorylation may have a strong association. In yeasts and animals, it is well known that DNA damage or DNA replication checkpoint targets CDC25 to arrest the cell cycle at G2/M. However, some of the upstream regulators such as p53 and Chk1 are not conserved in plants; this is an indication that a certain plant-specific pathway possibly controls the G2/M progression under various environmental conditions. Such mechanisms will be involved not only in cell cycle regulation but also in cell size control; thus, identification of upstream players will clarify the signal transduction pathway that governs cell morphogenesis during plant development.

Acknowledgments

We thank Masatoshi Yamaguchi for helping with the preparation of the manuscript. M. Yamaguchi, S. Matsubayashi, K. Bisova and C. Umeda-Hara are thanked for collaboratively studying plant CAKs with us. The CAK research was partly supported by a Grant-in-Aid for Scientific Research on Priority Areas (grant no. 17027007 and 18056006), a Grant-in-Aid for Scientific Research (B) (grant no. 16370019) and the Program for Promotion of Basic Research Activities for Innovative Biosciences (PROBRAIN).

References

Adamczewski, J.P., Rossignol, M., Tassan, J.P., Nigg, E.A., Moncollin, V. and Egly, J.M. (1996) MAT1, cdk7 and cyclin H form a kinase complex which is UV light-sensitive upon association with TFIIH. *EMBO J* **15**, 1877–1884.

Akoulitchev, S., Chuikov, S. and Reinberg, D. (2000) TFIIH is negatively regulated by cdk8-containing mediator complexes. *Nature* **407**, 102–106.

Bakó, L., Umeda, M., Tiburcio, A.F., Schell, J. and Koncz, C. (2003) The VirD2 pilot protein of *Agrobacterium*-transferred DNA interacts with the TATA box-binding protein and a nuclear protein kinase in plants. *Proc Natl Acad Sci USA* **100**, 10108–10113.

Barrôco, R.M., De Veylder, L., Magyar, Z., Engler, G., Inzé, D. and Mironov, V. (2003) Novel complexes of cyclin-dependent kinases and a cyclin-like protein from *Arabidopsis thaliana* with a function unrelated to cell division. *Cell Mol Life Sci* **60**, 401–412.

Bell, M.H., Halford, N.G., Ormrod, J.C. and Francis, D. (1993) Tobacco plants transformed with *cdc25*, a mitotic inducer gene from fission yeast. *Plant Mol Biol* **23**, 445–451.

Bellec, Y., Harrar, Y., Butaeye, C., Darnet, S., Bellini, C. and Faure, J.-D. (2002) *Pasticcino2* is a protein tyrosine phosphatase-like involved in cell proliferation and differentiation in *Arabidopsis*. *Plant J* **32**, 713–722.

Bleeker, P.M., Hakvoort, H.W.J., Bliek, M., Souer, E. and Schat, H. (2006) Enhanced arsenate reduction by a CDC25-like tyrosine phosphatase explains increased phytochelatin accumulation in arsenate-tolerant *Holcus lanatus*. *Plant J* **45**, 917–929.

Booher, R.N., Deshaies, R.J. and Kirschner, M.W. (1993) Properties of *Saccharomyces cerevisiae wee1* and its differential regulation of p34CDC28 in response to G1 and G2 cyclins. *EMBO J* **12**, 3417–3426.

Booher, R.N., Holman, P.S. and Fattaey, A. (1997) Human Myt1 is a cell cycle-regulated kinase that inhibits Cdc2 but not Cdk2 activity. *J Biol Chem* **272**, 22300–22306.

Bordo, D. and Bork, P. (2002) The rhodanese/Cdc25 phosphatase superfamily. Sequence-structure-function relations. *EMBO Rep* **3**, 741–746.

Buck, V., Russell, P. and Millar, J.B. (1995) Identification of a cdk-activating kinase in fission yeast. *EMBO J* **14**, 6173–6183.

Busso, D., Keriel, A., Sandrock, B., Poterszman, A., Gileadi, O. and Egly, J.M. (2000) Distinct regions of MAT1 regulate cdk7 kinase and TFIIH transcription activities. *J Biol Chem* **275**, 22815–22823.

Campbell, S.D., Sprenger, F., Edgar, B.A. and O'Farrell, P.H. (1995) *Drosophila* Wee1 kinase rescues fission yeast from mitotic catastrophe and phosphorylates *Drosophila* Cdc2 in vitro. *Mol Biol Cell* **6**, 1333–1347.

Chen, R., Yang, Z. and Zhou, Q. (2004) Phosphorylated positive transcription elongation factor b (P-TEFb) is tagged for inhibition through association with 7SK snRNA. *J Biol Chem* **279**, 4153–4160.

Cismowski, M.J., Laff, G.M., Solomon, M.J. and Reed, S.I. (1995) KIN28 encodes a C-terminal domain kinase that controls mRNA transcription in *Saccharomyces cerevisiae* but lacks cyclin-dependent kinase-activating kinase (CAK) activity. *Mol Cell Biol* **15**, 2983–2992.

Coin, F., Bergmann, E., Tremeau Bravard, A. and Egly, J.M. (1999) Mutations in XPB and XPD helicases found in xeroderma pigmentosum patients impair the transcription function of TFIIH. *EMBO J* **18**, 1357–1366.

Da Costa, M., Bach, L., Landrieu, I., *et al.* (2006) *Arabidopsis* PASTICCINO2 is an antiphosphatase involved in regulation of cyclin-dependent kinase A. *Plant Cell* **18**, 1426–1437.

Damagnez, V., Mäkelä, T.P. and Cottarel, G. (1995) *Schizosaccharomyces pombe* Mop1–Mcs2 is related to mammalian CAK. *EMBO J* **14**, 6164–6172.

Devault, A., Martinez, A.M., Fesquet, D., *et al.* (1995) MAT1 ('menage a trois') a new RING finger protein subunit stabilizing cyclin H-cdk7 complexes in starfish and *Xenopus* CAK. *EMBO J* **14**, 5027–5036.

Dewitte, W., Riou-Khamlichi, C., Scofield, S., *et al.* (2003) Altered cell cycle distribution, hyperplasia, and inhibited differentiation in Arabidopsis caused by the D-type cyclin CYCD3. *Plant Cell* **15**, 79–92.

Draetta, G.F. (1997) Cell cycle: will the real Cdk-activating kinase please stand up. *Curr Biol* **7**, R50–R52.

Drapkin, R., Le Roy, G., Cho, H., Akoulitchev, S. and Reinberg, D. (1996) Human cyclin-dependent kinase-activating kinase exists in three distinct complexes. *Proc Natl Acad Sci USA* **93**, 6488–6493.

Enke, D.A., Kaldis, P., Holmes, J.K. and Solomon, M.J. (1999) The CDK-activating kinase (Cak1p) from budding yeast has an unusual ATP-binding pocket. *J Biol Chem* **274**, 1949–1956.

Espinoza, F.H., Farrell, A., Erdjument-Bromage, H., Tempst, P. and Morgan, D.O. (1996) A cyclin-dependent kinase-activating kinase (CAK) in budding yeast unrelated to vertebrate CAK. *Science* **273**, 1714–1717.

Fabian-Marwedel, T., Umeda, M. and Sauter, M. (2002) The rice cyclin-dependent kinase-activating kinase R2 regulates S-phase progression. *Plant Cell* **14**, 197–210.

Faure, J.-D., Vittorioso, P., Santoni, V., *et al.* (1998) The *PASTICCINO* genes of *Arabidopsis thaliana* are involved in the control of cell division and differentiation. *Development* **125**, 909–918.

Faye, G., Simon, M., Valay, J.G., Fesquet, D. and Facca, C. (1997) Rig2, a RING finger protein that interacts with the Kin28/Ccl1 CTD kinase in yeast. *Mol Gen Genet* **255**, 460–466.

Feaver, W.J., Henry, N.L., Wang, Z., *et al.* (1997) Genes for Tfb2, Tfb3, and Tfb4 subunits of yeast transcription/repair factor IIH. Homology to human cyclin-dependent kinase activating kinase and IIH subunits. *J Biol Chem* **272**, 19319–19327.

Feaver, W.J., Svejstrup, J.Q., Henry, N.L. and Kornberg, R.D. (1994) Relationship of CDK-activating kinase and RNA polymerase II CTD kinase TFIIH/TFIIK. *Cell* **79**, 1103–1109.

Fisher, R.P. and Morgan, D.O. (1994) A novel cyclin associates with MO15/CDK7 to form the CDK-activating kinase. *Cell* **78**, 713–724.

Foe, V.E. (1989) Mitotic domains reveal early commitment of cells in *Drosophila* embryos. *Development* **107**, 1–22.

Frit, P., Bergmann, E. and Egly, J.M. (1999) Transcription factor IIH: a key player in the cellular response to DNA damage. *Biochimie* **81**, 27–38.

Fülöp, K., Pettkó-Szandtner, A., Magyar, Z., *et al.* (2005) The *Medicago* CDKC;1-CYCLINT;1 kinase complex phosphorylates the carboxy-terminal domain of RNA polymerase II and promotes transcription. *Plant J* **42**, 810–820.

Gonzalez, N., Hernould, M., Delmas, F., *et al.* (2004) Molecular characterization of a *WEE1* gene homologue in tomato (*Lycopersicon esculentum* Mill.). *Plant Mol Biol* **56**, 849–861.

Grafi, G. and Larkins, B.A. (1995) Endoreduplication in maize endosperm: involvement of M phase-promoting factor inhibition and induction of S phase-related kinases. *Science* **269**, 1262–1264.

Hanks, S.K. and Hunter, T. (1995) The eukaryotic protein kinase superfamily: kinase (catalytic) domain structure and classification. *FASEB J* **9**, 576–596.

Hata, S. (1991) cDNA cloning of a novel cdc2^{+}/CDC28-related protein kinase from rice. *FEBS Lett* **279**, 149–152.

Hermand, D., Pihlak, A., Westerling, T., *et al.* (1998) Fission yeast Csk1 is a CAK-activating kinase (CAKAK). *EMBO J* **17**, 7230–7238.

Hermand, D., Westerling, T., Pihlak, A., *et al.* (2001) Specificity of Cdk activation in vivo by the two Caks Mcs6 and Csk1 in fission yeast. *EMBO J* **20**, 82–90.

Igarashi, M., Nagata, A., Jinno, S., Suto, K. and Okayama, H. (1991) Wee1^{+}-like gene in human cells. *Nature* **353**, 80–83.

Joubès, J., Chevalier, C., Dudits, D., *et al.* (2000) CDK-related protein kinases in plants. *Plant Mol Biol* **43**, 607–620.

Kaldis, P. and Solomon, M.J. (2000) Analysis of CAK activities from human cells. *Eur J Biochem* **267**, 4213–4221.

Kaldis, P., Sutton, A. and Solomon, M.J. (1996) The Cdk-activating kinase (CAK) from budding yeast. *Cell* **86**, 553–564.

Kimmelman, J., Kaldis, P., Hengartner, C.J., *et al.* (1999) Activating phosphorylation of the Kin28p subunit of yeast TFIIH by Cak1p. *Mol Cell Biol* **19**, 4774–4787.

Krempler, A., Kartarius, S., Gunther, J. and Montenarh, M. (2005) Cyclin H is targeted to the nucleus by C-terminal nuclear localization sequences. *Cell Mol Life Sci* **62**, 1379–1387.

Labbé, J.C., Martinez, A.M., Fesquet, D., *et al.* (1994) p40^{MO15} associates with a p36 subunit and requires both nuclear translocation and Thr176 phosphorylation to generate cdk-activating kinase activity in *Xenopus oocytes*. *EMBO J* **13**, 5155–5164.

Landrieu, I., de Costa, M., De Veylder, L., *et al.* (2004) A small CDC25 dual-specificity tyrosine-phosphatase isoform in *Arabidopsis thaliana*. *Proc Natl Acad Sci USA* **101**, 13380–13385. (Erratum: *Proc Natl Acad Sci USA* (2004) **101**, 16391).

Larochelle, S., Pandur, J., Fisher, R.P., Salz, H.K. and Suter, B. (1998) Cdk7 is essential for mitosis and for in vivo Cdk-activating kinase activity. *Genes Dev* **12**, 370–381.

Laureys, F., Dewitte, W., Witters, E., Van Montagu, M., Inzé, D. and Van Onckelen, H. (1998) Zeatin is indispensable for the G2–M transition in tobacco BY-2 cells. *FEBS Lett* **426**, 29–32.

Leclerc, V., Raisin, S. and Leopold, P. (2000) Dominant-negative mutants reveal a role for the Cdk7 kinase at the mid-blastula transition in *Drosophila* embryos. *EMBO J* **19**, 1567–1575.

Leclerc, V., Tassan, J.P., O'Farrell, P.H., Nigg, E.A., Leopold, P. (1996) *Drosophila* Cdk8, a kinase partner of cyclin C that interacts with the large subunit of RNA polymerase II. *Mol Biol Cell* **7**, 505–513.

Lee, K.M., Saiz, J.E., Barton, W.A. and Fisher, R.P. (1999) Cdc2 activation in fission yeast depends on Mcs6 and Csk1, two partially redundant Cdk-activating kinases (CAKs). *Curr Biol* **9**, 441–444.

Liu, J. and Kipreos, E.T. (2002) The evolution of CDK-activating kinases. In: *CDK-Activating Kinase (CAK)* (ed. Kaldis, P.). Landes Bioscience, Georgetown, TX, pp. 99–111.

Liu, Y., Wu, C. and Galaktionov, K. (2004) p42, a novel cyclin-dependent kinase-activating kinase in mammalian cells. *J Biol Chem* **279**, 4507–4514.

Lundgren, K., Walworth, N., Booher, R., Dembski, M., Kirscner, M. and Beach, D. (1991) mik1 and wee1 cooperate in the inhibitory tyrosine phosphorylation of cdc2. *Cell* **64**, 1111–1122.

Mäkelä, T.P., Parvin, J.D., Kim, J., Huber, L.J., Sharp, P.A. and Weinberg, R.A. (1995) A kinase-deficient transcription factor TFIIH is functional in basal and activated transcription. *Proc Natl Acad Sci USA* **92**, 5174–5178.

Mäkelä, T.P., Tassan, J.P., Nigg, E.A., Frutiger, S., Hughes, G.J. and Weinberg, R.A. (1994) A cyclin associated with the CDK-activating kinase MO15. *Nature* **371**, 254–257.

Maldonado, E., Shiekhattar, R., Sheldon, M., *et al.* (1996) A human RNA polymerase II complex associated with SRB and DNA-repair proteins. *Nature* **381**, 86–89. (Erratum: *Nature* **384**, 384).

Marshall, N.F. and Price, D.H. (1995) Purification of P-TEFb, a transcription factor required for the transition into productive elongation. *J Biol Chem* **270**, 12335–12338.

McKibbin, R.S., Halford, N.G. and Francis, D. (1998) Expression of fission yeast *cdc25* alters the frequency of lateral root formation in transgenic tobacco. *Plant Mol Biol* **36**, 601–612.

Menges, M., de Jager, S.M., Gruissem, W. and Murray, J.A. (2005) Global analysis of the core cell cycle regulators of *Arabidopsis* identifies novel genes, reveals multiple and highly specific profiles of expression and provides a coherent model for plant cell cycle control. *Plant J* **41**, 546–566.

Menges, M., Hennig, L., Gruissem, W. and Murray, J.A. (2003) Genome-wide gene expression in an Arabidopsis cell suspension. *Plant Mol Biol* **53**, 423–442.

Mironov, V.V., De Veylder, L., Van Montagu, M. and Inzé, D. (1999) Cyclin-dependent kinases and cell division in plants – the nexus. *Plant Cell* **11**, 509–522.

Morgan, D.O. (1997) Cyclin-dependent kinases: engines, clocks, and microprocessors. *Annu Rev Cell Dev Biol* **13**, 261–291.

Mueller, P.R., Coleman, T.R., Kumagai, A. and Dunphy, W.G. (1995) Myt1: a membrane-associated inhibitory kinase that phosphorylates Cdc2 on both threonine-14 and tyrosine-15. *Science* **270**, 86–90.

Nagahara, H., Ezhevsky, S.A., Vocero-Akbani, A.M., Kaldis, P., Solomon, M.J. and Dowdy, S.F. (1999) Transforming growth factor beta targeted inactivation of cyclin E: cyclin-dependent kinase 2 (Cdk2) complexes by inhibition of Cdk2 activating kinase activity. *Proc Natl Acad Sci USA* **96**, 14961–14966.

Nigg, E.A. (1996) Cyclin-dependent kinase 7: at the cross-roads of transcription, DNA repair and cell cycle control? *Curr Opin Cell Biol* **8**, 312–317.

Nilsson, I. and Hoffmann, I. (2000) Cell cycle regulation by the Cdc25 phosphatase family. *Prog Cell Cycle Res* **4**, 107–114.

Nishinari, N. and Syono, K. (1986) Induction of cell division synchrony and variation of cytokinin contents through the cell cycle in tobacco cultured cells. *Plant Cell Physiol* **27**, 147–153.

Orchard, C.B., Siciliano, I., Sorrell, D.A., *et al.* (2005) Tobacco BY-2 cells expressing fission yeast *cdc25* bypass a G2/M block on the cell cycle. *Plant J* **44**, 290–299.

Pei, Y., Du, H., Singer, J., *et al.* (2006) Cyclin-dependent kinase 9 (Cdk9) of fission yeast is activated by the CDK-activating kinase Csk1, overlaps functionally with the TFIIH-associated kinase Mcs6, and associates with the mRNA cap methyltransferase Pcm1 in vivo. *Mol Cell Biol* **26**, 777–788.

Peng, C.Y., Graves, P.R., Thoma, R.S., Wu, Z., Shaw, A.S. and Piwnica-Worms, H. (1997) Mitotic and G2 checkpoint control: regulation of 14-3-3 protein binding by phosphorylation of Cdc25C on serine-216. *Science* **277**, 1501–1505.

Reardon, J.T., Ge, H., Gibbs, E., Sancar, A., Hurwitz, J. and Pan, Z.Q. (1996) Isolation and characterization of two human transcription factor IIH (TFIIH)-related complexes: ERCC2/CAK and TFIIH. *Proc Natl Acad Sci USA* **93**, 6482–6487. (Erratum: *Proc Natl Acad Sci USA* **93**, 10538).

Redig, P., Shaul, O., Inzé, D., Van Montagu, M. and Van Onckelen, H. (1996) Levels of endogenous cytokinins, indole-3-acetic acid and abscisic acid during the cell cycle of synchronized tobacco BY-2 cells. *FEBS Lett* **391**, 175–180.

Rickert, P., Seghezzi, W., Shanahan, F., Cho, H. and Lees, E. (1996) Cyclin C/CDK8 is a novel CTD kinase associated with RNA polymerase II. *Oncogene* **12**, 2631–2640.

Rossignol, M., Kolb-Cheynel, I. and Egly, J.M. (1997) Substrate specificity of the cdk-activating kinase (CAK) is altered upon association with TFIIH. *EMBO J* **16**, 1628–1637.

Roy, R., Adamczewski, J.P., Seroz, T., *et al.* (1994) The MO15 cell cycle kinase is associated with the TFIIH transcription-DNA repair factor. *Cell* **79**, 1093–1101.

Russell, P., Moreno, S. and Reed, S.I. (1989) Conservation of mitotic controls in fission and budding yeasts. *Cell* **57**, 295–303.

Russell, P and Nurse, P. (1986) $cdc25^{+}$ functions as an inducer in the mitotic control of fission yeast. *Cell* **45**, 145–153.

Sauter, M. (1997) Differential expression of a CAK (cdc2-activating kinase)-like protein kinase, cyclins and cdc2 genes from rice during the cell cycle and in response to gibberellin. *Plant J* **11**, 181–190.

Sauter, M., Mekhedov, S.L. and Kende, H. (1995) Gibberellin promotes histone H1 kinase activity and the expression of cdc2 and cyclin genes during the induction of rapid growth in deepwater rice internodes. *Plant J* **7**, 623–632.

Schaber, M., Lindgren, A., Schindler, K., Bungard, D., Kaldis, P. and Winter, E. (2002) CAK1 promotes meiosis and spore formation in *Saccharomyces cerevisiae* in a CDC28-independent fashion. *Mol Cell Biol* **22**, 57–68.

Schneider, E., Kartarius, S., Schuster, N. and Montenarh, M. (2002) The cyclin H/cdk7/Mat1 kinase activity is regulated by CK2 phosphorylation of cyclin H. *Oncogene* **21**, 5031–5037.

Schultz, P., Fribourg, S., Poterszman, A., Mallouh, V., Moras, D. and Egly, J.M. (2000) Molecular structure of human TFIIH. *Cell* **102**, 599–607.

Schuppler, U., He, P.-H., John, P.C. and Munns, R. (1998) Effect of water stress on cell division and cell-division-cycle 2-like cell-cycle kinase activity in wheat leaves. *Plant Physiol* **117**, 667–678.

Serizawa, H., Mäkelä, T.P., Conaway, J.W., Conaway, R.C., Weinberg, R.A. and Young, R.A. (1995) Association of Cdk-activating kinase subunits with transcription factor TFIIH. *Nature* **374**, 280–282.

Shiekhattar, R., Mermelstein, F., Fisher, R.P., *et al.* (1995) Cdk-activating kinase complex is a component of human transcription factor TFIIH. *Nature* **374**, 283–287.

Shimotohno, A., Matsubayashi, S., Yamaguchi, M., Uchimiya, H. and Umeda, M. (2003) Differential phosphorylation activities of CDK-activating kinases in *Arabidopsis thaliana*. *FEBS Lett* **534**, 69–74.

Shimotohno, A., Ohno, R., Bisova, K., *et al.* (2006) Diverse phosphoregulatory mechanisms controlling cyclin-dependent kinase-activating kinases in *Arabidopsis*. *Plant J* **47**, 701–710.

Shimotohno, A., Umeda-Hara, C., Bisova, K., Uchimiya, H. and Umeda, M. (2004) The plant-specific kinase CDKF;1 is involved in activating phosphorylation of cyclin-dependent kinase-activating kinases in *Arabidopsis*. *Plant Cell* **16**, 2954–2966.

Solomon, M.J., Harper, J.W. and Shuttleworth, J. (1993) CAK, the $p34^{cdc2}$ activating kinase, contains a protein identical or closely related to $p40^{MO15}$. *EMBO J* **12**, 3133–3142.

Solomon, M.J. and Kaldis, P. (1998) Regulation of cdks by phosphorylation. In: *Results and Problems in Cell Differentiation: Cell cycle control* (ed. Pagano, M.). Heidelberg, Springer, pp. 79–109.

Sorrell, D.A., Chrimes, D., Dickinson, J.R., Rogers, H.J. and Francis, D. (2005) The Arabidopsis CDC25 induces a short cell length when overexpressed in fission yeast: evidence for cell cycle function. *New Phytol* **165**, 425–428.

Sorrell, D.A., Marchbank, A.M., Chrimes, D.A., *et al.* (2003) The *Arabidopsis* 14-3-3 protein, GF14ω, binds to the *Schizosaccharomyces pombe* Cdc25 phosphatase and rescues checkpoint defects in the rad24$^-$ mutant. *Planta* **218**, 50–57.

Sorrell, D.A., Marchbank, A.M., McMahon, K., Dickinson, J.R., Rogers, H.J. and Francis, D. (2002) A *WEE1* homologue from *Arabidopsis thaliana*. *Planta* **215**, 518–522.

Suchomelová, P., Velgová, D., Masek, T., *et al.* (2004) Expression of the fission yeast cell cycle regulator *cdc25* induces de novo shoot formation in tobacco: evidence of a cytokinin-like effect by this mitotic activator. *Plant Physiol Biochem* **42**, 49–55.

Sun, Y., Dilkes, B.P., Zhang, C., *et al.* (1999) Characterization of maize (*Zea mays* L.) Wee1 and its activity in developing endosperm. *Proc Natl Acad Sci USA* **96**, 4180–4185.

Svejstrup, J.Q., Feaver, W.J. and Kornberg, R.D. (1996a) Subunits of yeast RNA polymerase II transcription factor TFIIH encoded by the CCL1 gene. *J Biol Chem* **271**, 643–645.

Svejstrup, J.Q., Vichi, P. and Egly, J.M. (1996b) The multiple roles of transcription/repair factor TFIIH. *Trends Biochem Sci* **21**, 346–350.

Tassan, J.P., Jaquenoud, M., Fry, A.M., Frutiger, S., Hughes, G.J. and Nigg, E.A. (1995) In vitro assembly of a functional human CDK7-cyclin H complex requires MAT1, a novel 36 kDa RING finger protein. *EMBO J* **14**, 5608–5617.

Thuret, J.Y., Valay, J.G., Faye, G. and Mann, C. (1996) Civ1 (CAK in vivo), a novel Cdk-activating kinase. *Cell* **86**, 565–576.

Tirode, F., Busso, D., Coin, F. and Egly, J.M. (1999) Reconstitution of the transcription factor TFIIH: assignment of functions for the three enzymatic subunits, XPB, XPD, and cdk7. *Mol Cell* **3**, 87–95.

Tsakraklides, V. and Solomon, M.J. (2002) Comparison of Cak1p-like cyclin-dependent kinase-activating kinase. *J Biol Chem* **277**, 33482–33489.

Umeda, M. (2002) CDK-activating kinases in higher plants. In: *CDK-Activating Kinase (CAK)* (ed. Kaldis, P.). Landes Bioscience, Georgetown, TX, pp. 55–64.

Umeda, M., Bhalerao, R.P., Schell, J., Uchimiya, H. and Koncz, C. (1998) A distinct cyclin-dependent kinase-activating kinase of *Arabidopsis thaliana*. *Proc Natl Acad Sci USA* **95**, 5021–5026.

Umeda, M., Shimotohno, A. and Yamaguchi, M. (2005) Control of cell division and transcription by cyclin-dependent kinase-activating kinases in plants. *Plant Cell Physiol* **46**, 1437–1442.

Umeda, M., Umeda-Hara, C. and Uchimiya, H. (2000) A cyclin-dependent kinase-activating kinase regulates differentiation of root initial cells in *Arabidopsis*. *Proc Natl Acad Sci USA* **97**, 13396–13400.

Vandepoele, K., Raes, J., De Veylder, L., Rouze, P., Rombauts, S. and Inzé, D. (2002) Genome-wide analysis of core cell cycle genes in *Arabidopsis*. *Plant Cell* **14**, 903–916.

Wang, G., Kong, H., Sun, Y., *et al.* (2004) Genome-wide analysis of the cyclin family in *Arabidopsis* and comparative phylogenetic analysis of plant cyclin-like proteins. *Plant Physiol* **135**, 1084–1099.

Wang, W. and Chen, X. (2004) HUA ENHANCER3 reveals a role for a cyclin-dependent protein kinase in the specification of floral organ identity in *Arabidopsis*. *Development* **131**, 3147–3156.

Weingartner, M., Pelayo, H.R., Binarova, P., *et al.* (2003) A plant cyclin B2 is degraded early in mitosis and its ectopic expression shortens G2-phase and alleviates the DNA-damage checkpoint. *J Cell Sci* **116**, 487–498.

Wyrzykowska, J., Pien, S., Shen, W.H. and Fleming, A.J. (2002) Manipulation of leaf shape by modulation of cell division. *Development* **129**, 957–964.

Yamaguchi, M., Fabian, T., Sauter, M., *et al.* (2000) Activation of CDK-activating kinase is dependent on interaction with H-type cyclins in plants. *Plant J* **24**, 11–20. (Erratum: *Plant J* **25**, 473).

Yamaguchi, M., Kato, H., Yoshida, S., Yamamura, S., Uchimiya, H. and Umeda, M. (2003) Control of in vitro organogenesis by cyclin-dependent kinase activities in plants. *Proc Natl Acad Sci USA* **100**, 8019–8023.

Yamaguchi, M., Umeda, M. and Uchimiya, H. (1998) A rice homolog of Cdk7/MO15 phosphorylates both cyclin-dependent protein kinases and the carboxy-terminal domain of RNA polymerase II. *Plant J* **16**, 613–619.

Yankulov, K.Y. and Bentley, D.L. (1997) Regulation of CDK7 substrate specificity by MAT1 and TFIIH. *EMBO J* **16**, 1638–1646.

Zhang, K., Diederich, L. and John, P.C.L. (2005) The cytokinin requirement for cell division in cultured *Nicotiana plumbaginifolia* cells can be satisfied by yeast Cdc25 protein tyrosine phosphatase. Implications for mechanisms of cytokinin response and plant development. *Plant Physiol* **137**, 308–316.

Zhang, K., Letham, D.S. and John, P.C. (1996) Cytokinin controls the cell cycle at mitosis by stimulating the tyrosine dephosphorylation and activation of $p34^{cdc2}$-like H1 histone kinase. *Planta* **200**, 2–12.

Zhang, S., Cai, M., Zhang, S., *et al.* (2002) Interaction of $p58^{PITSLRE}$, a G_2/M-specific protein kinase, with cyclin D3. *J Biol Chem* **277**, 35314–35322.

Zhou, Q., Chen, D., Pierstorff, E. and Luo, K. (1998) Transcription elongation factor P-TEFb mediates Tat activation of HIV-1 transcription at multiple stages. *EMBO J* **17**, 3681–3691.

6 E2F–DP transcription factors

Elena Ramirez-Parra, Juan Carlos del Pozo, Bénédicte Desvoyes, María de la Paz Sanchez and Crisanto Gutierrez

6.1 E2F–DP transcription factors: a historical perspective

Transcriptional regulation that controls cell cycle transitions is one revealing example of the differences between different eukaryotes: yeast, on one side, and plant and animal cells, on the other. Transcription is one of the major regulatory levels that control the availability of specific cellular factors whose temporal and spatial coordinated action is required for a variety of processes, e.g. cell cycle progression. In the case of transcriptional regulation during the G1/S phase, yeast rely on the function of the Swi4/Swi6 and Mbp1/Swi6 transcription factors (Breeden, 1996). In contrast, G1/S gene expression in animal cells is controlled to a large extent by an entirely different family of transcription factors, the E2F–DP family, whose activity is modulated by the retinoblastoma (RB) tumour suppressor protein (Korenjak and Brehm, 2005). These transcription factors were originally identified in studies with human adenoviruses that revealed that one of the virus's early promoters is activated by a cellular protein, E2F (adenovirus *E2* promoter-binding *f*actor; Helin *et al.*, 1992). Later, the E2F *d*imerization *p*artner, DP, was identified (Helin *et al.*, 1993).

About 10 years ago, a new step ahead in our understanding of plant cell cycle regulation was achieved. Three independent studies clearly pointed to the existence of G1/S regulators in plant cells that were structurally, and in many respects also functionally, remarkably similar to those present in animal cells but unrelated to those functioning in yeast. One was the identification and cloning of three D-type cyclins in *Arabidopsis thaliana*, related to the human counterparts, which possessed an LxCxE amino acid motif, which in human cells mediated their interaction with the RB protein (Soni *et al.*, 1995). Likewise, D-type cyclins were also identified in another plant species (Dahl *et al.*, 1995). From a different approach, a LxCxE motif was identified in RepA, a protein encoded by the small, circular genome of the wheat dwarf geminivirus, which conserved the function of interacting with heterologous proteins of the human RB family (Xie *et al.*, 1995). Interestingly, this viral strategy of interacting with the host RB protein is conserved with animal oncovirus such as SV40, adenovirus and human papilloma virus (Gutierrez, 2000).

These findings provided the strongest evidence at the time, supporting that not only should plants contain proteins homologous to the human tumour suppressor RB protein but, if so, they should rely on the E2F–DP family of transcription factors, described a few years earlier in human cells (Helin *et al.*, 1992, 1993). One of the components of this pathway, the plant RB-related (RBR) protein, and gene, was identified shortly afterwards (Grafi *et al.*, 1996; Xie *et al.*, 1996). Then, E2F (Ramirez-Parra *et al.*, 1999; Sekine *et al.*, 1999) and its dimerization partner DP (Magyar *et al.*, 2000; Ramirez-Parra *et al.*, 2000) was cloned. Together, these studies revealed that an RBR–E2F–DP pathway was active in plant cells (Shen, 2002). Here, we will review more recent efforts aimed at understanding the biology of plant E2F–DP transcription factors, including their structural organization, expression pattern, target genes and the impact of their function during plant growth and development.

6.2 Domain organization of E2F–DP proteins

Efforts that followed the initial identification of plant *E2F* (Ramirez-Parra *et al.*, 1999; Sekine *et al.*, 1999) and *DP* (Magyar *et al.*, 2000; Ramirez-Parra and Gutierrez, 2000) genes succeeded in identifying them in several plant species, including *A. thaliana* (de Jager *et al.*, 2001; Kosugi and Ohashi, 2002a; Mariconti *et al.*, 2002), carrot (Albani *et al.*, 2000), tobacco (Chaboute *et al.*, 2000), rice (Kosugi and Ohashi, 2002b) and, more recently, maize (Sabelli *et al.*, 2005), *Chlamydomonas reinhardtii* (Bisova *et al.*, 2005) and *Ostreococcus tauri* (Robbens *et al.*, 2005).

These studies have revealed that plants contain a complex set of E2F–DP family members, those of *A. thaliana* (six E2Fs and two DP) being the most intensely studied. Three Arabidopsis E2Fs, named AtE2Fa, b and c, share a common domain organization, including domains for DNA binding, dimerization with DP, interaction with the RBR protein and transcriptional regulation (Figure 6.1; de Jager *et al.*, 2001; Mariconti *et al.*, 2002). They are structurally related to human E2F1–5 family members. E2Fa and E2Fb present properties of transcriptional activators whereas E2Fc has a repressor nature. The other three Arabidopsis *E2F* genes were identified in several laboratories receiving different names: *ELP3*, *2* and *1* (de Jager *et al.*, 2001), *E2Fd*, *e* and *f* (Mariconti *et al.*, 2002), *DEL2*, *1* and *3* (Vandepoele *et al.*, 2002) and *E2L1*, *3* and *2* (Kosugi and Ohashi, 2002a), respectively. Here we will use the names *E2Fd/DEL2*, *E2Fe/DEL1* and *E2Ff/DEL3*, which has the advantage of maintaining the acronym E2F, which is especially useful in comparative studies with other species, e.g. mammalian cells, for which it has been maintained. These E2F proteins are atypical in that they have a duplicated DNA-binding domain and they function independently of DP (Figure 6.1). These properties are shared with mammalian E2F7 (Bracken *et al.*, 2004) and E2F8 (Christensen *et al.*, 2005; Maiti *et al.*, 2005), all of which were, remarkably, identified based on the information of atypical Arabidopsis *E2F* genes obtained earlier. Arabidopsis contains two DP proteins (DPa and DPb) which are also structurally related to the human DP1 protein (Figure 6.1) sharing a DNA-binding and a dimerization domain.

6.2.1 DNA-binding and dimerization domains

This is the most conserved domain in amino acid sequence among all plant E2F family members as well as among the animal E2F proteins. This high residue

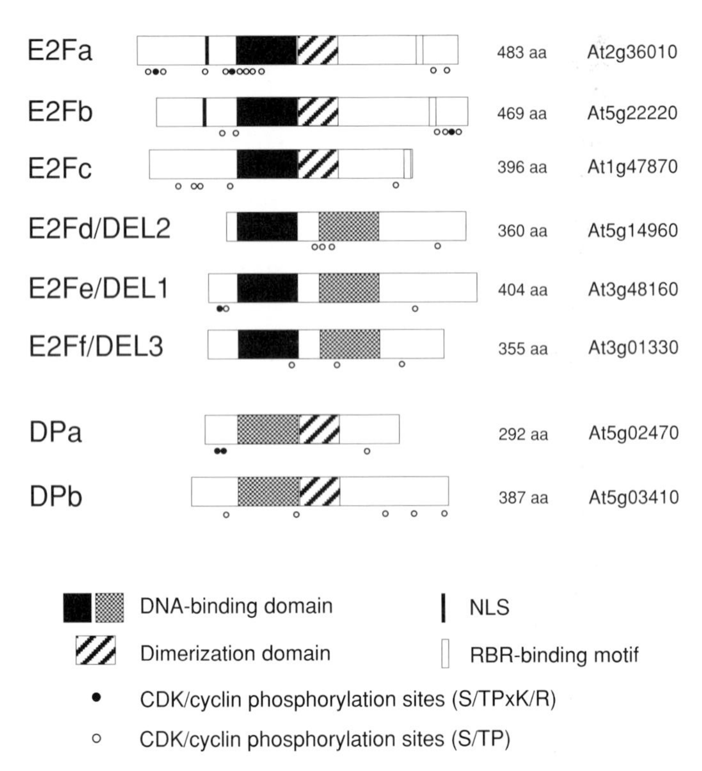

Figure 6.1 Domain organization and landmarks of the Arabidopsis E2F–DP family members.

conservation suggests that the folded protein structure is similar. In fact, the plant E2F DNA-binding domain can be modelled into the known crystal structure of the human E2F4–DP2 heterodimer that is of the winged-helix type (Zheng *et al.*, 1999). The DNA-binding domains present in E2F and DP are structurally related. In the case of the atypical E2Fs, both of plant and of animal origin, that contain two DNA-binding domains, the most N-terminal one is more related in amino acid sequence to that of E2F while the other is more similar to that of DP (Figure 6.1). This may explain, from a simple structural point of view, that atypical E2Fs do not need heterodimerization with DP for DNA binding, since the two domains that are required for high-affinity binding are provided by the same protein. The dimerization domain is also highly conserved at the amino acid level (Figure 6.1) in such a way that it can mediate heterologous interactions between plant DP and human E2F proteins (Ramirez-Parra and Gutierrez, 2000). This domain is absent in the atypical E2F, as mentioned above.

This structural similarity led support to the idea that plant E2F could also interact with the consensus DNA sequence identified for animal E2F. Human E2F transcription factors bind the consensus site, TTTSSCGS (S being a G or C nucleotide), in direct or reverse orientation without preference with respect to the transcriptional direction (Black and Azizkhan-Clifford, 1999; Zheng *et al.*, 1999). The amount of information as to how E2F factors regulate plant promoters is still comparatively more limited. However, plant E2F–DP proteins bind the same canonical DNA-binding site as their mammalian counterparts as revealed by electrophoresis mobility shift assay (EMSA; Albani *et al.*, 2000; Ramirez-Parra and Gutierrez, 2000; de Jager *et al.*, 2001), most likely using the same type of contacts with the DNA sequence as revealed for the human counterparts (Zheng *et al.*, 1999). Genome-wide analysis of putative plant E2F target genes have confirmed the *in vitro* observations, as discussed below (Ramirez-Parra *et al.*, 2003; Vandepoele *et al.*, 2005).

6.2.2 RBR-binding domain

The C-terminus of animal E2F1 (and related members) contains a 17 amino acid motif (*DY*x_7Ex_3*DLF*D; residues in italic are critical for interaction based on mutational studies) that mediates its interaction with the RB family proteins (Xiao *et al.*, 2003; Rubin *et al.*, 2005). The C-terminal moiety in all plant E2Fs is the less conserved at the amino acid level. However, most of them contain near their C-terminus a 17 amino acid sequence with the consensus *DYW*$x_{7/8}$S(M/I)*TD*(M/I)*W* (residues in italic indicate high conservation), which is related to that of the human counterparts. Detailed mutational studies of this region of plant E2F are lacking, but it is conceivable that it plays a similar role based on the conservation of the acidic and hydrophobic nature of residues at conserved positions and on the spacing between them. In support of this view, maize RBR1 (ZmRb1) is able to partially suppress an E2F-responsive promoter in transfected human cells (Huntley *et al.*, 1998). Likewise, expression of tobacco *E2F* is repressed by tobacco RBR (Uemukai *et al.*, 2005). The crystal structure of human RB with E2F1–DP1 also revealed additional interactions that stabilize the ternary complex between the C-terminal domain of RB and residues in the marked box domain of E2F1, a region whose functional relevance is poorly understood. Plant E2Fs also show conservation of a group of residues near the dimerization domain, which may be functionally equivalent to the marked box.

6.3 Transcriptional and post-translational regulation of E2F

The regulatory function of E2F–DP proteins must be strictly controlled in order to obtain the adequate response in terms of amplitude and cell/organ specificity. To achieve this, the activity of different E2F–DP is regulated at various levels, such as transcription, post-translational modifications and specific proteolysis.

6.3.1 Transcription

At present, scattered expression data for the majority of the *E2F* and *DP* family genes have been obtained, although detailed information of the spatio-temporal expression pattern has been reported only for some of them.

Microarray analyses of cultured cells synchronized by aphidicolin treatment or sucrose deprivation show that expression of several *E2F* genes is cell cycle regulated (Hennig *et al.*, 2003; Menges *et al.*, 2003, 2005). Thus, *E2Fb* and *E2Fc* are upregulated during the S phase, while *E2Fc* and *E2Ff/DEL3* levels remain high during late S/G2. *E2Fa*, *DPa* and *DPb* do not seem to have such strict cell-cycle-regulated expression.

The *E2F–DP* gene expression pattern seems to be complex also during seed germination (Masubelele *et al.*, 2005). First, it is worth noting that dry seeds contain very low levels of *E2Fa* and *E2Ff/DEL3* and high levels of *E2Fc* and *E2Fd/DEL2*, suggesting that the balanced activity of several E2Fs could be important for the maintenance of seed dormancy. Then, upon imbibition, *E2Fa* and *E2Fc* follow opposed patterns, consistent with their proposed activator and repressor major roles.

The expression patterns of *E2Fc*, *E2Ff/DEL3* and *E2Fb* genes have been obtained in plants expressing the *uidA* (*GUS*) reporter gene under the control of their promoters (del Pozo *et al.*, 2002a; Ramirez-Parra *et al.*, 2004; Sozzani *et al.*, 2006). These genes are highly expressed in young seedlings, with a strong expression in the meristems, except for *E2Ff*, which is undetectable in the root apical meristem. In the case of *E2Fa*, the *GUS* expression pattern has been confirmed by *in situ* hybridization (De Veylder *et al.*, 2002). In older plants, the expression of *E2Fc* is still high in leaves, while the expression levels of *E2Ff* and *E2Fb* are limited to the vascular strands. In flowers, these genes are also expressed although *E2Fc* levels are higher, except in pollen and anthers, where they are strongly expressed. These expression patterns suggested that these genes might have a function not only in cell division but also in other processes in differentiated cells.

A detailed picture of the overall expression patterns of *E2F–DP* genes can be obtained by using the Genevestigator® tool package (Zimmermann *et al.*, 2004). First conclusion after inspecting these data is that, frequently, expression of several *E2F* genes follows similar patterns in different organs, again strongly suggesting that it is the balanced activity of several E2F proteins that determines the final transcriptional regulation of particular target genes. Thus, as an example, stamen and pollen, in particular, show high levels of *E2Fc*, *E2Fd/DEL2* and *E2Ff/DEL3*, and comparatively much lower levels of *E2Fa*. However, *E2Fd/DEL2* and *E2Ff/DEL3*, together with *E2Fb*, are expressed at low levels in petals and sepals, while *E2Fa* expression is high in these tissues. *E2Fa* and *E2Fc* are high in both cauline and rosette leaves, but *E2Ff/DEL3* is high in cauline leaves and very low in rosette leaves. If we consider that different cell types are making up all plant organs it would be necessary to have detailed information of *E2F–DP* gene expression in individual cell types before we can finally identify the complete set of putative target genes regulated in each cell type at particular developmental stages.

Taken together, the wide expression of the *E2F–DP* genes, supported by functional data, indicates that this pathway plays important roles not only in controlling cell division but also in cell differentiation, growth and metabolism (see below). Further analyses of overexpressing and knockout plants will help us to understand the function of the RBR–E2F pathway during plant development and during the response to environmental challenges.

6.3.2 Phosphorylation

In mammals, phosphorylation of E2F and DP proteins by CDK/cyclin complexes is a widespread way to regulate the activity of these proteins (Dynlacht *et al.*, 1997). All plant E2F and DP proteins contain several minimal CDK consensus phosphorylation sites (S/TP), but only E2Fa, E2Fb, E2Fe/DEL1 and DPa contain the full sequence (S/TPxK/R; Figure 6.1). Studies of E2F–DP phosphorylation are still limited. Using *in vitro* phosphorylation assays, it has been shown that E2Fc can be phosphorylated, at least, by the CDKA/CYCA2;2 or the CDKA/CYCD2;2 complexes (del Pozo *et al.*, 2002a). Similar experiments were carried out to analyze the phosphorylation of other members of the E2F family, finding that E2Fa, E2Fb, *E2Fe/DEL1* and *E2Fd/DEL2* are also phosphorylated by the CDKA/CYCA2;2 kinase *in vitro* (J.C. del Pozo and C. Gutierrez, unpublished). Although all these results are obtained in *in vitro* assays, they suggest that CDK-dependent phosphorylation also regulates the E2F activity in plants. In fact, Espinosa-Ruiz *et al.* (2004) showed that phosphorylation of a hybrid aspen E2F protein by CDK reduces the DNA-binding activity of E2F. Similar results are found for E2Fc, which also reduces its affinity for DNA when phosphorylated by CDKA/CYCA2;2 complex *in vitro* (del Pozo and Gutierrez, unpublished).

6.3.3 Subcellular localization

Only E2Fa and E2Fb show typical nuclear localization signals (NLS; Figure 6.1), and studies aimed at defining the subcellular localization properties of different endogenous E2F are lacking. However, transient expression of Arabidopsis E2F in tobacco cultured cells revealed that a complex nuclear localization control appears to operate. Thus, in this study, E2Fa and E2Fc were translocated to the nucleus only in the presence of DPa, but not DPb, while E2Fb always appeared to be cytoplasmic (Kosugi and Ohashi, 2002c). Whether these interactions also occur *in planta* is not known but they seem to be somehow different based on the functional interaction observed in transgenic Arabidopsis plants (De Veylder *et al.*, 2002; Rossignol *et al.*, 2002; del Pozo *et al.*, 2006; Sozzani *et al.*, 2006).

6.3.4 Selective proteolysis of E2F and DP

Programmed proteolysis of cell proliferation regulators is a well-known system to control cell cycle progression in multicellular organisms. In mammals, the ubiquitin-proteasome pathway regulates the availability of a variety of proteins such as cyclins, RB, E2F1 or DP1, among others (Yamasaki and Pagano, 2004). The stability of human E2F1 is regulated by the SCF^{SKP2} complex in a phosphorylation-dependent manner (Marti *et al.*, 1999), although this is not the only degradation pathway for human E2F1 (Ohta and Xiong, 2001). In Arabidopsis, the regulation of E2Fc through the ubiquitin-proteasome pathway has been documented (del Pozo *et al.*, 2002a). Interestingly, the SCF^{SKP2A} complex targets E2Fc in a CDK-dependent manner. In addition, E2Fc accumulated in the auxin-response mutant *axr1*, which is deficient in the RUB–CULLIN signalling pathway (del Pozo *et al.*, 2002b). It is remarkable that E2Fc protein is degraded during the transition from the skoto-morphogenetic to the photo-morphogenetic development. Thus, high level of E2Fc accumulates in hypocotyls of dark-grown seedlings and becomes undetectable just a few hours after transferring plants to light. This suggests that degradation of E2Fc is a prerequisite to trigger cell proliferation in arrested cells (see also the discussion on lateral root initiation). Because of the repressor function of E2Fc, it was suggested that the degradation of E2Fc during this transition activates cell proliferation in response to light. Magyar *et al.* (2005) showed that E2Fb was an unstable protein and that the presence of auxin in the medium stabilized it. However, at present the pathway involved in the degradation of E2Fb or the proteins/complexes required for its proteolysis is unknown. DPb protein, which interacts with E2Fc *in vivo*, is also regulated through the ubiquitin-proteasome pathway and similarly to E2Fc, the SCF^{SKP2A} complex targets DPb for degradation (del Pozo *et al.*, 2006). From these studies, it is tempting to speculate that the SCF^{SKP2A} complex plays a central role possibly regulating a large number of cell cycle proteins.

6.4 E2F–DP target genes

Mammalian and Drosophila E2F transcription factors regulate genes expressed at the G1/S boundary of the cell division cycle in higher eukaryotes. Genes regulated by E2F include those whose products are part of the enzymatic machinery of DNA replication, such as dihydrofolate reductase and DNA polymerase α (reviewed in Herwig and Strauss, 1997; Helin, 1998), as well as components implicated in the initiation of DNA replication (Figure 6.2), including ORC1 (Ohtani *et al.*, 1996), CDC6 (Hateboer *et al.*, 1998) and the minichromosome maintenance (MCM) proteins (Leone *et al.*, 1998). In addition, regulators of the cell cycle progression, such as cyclin E, cyclin A, E2F1, E2F2, CDK1 (reviewed in Herwig and Strauss, 1997; Helin, 1998) and CDC25 (Vigo *et al.*, 1999), are also regulated by E2Fs. Other physiologically important targets of the mammalian E2F transcription factors have been identified by microarray experiments, chromatin immunoprecipitation and computer-assisted predictions (Ishida *et al.*, 2001; Kel *et al.*, 2001; Müller *et al.*, 2001; Weinmann *et al.*, 2001; Ren *et al.*, 2002). Thus, genes involved in cell division, DNA repair and replication, mitotic progression, apoptosis and differentiation have been identified as E2F targets.

During the last years, consensus E2F-binding sites have been found in the promoters of numerous plant genes and experimentally demonstrated to mediate transcriptional regulation (Table 6.1). To date, most of them are cell-cycle-related genes but we now know that some plant-specific pathways such as cell wall biogenesis and expansion, nitrogen assimilation or photosynthesis also depend on genes whose transcriptional regulation is mediated by E2F family members. More recently, the use of genome-wide approaches that integrate microarray information with bioinformatic analysis has led to the identification of a broad number of potential plant E2F targets, which are just beginning to be studied in detail.

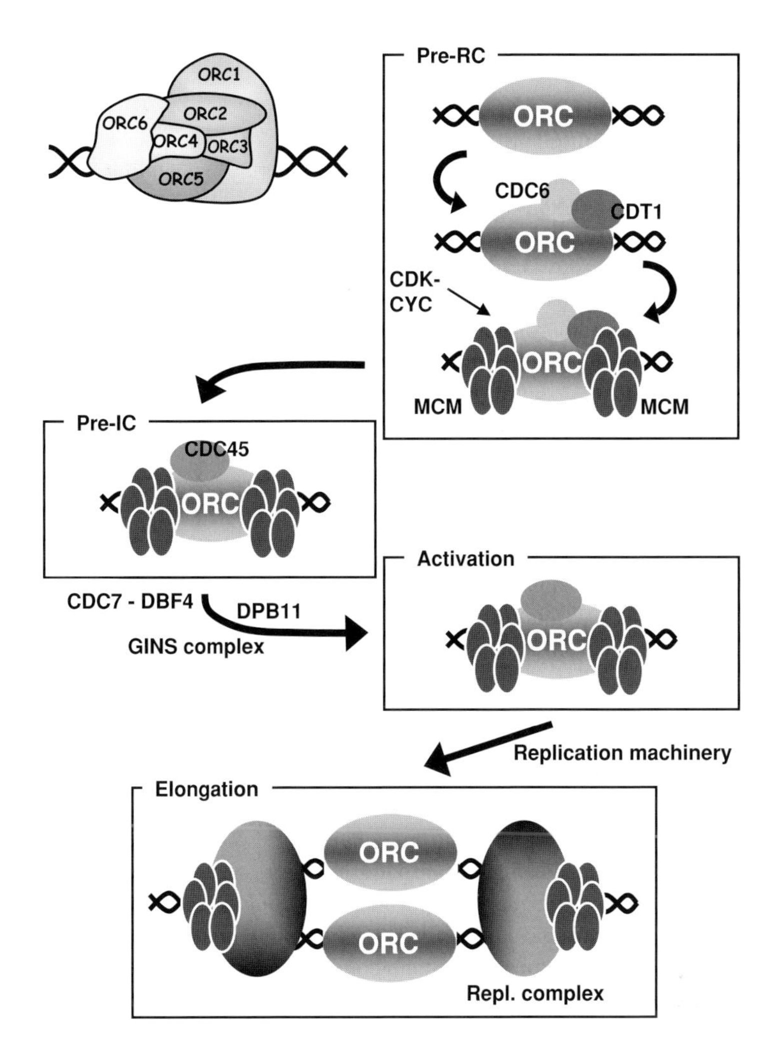

Figure 6.2 Proteins involved in the initial steps of DNA replication. Top left: schematic of the six subunits that constitute the origin recognition complex (ORC). Note that the ORC2-3-4-5 subunits make up the ORC core, whereas the ORC1 and ORC6 subunits seem to be more labile in the complex. Rest of the scheme: sequence of the major events that occur from the establishment of pre-replication complexes (pre-RC) to the elongation phase. The initial requirement for pre-RC is that once ORC binds to a DNA site, CDC6 and CDT1 proteins interact with the complex. Then, CDK/cyclin complexes modify various pre-RC components allowing the final incorporation of the six-subunit minichromosome maintenance (MCM) complex. Pre-initiation complexes are formed after CDC45 protein interacts with pre-RC. Subsequent activity of CDC7–DBF4, Dpb11 and the GINS complex (that consists of the SLD5, PSf1, PSF2 and PSF3 proteins) leads to activation of the origin of DNA replication. Further incorporation of the DNA replication machinery, including several DNA polymerases and accessory factors, e.g. PCNA and RFC, among others, to each of the two divergent replication forks allows initiation of DNA replication and the elongation process to occur. Note that (i) this scheme is a composite derived from information gathered from different model systems where specific regulatory steps occur, but they have not been included here for simplicity; (ii) since most, if not all, of these replication factors are largely conserved throughout evolution, including plants (Gutierrez, 2006), it is likely that the basic molecular events are also conserved; (iii) several of the genes encoding these replication factors have been shown to be E2F targets in Arabidopsis, as discussed in the text.

Table 6.1 Plant E2F target genes

Gene	Species	E2F–binding site(s)	Position[a]	Sense	Cell cycle regulation	Experimental approach	Reference[b]	E2F regulation in animal homologues
CDC6a	*Arabidopsis thaliana* (At2g29680)	TTTCCCGC	−182	+	G1/S	EMSA, deregulation in E2F transgenic	1–4	Y
CDT1	*A. thaliana* (At2g31270)	TTTCGCGG TTTGGCGC	−186–113	+−		EMSA, deregulation in E2F transgenic	5	Y
ORC1a	*A. thaliana* (At4g14700)	TTTCCCGC TTTGGCGG	–75–65	+−	G1/S	EMSA, deregulation in E2F transgenic	6	Y
ORC1b	*A. thaliana* (At4g12620)	TTTCCCGC GTTGGCGG	–111–101	+−	G1/S	EMSA, deregulation in E2F transgenic	6	Y
ORC2	*A. thaliana* (At2g37560)	TTTCCCGG TTTCCCGC	–80[a]–65[a]	−−	G1/S	EMSA, deregulation in E2F transgenic	6	N
ORC3	*A. thaliana* (At5g16690)	TTTCCCGC TTTGGCGG	–93–83	+−	G1/S	EMSA, deregulation in E2F transgenic	6	
ORC4	*A. thaliana* (At2g01120)	TTTCCCGC	–18	−	G1/S	EMSA, deregulation in E2F transgenic	6	N
ORC6	*A. thaliana* (At1g26840)	ATTCGCGG TTTGGCGG	–164–121	−−	G1/S	EMSA, deregulation in E2F transgenic	6	Y
MCM3	*A. thaliana* (At5g46280)	TTTGGCGC TTTGGCGG	–272–99	−−	G1/S	EMSA, deregulation in E2F transgenic, promoter mutational studies	7	Y
CDKB1;1	*A. thaliana* (At3g54180)	TTTCCCGC	–151	+	G1/S/G2	Deregulation in E2F transgenic; promoter mutational studies	8	N
E2Fb	*A. thaliana* (At5g22220)	CTTCCCGG ATTCCCGC	–140–125	++	S	Deregulation in E2F transgenic; chromatin immunoprecipitation	9	Y

(*Continued*)

Table 6.1 Plant E2F target genes (*Continued*)

Gene	Species	E2F–binding site(s)	Position[a]	Sense	Cell cycle regulation	Experimental approach	Reference[b]	E2F regulation in animal homologues
EXP3	*A. thaliana* (At2g37640)	CTTCCCGC	–52	+		Deregulation in E2F transgenic; chromatin immunoprecipitation	10	N
EXP7	*A. thaliana* (At1g12560)	TCTCCCGC	–575	+		Deregulation in E2F transgenic; chromatin immunoprecipitation	10	N
EXP9	*A. thaliana* (At5g02260)	TTCGCCGC ATTCGCGG TTCGCCGC TATGGCGC	–700–690 +4+24	−+ −+		Deregulation in E2F transgenic; chromatin immunoprecipitation	10	N
UGT	*A. thaliana* (At1g22400)	TTTCGCGC	–21	+		Deregulation in E2F transgenic; chromatin immunoprecipitation	10	N
RNR1b	*Nicotiana tabacum*	TTTCCCGC	–177	+	S	*In vivo* footprint; EMSA; promoter mutational studies	11	N
RNR2	*N. tabacum*	TTTCCCGC TTTGCCGC	–355–294	+−	G1/S	*In vivo* footprint; EMSA; promoter mutational studies	12	N
PCNA	*Nicotiana benthamiana*	TTTCCCGC ATTCCCGC	–115–77	−+	G1/S	Promoter mutational studies	13–14	Y
PCNA	*Oryza sativa*	TTTCCCGC	–135	−	G1/S	Promoter mutational studies	15	Y
RBR3	*Zea mays*	TTTGGCGG TCTCCCGC	–705 –648	−+		EMSA	16	N

[a]Position refers to nucleotide position, relative to the initiator ATG, the A residue being +1, of the first residue of the binding site, except for the case of ORC2 where it refers to the transcription start site.

[b]References are as follows: 1, Castellano *et al.*, 2001; 2, de Jager *et al.*, 2001; 3, De Veylder *et al.*, 2002; 4, del Pozo *et al.*, 2002a; 5, Castellano *et al.*, 2004; 6, Diaz-Trivino *et al.*, 2005; 7, Stevens *et al.*, 2002; 8, Bouldolf *et al.*, 2004; 9, Sozzani *et al.*, 2006; 10, Ramirez-Parra *et al.*, 2004; 11, Chaboute *et al.*, 2002; 12, Chaboute *et al.*, 2000; 13, Egelkrout *et al.*, 2001; 14, Egelkrout *et al.*, 2002; 15, Kosugi and Ohashi, 2002b; 16, Sabelli *et al.*, 2005.

6.4.1 DNA replication genes

Completion of the *A. thaliana* genome sequence has facilitated the identification of genes involved in DNA replication which are largely conserved among all multicellular eukaryotes, including plants (Gutierrez, 2006). The mechanisms controlling the availability of DNA replication proteins, e.g. transcriptional regulation, post-translational modifications, subcellular localization and proteolysis, show specific features in different model organisms, which are beginning to be elucidated. However, it is likely that the major protein–protein interactions that occur during initiation and elongation of DNA replication have been largely conserved throughout evolution. Thus, it is conceivable that the initial series of events involved in initiation of DNA replication is similar to that depicted in Figure 6.2. As discussed below, genes encoding many DNA replication proteins are E2F targets.

The ORC complex is made up of six subunits that mark the origin of replication (Figure 6.2). All Arabidopsis *ORC* genes, except *ORC5*, contain at least one consensus E2F-binding site, the most common being TTTCCCGC at optimal positions from the translation start site (Table 6.1). Consistent with this, expression of *ORC* genes containing E2F-binding sites peaks at the G1/S phase of cell cycle as deduced from analysis in Arabidopsis synchronized cells (Diaz-Trivino *et al.*, 2005). EMSA using the E2F consensus sequences found in the *ORC* gene promoters indicates specific binding of E2F. Transgenic plants expressing a dominant-negative version of a DP protein that inhibits binding of E2F to DNA (Ramirez-Parra *et al.*, 2003) show decreased expression levels in most of the *ORC* genes that contain E2F sites. A complementary analysis using transgenic plants where E2F activity was increased by inducible inactivation of the RBR protein shows an increased expression of all *ORC* genes, except for *ORC5* (Desvoyes *et al.*, 2006). Therefore, these *in vitro*, *in silico* and *in planta* data together support the conclusion that expression of all *ORC* genes, except *ORC5*, is regulated by E2F (Diaz-Trivino *et al.*, 2005). It must be emphasized that this situation is different from that found in animal cells where only the Orc1 has been demonstrated to respond to E2F (Ohtani *et al.*, 1996), suggesting a different regulation of the complex in plants.

In Arabidopsis, as it occurs in yeast and animal cells, CDC6 is a protein crucial for the initiation of DNA replication which allows the loading of MCMs to the DNA replication origins (Figure 6.2). Consistent with this role, Arabidopsis *CDC6a* gene expression is upregulated in the S phase of cell cycle (Table 6.1). The analysis of its promoter indicates the presence of an E2F consensus site at an optimal position with respect to the translation initiation site (TTTCCCGC; −182 upstream from the ATG in sense orientation). Using EMSA, it has been proved that purified plant E2F protein specifically binds *AtCDC6* promoter sequences and that a point mutation in the E2F site abolishes binding (Castellano *et al.*, 2001; de Jager *et al.*, 2001). In this context, it is worth mentioning that Arabidopsis plants overexpressing the activator E2Fa–DPa have elevated levels of *CDC6* (De Veylder *et al.*, 2002) while overexpression of the repressor E2Fc leads to downregulation of this gene (del Pozo *et al.*, 2002a).

CDT1 cooperates with CDC6 in the loading of MCMs to the DNA replication origins (Figure 6.2). The Arabidopsis *CDT1a* gene promoter contains E2F consensus binding sites (Table 6.1; TTTCGCGG and TTTGGCGCG, at positions −186 and −113 bp relative to the ATG, in direct and reverse orientation, respectively). These are functional in *in vitro* EMSA analysis (E. Caro, unpublished data). In addition, Arabidopsis plants expressing a dominant-negative version of a DP protein, which prevents E2F–DNA interactions (Ramirez-Parra *et al.*, 2003), show decreased *CDT1a* expression levels (Castellano *et al.*, 2004), strongly suggesting that it is an E2F target. Interestingly, a similar regulation has been reported for the Drosophila Cdt1 homologue (also known as 'Double parked'), based on the reduced Cdt1 expression levels observed in Drosophila E2F mutants (Whittaker *et al.*, 2000), as well as for human Cdt1, where its expression is activated in E2F1-expressing cells (Yoshida and Inoue, 2004).

The Arabidopsis MCM3 homologue forms part of the minichromosome maintenance (MCM) complex involved in the initiation of DNA replication at the G1/S transition (Figure 6.2). Consistent with its role, the *MCM3* gene is transcriptionally regulated at S phase of cell cycle (Stevens *et al.*, 2002). Two E2F consensus binding sites have been found in its promoter (Table 6.1; TTTGGCGCG and TTTGGCGGG, at −272 and −99 bp respectively, both in reverse orientation), and EMSA using Arabidopsis E2F indicated that it specifically binds these two E2F sites in the *MCM3* promoter. Transient expression studies using tobacco BY2 cells expressing the fusion of *GUS* gene to the *MCM3* promoter indicate that it is transcriptionally regulated at G1/S phase (Stevens *et al.*, 2002). Deletion of the first E2F-binding site leads to loss of the S-phase regulation, by a loss of repression during G2 in synchronized cell suspensions. In contrast, deletion of the second site leads to S-phase transcriptional regulation similar to the wild-type promoter, but transcript levels are much lower. The double mutant in both E2F sites no longer shows cell-cycle-regulated expression and loses the transcriptional downregulation observed during G2 in the wild-type promoter. These data indicate that in cell suspension cultures the first E2F site represses the G2 inhibition of *MCM3* promoter activity, whereas the second E2F site may affect the level of promoter activity in S phase. Studies *in planta* revealed that the wild-type *MCM3* promoter was active in meristematic regions, in particular in secondary root tips and the proliferating regions at the base of leaves. These studies showed that two highly similar E2F-binding sites in the promoter of the *MCM3* gene are responsible for different cell cycle regulation or developmental expression patterns depending on the cellular environment.

PCNA, originally characterized as an auxiliary protein of DNA polymerase δ, functions in many events including DNA replication, DNA repair, chromatin assembly and cell cycle control (Jonsson and Hübscher, 1997; Kelman, 1997; Tsurimoto, 1998). Consistent with this role, its expression has a cell cycle regulation peaking at the G1/S phase. Tobacco *PCNA* promoter contains two E2F sites (Table 6.1; TTTCCCGCG and ATTCCCGC, at −122 and −99 bp position, in reverse and direct orientation, respectively). These two E2F sites are specifically recognized in EMSA using plant nuclear extracts or purified Arabidopsis E2F and DP proteins.

Cultured cells transfected with a *PCNA* promoter mutated at the two E2F sites fused to GUS reporter had only about 30% of the activity of the wild-type promoter in the transgenic cells (Kosugi and Ohashi, 2002b). Mature leaves of *Nicotiana benthamiana* transgenic plants that expressed luciferase reporter gene (*LUC*) fused to the *PCNA* promoter, had a significant increase in reporter activity using constructs with mutations in either or both E2F sites. These data indicate an E2F-mediated partial loss of repression. However, in young leaves mutation in the first E2F consensus site produces a decrease in activity compared with that in the wild-type *PCNA* promoter and mutation in the second or in both E2F consensus sites presents an activity similar to those of the wild type. These results demonstrate that both E2F elements contribute to the repression of the *PCNA* promoter in mature leaves, whereas the first E2F site counters the repression activity of the second E2F site in young leaves (Egelkrout *et al.*, 2001, 2002). The different results in young versus mature tissues may reflect developmental regulation of the factors that bind to the two E2F sites. This complex situation reveals that the regulation of *PCNA* transcription in plants is mediated by an intricate network of interactions that probably involve different E2F complexes as well as other transcription factors (Kosugi and Ohashi, 1997) that act in a concerted manner during development.

A putative E2F site has been also found in rice *PCNA* promoter (Table 6.1; TTTC-CCGC, at −142 in reverse orientation). Recombinant rice E2F protein specifically binds this E2F site. The rice *PCNA* promoter mutated at the E2F site fused to the *GUS* reporter gene introduced into cultured tobacco cells sites had a reduction of about half of the activity of the wild-type promoter in the transgenic cells (Kosugi and Ohashi, 2002b). These results indicate that the E2F sites in the tobacco and rice *PCNA* promoter function as positive *cis*-elements responsible for the expression in actively dividing cells. Several E2F sites are also present in the proximal promoter region of the two Arabidopsis *PCNA* genes, and E2F elements have been identified in Drosophila, human, and murine *PCNA* genes, suggesting that E2F may regulate all eukaryotic *PCNA* genes. However, *PCNA* promoter function is complex, involving a variety of transcription factors, and E2F can activate as well as repress its activity (Yamaguchi *et al.*, 1995; Huang and Prystowsky, 1996; Liu *et al.*, 1998; Hayashi *et al.*, 1999).

Ribonucleotide reductase (RNR) is a key enzyme involved in the DNA synthesis pathway. It catalyzes the reduction of ribonucleoside diphosphates (NDPs) in deoxyribonucleoside diphosphates (dNDPs), a rate-limiting step in DNA synthesis. In eukaryotes, ribonucleotide reductases belong to the RNR class I in which the active enzyme consists of two large (R1) and two small (R2) subunits encoded by *RNR1* and *RNR2* genes, respectively. There are at least two *RNR1* genes (*RNR1a* and *RNR1b*) and one *RNR2* gene in tobacco. The *RNR* genes are cell cycle regulated and specifically expressed in S phase. Two E2F-binding consensus sites have been identified in tobacco *RNR2* promoter (Table 6.1; TTTCCCGC and TTTGCCGC, at −355 and −294 bp, in sense and antisense orientation, respectively). *In vivo* footprinting assays of aphidicolin-synchronized BY2 cells show the *in vivo* occupancy of these E2F consensus sites in a cell-cycle-dependent manner, thereby suggesting that they have a functional significance as regulators of *RNR2* gene expression

(Chabouté *et al.*, 2000). In fact, specific interaction of nuclear complexes and purified tobacco E2F protein with the two E2F sites was demonstrated by EMSA. Additional analyses in transgenic synchronized transformed BY2 cells indicated that the two E2F elements were involved synergistically in upregulation of the promoter at the G1/S transition, and mutation of both elements prevented any significant induction of the *RNR* promoter. The first E2F site acts essentially as an activator while the second behaves as a repressor outside of the S phase (Chabouté *et al.*, 2000).

One E2F-binding site has been found in tobacco *RNR1b* gene promoter (Table 6.1; TTTCCCGC at −258 bp, in sense orientation). Gel-shift assays show that a nuclear protein complex as well as purified tobacco E2F protein associate specifically with the E2F site in the *RNR1b* promoter. In a way similar to that for *RNR2*, the tobacco *RNR1b* promoter activity was detected during S phase in synchronized cells, and mutation of the E2F site substantially reduced this regulation (Chabouté *et al.*, 2002). Transgenic plants expressing the *RNR1b* promoter fused to the *GUS* reporter gene show active expression in root meristems as well as in axillary meristems of developing plantlets, but not in the shoot apical meristem. However, no GUS activity was detected in transgenic seedlings harbouring the same construction with a mutated E2F site. Arabidopsis and rice *RNR* genes seem to have a similar E2F-dependent regulation, as indicated by the presence of E2F consensus sites in their promoters, suggesting that E2F regulation of *RNR* promoters could be a general rule in plants. In contrast, this seems to be different in the mammalian *RNR* gene promoters, which lack E2F sites (Filatov and Thelander, 1995).

6.4.2 Cell cycle genes

Arabidopsis plants overexpressing the activator E2Fa–DPa have elevated levels of *CDKB1;1* (Vlieghe *et al.*, 2003), as revealed by microarray and Northern blot analysis. The promoter analysis indicates the presence of an E2F consensus site (Table 6.1; TTTCCCGC, at −151 bp downstream of the start codon). Tobacco BY2 cells that express the GUS reporter gene fused to the *CDKB1;1* promoter showed that GUS activity was ∼30-fold higher in the calli containing the wild-type promoter construct than in those harbouring the mutant promoter without the E2F site. These complementary data indicate an E2F direct regulation of *CDKB1;1* gene. In addition, transgenic plants expressing a *CDKB1;1*-promoter:*GUS* construct in an *E2Fa–DPa*-overexpressing background show increased GUS staining in leaf primordia and stomatal cells (Boudolf *et al.*, 2004). The E2F regulation of *CDKB1;1* transcription was surprising, because E2F–DP transcription factors were presumed to operate at the G1/S transition, whereas CDKB1;1 kinase activity peaks at G2/M. However, *CDKB1;1* transcript levels increase from the G1/S transition onwards, followed by an increase in CDKB1;1 protein and associated activity during S phase (Porceddu *et al.*, 2001; Sorrell *et al.*, 2001; Breyne *et al.*, 2002; Menges *et al.*, 2002). These findings illustrate a crosstalk mechanism between the G1/S and G2/M transition points, in which upregulation of *CDKB1;1* by E2F might be a mechanism linking DNA replication with the following mitosis. Similar E2F-mediated links between

G1/S and G2/M transcription have been observed in mammalian cells (Zhu *et al.*, 2004).

Three *RBR* genes have been identified in maize and one of them, *RBR3*, contains two consensus E2F-binding sites in its promoter which actually bound E2F–DP-containing complexes (Table 6.1; Sabelli *et al.*, 2005). Furthermore, expression of the wheat dwarf geminivirus protein which targets RBR proteins (Grafi *et al.*, 1996; Xie *et al.*, 1996) and inactivates them (Desvoyes *et al.*, 2006), and RBR1 in particular, upregulates *RBR3*, but not other maize *RBR* genes (Sabelli *et al.*, 2005). Whether *RBR* genes in other species are also E2F targets is not presently known, although consensus E2F sites are present in Arabidopsis *RBR* (B. Desvoyes and C. Gutierrez, unpublished). Together, these data revealed that *RBR3* expression is controlled by RBR1 and E2F–DP and that the situation in grasses seems to be complex (Sabelli and Larkins, 2006).

The Arabidopsis *E2Fb* gene contains E2F sites in its promoter (Table 6.1; CTTC-CCGG and ATTCCCGC, at −140 and −125 bp from the ATG, both in direct orientation). The E2Fb protein accumulates in *E2Fa–DPa*-overexpressing plants and its promoter binds E2F *in planta*, as demonstrated by chromatin immunoprecipitation experiments (Sozzani *et al.*, 2006).

6.4.3 E2F targets in differentiated cells

Previous studies have revealed the presence of E2F-binding sites in promoters of cell cycle and non-cell-cycle genes (Ramirez-Parra *et al.*, 2003; Vlieghe *et al.*, 2003). Defects in cell expansion in root and hypocotyls observed in transgenic plants that overexpress *E2Ff/DEL3* indicated the possible implication of E2F in regulation of genes involved in cell wall biosynthesis, in which expansins play important roles (Vissenberg *et al.*, 2000; Li *et al.*, 2003). Several expansin genes (Table 6.1), including *EXP3* (CTTCCCGC at −52 bp from the putative ATG), *EXP7* (TCTCCCGC at −575 bp) and *EXP9* (TATGGCGG at +24, TTCGC-CGC at +4, ATTCGCGG at −690, and TTCGCCGC at −700 bp), contain E2F-binding sites in their putative promoters. Furthermore, the UDP-glucose-glycosyl transferase gene or *UGT* also contains one E2F site (TTTCGCGC at −21 bp). All these genes are upregulated in *E2Fa–DPa* overexpressing plants (Vlieghe *et al.*, 2003) and downregulated in *E2Ff/DEL3*-overexpressing plants (Ramirez-Parra *et al.*, 2004). Chromatin immunoprecipitation shows that promoter fragments of the *EXP3*, *EXP7*, *EXP9* and *UGT* genes were specifically amplified from *E2Ff/DEL3*-overexpressing transgenic plants. These data indicate that a set of cell wall biogenesis genes likely are direct E2F targets *in vivo* and uncover a role of E2Ff/DEL3 in repressing cell wall biosynthesis genes in differentiated cells (Ramirez-Parra *et al.*, 2004).

E2Fa–DPa have been also implicated in other processes typical of differentiated cells, such as nitrogen metabolism or photosynthesis, although the specific binding of E2F to the relevant promoters has not been yet demonstrated (Vlieghe *et al.*, 2003).

6.4.4 Genome-wide approaches to identify E2F target genes

Genomic approaches to identify E2F targets in the human genome have been reported (Ishida *et al.*, 2001; Kel *et al.*, 2001; Müller *et al.*, 2001; Ren *et al.*, 2002; Wells *et al.*, 2002). In a similar way, the availability of Arabidopsis genomic resources and the microarray databases are facilitating enormously genome-wide approaches aimed at identifying novel plant E2F target genes.

A genome-wide search for direct E2F–DP target genes within the Arabidopsis genome was carried out by computational analysis. Using one typical E2F-binding site (TTTCCCGCC) and the variant site (TCTCCCGCC) as search criteria allowed the identification of over 180 potential E2F target genes (Ramirez-Parra *et al.*, 2003). Interestingly, the distribution of E2F sites in the promoters was not random, because most of the identified genes contained the E2F site within a 400-bp-long region, upstream from the putative ATG, a characteristic feature of functionality, as previously reported in human promoters containing E2F elements (Kel *et al.*, 2001). Among them, and in addition to cell cycle and DNA-replication-related genes, genes belonging to other functional categories, such as transcription, stress, defence, signal transduction, cellular biogenesis and protein fate, were also identified. Most potential E2F targets identified *in silico* show a cell-cycle-regulated expression pattern with a peak in early/mid S phase in synchronized cells. In addition, the expression of a large number of these potential E2F targets was decreased in transgenic plants expressing a truncated DP protein that acts with dominant-negative effect over E2F activity. These data strongly support that the E2F activity is crucial for the expression of a large variety of genes (Ramirez-Parra *et al.*, 2003).

Genome-wide transcriptome analysis of transgenic Arabidopsis plants overexpressing *E2Fa–DPa* genes was done by combining microarray analysis and bioinformatics tools in order to identify novel plant E2F-responsive genes. Promoter regions of genes that were transcriptionally induced were searched for the presence of E2F-binding sites, and over 180 genes were identified as potential direct E2F targets (Vandepoele *et al.*, 2005). A comparison of these data with available microarray datasets of synchronized cell suspensions revealed that the E2F target genes that were upregulated in the array were expressed almost exclusively during G1 and S phases and activated upon re-entry of quiescent cells into the cell cycle (Menges *et al.*, 2002). The most abundant E2F DNA-binding motif was *TTTCCCGC*, and it was localized within 400 bp upstream from the putative ATG, in agreement with other studies (Ramirez-Parra *et al.*, 2003). The genes identified included previously reported E2F targets (*PCNA*, *RNR*, *CDC6* and *MCM3*), but in addition, other genes involved in initiation of DNA replication (*CDC45*), elongation of DNA replication (DNA polymerases and DNA primase), cell cycle (*E2Fb*, *E2Fc*, *RBR1*, *E2Ff/DEL3* and *CYCA3;2*), chromatin dynamics (*MSI3*, *CMT3* and *trihorax-like protein* gene) and DNA repair (*RAD17*, *MSH6-1* and *UVR3*) (Vandepoele *et al.*, 2005) were also identified. The information obtained in Arabidopsis was useful for extrapolation to other plant species, and E2F sites in the promoters of ~70 rice (*Oryza sativa*) orthologue genes have been found (Vandepoele *et al.*, 2005).

6.5 Functional relevance of E2F–DP in development

Transcriptional regulation of E2F target genes is crucial for cell cycle progression as well as for initiating and maintaining specific differentiation states. Beyond the cellular level, E2F regulatory networks also impact different aspects of organogenesis at different stages of plant development. Very importantly, the balance between the size of different cellular pools, such as proliferating cells and endoreplicating cells, from which a proportion is recruited to differentiation, also seems to depend on E2F function (Gutierrez, 2005). Accumulating evidence discussed below clearly points to a complex interplay that is being elucidated by using overexpression and loss-of-function approaches. However, much effort is still needed to define more precisely the function of different E2F members, both individually and in cooperation, by selectively modifying their levels in specific plant cells and tissues.

E2Fa associates preferentially with DPa as shown *in vitro* (Kosugi and Ohashi, 2002c; Mariconti *et al.*, 2002) and *in vivo* (De Veylder *et al.*, 2002). When *E2Fa* and *DPa* are constitutively co-expressed, they induce cell division and endoreplication (De Veylder *et al.*, 2002; Rossignol *et al.*, 2002). The E2Fa–DPa-mediated hyperplasia is inhibited by co-expressing a dominant-negative mutant of *CDKB1;1*, while the endoreplication potential is enhanced (Boudolf *et al.*, 2004).

E2Fb also prefers DPa for heterodimerization (Kosugi and Ohashi, 2002c; Magyar *et al.*, 2005). It seems to play a role in regulating proliferation in coordination with phytohormone signalling. Thus, coexpression of *E2Fb*, but not *E2Fa*, with *DPa* in tobacco cells stimulates cell proliferation in the absence of auxin (Magyar *et al.*, 2005). This study also demonstrated that E2Fb is a key factor controlling the balance between cell proliferation and endoreplication. High levels of E2Fb *in planta* also lead to phenotypes in roots, leaves and cotyledons consistent with hyperplasia, possibly as a consequence of the upregulation of a variety of both G1/S and G2/M target genes (Sozzani *et al.*, 2006).

E2Fc is a transcriptional repressor that is abundant in arrested cells and targeted to the proteosome after CDK phosphorylation by the SCF^{SKP2} complex upon stimulation of cell proliferation (del Pozo *et al.*, 2002a). E2Fc and DPb associate *in vivo* (del Pozo *et al.*, 2006). Overexpression of E2Fc impairs formation of leaf primordia (del Pozo *et al.*, 2002a) while strong reduction of E2Fc mRNA levels by RNAi produces leaves with a reduced ploidy level, suggesting that E2Fc–DPb regulates the switch from proliferation to the endocycle programme during leaf development (del Pozo *et al.*, 2006).

Alteration in RBR-dependent E2F activity has been also achieved by various ways of inactivating RBR function. Virus-induced transcriptional silencing of the tobacco *RBR* gene prolongs cell proliferation activity and produces the occurrence of extra endocycles in leaves (Park *et al.*, 2005). Similar phenotypes are observed in Arabidopsis after inactivation of RBR by expressing the geminivirus RepA protein (Desvoyes *et al.*, 2006). Furthermore, the consequences of RBR inactivation depend on the cell type and the developmental stage, e.g., hyperplasia in young leaves and stimulation of the endocycle programme in older leaves (Desvoyes *et al.*, 2006). In

Arabidopsis these effects are associated with an increased level of both E2Fa and E2Fc activities, suggesting that these two proteins act in coordination to regulate the transcriptional programme during development. Local inactivation of RBR in Arabidopsis roots increases the amount of stem cells without affecting cell cycle duration, as it occurs in plants overexpressing *CYCD3;1* or *E2Fa–DPa* (Wildwater *et al.*, 2005).

E2Fe/DEL1 is expressed in proliferating cells and its overexpression produces a significant reduction in ploidy level, whereas reduced levels of *E2Fe/DEL1* causes an increase in the endoreplication level (Vlieghe *et al.*, 2005). Furthermore, the endoreplication phenotype, but not the hyperplasia, of the E2Fa–DPa-overexpressing plants is reduced by half when E2Fe/DEL1 is overexpressed (Vlieghe *et al.*, 2005). These studies suggest that several interconnected pathways act to regulate the switch from proliferation to endoreplication during development.

A previously unforeseen role of E2Ff/DEL3 in controlling cell expansion, more clearly observed in hypocotyl cells, has been reported (Ramirez-Parra *et al.*, 2004). As discussed above, this E2F member seems to play a role in differentiated cells by negatively regulating the expression of a set of genes involved in cell wall biosynthesis without having any detectable impact in ploidy level.

Together, the studies of the role played by different E2F family members during development need to be expanded to obtain information about the transcriptional regulatory potential of each of them and to determine whether and how they act in coordination to control different processes in a cell-type-specific manner.

6.6 E2F and epigenetic regulation of gene expression

The main repressor function of RBR on E2F is to directly mask its transactivation domain, thus blocking E2F activity. However, recent findings in animal cells indicate that the RB/E2F complex actively represses transcription by interacting with different chromatin-remodelling factors. Binding of RB to E2F target promoters permits the direct recruitment of chromatin factors that modulate epigenetically the transcription of these genes. Thus, these activities can actively repress the transcription of E2F-regulated genes by blocking the access of other transcription factors in the proximity of the E2F-binding sites, or by direct nucleosome modification on these loci (Zhang and Dean, 2001).

The regulation of nucleosome structure is an important mechanism of gene transcription, and SWI/SNF-like complexes are one of the main modulators involved in the chromatin-remodelling processes, although the mechanism by which SWI2-related complexes reorganize nucleosomal interactions remains unknown. SWI2 (also called SNF2 in yeast) is a DNA-dependent ATPase that is the central component of the SWI/SNF complex and can alter chromatin structure in the absence of the other subunits. Two SWI2 homologues have been characterized in human cells, BRM and BRG1, and both can interact with RB protein. Recently, over 40 SWI2-related genes have been identified in Arabidopsis, and some of them have been shown to be essential for Arabidopsis development (reviewed in Reyes, 2006).

Thus, *PHOTOPERIOD-INDEPENDENT EARLY FLOWERING* (*PIE1*) gene controls flowering time and *PICKLE* (*PKL*) belongs to the CHD subfamily and is involved in the suppression of embryonic and meristematic characteristics during development (Noh and Amasino, 2003). The Arabidopsis homologue of Drosophila Brahma, *BRM*, controls shoot development and flowering (Farrona *et al.*, 2004), and its closest homologue, *SPLAYED* (*SYD*), seems to be involved in the repression of meristem phase transition (Wagner and Meyerowitz, 2002). However, until now, a direct *in vivo* interaction of these proteins with plant RBR is still lacking.

Another important mechanism of E2F/RB-mediated epigenetic control is the recruitment of histone deacetylases. The most frequent mechanism of histone covalent modification is the acetylation of lysines of the histone N-terminal mediated by histone acetyl transferases (HATs), causing an open chromatin conformation that activates transcription of E2F target genes. Subsequent deacetylation by histone deacetylases (HDACs) reverts this process. In mammals, it has been reported that RB recruits HDACs through the Rb-associated protein RbAp48. Recently, this interaction has been described between the tomato and maize RbAp48 proteins and the maize RBR, suggesting that it might also recruit the RbAp48-associated HDACs (Ach *et al.*, 1997; Nicolas *et al.*, 2001; Rossi *et al.*, 2001; Rossi and Varotto, 2002).

A third mechanism that involves covalent modification of histones is methylation of lysine residues by histone methyl transferases (HMTases). In human cells, RB can recruit the Suv39H1 HMTase through its SET-domain, which methylates specifically the lysine 9 of histone H3, producing transcriptional silencing. As in heterochromatin silencing, transcriptional repression of E2F target genes by Suv39H1 is mediated through the binding of heterochromatin protein 1 (HP1) to the methylated histone (Nielsen *et al.*, 2001; Vandel *et al.*, 2001; Vaute *et al.*, 2002; Ait-Si-Ali *et al.*, 2004). The Arabidopsis genome contains at least 29 genes encoding SET-domain proteins, and two of them present chromatin-associated HMTase activity (Baumbusch *et al.*, 2001). HP1 homologues have also been identified from several plant species. *TERMINAL FLOWER2* (*TFL2*), the only homologue of HP1 in Arabidopsis, silences genes within the euchromatic region but not genes positioned in heterochromatin (Nakahigashi *et al.*, 2005). In addition, the Arabidopsis SET-domain protein *CURLY LEAF* (*CLF*) has been shown to bind both the maize and human RB proteins (Williams and Grafi, 2000; Gaudin *et al.*, 2001), and maize RBR (ZmRb1) interacts with the polycomb protein FIE (Mosquna *et al.*, 2004). Altogether, these data suggest that RBR/E2F/HMTase/HP1 complexes could also function in E2F-mediated gene expression control in plants.

DNA methylation is another mechanism of transcriptional repression that is conserved between animals and plants (Martienssen and Colot, 2001). DNMT1 has been found to form a complex with mammalian RB, E2F1 and HDAC1 and represses transcription from E2F-responsive promoters (Robertson *et al.*, 2000). A *DNMT1* homologous gene is also present in the Arabidopsis genome, suggesting that this mechanism may also operate in plants. Recent findings indicate that human complexes of HDAC/Rb/E2F also contain hSWI/SNF complexes and DNMT1, and HMT Suv39H1 interacts with histone deacetylases (Robertson *et al.*, 2000; Zhang *et al.*, 2000; Vaute *et al.*, 2002), indicating that different types of modifications might communicate with each other.

6.7 Concluding remarks: complexity of E2F-dependent regulation of gene expression

The recent use of genome-wide approaches to discover novel E2F target genes allowed the identification of several hundred such genes that are involved not only in DNA replication and cell cycle progression, but also in DNA damage repair, apoptosis, differentiation and development (Bracken *et al.*, 2004). These new findings that can be extrapolated to plants (Ramirez-Parra *et al.*, 2003; Vandepoele *et al.*, 2005) have greatly enriched our understanding of how E2F controls transcription and cellular homeostasis. However, and surprisingly, some E2F target genes regulate novel pathways not previously reported in mammals, such as cell expansion, nitrogen assimilation or photosynthesis, which seem to be plant-specific processes.

In addition, here we have discussed the complex regulation of E2F target genes as *MCM3*, *PCNA* and *RNR*, in which E2F-binding sites can have very different roles in transcriptional regulation in different cell cycle phases or developmental stages depending on the sequence context. Thus, it is likely that the activity of a site is influenced by the surrounding promoter structure and the cellular environment or tissue type (van Ginkel *et al.*, 1997). This fact may promote that two or more E2F sites act either synergistically or antagonistically to regulate transcription of a particular gene under different cellular settings. In addition, we cannot exclude the possibility that different E2F sites may interact with different E2F–DP complexes, as previously reported in animals (Zhu *et al.*, 1995; Di Fiore *et al.*, 1999), although this type of selectivity has not been described for plant E2F elements, until now. However, this idea is coherent with the distinct developmental and tissue-specific expression patterns of the different E2F and DP family members.

Transcriptional regulation by E2F family members is being shown to have an unprecedented complexity. On one side, different pathways that impinge on numerous cellular processes, both in proliferating and in differentiated cells, are affected by E2F. This implies that the same E2F site may be acting in different ways depending on the cellular status. On another, recent evidence indicates that increasing the level of a particular E2F member, which is considered as a typical activator of gene expression, produces downregulation of some genes, bearing E2F sites in their promoters. The reverse, that is increasing the level of E2F members that produce repression of certain E2F targets, increases the level of others. These observations clearly indicate that caution has to be taken in defining individual E2F members as activators or repressors, since this may depend on the cellular and/or developmental context. Finally, different E2F members may contribute to transcriptional regulation of a particular target, again depending on availability of the various E2Fs at a given time or cell type. Therefore, the challenge ahead is to define as precisely as possible the intricate regulatory gene network controlled by E2F family members in a cell type/status-, organ- and developmental-stage-specific manner.

Acknowledgments

Authors are indebted to C. Vaca for technical assistance. This work has been partially supported by grant BMC2003-2131 and BFU2006-5602 (Spanish Ministry of Science and Technology) and by an institutional grant from Fundación Ramon Areces.

References

Ach, R.A., Taranto, P. and Gruissem, W. (1997) A conserved family of WD-40 proteins binds to the retinoblastoma protein in both plants and animals. *Plant Cell* **9**, 1595–1606.

Ait-Si-Ali, S., Guasconi, V., Fritsch, L., *et al.* (2004) A Suv39h-dependent mechanism for silencing S-phase in differentiated but in cycling cells. *EMBO J* **23**, 605–615.

Albani, D., Mariconti, L., Ricagno, S., *et al.* (2000) DcE2F, a functional plant E2F-like transcriptional activator from Daucus carota. *J Biol Chem* **275**, 19258–19267.

Baumbusch, L.O., Thorstensen, T., Krauss, V., *et al.* (2001) The *Arabidopsis thaliana* genome contains at least 29 active genes encoding SET domain proteins that can be assigned to four evolutionarily conserved classes. *Nucleic Acids Res* **29**, 4319–4333.

Bisova, K., Krylov, D.M. and Umen, J.G. (2005) Genome-wide annotation and expression profiling of cell cycle regulatory genes in *Chlamydomonas reinhardtii*. *Plant Physiol* **137**, 475–491.

Black, A.R. and Azizkhan-Clifford, J. (1999) Regulation of E2F: a family of transcription factors involved in proliferation control. *Gene* **237**, 281–302.

Boudolf, V., Vlieghe, K., Beemster, G.T.S., *et al.* (2004) The plant-specific cyclin-dependent kinase CDKB1;1 and transcription factor E2Fa-DPa control the balance of mitotically dividing and endoreduplicating cells in Arabidopsis. *Plant Cell* **16**, 2683–2692.

Bracken, A.P., Ciro, M., Cocito, A. and Helin, K. (2004) E2F target genes: unraveling the biology. *Trends Biochem Sci* **29**, 409–417.

Breeden, L. (1996) Start-specific transcription in yeast. *Curr Topics Microbiol Immunol* **208**, 95–127.

Breyne, P., Dreesen, R., Vandepoele, K., *et al.* (2002) Transcriptome analysis during cell division in plants. *Proc Natl Acad Sci USA* **99**, 14825–14830.

Castellano, M.M., Boniotti, M.B., Caro, E., Schnittger, A. and Gutierrez, C. (2004) DNA replication licensing affects cell proliferation or endoreplication in a cell type-specific manner. *Plant Cell* **16**, 2380–2393.

Castellano, M.M., del Pozo, J.C., Ramirez-Parra, E., Brown, S. and Gutierrez, C. (2001) Expression and stability of Arabidopsis CDC6 are associated with endoreplication. *Plant Cell* **13**, 2671–2686.

Chabouté, M.-E., Clément, B. and Philipps, G. (2002) S phase and meristem-specific expression of the tobacco RNR1b gene is mediated by an E2F element located in the 50 leader sequence. *J Biol Chem* **277**, 17845–17851.

Chabouté, M.-E., Clement, B., Sekine, M., Philipps, G. and Chaubet-Gigot, N. (2000) Cell cycle regulation of the tobacco ribonucleotidereductase small subunit gene is mediated by E2F-like elements. *Plant Cell* **12**, 1987–2000.

Christensen, J., Cloos, P., Toftegaard, U., *et al.* (2005) Characterization of E2F8, a novel E2F-like cell-cycle regulated repressor of E2F-activated transcription. *Nucleic Acids Res* **33**, 5458–5470.

Dahl, M., Meskiene, I., Bogre, L., *et al.* (1995) The D-type alfalfa cyclin gene cycMs4 complements G1 cyclin-deficient yeast and is induced in the G1 phase of the cell cycle. *Plant Cell* **7**, 1847–1857.

de Jager, S.M., Menges, M., Bauer, U.M. and Murray, J.A.H. (2001) Arabidopsis E2F1 binds a sequence present in the promoter of S-phase-regulated gene AtCDC6 and is a member of a multigene family with differential activities. *Plant Mol Biol* **47**, 555–568.

De Veylder, L., Beeckman, T., Beemster, G.T.S., *et al.* (2002) Control of proliferation, endoreduplication and differentiation by the Arabidopsis E2Fa-DPa transcription factor. *EMBO J* **21**, 1360–1368.

del Pozo, J.C., Boniotti, M.B. and Gutierrez, C. (2002a) Arabidopsis E2Fc functions in cell division and is degraded by the ubiquitin-SCF^{AtSKP2} pathway in response to light. *Plant Cell* **14**, 3057–3071.

del Pozo, J.C., Dharmasiri, S., Hellmann, H., Walker, L., Gray, W.M. and Estelle, M. (2002b) AXR1-ECR1-dependent conjugation of RUB1 to the Arabidopsis Cullin AtCUL1 is required for auxin response. *Plant Cell* **14**, 421–433.

del Pozo, J.C., Diaz-Trivino, S., Cisneros, N. and Gutierrez, C. (2006) The balance between cell division and endoreplication depends on E2Fc-DPb, transcription factors regulated by the ubiquitin-SCF^{SKP2A} pathway. *Plant Cell* **18**, 2224–2235.

Desvoyes, B., Ramirez-Parra, E., Xie, Q., Chua, N.H. and Gutierrez, C. (2006) Cell type-specific role of the retinoblastoma/E2F pathway during Arabidopsis leaf development. *Plant Physiol* **140**, 67–80.

Di Fiore, B., Guarguaglini, G., Palena, A., Kerkhoven, R.M., Bernards, R. and Lavia, P. (1999) Two E2F sites control growth-regulated and cell cycle-regulated transcription of the Htf9-a/RanBP1 gene through functionally distinct mechanisms. *J Biol Chem* **274**, 10339–10348.

Diaz-Trivino, S., del Mar Castellano, M., de la Paz Sanchez, M., Ramirez-Parra, E., Desvoyes, B. and Gutierrez, C. (2005) The genes encoding Arabidopsis ORC subunits are E2F targets and the two ORC1 genes are differently expressed in proliferating and endoreplicating cells. *Nucleic Acids Res* **33**, 5404–5414.

Dynlacht, B.D., Moberg, K., Lees, J.A., Harlow, E. and Zhu, L. (1997) Specific regulation of E2F family members by cyclin-dependent kinases. *Mol Cell Biol* **17**, 3867–3875.

Egelkrout, E.M., Mariconti, L., Settlage, S.B., Cella, R., Robertson, D. and Hanley-Bowdoin, L. (2002) Two E2F elements regulate the proliferating cell nuclear antigen promoter differently during leaf development. *Plant Cell* **14**, 3225–3236.

Egelkrout, E.M., Robertson, D. and Hanley-Bowdoin, L. (2001) Proliferating cell nuclear antigen transcription is repressed through an E2F consensus element and activated by geminivirus infection in mature leaves. *Plant Cell* **13**, 1437–1452.

Espinosa-Ruiz, A., Saxena, S., Schmidt, J., *et al.* (2004) Differential stage-specific regulation of cyclin-dependent kinases during cambial dormancy in hybrid aspen. *Plant J* **38**, 603–615.

Farrona, S., Hurtado, L., Bowman, J.L. and Reyes, J.C. (2004) The *Arabidopsis thaliana* SNF2 homolog AtBRM controls shoot development and flowering. *Development* **131**, 965–975.

Filatov, D. and Thelander, L. (1995) Role of a proximal NF-Y binding promoter element in S phase-specific expression of mouse ribonucleotide reductase R2 gene. *J Biol Chem* **270**, 25239–25243.

Gaudin, V., Libault, M., Pouteau, S., *et al.* (2001) Mutations in like heterochromatin protein 1 affect flowering time and plant architecture in Arabidopsis. *Development* **128**, 4847–4858.

Grafi, G., Burnett, R.J., Helentjaris, T., *et al.* (1996) A maize cDNA encoding a member of the retinoblastoma protein family: involvement in endoreduplication. *Proc Natl Acad Sci USA* **93**, 8962–8967.

Gutierrez, C. (2000) DNA replication and cell cycle in plants: learning from geminiviruses. *EMBO J* **19**, 792–799.

Gutierrez, C. (2005) Coupling cell proliferation and development in plants. *Nat Cell Biol* **7**, 535–541.

Gutierrez, C. (2006) Plant cells and viruses. In: *DNA Replication and Human Disease* (ed. DePamphilis, M.L.). Cold Spring Harbor Laboratory Press, Cold Spring Harbor, New York, pp. 257–272.

Hateboer, G., Wobst, A., Petersen, B.O., *et al.* (1998) Cell cycle-regulated expression of mammalian CDC6 is dependent on E2F. *Mol Cell Biol* **18**, 6679–6697.

Hayashi, Y., Yamagishi, M., Nishimoto, Y., Taguchi, O., Matsukage, A. and Yamaguchi, M. (1999) A binding site for the transcription factor Grainyhead/Nuclear transcription factor-1 contributes to regulation of the Drosophila proliferating cell nuclear antigen gene promoter. *J Biol Chem* **274**, 35080–35088.

Helin, K. (1998) Regulation of cell proliferation by the E2F transcription factors. *Curr Opin Genet Dev* **8**, 28–35.

Helin, K., Lees, J.A., Vidal, M., Dyson, N., Harlow, E. and Fattaey, A. (1992) A cDNA encoding a pRB-binding protein with properties of the transcription factor E2F. *Cell* **70**, 337–350.

Helin, K., Wu, C.L., Fattaey, A.R., *et al.* (1993) Heterodimerization of the transcription factors E2F-1 and DP-1 leads to cooperative trans-activation. *Genes Dev* **7**, 1850–1861.

Hennig, L., Taranto, P., Walser, M., Schonrock, N. and Gruissem, W. (2003) Arabidopsis MSI1 is required for epigenetic maintenance of reproductive development. *Development* **130**, 2555–2565.

Herwig, S. and Strauss, M. (1997) The retinoblastoma protein: a master regulator of cell cycle, differentiation and apoptosis. *Eur J Biochem* **246**, 581–601.

Huang, D.Y. and Prystowsky, M.B. (1996) Identification of an essential cis-element near the transcription start site for transcriptional activation of the proliferating cell nuclear antigen gene. *J Biol Chem* **271**, 1218–1225.

Huntley, R., Healy, S., Freeman, D., *et al.* (1998) The plant retinoblastoma protein homologue ZmRb1 is regulated during leaf development and displays conserved interactions with G1/S regulators and plant cyclin D (CycD) proteins. *Plant Mol Biol* **37**, 155–169.

Ishida, S., Huang, E., Zuzan, H., *et al.* (2001) Role for E2F in control of both DNA replication and mitotic functions as revealed from DNA microarrayanalysis. *Mol Cell Biol* **21**, 4684–4699.

Jonsson, Z.O. and Hübscher, U. (1997). Proliferating cell nuclear antigen: more than a clamp for DNA polymerases. *Bioessays* **1**, 967–975.

Kel, A.E., Kel-Margoulis, O.V., Farnham, P.J., Bartley, S.M., Wingender, E. and Zhang, M.Q. (2001) Computer-assisted identification of cell cycle-related genes: new targets for E2F transcription factors. *J Mol Biol* **309**, 99–120.

Kelman, Z. (1997) PCNA: structure, functions and interactions. *Oncogene* **14**, 629–640.

Korenjak, M. and Brehm, A. (2005) E2F-Rb complexes regulating transcription of genes important for differentiation and development. *Curr Opin Genet Dev* **15**, 520–527.

Kosugi S. and Ohashi, Y. (1997) PCF1 and PCF2 specifically bind to cis elements in the rice proliferating cell nuclear antigen gene. *Plant Cell* **9**, 1607–1619.

Kosugi, S. and Ohashi, Y. (2002a) E2Ls, E2F-like repressors of Arabidopsis that bind to E2F sites in a monomeric form. *J Biol Chem* **277**, 16553–16558.

Kosugi, S. and Ohashi, Y. (2002b) E2F sites that can interact with E2F proteins cloned from rice are required for meristematic tissue-specific expression of rice and tobacco proliferating cell nuclear antigen promoters. *Plant J* **29**, 45–59.

Kosugi, S. and Ohashi, Y. (2002c) Interaction of the Arabidopsis E2F and DP proteins confers their concomitant nuclear translocation and transactivation. *Plant Physiol* **128**, 833–843.

Leone, G., DeGregori, J., Yan, Z., *et al.* (1998) E2F3 activity is regulated during the cell cycle and is required for the induction of S phase. *Genes Dev* **12**, 2120–2130.

Li, L.-C., Bedinger, P.A., Volk, C., Jones, A.D. and Cosgrove, D.J. (2003) Purification and characterization of four β-expansins (Zea m 1 isoforms) from maize pollen. *Plant Physiol* **132**, 2073–2085.

Liu, Y.C., Chang, H.W., Lai, Y.C., Ding, S.T. and Ho, J.L. (1998) Serum responsiveness of the rat PCNA promoter involves the proximal ATF and AP-1 sites. *FEBS Lett* **441**, 200–204.

Magyar, Z., Atanassova, A., De Veylder, L., Rombauts, S. and Inzé, D. (2000) Characterization of two distinct DP-related genes from *Arabidopsis thaliana*. *FEBS Lett* **486**, 79–87.

Magyar, Z., De Veylder, L., Atanassova, A., Bakó, L., Inzé, D. and Bögre, L. (2005) The role of the Arabidopsis E2FB transcription factor in regulating auxin-dependent cell division. *Plant Cell* **17**, 2527–2541.

Maiti, B., Li, J., de Bruin, A., *et al.* (2005) Cloning and characterization of mouse E2F8, a novel mammalian E2F family member capable of blocking cellular proliferation. *J Biol Chem* **280**, 18211–18220.

Mariconti, L., Pellegrini, B., Cantoni, R., *et al.* (2002) The E2F family of transcription factors from *Arabidopsis thaliana*. Novel and conserved components of the retinoblastoma pathway in plants. *J Biol Chem* **277**, 9911–9919.

Marti, A., Wirbelauer, C., Scheffner, M. and Krek, W. (1999) Interaction between ubiquitin-protein ligase SCFSKP2 and E2F-1 underlies the regulation of E2F-1 degradation. *Nat Cell Biol* **1**, 14–19.

Martienssen, R.A. and Colot, V. (2001) DNA methylation and epigenetic inheritance in plants and filamentous fungi. *Science* **293**, 1070–1074.

Masubelele, N.H., Dewitte, W., Menges, M., *et al.* (2005) D-type cyclins activate division in the root apex to promote seed germination in Arabidopsis. *Proc Natl Acad Sci USA* **102**, 15694–15699.

Menges, M., de Jager, S.M., Gruissem, W. and Murray, J.A. (2005) Global analysis of the core cell cycle regulators of Arabidopsis identifies novel genes, reveals multiple and highly specific profiles of expression and provides a coherent model for plant cell cycle control. *Plant J* **41**, 546–566.

Menges, M., Hennig, L., Gruissem, W. and Murray, J.A. (2002) Cell cycle-regulated gene expression in Arabidopsis. *J Biol Chem* **277**, 41987–42002.

Menges, M., Hennig, L., Gruissem, W. and Murray, J.A. (2003) Genome-wide gene expression in an Arabidopsis cell suspension. *Plant Mol Biol* **53**, 423–442.

Mosquna, A., Katz, A., Shochat, S., Grafi, G. and Ohad, N. (2004) Interaction of FIE, a polycomb protein, with pRb: a possible mechanism regulating endosperm development. *Mol Genet Genomics* **271**, 651–657.

Müller, H., Bracken, A.P., Vernell, R., *et al.* (2001) E2Fs regulate the expression of genes involved in differentiation, development, proliferation and apoptosis. *Genes Dev* **15**, 267–285.

Nakahigashi, K., Jasencakova, Z., Schubert, I. and Goto K. (2005) The Arabidopsis heterochromatin protein1 homolog (TERMINAL FLOWER2) silences genes within the euchromatic region but not genes positioned in heterochromatin. *Plant Cell Physiol* **46**, 1747–1756.

Nicolas, E., Ait-Si-Ali, S. and Trouche, D. (2001) The histone deacetylase HDAC3 targets RbAp48 to the retinoblastoma protein. *Nucleic Acids Res* **29**, 3131–3136.

Nielsen, S.J., Schneider, R., Bauer, U.M., *et al.* (2001) Rb targets histone H3 methylation and HP1 to promoters. *Nature* **412**, 561–565.

Noh, Y.S. and Amasino, R.M. (2003) PIE1, an ISWI family gene, is required for FLC activation and floral repression in Arabidopsis. *Plant Cell* **15**, 1671–1682.

Ohta, T. and Xiong, Y. (2001) Phosphorylation- and Skp1-independent *in vitro* ubiquitination of E2F1 by multiple ROC-cullin ligases. *Cancer Res* **61**, 1347–1353.

Ohtani, K., DeGregori, J., Leone, G., Herendeen, D.R., Kelly, T.J. and Nevins, J.R. (1996) Expression of the HsOrc1 gene, a human ORC1 homolog, is regulated by cell proliferation via the E2F transcription factor. *Mol Cell Biol* **16**, 6977–6984.

Park, J.A., Ahn, J.W., Kim, Y.K., *et al.* (2005) Retinoblastoma protein regulates cell proliferation, differentiation and endoreduplication in plants. *Plant J* **42**, 153–163.

Porceddu, A., Stals, H., Reichheld, J.-P., *et al.* (2001) A plant-specific cyclin-dependent kinase is involved in the control of G_2/M progression in plants. *J Biol Chem* **276**, 36354–36360.

Ramirez-Parra, E., Frundt, C. and Gutierrez, C. (2003) A genome-wide identification of E2F-regulated genes in Arabidopsis. *Plant J* **33**, 801–811.

Ramirez-Parra, E. and Gutierrez, C. (2000) Characterization of wheat DP, a heterodimerization partner of the plant E2F transcription factor which stimulates E2F-DNA binding. *FEBS Lett* **486**, 73–78.

Ramirez-Parra, E., Lopez-Matas, M.A., Frundt, C. and Gutierrez, C. (2004) Role of an atypical E2F transcription factor in the control of Arabidopsis cell growth and differentiation. *Plant Cell* **16**, 2350–2363.

Ramirez-Parra, E., Xie, Q., Boniotti, M.B. and Gutierrez, C. (1999) The cloning of plant E2F, a retinoblastoma-binding protein, reveals unique and conserved features with animal G(1)/S regulators. *Nucleic Acids Res* **27**, 3527–3533.

Ren, B., Cam, H., Takahashi, Y., *et al.* (2002). E2F integrates cell cycle progression with DNA repair, replication and G2/M checkpoints. *Genes Dev* **16**, 245–256.

Reyes, J.C. (2006) Chromatin modifiers that control plant development. *Curr Opin Plant Biol* **9**, 21–27.

Robbens, S., Khadaroo, B., Camasses, A., *et al.* (2005) Genome-wide analysis of core cell cycle genes in the unicellular green alga *Ostreococcus tauri*. *Mol Biol Evol* **22**, 589–597.

Robertson, K.D., Ait-Si-Ali, S., Yokochi, T., Wade, P.A., Jones, P.L. and Wolffe, A.P. (2000) DNMT1 forms a complex with Rb, E2F1 and HDAC1 and represses transcription from E2F-responsive promoters. *Nat Genet* **25**, 338–342.

Rossi, V. and Varotto, S. (2002) Insights into the G1/S transition in plants. *Planta* **215**, 345–356.

Rossi, V., Varotto, S., Locatelli, S., *et al.* (2001) The maize WD-repeat gene ZmRbAp1 encodes a member of the MSI/RbAp sub-family and is differentially expressed during endosperm development. *Mol Genet Genomics* **265**, 576–584.

Rossignol, P., Stevens, R., Perennes, C., *et al.* (2002) AtE2F-a and AtDP-a, members of the E2F family of transcription factors, induce Arabidopsis leaf cells to re-enter S-phase. *Mol Gen Genet* **266**, 995–1003.

Rubin, S.M., Gall, A.L., Zheng, N. and Pavletich, N.P. (2005) Structure of the Rb C-terminal domain bound to E2F1-DP1: a mechanism for phosphorylation-induced E2F release. *Cell* **123**, 1093–1106.

Sabelli, P.A., Dante, R.A., Leiva-Neto, J.T., Jung, R., Gordon-Kamm, W.J. and Larkins, B.A. (2005) RBR3, a member of the retinoblastoma-related family from maize, is regulated by the RBR1/E2F pathway. *Proc Natl Acad Sci USA* **102**, 13005–13012.

Sabelli, P.A. and Larkins, B.A. (2006) Grasses like mammals? Redundancy and compensatory regulation within the retinoblastoma protein family. *Cell Cycle* **5**, 352–355.

Sekine, M., Ito, M., Uemukai, K., Maeda, Y., Nakagami, H. and Shinmyo, A. (1999) Isolation and characterization of the E2F-like gene in plants. *FEBS Lett* **460**, 117–122.

Shen, W.H. (2002) The plant E2F-Rb pathway and epigenetic control. *Trends Plant Sci* **7**, 505–511.

Soni, R., Carmichael, J.P., Shah, Z.H. and Murray, J.A. (1995) A family of cyclin D homologs from plants differentially controlled by growth regulators and containing the conserved retinoblastoma protein interaction motif. *Plant Cell* **7**, 85–103.

Sorrell, D.A., Menges, M., Healy, J.M., *et al.* (2001) Cell cycle regulation of cyclin-dependent kinases in tobacco cultivar Bright Yellow-2 cells. *Plant Physiol* **126**, 1214–1223.

Sozzani, R., Maggio, C., Varotto, S., *et al.* (2006) Interplay between Arabidopsis activating factors E2Fb and E2Fa in cell cycle progression and development. *Plant Physiol* **140**, 1355–1366.

Stevens, R., Mariconti, L., Rossignol, P., Perennes, C., Cella, R. and Bergounioux, C. (2002) Two E2F sites in the Arabidopsis MCM3 promoter have different roles in cell cycle activation and meristematic expression. *J Biol Chem* **277**, 32978–32984.

Tsurimoto, T. (1998) PCNA, a multifunctional ring on DNA. *Biochim Biophys Acta* **1443**, 23–39.

Uemukai, K., Iwakawa, H., Kosugi, S., *et al.* (2005) Transcriptional activation of tobacco E2F is repressed by co-transfection with the retinoblastoma-related protein: cyclin D expression overcomes this repressor activity. *Plant Mol Biol* **57**, 83–100.

van Ginkel, P.R., Hsiao, K.M., Schjerven, H. and Farnham, P.J. (1997) E2F-mediated growth regulation requires transcription factor cooperation. *J Biol Chem* **272**, 18367–18374.

Vandel, L., Nicolas, E., Vaute, O., Ferreira, R., Ait-Si-Ali, S. and Trouche, D. (2001) Transcriptional repression by the retinoblastoma protein through the recruitment of a histone methyltransferase. *Mol Cell Biol* **21**, 6484–6494.

Vandepoele, K., Raes, J., De Veylder, L., Rouzé, P., Rombauts, S. and Inzé, D. (2002) Genome-wide analysis of core cell cycle genes in Arabidopsis. *Plant Cell* **14**, 903–916.

Vandepoele, K., Vlieghe, K., Florquin, K., e*t al.* (2005) Genome-wide identification of potential plant E2F target genes. *Plant Physiol* **139**, 316–328.

Vaute, O., Nicolas, E., Vandel, L. and Trouche, D. (2002) Functional and physical interaction between the histone methyl transferase Suv39H1 and histone deacetylases. *Nucleic Acids Res* **30**, 475–481.

Vigo, E., Muller, H., Prosperini, E., *et al.* (1999) CDC25A phosphatase is a target of E2F and is required for efficient E2F-induced S phase. *Mol Cell Biol* **19**, 6379–6395.

Vissenberg, K., Martinez-Vilchez, I.M., Verbelen, J.P., Miller, J.G. and Fry, S.C. (2000) *In vivo* colocalization of xyloglucan endotransglycosylase activity and its donor substrate in the elongation zone of Arabidopsis roots. *Plant Cell* **12**, 1229–1237.

Vlieghe, K., Boudolf, V., Beemster, G.T.S., *et al.* (2005) The DP-E2F-like gene DEL1 controls the endocycle in *Arabidopsis thaliana*. *Curr Biol* **15**, 59–63.

Vlieghe, K., Vuylsteke, M., Florquin, K., e*t al.* (2003) Microarray analysis of E2Fa-DPa-overexpressing plants uncovers a cross-talking genetic network between DNA replication and nitrogen assimilation. *J Cell Sci* **116**, 4249–4259.

Wagner, D. and Meyerowitz, E.M. (2002) SPLAYED, a novel SWI/SNF ATPase homolog, controls reproductive development in Arabidopsis. *Curr Biol* **12**, 85–94.

Weinmann, A.S., Bartley, S.M., Zhangm, T., Zhangm, M.Q. and Farnham, P.J. (2001) Use of chromatin immunoprecipitation toclone novel E2F target promoters. *Mol Cell Biol* **21**, 6820–6832.

Wells, J., Graveel, C.R., Bartley, S.M., Madore, S.J. and Farnham, P.J (2002) The identification of E2F1-specific target genes. *Proc Natl Acad Sci USA* **99**, 3890–3895.

Whittaker, A.J., Royzman, I. and Orr-Weaver, T.L. (2000) Drosophila double parked: a conserved, essential replication protein that colocalizes with the origin recognition complex and links DNA replication with mitosis and the down-regulation of S phase transcripts. *Genes Dev* **14**, 1765–1776.

Wildwater, M., Campilho, A., Perez-Perez, J.M., *et al.* (2005) The retinoblastoma-related gene regulates stem cell maintenance in Arabidopsis roots. *Cell* **123**, 1337–1349.

Williams, L. and Grafi, G. (2000) The retinoblastoma protein – a bridge to heterochromatin. *Trends Plant Sci* **5**, 239–240.

Xiao, B., Spencer, J., Clements, A., *et al.* (2003) Crystal structure of the retinoblastoma tumor suppressor protein bound to E2F and the molecular basis of its regulation. *Proc Natl Acad Sci USA* **100**, 2363–2368.

Xie, Q., Sanz-Burgos, A., Hannon, G.J. and Gutierrez, C. (1996) Plant cells contain a novel member of the retinoblastoma family of growth regulatory proteins. *EMBO J* **15**, 4900–4908.

Xie, Q., Suárez-López, P. and Gutierrez, C. (1995) Identification and analysis of a retinoblastoma-binding motif in the replication protein of a plant DNA virus: requirement for efficient viral DNA replication. *EMBO J* **14**, 4073–4082.

Yamaguchi, M., Hayashi, Y. and Matsukage, A. (1995) Essential role of E2F recognition sites in regulation of the proliferating cell nuclear antigen gene promoter during Drosophila development. *J Biol Chem* **270**(42), 25159–25165.

Yamasaki, L. and Pagano, M. (2004) Cell cycle, proteolysis and cancer. *Curr Opin Cell Biol* **16**, 623–628.

Yoshida, K. and Inoue, I. (2004) Regulation of Geminin and Cdt1 expression by E2F transcription factors. *Oncogene* **23**, 3802–3812.

Zhang, H.S. and Dean, D.C. (2001) Rb-mediated chromatin structure regulation and transcriptional repression. *Oncogene* **20**, 3134–3138.

Zhang, H.S., Gavin, M., Dahiya, A., *et al.* (2000) Exit from G1 and S phase of the cell cycle is regulated by repressor complexes containing HDAC-Rb-hSWI/SNF and Rb-hSWI/SNF. *Cell* **101**, 79–89.

Zheng, N., Fraenkel, E., Pabo, C.O. and Pavletich, N.P. (1999) Structural basis of DNA recognition by the heterodimeric cell cycle transcription factor E2F–DP. *Genes Dev* **13**, 666–674.

Zhu, L., Xie, E. and Chang, L.S. (1995) Differential roles of two tandem E2F sites in repression of the human p107 promoter by retinoblastoma and p107 proteins. *Mol Cell Biol* **15**, 3552–3562.

Zhu, W., Giangrande, P.H. and Nevins, J.R. (2004) E2Fs link the control of G1/S and G2/M transcription. *EMBO J* **23**, 4615–4626.

Zimmermann, P., Hirsch-Hoffmann, M., Hennig, L., Gruissem, W. (2004) GENEVESTIGATOR. Arabidopsis Microarray Database and Analysis Toolbox. *Plant Physiol* **136**, 2621–2632.

7 Function of the retinoblastoma-related protein in plants

Wilhelm Gruissem

7.1 Introduction

Embryonic and postembryonic development of plants depends on the tight coupling of cell cycle activity with temporal and spatial cell differentiation processes to control form and function of the organism. Postembryonic plant development is directed by stem cells in shoot and root apical meristems that are maintained throughout the life of the plant (reviewed in Williams and Fletcher, 2005; Fleming, 2006). While many proteins that regulate programmed gene expression during the plant cell cycle, stem cell maintenance and development have now been identified, their concerted action in integrating cell proliferation, differentiation and morphogenesis is still poorly understood (Gutierrez, 2005; Inzé, 2005). Disruption of these processes results in developmental abnormalities, and in animals it can lead to unscheduled cell proliferation and tumor development.

Perhaps the most fundamental decision facing each cell exiting a stem cell niche in a multicellular organism is whether to commit to another round of DNA replication or exit the cell cycle and undergo differentiation. This decision is governed by complex regulatory mechanisms, which have evolved to ensure that cells are able to properly interpret various cues and translate the information to the basic cell cycle machinery. In animals, many of the key regulators have first been identified as oncogenes or tumor suppressor genes that are associated with various types of cancer. We are now beginning to appreciate that genes of both classes have homologs in plants, suggesting that the pathways in which they function have evolved early during the emergence of eukaryotic organisms. One such example was the discovery of genes in plants that encode proteins with homology to the mammalian retinoblastoma tumor suppressor protein pRB, which has emerged as a key regulator at the interface of cell cycle regulation and differentiation. This chapter will provide a brief introduction to the prevailing view of animal pRB function and relate this to information that has been gained from the genetic and biochemical analysis of plant retinoblastoma-related (RBR) proteins in cell cycle regulation and development.

7.2 Retinoblastoma proteins and the tumor suppressor concept

Nearly 20 years ago, the retinoblastoma tumor suppressor gene (*RB*) was identified and cloned (Friend *et al.*, 1986; Fung *et al.*, 1987; Lee *et al.*, 1987). In mammals,

RB is part of a small family of genes that includes the genes for p107 and p130, whose functions are partially redundant with RB (Classon and Dyson, 2001; Claudio *et al.*, 2002). These regulatory proteins are also known as 'pocket proteins' due to the structure of their so-called A- and B-domains that make up the C-terminal half of the proteins (Livingston *et al.*, 1993). Products of the mammalian RB tumor suppressor gene family have been implicated in several different cellular processes, but their functions are currently best understood as important regulators of the cell cycle, differentiation and apoptosis. In humans, mutational inactivation of both *RB* alleles is necessary and sufficient for both hereditary and sporadic retinoblastoma, validating Knudson's two-hit hypothesis for tumor formation (Knudson, 1971). Subsequent examination has shown that *RB* is functionally inactivated in nearly every human tumor (Weinberg, 1995; Zheng and Lee, 2001; Sherr and McCormick, 2002), suggesting that elimination of pRB function is a prerequisite of human cancer development (Sherr, 2000).

Efforts to clearly establish the mechanism by which pRB acts as a tumor suppressor in mammals, however, have been complicated in part by functional redundancies with p107 and p130 (Classon and Dyson, 2001). In addition to its role in cell cycle regulation, pRB has also been implicated in a wide variety of cellular processes. It is now accepted that pRB exerts broader control in the balance of cell proliferation, differentiation and apoptosis (Lipinski and Jacks, 1999; Classon and Harlow, 2002; Hickman *et al.*, 2002). In many cell types, pRB promotes terminal differentiation by inducing cell cycle exit and tissue-specific gene expression (Liu *et al.*, 2004). It is therefore easy to imagine that a decrease in differentiation potential and increase in cell proliferation rates, which are found for RB-deficient cells, contribute to tumor formation. As a result, the current view is shifting to a major role of pRB in the maintenance of a differentiated state, which is consistent with the escape of differentiated cells from this control when pRB is mutated.

7.3 The retinoblastoma pathway is conserved in animals and plants

Although it was initially thought that pRB-family proteins were peculiar to vertebrates, it is now clear that orthologs are found in eukaryotic cells of different phyla and kingdoms (Durfee *et al.*, 2000; Claudio *et al.*, 2002; Whyatt and Grosveld, 2002), including yeast (Costanzo *et al.*, 2004; de Bruin *et al.*, 2004). *RB* homologs from plants were first cloned from maize (Grafi *et al.*, 1996; Xie *et al.*, 1996; Ach *et al.*, 1997a) and later from tobacco (Nakagami *et al.*, 1999) and Arabidopsis (Kong *et al.*, 2000), concomitantly with the identification of the genes in *Drosophila melanogaster* and *Caenorhabditis elegans* (Du *et al.*, 1996; Lu and Horvitz, 1998). It is now clear that *RB*-related genes are found widely in different phyla and kingdoms, including single-cell algae (Durfee *et al.*, 2000; Umen and Goodenough, 2001; Claudio *et al.*, 2002; Sabelli and Larkins, 2006). The highest level of identity between plant and human RB-like proteins (20–35%) is found in the A- and B-domains (de Jager and Murray, 1999; Durfee *et al.*, 2000). Considering the amino acid homology observed between the animal RB and plant RBR proteins, it was not unexpected that they also have similar biochemical properties (Ach *et al.*, 1997a). Similar to genes for the mammalian pocket proteins, plant *RBR* genes are ubiquitously transcribed but also show temporal and spatial differences in their expression patterns (Ach *et al.*, 1997a; Huntley *et al.*, 1998; Sabelli *et al.*, 2005). Together, the discovery of RB-like proteins outside of metazoans confirms that they are more highly conserved in eukaryotes than initially thought and suggests that the RB pathway may have an important role in the development of all multicellular eukaryotes, not just animals.

Plant cyclin-D proteins also bind to RBR, and the LxCxE motif is necessary for those interactions (Meijer and Murray, 2000). The mitotic cyclin, CycA, may have some weak affinity for RBR as well. Finally, the E2F/DP family of transcription factors, which interact with RB in animals, is conserved in plants (De Veylder *et al.*, 2002; Boudolf *et al.*, 2004; Vlieghe *et al.*, 2005). Together, while animals and plants clearly evolved distinct development strategies, growing evidence suggests that key mechanisms that control the cell cycle, proliferation and differentiation are highly conserved (Nakagami *et al.*, 2002; Beemster *et al.*, 2003; De Veylder *et al.*, 2003; Barrôco *et al.*, 2005; Inzé, 2005). Unlike mammalian cells in which pRB function is partially redundant with p107 and p130, however, Arabidopsis has only a single gene for an RBR protein (Kong *et al.*, 2000). This, together with the unique postembryonic development of plants, will facilitate the functional analysis of the RB pathway in plants, as will be discussed below.

7.4 Retinoblastoma proteins form complexes with E2F transcription factors to control entry into the cell cycle

A hallmark of pRB function is the regulation of the downstream E2F and DP families of transcription factors, which are also conserved in plants (Sekine *et al.*, 1999; Albani *et al.*, 2000; Magyar *et al.*, 2000; Ramirez-Parra and Gutierrez, 2000; de Jager *et al.*, 2001; Mariconti *et al.*, 2002; Inzé, 2005). This aspect of the RBR pathway in plants is discussed in more detail in Chapter 6. It is now well established that mammalian E2F1–E2F6 require heterodimerization with DP1 or DP2 for their function. Functional cooperation of plant E2F and DP proteins has been confirmed in transgenic Arabidopsis expressing both proteins (De Veylder *et al.*, 2002; Rossignol *et al.*, 2002), but interactions between all E2F and DP members have not been analyzed in detail. In animals, E2F1, E2F2 and E2F3a function as activators while E2F3b, E2F4, E2F5, E2F6 and E2F7 function as repressors. The size of the E2F family of transcription factors is similar in plants and E2Fa as well as E2Fb can clearly function as activators of the cell cycle (De Veylder *et al.*, 2002; Rossignol *et al.*, 2002; Sozzani *et al.*, 2006), but the function of the other family members is largely unknown.

The discovery that pRB controls the G1 restriction point, which inhibits S-phase entry and progression of DNA replication, has led to an early model for cell cycle control (Figure 7.1). This model suggests that G1 control depends on the interaction of pRB with the E2F/DP family of transcription factors (Harbour and Dean, 2000; Müller and Helin, 2000; Attwooll *et al.*, 2004; Blais and Dynlacht,

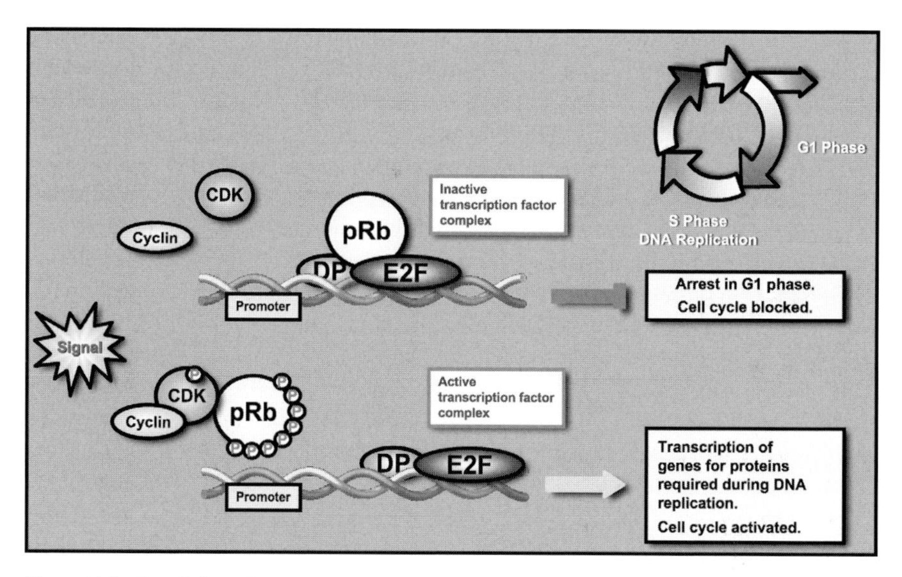

Figure 7.1 Regulation of E2F-dependent promoters by pRB. The model reflects our current understanding of pRB, E2F and DP functions in G1 control and activation of genes that mediate S-phase entry. In noncycling or quiescent cells, hypophosphorylated pRB inhibits proliferation through association with other proteins. A principal target is the E2F transcription factor, which together with pRB and DP forms a repressive complex on promoters containing E2F binding sites. Following activation by mitogenic signal transduction pathways, phosphorylation of pRB and RBR by CDKs requires D- and E-type cyclins in animals and D- and A-type cyclins in plants. Hyperphosphorylated pRB dissociates from the DP/E2F complex to release transcriptional arrest. This model illustrates how mutation or loss of *RB* function causes deregulation of E2F and inappropriate cell proliferation. The model also explains how other proteins, such as viral proteins that bind pRB and displace E2F (Nevins, 2001), can deregulate E2F through their effect on pRB. With the discovery that pRB, E2F and DP are members of protein families, which consist of the pRB, p107 and p130 pocket proteins, E2F1–7, and the DP1 and DP2 binding partners of E2F1–6 (Dimova and Dyson, 2005), the model of the pRB regulatory pathway has become more complex. It is now clear that pRB interacts with many other proteins and transcription factors in regulating fundamental aspects of cell growth and differentiation (Morris and Dyson, 2001). While plant RBR has not been investigated at this level of detail, conserved biochemical properties of RBR and the presence of the E2F/DP family of transcription factors suggest that the RBR–E2F has similar functions in plants.

2004; Bracken *et al.*, 2004; Dimova and Dyson, 2005). Together they form repressive complexes on promoters of cell cycle regulatory genes such as *cyclin E*, many S-phase-specific genes, and the *E2F* genes themselves (Stevaux and Dyson, 2002). Following mitogenic stimulation of the cell and rise of CDK activity, this repressive function is relieved by phosphorylation of pRB (see below), leading to S-phase entry and DNA synthesis. This model explains the tumor suppressor activity of pRB and how loss of pRB function leads to deregulation of E2F and unscheduled cell proliferation.

While the prevailing model of pRB/E2F function is useful to conceptualize the core cell cycle function of pRB, it is now becoming increasingly clear that pRB and the pRB/E2F pathway have much broader control functions. Gene expression profiling experiments have revealed that the gene network controlled by the E2F family of transcription factors in animals and plants is complex (Ramirez-Parra *et al.*, 2003; Blais and Dynlacht, 2004; Bracken *et al.*, 2004; Dimova and Dyson, 2005; Vandepoele *et al.*, 2005). The complexity of the E2F transcriptional network in terms of number of possible transcriptional complexes and number of putative gene targets makes it difficult to understand the concerted actions of E2F beyond the mechanism of transcriptional activation and repression by individual complexes. The problem is further confounded by the realization that animal E2F transcription factors have specialized functions in development beyond establishing a cell cycle arrest that is permissive for cell differentiation (DeGregori, 2002; Attwooll *et al.*, 2004; Dimova and Dyson, 2005). Although this aspect has not been broadly investigated in plants (Ramirez-Parra *et al.*, 2004), given a similar complexity of the E2F family of transcription factors it can be expected that E2Fs in plants will also show functional specialization.

7.5 G1 restriction point control is mediated by retinoblastoma protein phosphorylation

Early studies in mammalian cells revealed that hypophosphorylated pRB interacts with the E2F/DP transcription factors in G0 or G1 cells to form repressive complexes by inhibiting the E2F transactivation domain (Figure 7.1). During G1 progression, pRB becomes increasingly phosphorylated on its 16 potential Ser/Thr-Pro phosphoacceptor sites (Mittnacht, 1998), causing hyperphosphorylated pRB to dissociate from the E2F/DP complex to allow the transcription of genes that mediate S-phase entry (Weinberg, 1995). In response to mitogenic signals, the cyclin-dependent kinases Cdk4 and Cdk6 together with D cyclins begin phosphorylation of pRB in early and mid-G1, followed by Cdk2 together with E cyclins later in G1 (Sherr, 1996). Although the functional significance of the different pRB phosphoacceptor sites remains largely obscure, there is now increasing evidence that Cdk2, Ckd4 and Cdk6 phosphorylate pRB at different sites (Kitagawa *et al.*, 1996; Zarkowska and Mittnacht, 1997; Takaki *et al.*, 2005).

Several aspects of the mammalian pRB/E2F regulatory network are conserved in other animals and plants as well. While plants do not appear to have orthologs of mammalian Cdk4 and Cdk6 (Inzé, 2005), they contain two classes of cyclin-dependent kinases (CDKA and CDKB) that can potentially phosphorylate RBR. D-type cyclins are conserved in plants (Oakenfull *et al.*, 2002) and their expression is associated with cell proliferation and responds strongly to mitogenic signals such as sucrose or cytokinin (Riou-Khamlichi *et al.*, 2000; Menges *et al.*, 2002; Oakenfull *et al.*, 2002; Menges *et al.*, 2005). Plant CYCD proteins interact with RBR through a conserved LxCxE motif in their N-terminal domain (Huntley *et al.*, 1998), and isolated CYCD/cyclin-dependent kinase complexes can phosphorylate RBR *in vitro* (Nakagami *et al.*, 1999, 2002; Boniotti and Gutierrez, 2001). Consistent with these interactions, CYCD can accelerate the cell cycle and overcome the repressive activity of the RBR/E2F complex in plant tissue culture cells (Koroleva *et al.*, 2004; Uemukai *et al.*, 2005; Menges *et al.*, 2006). During seed germination in Arabidopsis elevated *CYCD* expression precedes root emergence, and loss-of-function mutants for D-type cyclins that are expressed early showed a significant delay in the onset of cell proliferation (Masubelele *et al.*, 2005). These experiments place CYCD upstream of the RBR/E2F pathway and suggest that plant CYCDs, like animals D cyclins, have a key role in transmitting mitogenic signals to G1/S control.

7.6 Animal and plant DNA viruses target retinoblastoma proteins to induce host DNA replication

In animal cells, small DNA viruses such as simian virus 40 (SV40), papilloma viruses, and adenoviruses encode oncoproteins that target pRB for induction of host cell DNA replication (for a recent review see Lee and Cho, 2002). The binding of these proteins inactivates the function of pRB to control the cell cycle, which often results in tumor formation induced by viral infection.

Like animal pRB, plant RBR also binds to viral proteins (Ach *et al.*, 1997a, b). Remarkably, RBR can associate with mammalian viral oncoproteins, SV40 T-antigen, adenovirus E1A, and HPV E, as well as plant virus proteins, as was shown for the RepA polypeptide of wheat dwarf virus (WDV) and the tomato golden mosaic virus (TGMV) replication protein AL1. These associations require a conserved LxCxE motif in the viral proteins (Grafi *et al.*, 1996; Horváth *et al.*, 1998; Gordon-Kamm *et al.*, 2002), with the exception of the AL1 interaction, which occurs through a novel motif (Kong *et al.*, 2000). Together, it is now well established that DNA geminiviruses encode proteins that can bind to RBR through the conserved LxCxE motif in the mastrevirus RepA protein and through a novel amino acid sequence motif in Rep proteins of the begomovirus and curtovirus subgroups. Expression of RepA in transformed plant tissue culture cells stimulates cell division and callus growth (Gordon-Kamm *et al.*, 2002). This is consistent with a model that after infection geminivirus proteins bypass the normal control of RBR phosphorylation and release of E2F activity to induce host cell DNA replication.

7.7 Information on retinoblastoma protein function in animal development is still incomplete

While many aspects of cell cycle regulation are now well understood, the precise mechanisms that connect the cell cycle to cell differentiation and development, as well as the role of pocket proteins in this process, remain largely obscure. Most of the biochemical research on pRB, p107 and p130 focused initially on the role of these proteins in the control of the G1 restriction point during the cell cycle, but it is now becoming increasingly clear that the pocket proteins coordinate a multitude of regulatory processes that influence transitions between cell proliferation and differentiation. Early lessons learned from knockout mouse models suggested that mammalian *RB* is essential during embryo development. $RB^{-/-}$ embryos die between day 13 and day 15 of gestation with defects in erythroid, neuronal and lens development (Clarke *et al.*, 1992; Jacks *et al.*, 1992; Lee *et al.*, 1992). Later studies using conditional *RB* loss-of-function alleles corrected some of the earlier observations (Wu *et al.*, 2003), but also confirmed that pRB has critical cell intrinsic functions in promoting cell cycle exit and terminal differentiation of diverse cell types, including neurons, muscle fibers, lens fibers, keratinocytes, adipocytes and trophoblast stem cells (Liu *et al.*, 2004). Since the inactivation of pRB is a hallmark of most human cancers (Weinberg, 1995), the differentiation-promoting function of pRB and its ability to induce tissue-specific gene expression contribute to the tumor suppressor function of pRB. The situation in mammalian cell is complex, however, because p107 and p130 may contribute to growth arrest and have substantial functional overlap with pRB during development (Classon and Dyson, 2001; Claudio *et al.*, 2002). For example, cultured mouse embryo fibroblasts (MEF) in which *RB* was lost by Cre-lox-mediated recombination can reenter the cell cycle from either the differentiated or senescent stage (Dannenberg *et al.*, 2000; Sage *et al.*, 2000). This is in contrast to germline $RB^{-/-}$ MEFs, in which p107 expression is upregulated and appears to compensate for the loss of pRB (Sage *et al.*, 2003). Together, these experiments suggest that pRB alone is sufficient to maintain cells in a state of growth arrest.

Embryos deficient in pRB/p107 or pRB/p130 have phenotypes similar to $RB^{-/-}$ embryos, but they die two days earlier and $RB^{-/-}/p107^{-/-}$ embryos have more extensive apoptosis in their central nervous system and hemapoetic cells than $RB^{-/-}$ embryos (E11–E13; Lee *et al.*, 1996; Lipinski and Jacks, 1999). As for MEFs, this suggests that p107 and p130 can partially substitute for pRB in certain developmental processes. Moreover, $p107^{-/-}$ and $p130^{-/-}$ mice in the same genetic background as $RB^{-/-}$ mice are viable and have no overt adult phenotypes. Mice deficient for both p107 and p130 die shortly after birth, confirming that both proteins have essential but overlapping functions (Cobrinik *et al.*, 1996). Analysis of chimeric mice generated using embryonic stem cells deficient for either pRB or pRB/p107 showed that in $RB^{+/+}/RB^{-/-}$ chimeras $RB^{-/-}$ cells contribute to most adult organs, while $RB^{-/-}/p107^{-/-}$ fail to make a significant contribution to most adult tissues (Robanus-Maandag *et al.*, 1998). These results imply that pRB has a nonautonomous function, because the intrinsic requirement for pRB function in the differentiation of certain cell types, such as erythroid or neuronal cells, can be bypassed by homotypic signaling from wild-type cells (Whyatt and Grosveld, 2002).

The functional compensation between mammalian pocket proteins makes it difficult to establish the direct mechanistic function of pRB in cell differentiation and development. In contrast to vertebrates, other animals and plants have single *RBR* genes. Genetic studies in *C. elegans* demonstrated that *lin-35* (the *C. elegans* homolog of *RB*) together with *efl-1* and *dpl-1* (*E2F* and *DP* homologs) function in a pathway that regulates vulva development (Lu and Horvitz, 1998; Ceol and Horvitz, 2001). Recent results from mosaic analysis have shown that the single *lin-35* gene function is acting in the hypodermal syncytium (hyp) cells to counteract the

Ras-MAPK (mitogen-activated protein kinase) signaling pathway, which in vulva precursor cells (VPCs) coordinates a defined program of cell divisions and differentiation that gives rise to single vulva (Myers and Greenwald, 2005). Since LIN-35 is active in hyp cells but does not seem to be necessary in VPCs, it is likely that its antagonizing function on Ras-MAPK signaling is indirect and involves currently unknown cell–cell signaling mechanisms.

7.8 Retinoblastoma proteins may have conserved functions in germline development

Genetic analysis of plant retinoblastoma homologs has been complicated at first because in grasses such as maize or rice RBR is encoded by a small family of genes, which could have redundant or compensatory functions (Ach *et al.*, 1997a; Durfee *et al.*, 2000; Sabelli and Larkins, 2006). The dicotyledonous plant Arabidopsis has only a single *RBR* gene (Kong *et al.*, 2000), and mutants with *rbr* null alleles can be maintained as heterozygous plants that develop normally from seedling to mature stages. Arabidopsis *RBR/rbr* flowers, however, produce siliques with 50% aborted ovules. Reciprocal crosses to wild-type Arabidopsis showed that the *rbr* mutant allele could only be transmitted paternally although at a very low frequency, suggesting strong maternal control and an effect of the *rbr* mutation on microspore development. Analysis of the aborted ovules revealed that inactivation of the *RBR* gene in Arabidopsis disrupts development of the female gametophyte (Ebel *et al.*, 2004). Thus, unlike *RB* in mice, plant *RBR* has an essential function already prior to zygotic development.

During normal female gametophyte development, the meiotic megaspore undergoes three mitotic divisions, resulting in eight nuclei (Plate 7.1). Six of these nuclei become cellularized to form the three cells of the egg cell apparatus, which consists of the egg cell and two synergids, and the three antipodal cells that later undergo apoptosis. The two remaining nuclei fuse to form the diploid nucleus of the central cell that after fertilization gives rise to the endosperm (Buchanan *et al.*, 2000). This reproductive structure of seven cells is called the embryo sac. Unfertilized ovules with *rbr* female gametophytes have embryo sacs with supernumerary nuclei and partially completed cell divisions instead of the distinct egg cell apparatus (Plate 7.1). Since the three antipodal cells were correctly positioned, it appears that the three scheduled mitotic divisions occur normally during early female gametophyte development. The nuclei that give rise to the egg cell apparatus, however, do not arrest in the absence of fertilization, which leads to the accumulation of supernumerary nuclei at the micropylar end of the ovule. Unfertilized *rbr* female gametophytes express the *CYCB1;1-promoter:GUS* marker gene, which is not expressed during the G1 and S phases of the cell cycle (Colón-Carmona *et al.*, 1999; Menges *et al.*, 2005) (Plate 7.1). This suggests that nuclei of the egg cell apparatus had continued in the cell cycle and that RBR is required to arrest mitotic activity in the female gametophyte prior to fertilization.

The roles of animal retinoblastoma family proteins in germ cell proliferation and differentiation are currently not well understood. Although expression levels and phosphorylation status of pRB, p107 and p130 in rat testis are modulated during germ cell cycle progression and apoptosis, the developmental significance of this specific regulation remains unknown (Toppari *et al.*, 2003). Recent screens for genetic modifiers of *C. elegans lin-35/RB* have identified a mutation in *xnp-1*, which encodes a homolog of human ATR-X, a member of the Swi2/Snf2 family of ATP-dependent DEAD/DEAH box helicases that function in nucleosome remodeling and transcriptional regulation. Although it was shown that *lin-35* and *xnp-1* function redundantly during *C. elegans* postembryonic development to control the execution of cell lineages that are required for somatic gonad development, it remains unknown if *lin-35* is required directly for germ cell differentiation (Bender *et al.*, 2004). More detailed analysis of the *lin-35* mutant showed that somatic cells express P-granule-like structures, which are normally assembled in the germline blastomers from maternally expressed gene products and segregated exclusively to the germline lineage during development (Wang *et al.*, 2005). Together, *RB* homologs may have a role in germline development that is conserved in animals and plants.

7.9 Retinoblastoma proteins connect stem cell maintenance to cell proliferation and differentiation

As discussed, conditional targeting of animal retinoblastoma proteins in mouse (pRB), Drosophila (RBF) and *C. elegans* (lin-35) is now revealing important functions of the proteins in the terminal differentiation of many tissues (Liu *et al.*, 2004; Korenjak and Brehm, 2005). Plants differ from animals, however, in that they produce simple embryos with few organs. A large part of plant development occurs postembryonically and depends on environmental cues. All organs of the growing plant are derived from stem cells in the shoot and root apical meristem, which are established early during embryogenesis (Takada and Tasaka, 2002). Plant meristems have high rates of cell proliferation, and pluripotent meristem cells become incorporated into leaves and flower organs in the shoot or committed to specific developmental fates in the root. Robust genetic circuits maintain meristem stem cell populations, and several regulatory genes have been identified that control cell fates in different regions of the meristems (Weigel and Jürgens, 2002; Bäurle and Laux, 2003; Byrne *et al.*, 2003; Sharma *et al.*, 2003; Vernoux and Benfey, 2005). As cells become displaced from the meristem, they acquire specific fates, their rate of cell division increases, and they begin to enlarge via vacuolation. Cells that are being displaced from the shoot apical meristem rapidly gain photosynthetic capacity, as is evident from the activation of genes required for photosynthesis (Fleming *et al.*, 1996; Fleming, 2006). Recent work is now revealing that the transition from stem cells to pluripotent meristem cells and organ patterning is orchestrated by auxin and cytokinin in cooperation with a network of regulatory genes (Fleming, 2005a; Castellano and Sablowski, 2005; Jiang and Feldman, 2005; Kepinski and Leyser,

2005; Leibfried *et al.*, 2005; Reddy and Meyerowitz, 2005; Barbier de Reuille *et al.*, 2006; Jönsson *et al.*, 2006; Smith *et al.*, 2006). But the mechanisms that connect meristem stem cell maintenance with cell proliferation and differentiation during plant development have remained elusive.

The unscheduled mitotic activity and disruption of normal female gametophyte development caused by Arabidopsis *rbr* loss-of-function alleles suggest that RBR may influence the switch from cell proliferation to differentiation (Ebel *et al.*, 2004). The gametophytic lethality of Arabidopsis *rbr* mutants, however, precludes direct testing of this hypothesis in postembryonic meristems. Even rare fertilization of *rbr* female gametophytes produces only abnormal embryos that abort very early in development, often with multinucleate cells (Plate 7.1). One approach to circumvent this problem is to use RNA interference to perturb the RBR pathway in meristems (Dinneny and Benfey, 2005; Wildwater *et al.*, 2005). This strategy revealed that suppression of *RBR* transcript accumulation in the root meristem of *RBR–RNAi* plants (rRBr) early after embryogenesis resulted in the production of several extra layers of undifferentiated cells in the columella root cap. Unlike mature columella cells, these cells did not accumulate amyloplasts, suggesting that they retained stem cell identity. This could be confirmed by ablation of the quiescent center (QC) cells, which in wild type results in differentiation of the adjacent columella root cap stem cells. After QC ablation the supernumerary undifferentiated root cap cell layers in rRBr plants immediately differentiate, indicating that suppression of RBR function in the root meristem allows expansion of the stem cell domain and that QC signaling prolongs maintenance of stem cell identity. Genetic interaction studies revealed that genes, which are necessary for QC specification, interact closely with RBR to maintain the root meristem stem cell niche and to prevent premature differentiation of stem cells before they exit the stem cell niche. Consistent with this view, ectopic expression of RBR in the root meristem results in rapid differentiation of stem cells (Wildwater *et al.*, 2005). Although modulation of other cell cycle genes can influence cell proliferation and growth, they do not affect meristem function in a way similar to what has been observed for RBR (Cockcroft *et al.*, 2000; Wang *et al.*, 2000; De Veylder *et al.*, 2001, 2002). Together, the results place RBR at a node that connects stem cell niche patterning and stem cell maintenance with cell proliferation and differentiation in plants.

There is now increasing evidence that pRB is also required for maintenance of mammalian stem cells (Liu *et al.*, 2004). Conditional targeting of *RB* loss to the mouse epidermis using the *keratin-14* promoter to drive Cre recombinase disrupts the maintenance of the epidermal stem cell pool and the regulation of the homeostatic balance between cells in the stem cell pool, differentiating and postmitotic keratinocytes (Ruiz *et al.*, 2004). In the absence of pRB, cell proliferation and differentiation become uncoupled, leading to epidermal hyperplasia and sebaceous gland carcinoma. Similar observations were made for p107 in the regulation of neural precursor cells in the mammalian brain (Vanderluit *et al.*, 2004). Moreover, stem cell RNA profiling data reveal that specific cell cycle genes (e.g. *CYCD*) and genes encoding proteins that have been reported to interact with pRB are upregulated in mouse stem cells (Ivanova *et al.*, 2002; Ramalho-Santos *et al.*, 2002). But the mechanisms by which pRB connects mammalian stem cell maintenance to proliferation and differentiation are still elusive.

7.10 Perturbation of RBR during leaf development affects cell proliferation and control of DNA replication

Leaf formation is a basic aspect of plant development and initiates with the recruitment of proliferating cells that have moved to the peripheral region of the SAM after exiting the stem cell niche. The initial stage of leaf formation involves a complex transcriptional network that switches cells in the primordia from a pluripotent meristem fate to a determinate, nonmeristem fate. The newly formed primordium then undergoes lateral growth and growth along the proximal–distal axis that is controlled by a set of specific transcription factor activities. Subsequent to the events that direct leaf polarity, the elaboration of final leaf shape and size results from coordinated cell proliferation and expansion along the length and the width of the leaf. While significant progress has been made in dissecting the genetic networks and regulatory components that control leaf shape and size, the molecular mechanisms that connect leaf cell proliferation to tissue differentiation and acquisition of cell fate remain largely unknown (Byrne, 2005; Fleming, 2005b).

Attempts have been made to understand the function of RBR in leaf development (Park *et al.*, 2005). Using virus-induced gene silencing (VIGS) to downregulate the *NbRBR1* gene in *Nicotiana benthamiana* results in growth retardation and development of small, curled and distorted leaves, although the typical dorsoventral organization of the leaf is not affected. The abnormal leaf development can be explained by prolonged cell proliferation, which produces a significantly increased number of smaller cells. This unscheduled cell proliferation affects stomata patterning in the epidermis, but mature stomata appear to differentiate normally. A different strategy to target the RBR pathway during Arabidopsis leaf development has used inducible expression of a viral RBR binding protein (geminivirus RepA). Ectopic expression of RepA appears to restrict cell division in the mesophyll and palisade tissues during early leaf development, but increases cell proliferation in the epidermis (Desvoyes *et al.*, 2006). In both cases trichome development is affected, and cells show enhanced endoreplication later in leaf development. At first sight, these results suggest that in plants RBR functions as a negative regulator of cell proliferation during leaf development and perhaps connects cell proliferation to specific cell differentiation processes, as may be the case in trichome development (Schnittger and Hülskamp, 2002; Schellmann and Hülskamp, 2005). But the different effects of downregulation of RBR function on cell proliferation are difficult to reconcile and could be explained by the specific experimental strategies. For example, VIGS in *N. benthamiana* may not result in complete inactivation of RBR expression, and expression of the viral RepA in Arabidopsis may affect specific RBR functions that are targeted by the virus to initiate viral DNA replication (Kong *et al.*, 2000), but that are not relevant for normal function of RBR during leaf development.

7.11 Roles of retinoblastoma proteins in transcription activation and repression

The examples discussed above are consistent with a broad role of the animal pRB and plant RBR pathways in developmental control. But precisely how pRB-family proteins or RBR regulate the transcription of genes to control cell proliferation and differentiation is not completely understood at present, considering that they interact with a broad range of transcription factors and chromatin remodeling complexes in animals (Morris and Dyson, 2001; Frolov and Dyson, 2004), and most likely in plants as well. The list of more than 100 unique pRB binding proteins in mammals includes over 70 transcription factors, almost all of which are differentiation regulators that interact with the unphosphorylated form of pRB. In addition to the interaction with E2F and DP, pRB therefore appears to engage in regulatory complexes with transcription factors to restrict cell proliferation and to promote differentiation. For example, pRB cooperates with the bHLH transcription factor MyoD to promote muscle differentiation by activating muscle-specific gene expression and myogenic conversion (Novitch *et al.*, 1996, 1999). How this cooperative transactivation of specific promoters is achieved remains unclear, however, because several studies have failed to show direct interactions between pRB and MyoD (Smialowski *et al.*, 2005). It is more likely that MyoD cooperates with pRB in a larger complex that includes E2F1 (Li *et al.*, 2000) and CDK4, which is inactivated by MyoD to prevent phosphorylation of pRB (Zhang *et al.*, 1999).

Considering the different transcription factors and other proteins that have been shown to bind pRB, their biochemical interactions suggest that pRB can repress transcription in at least three distinct ways. As discussed above and elsewhere, one involves binding to activation domain for E2F, thereby blocking the activity of E2F (Dimova and Dyson, 2005). This mechanism may also be conserved for RBR in plants, although direct evidence is still missing (Ramirez-Parra *et al.*, 2006, this volume). In other cases, recruitment of pRB to a promoter may block the binding of other transcription factors and the assembly of transcription preinitiation complexes. Finally, interaction domains in pRB outside of the E2F binding domain could associate with protein complexes that modify chromatin (Frolov and Dyson, 2004). For example, pRB has a long N-terminal domain of approximately 40 kD that is conserved throughout vertebrate evolution, but the function of this domain is still poorly understood. Mouse models have been used to test the requirement for the N-terminal domain for pRB function, but the results from these studies have been inconclusive. On the other hand, specific mutations in the N-terminal domain have been identified in retinoblastoma families, and the N-terminal domain is important for the interaction of pRB with several proteins (Goodrich, 2003). All plant RBR proteins identified to date contain a similar long N-terminal domain containing regions that have strong amino acid sequence similarity with pRB (Durfee *et al.*, 2000).

Increasing evidence suggests that the recruitment of histone deacetylases (HDAC) and histone acetyltransferases (HAT) to E2F-regulated promoters is a key regulatory mechanism for their activation or inactivation. pRB binds to class I HDACs *in vitro* and associates with HDAC activity *in vivo*, suggesting that pRB–HDAC complexes could be directly involved in the repression of promoters (Brehm and Kouzarides, 1999; Frolov and Dyson, 2004). Although HATs have been implicated in the selective activation of E2F-regulated promoters, these results are difficult to interpret at present because both E2F and pRB can be acetylated (Martínez-Balbás *et al.*, 2000; Chan *et al.*, 2001), and at least for pRB it has been suggested that acetylation regulates its differentiation-specific function (Nguyen *et al.*, 2004). Several HDACs and HATs have been identified in plants as well that are required for normal development (Reyes *et al.*, 2002), but their relationship to the plant RBR pathway has not been clarified.

7.12 Retinoblastoma proteins interact with polycomb group complexes in controlling gene expression

The discovery that pRB interacts with the histone methyltransferase SUV39H1 focused the attention on a role of pRB in long-term chromatin silencing (Nielsen *et al.*, 2001; Vandel *et al.*, 2001). There is now increasing evidence that crosstalk between the pRB pathway, developmental regulators and epigenetic silencers may be key for cell-fate specification and cell differentiation (Cui *et al.*, 2006; Ferres-Marco *et al.*, 2006).

In view of these recent discoveries, how can the function of RBR be explained at a mechanistic level in diverse processes such as plant stem cell maintenance and differentiation as well as gametophyte and leaf development? One possibility is that RBR acts through the E2F pathway by controlling the expression of target genes that are involved in differentiation and development, similar to mechanisms identified in animal cells. For example, genetic and biochemical studies in Drosophila and *C. elegans* revealed that complexes of their pRB and E2F homologs together with chromatin-modifying enzymes control the expression of genes in specific developmental pathways (Frolov and Dyson, 2004; Korenjak and Brehm, 2005). Similar functions are currently difficult to assign to RBR/E2F complexes in plants because a systematic analysis of mutants in the E2F family of transcription factors has not been reported. Overexpression of E2Fa and DPa synergistically induces ectopic cell divisions that result in accumulation of small cells and strong growth retardation; however, it does not appear to have a significant effect on the plant development program (De Veylder *et al.*, 2002). In contrast, ectopic expression of RBR results in cell cycle arrest and differentiation of stem cells. This suggests that RBR can override stem cell maintenance to initiate a differentiation program (Wildwater *et al.*, 2005), perhaps by interaction with other proteins in addition to E2F when the balance of proteins in the RBR/E2F pathway is perturbed.

Other insights into possible mechanistic functions of RBR come from the identification of RBR and E2F target genes in the Arabidopsis gametophytic-lethal *rbr* mutants and *E2Fa/DPa* overexpression plants (Ebel *et al.*, 2004; Vandepoele *et al.*, 2005). Conserved plant E2F target genes encode proteins involved in cell cycle regulation, DNA replication and, interestingly, chromatin remodeling, including members of the RBAp48/MSI1 family of WD-40 proteins (Hennig *et al.*, 2005; Schönrock *et al.*, 2006) and p150/FAS1, a subunit of the chromatin assembly factor-1 (CAF-1) (Schönrock *et al.*, 2006). RBAp48/MSI1 proteins interact with mammalian pRB and plant RBR (Ach *et al.*, 1997b), and in Arabidopsis *MSI1* is required for the epigenetic maintenance of reproductive development and the arrest of the endosperm nucleus prior to fertilization (Hennig *et al.*, 2003; Köhler *et al.*, 2003; Guitton *et al.*, 2004). Loss of Arabidopsis *RBR* function also results in autonomous proliferation of the endosperm nucleus (Ebel *et al.*, 2004), similar to other mutants collectively known as *fertilization-independent seed* (*fis*) mutants. All *FIS* genes encode subunits of the PRC2-type polycomb group (PcG) complex (Figure 7.2). They participate in the genetic pathway controlled by the MEA–FIE complex that represses central cell development in the absence of fertilization (Köhler and Grossniklaus, 2002; Hsieh *et al.*, 2003; Guitton and Berger, 2005). Animal PcG proteins control the expression of homeotic genes (*HOX*) that regulate correct body pattern formation. They were first discovered in Drosophila where they maintain the segment-specific expression pattern of *HOX* genes that are established by the segmentation genes through subsequent cell divisions (Jürgens, 1985). Therefore, animal and plant PcG proteins are part of a cellular memory mechanism that ensures the epigenetic inheritance of gene expression states. Targeted disruption of PcG complex subunit genes has shown, however, that PcG proteins are also involved in the regulation of cell cycle and proliferation of certain cell types (Pasini *et al.*, 2004). Recently, PcG complexes have emerged as major players in establishing and interpretation of the histone code that may direct the function of chromatin remodeling machines and transcription factors (Strahl, 2000; Turner, 2000; Jenuwein and Allis, 2001). In plants, chromatin remodeling is important for genome integrity, epigenetic inheritance and during development (Martienssen and Colot, 2001; Goodrich and Tweedie, 2002).

In Drosophila and mammalian cells, two distinct PcG complexes have been identified. PRC1 is required for the continued maintenance of repression of gene activity by restricting access of ATP-dependent chromatin remodeling complexes to nucleosome arrays (Simon and Tamkun, 2002). Of the distinct PCR1 and PCR2 PcG complexes that are conserved in fly and mammals, only genes for subunits of the PCR2 complex have been identified in plants (Figure 7.2). Together with MSI1, MEDEA (MEA), fertilization-independent endosperm (FIE) and fertilization-independent seed 2 (FIS2) form a PCR2 PcG complex (FIE–MEA complex) that could be identified in chromatin immunoprecipitation experiments (Hsieh *et al.*, 2003; Köhler *et al.*, 2003; Guitton and Berger, 2005). It has been shown that RBR interacts with MSI1 and FIE (Ach *et al.*, 1997b; Mosquna *et al.*, 2004), suggesting that RBR could recruit the FIE–MEA complex to specific target genes to maintain their transcriptional repression. Such a function of RBR would be consistent with the early *fis* phenotype of *rbr* mutants (Ebel *et al.*, 2004), but it does not explain the hyperproliferation of nuclei in the egg cell apparatus of *rbr* mutants, which is not observed in *fis* mutants. Thus, RBR has two distinct maternal functions during gametophyte development, although the mechanism by which RBR controls differentiation and mitotic arrest of the egg cell apparatus prior to fertilization is currently unknown.

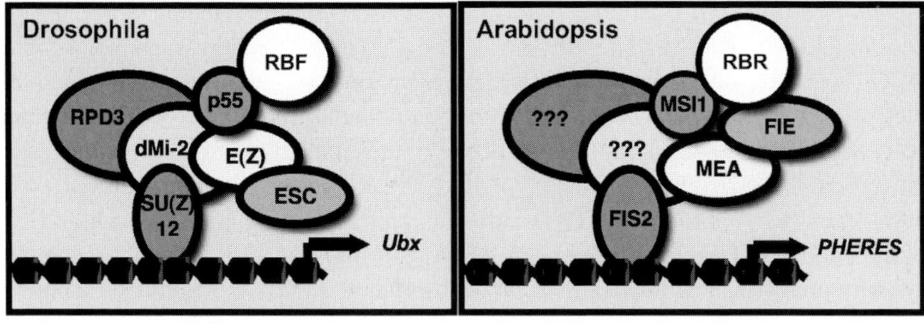

- **Regulation of *HOX* genes**
- **Regulation of cell proliferation**

- **Regulation of cell proliferation**
- **Endosperm development**

Figure 7.2 Retinoblastoma proteins interact with polycomb group (PcG) complexes that control cell proliferation and developmental processes in animals and plants. PcG proteins were discovered in Drosophila as transcriptional repressors of homeotic genes. For example, they maintain the segment-specific expression pattern of *HOX* genes that were initially established by the segmentation genes. In recent years, it has been found that PcG proteins are conserved in mammals and plants, demonstrating their important function in maintaining the transcriptional states of regulatory genes during development (Köhler and Grossniklaus, 2002; Hsieh *et al.*, 2003). Molecular cloning of the *FIS* genes revealed that they encode PcG homologs of the 600 kDa polycomb repressive complex 2 (PCR2), which was first purified from Drosophila. The ESZ–E(Z) complex has a histone methyltransferase activity that is required for the maintenance of *HOX* gene repression (Czermin *et al.*, 2002; Müller *et al.*, 2002). MEA is a SET-domain protein that is closely related to Drosophila E(Z), the enhancer of zeste protein. SET-domain proteins are implicated in histone H3 lysine methylation (H3K27 trimethylation) and possibly H1K26 methylation. E(Z)-based complexes can also methylate H3K9 *in vitro*, although the *in vivo* significance of this activity is not known. FIE, a homolog of Drosophila extra sex combs (ESC), is a WD-40 protein that binds to MEA and RBR (Mosquna *et al.*, 2004). FIS2 is a C2H2-type zinc finger protein that is homologous to Drosophila SU(Z)12 (suppressor of zeste 12; (Luo *et al.*, 1999; Birve *et al.*, 2001). Drosophila p55 and Arabidopsis MSI1 (multicopy suppressor of IRA1) are WD-40 proteins that are subunits of PCR2 (Tie *et al.*, 2001; Köhler *et al.*, 2003). In Drosophila, the complex consisting of ESC, E(Z) and p55 is sufficient for methyltransferase activity *in vitro* (Czermin *et al.*, 2002), although similar activity has not been demonstrated for the Arabidopsis MEA–FIE–MSI1 complex (Köhler *et al.*, 2003). Genetic and biochemical studies have shown that p55 and MSI1 interact with RBF (the pRB homolog in Drosophila) and RBR, respectively, but the pRB-related proteins in Drosophila and Arabidopsis do not copurify with the ESC–E(Z) or MEA–FIE PCR2 complex (Ach *et al.*, 1997b; Taylor-Harding *et al.*, 2004). Presumably RBF and RBR interact only transiently with PCR2 complexes to direct them to target genes. The ESC–E(Z) complex also associates with the histone deacetylase (HDAC) RPD3, although this interaction may be weak and no HDAC activity has been reported (Müller *et al.*, 2002). Although only the MEA–FIE complex has been biochemically isolated from female gametophytes and endosperm (Köhler *et al.*, 2003), the presence of other PcG-related genes in Arabidopsis suggests that distinct PCR2 complexes can exist during development.

7.13 Conclusion

Dissection of the pRB and RBR pathways in animal and plants is now revealing important new insights into the coordination of cell proliferation and differentiation. While many aspects of pRB and RBR regulation appear to be surprisingly similar between animals and plants considering their different developmental strategies, there are clearly important differences that will uncover new functions of these proteins. Unlike animals, plant cells are constrained by cell walls, which would confine uncontrolled cell proliferation caused by loss of RBR function in mature organs to specific cell types or tissues. Although RBR function is required early during specific developmental processes, it is therefore not immediately obvious why RBR expression is maintained in mature plant organs (Ach *et al.*, 1997a). Also, while many proteins that interact with pRB and RBR are similar in animals and plants, other proteins that associate with pRB in animals are not found in the plant genome. Thus, it can be expected that further analysis of RBR will reveal novel protein binding partners that may point to plant-specific mechanisms of RBR pathway regulation. Even after nearly 20 years of research on the retinoblastoma protein, the challenge remains to understand the regulatory network in which pRB and RBR function to control the gene expression program that governs the integration of cell proliferation, differentiation and morphogenesis in animals and plants.

Acknowledgments

Research on the Arabidopsis RBR pathway in the W.G. laboratory is supported by the Swiss National Science Foundation and the European Union Marie-Curie Training Program (SY-STEM).

References

Ach, R.A., Durfee, T., Miller, A.B., *et al.* (1997a) *RRB1* and *RRB2* encode maize retinoblastoma-related proteins that interact with a plant D-type cyclin and geminivirus replication protein. *Mol Cell Biol* **17**, 5077–5086.

Ach, R.A., Taranto, P. and Gruissem, W. (1997b) A conserved family of WD-40 proteins binds to the retinoblastoma protein in both plants and animals. *Plant Cell* **9**, 1595–1606.

Albani, D., Mariconti, L., Ricagno, S., *et al.* (2000) DcE2F, a functional plant E2F-like transcriptional activator from *Daucus carota*. *J Biol Chem* **275**, 19258–19267.

Attwooll, C., Lazzerini Denchi, E. and Helin, K. (2004) The E2F family: specific functions and overlapping interests. *EMBO J* **23**, 4709–4716.

Barbier de Reuille, P., Bohn-Courseau, I., Ljung, K., *et al.* (2006) Computer simulations reveal properties of the cell–cell signaling network at the shoot apex in *Arabidopsis*. *Proc Natl Acad Sci USA* **103**, 1627–1632.

Barrôco, R.M., Van Poucke, K., Bergervoet, J.H.W., *et al.* (2005) The role of the cell cycle machinery in resumption of postembryonic development. *Plant Physiol* **137**, 127–140.

Bäurle, I. and Laux, T. (2003) Apical meristems: the plant's fountain of youth. *Bio Essays* **25**, 961–970.

Beemster, G.T.S., Fiorani, F. and Inzé, D. (2003) Cell cycle: the key to plant growth control? *Trends Plant Sci* **8**, 154–158.

Bender, A.M., Wells, O. and Fay, D.S. (2004) *lin-35/Rb* and *xnp-1*/ATR-X function redundantly to control somatic gonad development in *C. elegans*. *Dev Biol* **273**, 335–349.

Birve, A., Sengupta, A.K., Beuchle, D., *et al.* (2001) *Su(z)12*, a novel *Drosophila* Polycomb group gene that is conserved in vertebrates and plants. *Development* **128**, 3371–3379.

Blais, A. and Dynlacht, B.D. (2004) Hitting their targets: an emerging picture of E2F and cell cycle control. *Curr Opin Genet Dev* **14**, 527–532.

Boniotti, M.B. and Gutierrez, C. (2001) A cell-cycle-regulated kinase activity phosphorylates plant retinoblastoma protein and contains, in *Arabidopsis*, a CDKA/cyclin D complex. *Plant J* **28**, 341–350.

Boudolf, V., Vlieghe, K., Beemster, G.T.S., *et al.* (2004) The plant-specific cyclin-dependent kinase CDKB1;1 and transcription factor E2Fa-DPa control the balance of mitotically dividing and endoreduplicating cells in *Arabidopsis*. *Plant Cell* **16**, 2683–2692.

Bracken, A.P., Ciro, M., Cocito, A. and Helin, K. (2004) E2F target genes: unraveling the biology. *Trends Biochem Sci* **29**, 409–417.

Brehm, A. and Kouzarides, T. (1999) Retinoblastoma protein meets chromatin. *Trends Biochem Sci* **24**, 142–145.

Buchanan, B.B., Gruissem, W. and Jones, R.L. (2000) *Biochemistry and Molecular Biology of Plants*. American Society of Plant Physiologists, Rockville, MD.

Byrne, M.E. (2005) Networks in leaf development. *Curr Opin Plant Biol* **8**, 59–66.

Byrne, M.E., Kidner, C.A. and Martienssen, R.A. (2003) Plant stem cells: divergent pathways and common themes in shoots and roots. *Curr Opin Genet Dev* **13**, 551–557.

Castellano, M.M. and Sablowski, R. (2005) Intercellular signalling in the transition from stem cells to organogenesis in meristems. *Curr Opin Plant Biol* **8**, 26–31.

Ceol, C.J. and Horvitz, H.R. (2001) *dpl-1* DP and *efl-1* E2F act with *lin-35* Rb to antagonize Ras signaling in *C. elegans* vulval development. *Mol Cell* **7**, 461–473.

Chan, H.M., Krstic-Demonacos, M., Smith, L., Demonacos, C. and La Thangue, N.B. (2001) Acetylation control of the retinoblastoma tumour-suppressor protein. *Nat Cell Biol* **3**, 667–674.

Clarke, A.R., Robanus Maandag, E., Van Roon, M., *et al.* (1992) Requirement for a functional *Rb-1* gene in murine development. *Nature* **359**, 328–330.

Classon, M. and Dyson, N. (2001) p107 and p130, versatile proteins with interesting pockets. *Exp Cell Res* **264**, 135–147.

Classon, M. and Harlow, E. (2002) The retinoblastoma tumour suppressor in development and cancer. *Nat Rev Cancer* **2**, 910–917.

Claudio, P.P., Tonini, T. and Giordano, A. (2002) The retinoblastoma family: twins or distant cousins? *Genome Biol* **3**, reviews 3012.1–3012.9.

Cobrinik, D., Lee, M.H., Hannon, G., *et al.* (1996) Shared role of the pRB-related p130 and p107 proteins in limb development. *Genes Dev* **10**, 1633–1644.

Cockcroft, C.E., den Boer, B.G.W., Healy, J.M.S. and Murray J.A.H. (2000) Cyclin D control of growth rate in plants. *Nature* **405**, 575–579.

Colón-Carmona, A., You, R., Haimovitch-Gal, T. and Doerner, P. (1999) Spatio-temporal analysis of mitotic activity with a labile cyclin-GUS fusion protein. *Plant J* **20**, 503–508.

Costanzo, M., Nishikawa, J.L., Tang, X., *et al.* (2004) CDK activity antagonizes Whi5, an inhibitor of G1/S transcription in yeast. *Cell* **117**, 899–913.

Cui, M., Kim, E.B. and Han, M. (2006) Diverse chromatin remodeling genes antagonize the Rb-involved SynMuv pathways in *C. elegans*. *PLoS Genet* **2**(e74), 0719–0732.

Czermin, B., Melfi, R., McCabe, D., Seitz, V., Imhof, A. and Pirrotta, V. (2002) *Drosophila* enhancer of Zeste/ESC complexes have a histone H3 methyltransferase activity that marks chromosomal Polycomb sites. *Cell* **111**, 185–196.

Dannenberg, J.-H., van Rossum, A., Schuijff, L. and te Riele, H. (2000) Ablation of the retinoblastoma gene family deregulates G_1 control causing immortalization and increased cell turnover under growth-restricting conditions. *Genes Dev* **14**, 3051–3064.

de Bruin, R.A.M., McDonald, W.H., Kalashnikova, T.I., Yates, J., III and Wittenberg, C. (2004) Cln3 activates G1-specific transcription via phosphorylation of the SBF bound repressor Whi5. *Cell* **117**, 887–898.

DeGregori, J. (2002) The genetics of the E2F family of transcription factors: shared functions and unique roles. *Biochim Biophys Acta* **1602**, 131–150.

de Jager, S.M. and Murray, J.A.H. (1999) Retinoblastoma proteins in plants. *Plant Mol Biol* **41**, 295–299.

de Jager, S.M., Menges, M., Bauer, U.-M. and Murray, J.A.H. (2001) *Arabidopsis* E2F1 binds a sequence present in the promoter of S-phase-regulated gene *AtCDC6* and is a member of a multigene family with differential activities. *Plant Mol Biol* **47**, 555–568.

Desvoyes, B., Ramirez-Parra, E., Xie, Q., Chua, N.-H. and Gutierrez, C. (2006) Cell type-specific role of the retinoblastoma/E2F pathway during *Arabidopsis* leaf development. *Plant Physiol* **140**, 67–80.

De Veylder, L., Beeckman, T., Beemster, G.T.S., *et al.* (2001) Functional analysis of cyclin-dependent kinase inhibitors of *Arabidopsis*. *Plant Cell* **13**, 1653–1667.

De Veylder, L., Beeckman, T., Beemster, G.T.S., *et al.* (2002) Control of proliferation, endoreduplication and differentiation by the *Arabidopsis* E2Fa/DPa transcription factor. *EMBO J* **21**, 1360–1368.

De Veylder, L., Joubès, J. and Inzé, D. (2003) Plant cell cycle transitions. *Curr Opin Plant Biol* **6**, 536–543.

Dimova, D.K. and Dyson, N.J. (2005) The E2F transcriptional network: old acquaintances with new faces. *Oncogene* **24**, 2810–2826.

Dinneny, J.R. and Benfey, P.N. (2005) Stem cell research goes underground: the *retinoblastoma-related* gene in root development. *Cell* **123**, 1180–1182.

Durfee, T., Feiler, H.S. and Gruissem, W. (2000) Retinoblastoma-related proteins in plants: homologues or orthologues of their metazoan counterparts. *Plant Mol Biol* **43**, 635–642.

Du, W., Vidal, M., Xie, J.E. and Dyson, N. (1996) RBF, a novel RB-related gene that regulates E2F activity and interacts with cyclin E in Drosophila. *Genes Dev* **10**, 1206–1218.

Ebel, C., Mariconti, L. and Gruissem, W. (2004) Plant retinoblastoma homologues control nuclear proliferation in the female gametophyte. *Nature* **429**, 776–780.

Ferres-Marco, D., Gutierrez-Garcia, I., Vallejo, D.M., Bolivar, J., Gutierrez-Aviño, F.J. and Dominguez, M. (2006) Epigenetic silencers and Notch collaborate to promote malignant tumours by *Rb* silencing. *Nature* **439**, 430–436.

Fleming, A.J. (2005a) Formation of primordia and phyllotaxy. *Curr Opin Plant Biol* **8**, 53–58.

Fleming, A.J. (2005b) The control of leaf development. *New Phytol* **166**, 9–20.

Fleming, A.J. (2006) The co-ordination of cell division, differentiation and morphogenesis in the shoot apical meristem: a perspective. *J Exp Bot* **57**, 25–32.

Fleming, A.J., Manzara, T., Gruissem, W. and Kuhlemeier, C. (1996) Fluorescent imaging of GUS activity and RT-PCR analysis of gene expression in the shoot apical meristem. *Plant J* **10**, 745–754.

Friend, S.H., Bernards, R., Rogelj, S., *et al.* (1986) A human DNA segment with properties of the gene that predisposes to retinoblastoma and osteosarcoma. *Nature* **323**, 643–646.

Frolov, M.V. and Dyson, N.J. (2004) Molecular mechanisms of E2F-dependent activation and pRB-mediated repression. *J Cell Sci* **117**, 2173–2181.

Fung, Y.K., Murphree, A.L., T'ang, A., Qian, J., Hinrichs, S.H. and Benedict, W.F. (1987) Structural evidence for the authenticity of the human retinoblastoma gene. *Science* **236**, 1657–1661.

Goodrich, D.W. (2003) How the other half lives, the amino-terminal domain of the retinoblastoma tumor suppressor protein. *J Cell Physiol* **197**, 169–180.

Goodrich, J. and Tweedie, S. (2002) Remembrance of things past: chromatin remodeling in plant development. *Annu Rev Cell Dev Biol* **18**, 707–746.

Gordon-Kamm, W., Dilkes, B.P., Lowe, K., *et al.* (2002) Stimulation of the cell cycle and maize transformation by disruption of the plant retinoblastoma pathway. *Proc Natl Acad Sci USA* **99**, 11975–11980.

Grafi, G., Burnett, R.J., Helentjaris, T., *et al.* (1996) A maize cDNA encoding a member of the retinoblastoma protein family: involvement in endoreduplication. *Proc Natl Acad Sci USA* **93**, 8962–8967.

Guitton, A.-E. and Berger, F. (2005) Control of reproduction by Polycomb Group complexes in animals and plants. *Int J Dev Biol* **49**, 707–716.

Guitton, A.-E., Page, D.R., Chambrier, P., *et al.* (2004) Identification of new members of Fertilisation Independent Seed Polycomb Group pathway involved in the control of seed development in *Arabidopsis thaliana*. *Development* 131:, 2971–2981.

Gutierrez, C. (2005) Coupling cell proliferation and development in plants. *Nat Cell Biol* **7**, 535–541.

Harbour, J.W. and Dean, D.C. (2000) The Rb/E2F pathway: expanding roles and emerging paradigms. *Genes Dev* **14**, 2393–2409.

Hennig, L., Bouveret, R. and Gruissem, W. (2005) MSI1-like proteins: an escort service for chromatin assembly and remodeling complexes. *Trends Cell Biol* **15**, 295–302.

Hennig, L., Taranto, P., Walser, M., Schönrock, N. and Gruissem, W. (2003) *Arabidopsis* MSI1 is required for epigenetic maintenance of reproductive development. *Development* **130**, 2555–2565.

Hickman, E.S., Moroni, M.C. and Helin, K. (2002) The role of p53 and pRB in apoptosis and cancer. *Curr Opin Genet Dev* **12**, 60–66.

Horváth, G.V., Pettkó-Szandtner, A., Nikovics, K., *et al.* (1998) Prediction of functional regions of the maize streak virus replication-associated proteins by protein–protein interaction analysis. *Plant Mol Biol* **38**, 699–712.

Hsieh, T.-F., Hakim, O., Ohad, N. and Fischer, R.L. (2003) From flour to flower, how Polycomb group proteins influence multiple aspects of plant development. *Trends Plant Sci* **8**, 439–445.

Huntley, R., Healy, S., Freeman, D., *et al.* (1998) The maize retinoblastoma protein homologue ZmRb-1 is regulated during leaf development and displays conserved interactions with G1/S regulators and plant cyclin D (CycD) proteins. *Plant Mol Biol* **37**, 155–169.

Inzé, D. (2005) Green light for the cell cycle. *EMBO J* **24**, 657–662.

Ivanova, N.B., Dimos, J.T., Schaniel, C., Hackney, J.A., Moore, K.A. and Lemischka, I.R. (2002) A stem cell molecular signature. *Science* **298**, 601–604.

Jacks, T., Fazeli, A., Schmitt, E.M., Bronson, R.T., Goodell, M.A. and Weinberg, R.A. (1992) Effects of an *Rb* mutation in the mouse. *Nature* **359**, 295–300.

Jenuwein, T. and Allis, C.D. (2001) Translating the histone code. *Science* **293**, 1074–1080.

Jiang, K. and Feldman, L.J. (2005) Regulation of root apical meristem development. *Annu Rev Cell Dev Biol* **21**, 485–509.

Jönsson, H., Heisler, M.G., Shapiro, B.E., Meyerowitz, E.M. and Mjolsness, E. (2006) An auxin-driven polarized transport model for phyllotaxis. *Proc Natl Acad Sci USA* **103**, 1633–1638.

Jürgens, G. (1985) A group of genes controlling the spatial expression of the bithorax complex in Drosophila. *Nature* **316**, 153–155.

Kepinski, S. and Leyser, O. (2005) Plant development: auxin in loops. *Curr Biol* **15**, R208–R210.

Kitagawa, M., Higashi, H., Jung, H.-K., *et al.* (1996) The consensus motif for phosphorylation by cyclin D1-Cdk4 is different from that for phosphorylation by cyclin A/E-Cdk2. *EMBO J* **15**, 7060–7069.

Knudson, A.G. Jr (1971) Mutation and cancer: statistical study of retinoblastoma. *Proc Natl Acad Sci USA* **68**, 820–823.

Köhler, C. and Grossniklaus, U. (2002) Epigenetic inheritance of expression states in plant development: the role of *Polycomb* group proteins. *Curr Opin Cell Biol* **14**, 773–779.

Köhler, C., Hennig, L., Bouveret, R., Gheyselinck, J., Grossniklaus, U. and Gruissem, W. (2003) *Arabidopsis* MSI1 is a component of the MEA/FIE *Polycomb* group complex and required for seed development. *EMBO J* **22**, 4804–4814.

Kong, L.-J., Orozco, B.M., Roe, J.L., *et al.* (2000) A geminivirus replication protein interacts with the retinoblastoma protein through a novel domain to determine symptoms and tissue specificity of infection in plants. *EMBO J* **19**, 3485–3495.

Korenjak, M. and Brehm, A. (2005) E2F-Rb complexes regulating transcription of genes important for differentiation and development. *Curr Opin Genet Dev* **15**, 520–527.

Koroleva, O.A., Tomlinson, M., Parinyapong, P., *et al.* (2004) *CycD1*, a putative G1 cyclin from *Antirrhinum majus*, accelerates the cell cycle in cultured tobacco BY-2 cells by enhancing both G1/S entry and progression through S and G2 phases. *Plant Cell* **16**, 2364–2379.

Lee, C. and Cho, Y. (2002) Interactions of SV40 large T antigen and other viral proteins with retinoblastoma tumour suppressor. *Rev Med Virol* **12**, 81–92.

Lee, E.Y.-H., Chang, C.-Y., Hu, N., *et al.* (1992) Mice deficient for Rb are nonviable and show defects in neurogenesis and haematopoiesis. *Nature* **359**, 288–294.

Lee, M.H., Williams, B.O., Mulligan, G., *et al.* (1996) Targeted disruption of p107: functional overlap between p107 and Rb. *Genes Dev* **10**, 1621–1632.

Lee, W.H., Bookstein, R., Hong, F., Young, L.J., Shew J.Y. and Lee, E.Y. (1987) Human retinoblastoma susceptibility gene: cloning, identification, and sequence. *Science* **235**, 1394–1399.

Leibfried, A., To, J.P.C., Busch, W., *et al.* (2005) WUSCHEL controls meristem function by direct regulation of cytokinin-inducible response regulators. *Nature* **438**, 1172–1175.

Li, F.-Q., Coonrod, A. and Horwitz, M. (2000) Selection of a dominant negative retinoblastoma protein (RB) inhibiting satellite myoblast differentiation implies an indirect interaction between MyoD and RB. *Mol Cell Biol* **20**, 5129–5139.

Lipinski, M.M. and Jacks, T. (1999) The retinoblastoma gene family in differentiation and development. *Oncogene* **18**, 7873–7882.

Liu, H., Dibling, B., Spike, B., Dirlam, A. and Macleod, K. (2004) New roles for the RB tumor suppressor protein. *Curr Opin Genet Dev* **14**, 55–64.

Livingston, D.M., Kaelin, W., Chittenden, T. and Qin, X. (1993) Structural and functional contributions to the G1 blocking action of the retinoblastoma protein. *Br J Cancer* **68**, 264–268.

Lu, X. and Horvitz, H.R. (1998) *lin-35* and *lin-53*, two genes that antagonize a *C. elegans* Ras pathway, encode proteins similar to Rb and its binding protein RbAp48. *Cell* **95**, 981–991.

Luo, M., Bilodeau, P., Koltunow, A., Dennis, E.S., Peacock, W.J. and Chaudhury, A.M. (1999) Genes controlling fertilization-independent seed development in *Arabidopsis thaliana*. *Proc Natl Acad Sci USA* **96**, 296–301.

Magyar, Z., Atanassova, A., De Veylder, L., Rombauts, S. and Inzé, D. (2000) Characterization of two distinct DP-related genes from *Arabidopsis thaliana*. *FEBS Lett* **486**, 79–87.

Mariconti, L., Pellegrini, B., Cantoni, R., *et al.* (2002) The E2F family of transcription factors from *Arabidopsis thaliana*. Novel and conserved components of the retinoblastoma/E2F pathway in plants. *J Biol Chem* **277**, 9911–9919.

Martienssen, R.A. and Colot, V. (2001) DNA methylation and epigenetic inheritance in plants and filamentous fungi. *Science* **293**, 1070–1074.

Martínez-Balbás, M.A., Bauer, U.-M., Nielsen, S.J., Brehm, A. and Kouzarides, T. (2000) Regulation of E2F1 activity by acetylation. *EMBO J* **19**, 662–671.

Masubelele, N.H., Dewitte, W., Menges, M., *et al.* (2005) D-type cyclins activate division in the root apex to promote seed germination in *Arabidopsis*. *Proc Natl Acad Sci USA* **102**, 15694–15699.

Meijer, M. and Murray, J.A.H. (2000) The role and regulation of D-type cyclins in the plant cell cycle. *Plant Mol Biol* **43**, 621–633.

Menges, M., de Jager, S.M., Gruissem, W. and Murray, J.A.H. (2005) Global analysis of the core cell cycle regulators of *Arabidopsis* identifies novel genes, reveals multiple and highly specific profiles of expression and provides a coherent model for plant cell cycle control. *Plant J* **41**, 546–566.

Menges, M., Hennig, L., Gruissem, W. and Murray, J.A.H. (2002) Cell cycle-regulated gene expression in *Arabidopsis*. *J Biol Chem* **277**, 41987–42002.

Menges, M., Samland, A.K., Planchais, S. and Murray, J.A.H. (2006) The D-type cyclin CYCD3;1 is limiting for the G1-to-S-phase transition in *Arabidopsis*. *Plant Cell* **18**, 893–906.

Mittnacht, S. (1998) Control of pRB phosphorylation. *Curr Opin Genet Dev* **8**, 21–27.

Morris, E.J. and Dyson, N.J. (2001) Retinoblastoma protein partners. *Adv Cancer Res* **82**, 1–54.

Mosquna, A., Katz, A., Shochat, S., Grafi, G. and Ohad, N. (2004) Interaction of FIE, a Polycomb protein, with pRB: a possible mechanism regulating endosperm development. *Mol Genet Genomics* **271**, 651–657.

Müller, H. and Helin, K. (2000) The E2F transcription factors: key regulators of cell proliferation. *Biochim Biophys Acta* **1470**, M1–M12.

Müller, J., Hart, C.M., Francis, N.J., *et al.* (2002) Histone methyltransferase activity of a *Drosophila* Polycomb group repressor complex. *Cell* **111**, 197–208.

Myers, T.R. and Greenwald, I. (2005) *lin-35* Rb acts in the major hypodermis to oppose Ras-mediated vulval induction in *C. elegans*. *Dev Cell* **8**, 117–123.

Nakagami, H., Kawamura, K., Sugisaka, K., Sekine, M. and Shinmyo, A. (2002) Phosphorylation of retinoblastoma-related protein by the cyclin D/cyclin-dependent kinase complex is activated at the G1/S-phase transition in tobacco. *Plant Cell* **14**, 1847–1857.

Nakagami, H., Sekine, M., Murakami, H. and Shinmyo, A. (1999) Tobacco retinoblastoma-related protein phosphorylated by a distinct cyclin-dependent kinase complex with Cdc2/cyclin D *in vitro*. *Plant J* **18**, 243–252.

Nevins, J.R. (2001) The Rb/E2F pathway and cancer. *Hum Mol Genet* **10**, 699–703.

Nguyen, D.X., Baglia, L.A., Huang, S.-M., Baker, C.M. and McCance, D.J. (2004) Acetylation regulates the differentiation-specific functions of the retinoblastoma protein. *EMBO J* **23**, 1609–1618.

Nielsen, S.J., Schneider, R., Bauer, U.-M., *et al.* (2001) Rb targets histone H3 methylation and HP1 to promoters. *Nature* **412**, 561–565.

Novitch, B.G., Mulligan, G.J., Jacks, T. and Lassar, A.B. (1996) Skeletal muscle cells lacking the retinoblastoma protein display defects in muscle gene expression and accumulate in S and G_2 phases of the cell cycle. *J Cell Biol* **135**, 441–456.

Novitch, B.G., Spicer, D.B., Kim, P.S., Cheung, W.L. and Lassar, A.B. (1999) pRb is required for MEF2-dependent gene expression as well as cell-cycle arrest during skeletal muscle differentiation. *Curr Biol* **9**, 449–459.

Oakenfull, E.A., Riou-Khamlichi, C. and Murray, J.A.H. (2002) Plant D-type cyclins and the control of G1 progression. *Phil Trans R Soc Lond B* **357**, 749–760.

Park, J.-A., Ahn, J.-W., Kim, Y.-K., *et al.* (2005) Retinoblastoma protein regulates cell proliferation, differentiation, and endoreduplication in plants. *Plant J* **42**, 153–163.

Pasini, D., Bracken, A.P. and Helin, K. (2004) Polycomb group proteins in cell cycle progression and cancer. *Cell Cycle* **3**, 396–400.

Ramalho-Santos, M., Yoon, S., Matsuzaki, Y., Mulligan, R.C. and Melton, D.A. (2002) "Stemness": transcriptional profiling of embryonic and adult stem cells. *Science* **298**, 597–600.

Ramirez-Parra, E., Fründt, C. and Gutierrez, C. (2003) A genome-wide identification of E2F-regulated genes in *Arabidopsis*. *Plant J* **33**, 801–811.

Ramirez-Parra, E. and Gutierrez, C. (2000) Characterization of wheat DP, a heterodimerization partner of the plant E2F transcription factor which stimulates E2F–DNA binding. *FEBS Lett* **486**, 73–78.

Ramirez-Parra, E., López-Matas, M.A., Fründt, C. and Gutierrez, C. (2004) Role of an atypical E2F transcription factor in the control of *Arabidopsis* cell growth and differentiation. *Plant Cell* **16**, 2350–2363.

Reddy, G.V. and Meyerowitz, E.M. (2005) Stem-cell homeostasis and growth dynamics can be uncoupled in the *Arabidopsis* shoot apex. *Science* **310**, 663–667.

Reyes, J.C., Hennig, L. and Gruissem, W. (2002) Chromatin-remodeling and memory factors. New regulators of plant development. *Plant Physiol* **130**, 1090–1101.

Riou-Khamlichi, C., Menges, M., Healy, J.M.S. and Murray, J.A.H. (2000) Sugar control of the plant cell cycle: differential regulation of *Arabidopsis* D-type cyclin gene expression. *Mol Cell Biol* **20**, 4513–4521.

Robanus-Maandag, E., Dekker, M., Van Der Valk, M., *et al.* (1998) p107 is a suppressor of retinoblastoma development in pRb-deficient mice. *Genes Dev* **12**, 1599–1609.

Rossignol, P., Stevens, R., Perennes, C., *et al.* (2002) AtE2F-a and AtDP-a, members of the E2F family of transcription factors, induce *Arabidopsis* leaf cells to re-enter S phase. *Mol Genet Genomics* **266**, 995–1003.

Ruiz, S., Santos, M., Segrelles, C., *et al.* (2004) Unique and overlapping functions of pRb and p107 in the control of proliferation and differentiation in epidermis. *Development* **131**, 2737–2748.

Sabelli, P.A., Dante, R.A., Leiva-Neto, J.T., Jung, R., Gordon-Kamm, W.J. and Larkins, B.A. (2005) RBR3, a member of the retinoblastoma-related family from maize, is regulated by the RBR1/E2F pathway. *Proc Natl Acad Sci USA* **102**, 13005–13012.

Sabelli, P.A. and Larkins, B.A. (2006) Grasses like mammals? Redundancy and compensatory regulation within the retinoblastoma protein family. *Cell Cycle* **5**, 352–355.

Sage, J., Miller, A.L., Pérez-Mancera, P.A., Wysocki, J.M. and Jacks, T. (2003) Acute mutation of retinoblastoma gene function is sufficient for cell cycle re-entry. *Nature* **424**, 223–228.

Sage, J., Mulligan, G.J., Attardi, L.D., *et al.* (2000) Targeted disruption of the three Rb-related genes leads to loss of G_1 control and immortalization. *Genes Dev* **14**, 3037–3050.

Schellmann, S. and Hülskamp, M. (2005) Epidermal differentiation: trichomes in *Arabidopsis* as a model system. *Int J Dev Biol* **49**, 579–584.

Schnittger, A. and Hülskamp, M. (2002) Trichome morphogenesis: a cell-cycle perspective. *Philos Trans R Soc Lond B* **357**, 823–826.

Schönrock, N., Exner, V., Probst, A., Gruissem, W. and Hennig, L. (2006) Functional genomic analysis of CAF-1 mutants in *Arabidopsis thaliana*. *J Biol Chem* **281**, 9560–9568.

Sekine, M., Ito, M., Uemukai, K., Maeda, Y., Nakagami, H. and Shinmyo, A. (1999) Isolation and characterization of the E2F-like gene in plants. *FEBS Lett* **460**, 117–122.

Sharma, V.K., Carles, C. and Fletcher, J.C. (2003) Maintenance of stem cell populations in plants. *Proc Natl Acad Sci USA* **100** (Suppl. 1), 11823–11829.

Sherr, C.J. (1996) Cancer cell cycles. *Science* **274**, 1672–1677.

Sherr, C.J. (2000) The Pezcoller lecture: cancer cell cycles revisited. *Cancer Res* **60**, 3689–3695.

Sherr, C.J. and McCormick, F. (2002) The RB and p53 pathways in cancer. *Cancer Cell* **2**, 103–112.

Simon, J.A. and Tamkun, J.W. (2002) Programming off and on states in chromatin: mechanisms of Polycomb and trithorax group complexes. *Curr Opin Genet Dev* **12**, 210–218.

Smialowski, P., Singh, M., Mikolajka, A., *et al.* (2005) NMR and mass spectrometry studies of putative interactions of cell cycle proteins pRb and CDK6 with cell differentiation proteins MyoD and ID-2. *Biochim Biophys Acta* **1750**, 48–60.

Smith, R.S., Guyomarc'h, S., Mandel, T., Reinhardt, D., Kuhlemeier, C. and Prusinkiewicz, P. (2006) A plausible model of phyllotaxis. *Proc Natl Acad Sci USA* **103**, 1301–1306.

Sozzani, R., Maggio, C., Varotto, S., *et al.* (2006) Interplay between *Arabidopsis* activating factors E2Fb and E2Fa in cell cycle progression and development. *Plant Physiol* **140**, 1355–1366.

Stevaux, O. and Dyson, N.J. (2002) A revised picture of the E2F transcriptional network and RB function. *Curr Opin Cell Biol* **14**, 684–691.

Strahl, B.D. (2000) The language of covalent histone modifications. *Nature* **403**, 41–45.

Takada, S. and Tasaka, M. (2002) Embryonic shoot apical meristem formation in higher plants. *J Plant Res* **115**, 411–417.

Takaki, T., Fukasawa, K., Suzuki-Takahashi, I., *et al.* (2005) Preferences for phosphorylation sites in the retinoblastoma protein of D-type cyclin-dependent kinases, Cdk4 and Cdk6, *in vitro*. *J Biochem* **137**, 381–386.

Taylor-Harding, B., Binné, U.K., Korenjak, M., Brehm, A. and Dyson, N.J. (2004) p55, the *Drosophila* ortholog of RbAp46/RbAp48, is required for the repression of dE2F2/RBF-regulated genes. *Mol Cell Biol* **24**, 9124–9136.

Tie, F., Furuyama, T., Prasad-Sinha, J., Jane, E. and Harte, P.J. (2001) The *Drosophila* Polycomb Group proteins ESC and E(Z) are present in a complex containing the histone-binding protein p55 and the histone deacetylase RPD3. *Development* **128**, 275–286.

Toppari, J., Suominen, J.S. and Yan, W. (2003) The role of retinoblastoma protein family in the control of germ cell proliferation, differentiation and survival. *APMIS* **111**, 245–251.

Turner, B.M. (2000) Histone acetylation and an epigenetic code. *BioEssays* **22**, 836–845.

Uemukai, K., Iwakawa, H., Kosugi, S., *et al.* (2005) Transcriptional activation of tobacco E2F is repressed by co-transfection with the retinoblastoma-related protein: cyclin D expression overcomes this repressor activity. *Plant Mol Biol* **57**, 83–100.

Umen, J.G. and Goodenough, U.W. (2001) Control of cell division by a retinoblastoma protein homolog in *Chlamydomonas*. *Genes Dev* **15**, 1652–1661.

Vandel, L., Nicolas, E., Vaute, O., Ferreira, R., Ait-Si-Ali, S. and Trouche, D. (2001) Transcriptional repression by the retinoblastoma protein through the recruitment of a histone methyltransferase. *Mol Cell Biol* **21**, 6484–6494.

Vandepoele, K., Vlieghe, K., Florquin, K., *et al.* (2005) Genome-wide identification of potential plant E2F target genes. *Plant Physiol* **139**, 316–328.

Vanderluit, J.L., Ferguson, K.L., Nikoletopoulou, V., *et al.* (2004) p107 regulates neural precursor cells in the mammalian brain. *J Cell Biol* **166**, 853–863.

Vernoux, T. and Benfey, P.N. (2005) Signals that regulate stem cell activity during plant development. *Curr Opin Genet Dev* **15**, 388–394.

Vlieghe, K., Boudolf, V., Beemster, G.T.S., *et al.* (2005) The DP-E2F-like *DEL1* gene controls the endocycle in *Arabidopsis thaliana*. *Curr Biol* **15**, 59–63.

Wang, D., Kennedy, S., Conte, D., Jr., et al. (2005) Somatic misexpression of germline P granules and enhanced RNA interference in retinoblastoma pathway mutants. *Nature* **436**, 593–597.

Wang, H., Zhou, Y., Gilmer, S., Whitwill, S. and Fowke, L.C. (2000) Expression of the plant cyclin-dependent kinase inhibitor ICK1 affects cell division, plant growth and morphology. *Plant J* **24**, 613–623.

Weigel, D. and Jürgens, G. (2002) Stem cells that make stems. *Nature* **415**, 751–754.

Weinberg, R.A. (1995) The retinoblastoma protein and cell cycle control. *Cell* **81**, 323–330.

Whyatt, D. and Grosveld, F. (2002) Cell-nonautonomous function of the retinoblastoma tumour suppressor protein: new interpretations of old phenotypes. *EMBO Rep* **3**, 130–135.

Wildwater, M., Campilho, A., Perez-Perez, J.M., *et al.* (2005) The *retinoblastoma-related* gene regulates stem cell maintenance in *Arabidopsis* roots. *Cell* **123**, 1337–1349.

Williams, L. and Fletcher, J.C. (2005) Stem cell regulation in the *Arabidopsis* shoot apical meristem. *Curr Opin Plant Biol* **8**, 582–586.

Wu, L., de Bruin, A., Saavedra, H.I., *et al.* (2003) Extra-embryonic function of Rb is essential for embryonic development and viability. *Nature* **421**, 942–947.

Xie, Q., Sanz-Burgos, A.P., Hannon, G.J. and Gutiérrez, C. (1996) Plant cells contain a novel member of the retinoblastoma family of growth regulatory proteins. *EMBO J* **15**, 4900–4908.

Zarkowska, T. and Mittnacht, S. (1997) Differential phosphorylation of the retinoblastoma protein by G1/S cyclin-dependent kinases. *J Biol Chem* **272**, 12738–12746.

Zhang, J.-M., Zhao, X., Wei, Q. and Paterson, B.M. (1999) Direct inhibition of G_1 cdk kinase activity by MyoD promotes myoblast cell cycle withdrawal and terminal differentiation. *EMBO J* **18**, 6983–6993.

Zheng, L. and Lee, W.H. (2001) The retinoblastoma gene: a prototypic and multifunctional tumor suppressor. *Exp Cell Res* **264**, 2–18.

8 Auxin fuels the cell cycle engine during lateral root initiation

Steffen Vanneste, Dirk Inzé and Tom Beeckman

8.1 Introduction

Meristems represent sites of mitotic activity in plants. In roots, meristems can be found at the apex and at the tips of lateral roots (primary, secondary and tertiary). The lateral roots are major contributors to the overall root system. The more meristems a root system generates, the more efficiently the soil can be exploited. For a single 16-week-old winter rye (*Secale cereale*) plant, up to an astonishing 13 million branches have been reported, resulting in a total root length of more than 500 km packed in less than 0.05 m^3 of soil (Dittmer, 1937). Lateral root development can optimally be studied in a species with a simple diarch root system (having only two protoxylem poles), such as the model plant *Arabidopsis thaliana*.

In *Arabidopsis*, lateral roots originate from three files of xylem-pole-associated pericycle cells at each protoxylem pole. The onset of forming a novel lateral root coincides with restarting cell cycle activity in these formerly non-dividing pericycle cells. Lineage analysis has shown that the majority of cells in a lateral root primordium are derived from the central file of xylem pole pericycle cells (Kurup *et al.*, 2005). Initially, these pericycle cells acquire founder cell identity. After commitment, the founder cells divide asymmetrically resulting in a stage I primordium (Malamy and Benfey, 1997). Subsequently, strictly organized cell divisions result in periclinal growth of the primordium. Between the three- and five-layered stage, a lateral root primordium is capable of autonomous growth (Laskowski *et al.*, 1995) and evolves into a fully functional lateral root meristem after emergence (Malamy and Benfey, 1997) (Figure 8.1).

Physiological studies have identified a plethora of interactions between most plant hormones during lateral root formation. Among the different plant hormones, auxin stands out for its key role in many, if not all, developmental steps of lateral root development. Indeed, overwhelming evidence shows the involvement of auxin in founder cell specification, cell division and meristem organization, lateral root emergence and meristem activation. Interestingly, many of the other plant hormone response pathways converge, at least partially, to the modulation of auxin activity.

Here we will summarize the most recent insights into auxin-induced lateral root initiation as a model for cell cycle (re)-activation.

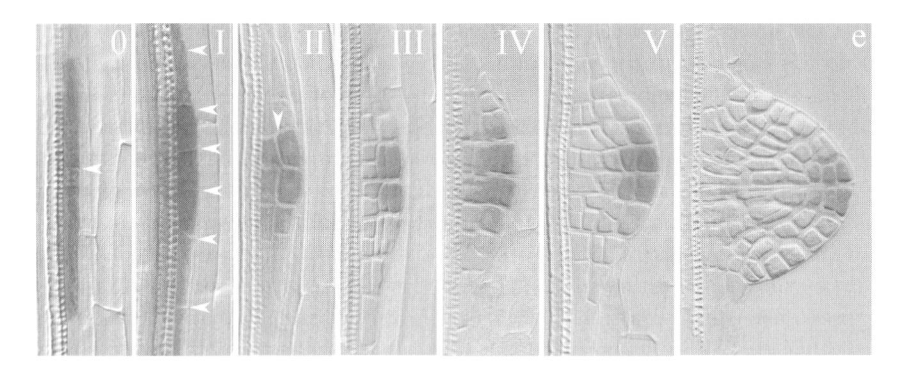

Figure 8.1 Stages of lateral root development and the organization of an auxin gradient. Darker staining representing DR5:GUS activity is indicative for auxin accumulation. At stage 0, prior to the asymmetric divisions hallmarking lateral root initiation, auxin accumulates in adjacent xylem pole pericycle cells. At stage I, the first anticlinal asymmetric divisions have occurred and the DR5:GUS activity is restricted to the central cells. At stage II, these cells undergo their first round of periclinal division and the DR5:GUS maximum becomes even more restricted. In the following stages (III to V) more rounds of anticlinal and pericyclinal divisions occur, while the DR5:GUS maximum becomes more and more restricted defining the future lateral root meristem stem cell niche. At emergence (e) a functional lateral root meristem is established. (Adapted, with permission from Elsevier, from Benková *et al.* (2003).)

8.2 Cell cycle regulation during lateral root development

Like most of the other cell types, xylem pole pericycle cells leave the meristem in G1 phase. Upon triggering into lateral root initiation, G1-to-S transition takes place. The E2F–RB pathway restricts this transition both in mammals and in plants (Shen, 2002). The key players in this pathway are conserved and fulfil similar functions. The RETINOBLASTOMA-RELATED (RBR) protein inhibits cell cycle progression, at least in part, through the obstruction of E2F-DP transcription factor complexes. D-type cyclin/CDK complexes phosphorylate RBR resulting in derepression of E2F-DP complexes followed by G1-to-S transition. In turn, cell cycle inhibitory proteins, called inhibitors of CDK/Kip-related proteins (ICKs/KRPs), can inhibit cyclin/CDK activity and thus affect cell cycle progression (reviewed in previous chapters). In roots, *CYCD3;1*, *CYCD3;2*, *RBR*, *E2Fa*, *E2Fc* and *DPa* were upregulated by auxin treatment, whereas *ICK1/KRP1* and *ICK2/KRP2* were downregulated (Himanen *et al.*, 2002; Vanneste *et al.*, 2005). In somatic tissues and cell suspensions, excessive cell proliferation can be induced by ectopic expression of *CYCD3;1* (Dewitte *et al.*, 2003), *E2Fa-DPa* (De Veylder *et al.*, 2002) and *E2Fb-DPa* (Magyar *et al.*, 2005). In contrast, ectopic overexpression of *ICK1/KRP1* or *ICK2/KRP2* strongly reduced cell cycle progression (Lui *et al.*, 2000; De Veylder *et al.*, 2001). *In situ* hybridizations showed that *ICK2/KRP2* is expressed in a specific pattern in the root and is highly expressed throughout the pericycle, except for sites of lateral root initiation. Furthermore, overexpression drastically reduced the number of lateral roots. Therefore, it is believed that the expression of *ICK2/KRP2* is involved in restricting sites of lateral root development (Himanen *et al.*, 2002).

Further progression through S phase can be inhibited by E2Fc, which is a repressor of E2F-DP-regulated transcription (del Pozo *et al.*, 2002a). Moreover, CYCA2–CDKA;1 protein complexes have been implicated in phosphorylation of E2Fc, targeting it for proteolysis (del Pozo *et al.*, 2002a). In roots, A-type cyclins *CYCA1;1*, *CYCA2;1* and *CYCA2;4* can be induced by auxin (Himanen *et al.*, 2002; Vanneste *et al.*, 2005), among which *CYCA2;4* expression may even be primary auxin responsive (Vanneste *et al.*, 2005). Therefore, it is plausible to assume that auxin-induced A2-type cyclins stimulate S-phase progression by modulating E2Fc stability in roots. Furthermore, E2Fa-DPa transcription factor complexes directly activate *CDKB1;1* expression (Boudolf *et al.*, 2004b), which is typically expressed from S phase to G2 to M transition (Segers *et al.*, 1996). In addition, *CDKB1;1* is expressed in xylem pole pericycle cells during lateral root initiation (Beeckman *et al.*, 2001; Vanneste *et al.*, 2005). Overexpression of a dominant-negative allele of *CDKB1;1* resulted in a reduced G2 to M transition and enhanced endoreduplication in all tissues tested, including roots (Boudolf *et al.*, 2004a). Consistently, a significant reduction in lateral root density was found in plants overexpressing a dominant-negative allele of CDKB1;1 (I. De Smet and T. Beeckman, personal communication). Furthermore, CDKB1;1 kinase was shown to regulate ICK2/KRP2 protein abundance (Verkest *et al.*, 2005), strongly suggesting that its kinase activity is correlated with lateral root initiation. *CDKB2;1*, another member of the same family and specific to the G2-to-M transition, was also identified as a potential primary auxin-responsive cell cycle gene (Vanneste *et al.*, 2005), supporting the idea of a specific auxin signal transduction pathway operating at this cell cycle transition. Still at the same cell cycle phase, the APC complex becomes active, regulating proteolysis of cell cycle regulators (see previous chapters). A specific APC complex subunit, HOBBIT/CDC25b, is involved in cell cycle progression and differentiation in the root meristem. Interestingly, the auxin response repressor protein, AXR3/IAA17, accumulates in *hobbit* mutants (Blilou *et al.*, 2002) suggesting that HOBBIT/CDC25b might mediate primary auxin response at the G2-to-M transition. It is tempting to speculate that HOBBIT/CDC25b would be involved in regulating AUX/IAA-dependent *CDKB2;1* expression at the onset of lateral root formation.

8.3 Stemness of the xylem-pole-associated pericycle

The terms *stem cell* and *stemness* are becoming increasingly popular in recent plant literature. Whilst no unambiguous definition for a ‘stem cell’ exists (Parker *et al.*, 2005), they are commonly defined as ‘pluripotent’ and ‘able to reconstitute entire tissues’. Usually, one discerns between ‘steady-state’ and ‘emergent’ stem cells (Shostak, 2006). The steady-state stem cells are self-renewing and give rise to complete tissues and organs through an iteration of asymmetric divisions. Emergent stem cells are transient and are most commonly found in developing tissues, such as developing embryos.

Within the root meristem, one can easily detect steady-state stem cells in a special microenvironment (‘stem cell niche’), contacting the quiescent centre

(Benfey and Scheres, 2000). These cells undergo asymmetric divisions, in which one daughter cell retains the parental cell fate, whilst the other gives rise to a 'proliferating precursor cell' of a particular tissue lineage. When such a stem cell is ablated, a neighbouring cell dedifferentiates and changes its cell fate to replace the ablated stem cell (Xu *et al.*, 2006), suggesting that 'stemness' diffuses within the meristem.

Beyond the root meristem, a population of 'quiescent' emergent stem cells can be found within the pericycle at the xylem poles. Upon stimulation, these cells, previously called 'progenitor cells' or 'founder cells', become 'true' emergent stem cells as they become part of the pool of cells participating in *de novo* development of a lateral root meristem.

Recently, it was found that modulation of the E2F–RB pathway in the root stem cell niche near the quiescence centre of *Arabidopsis* affects rate of differentiation rather than speed of cell division (Wildwater *et al.*, 2005). This is in contrast to the previously reported ectopic proliferation, in non-stem cells, induced by overexpression of stimulatory E2F–RB components (CYCD3;1 and E2Fa/DPa) (De Veylder *et al.*, 2002; Dewitte *et al.*, 2003). These data suggest that the output of the E2F–RB pathway depends on intrinsic levels of stemness of the target cells. Furthermore, the E2F–RB pathway is limiting to lateral root initiation as shown by overexpression of the E2F–RB inhibitory protein, ICK2/KRP2 (Himanen *et al.*, 2002; Vanneste *et al.*, 2005). As xylem pole pericycle cells have high levels of stemness, one might expect a specific output for the E2F–RB pathway in these cells. Overexpression of *CYCD3;1* or *E2Fa-DPa* did not boost lateral root initiation (I. De Smet and T. Beeckman, personal communication), suggesting that another signal needs to coincide with the activation of the E2F–RB pathway for lateral root initiation to occur. Concordantly, activation of the E2F–RB pathway in the lateral root-less auxin response mutant, *solitary root-1*, simply stimulated proliferative divisions in the xylem pole pericycle, rather than lateral root initiation (Figure 8.2). This implies that an auxin-derived signal shifts the output of the E2F–RB pathway from proliferation towards lateral root initiation. The nature of this auxin-derived signal remains elusive. However, the identification of novel downstream components of the auxin-signalling cascade will be essential to unmask this (these) mysterious component(s).

8.4 Auxin signalling during lateral root initiation

Decades of research have been dedicated to the elucidation of the modus operandi of early auxin signalling (Figure 8.3). Only recently, auxin receptors have been identified unambiguously (Dharmasiri *et al.*, 2005a; Kepinski and Leyser, 2005) and comprise a small family AUXIN SIGNALING F-BOX proteins (AFB) (Dharmasiri *et al.*, 2005b), of which TIR1 is the best characterized. These AFBs occur in complexes forming functional $SCF^{TIR1/AFB1/2/3}$ E3 ligases (Gray *et al.*, 1999; Dharmasiri *et al.*, 2005b). Most of the tested components and regulators of the $SCF^{TIR1/AFB1/2/3}$ complex are found to be highly expressed at sites of lateral root formation, and mutations result in a reduction in lateral root density (del Pozo *et al.*, 2002b;

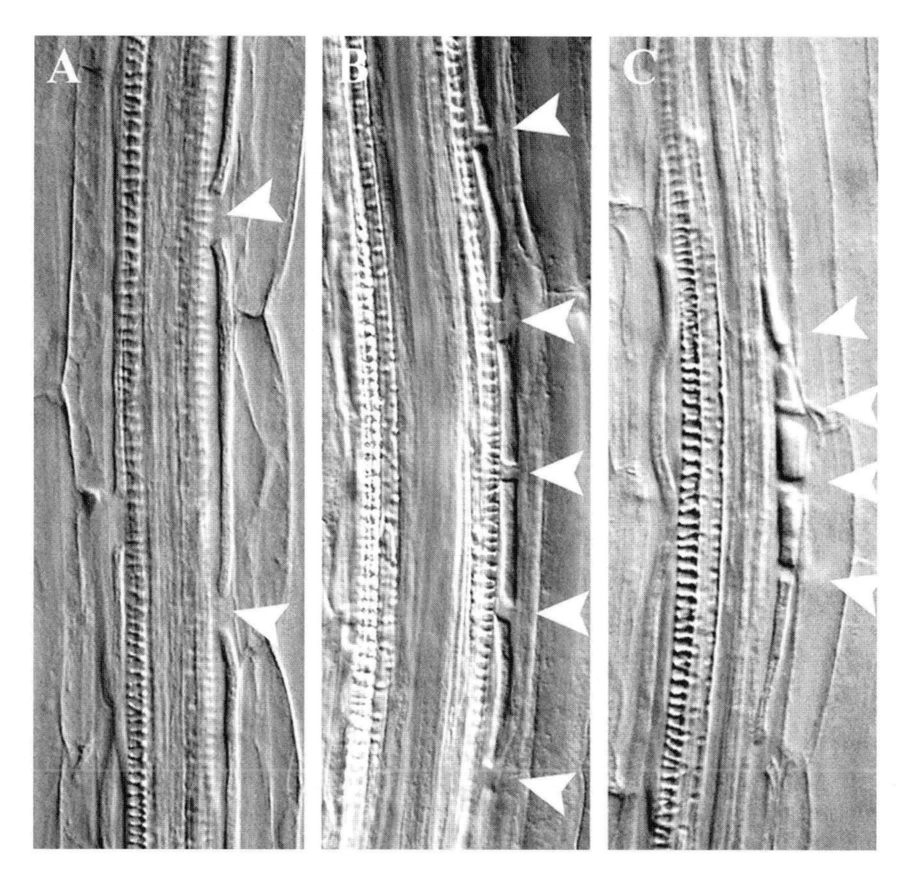

Figure 8.2 Proliferation versus asymmetric division in the xylem pole pericycle. (A) Non-dividing mature pericycle cell. (B) Proliferation of xylem pole pericycle cells upon stimulation of the E2F–RB pathway in the auxin signalling mutant *solitary root-1*. (C) Formation of a stage I primordium after asymmetric division in xylem pole pericycle cells. (Adapted, with permission from American Society of Plant Biologists, from Vanneste *et al.* (2005).)

Dharmasiri *et al.*, 2003, 2005b; Gray *et al.*, 2003) implicating that $SCF^{TIR1/AFB1/2/3}$-mediated ubiquitination is an essential regulatory component for lateral root initiation.

The most notorious targets for $SCF^{TIR1/AFB1/2/3}$-mediated ubiquitination are proteins of the AUX/IAA family (Gray *et al.*, 2001; Dharmasiri *et al.*, 2005b). Direct binding of auxin to $SCF^{TIR1/AFB1/2/3}$ enhances the affinity for AUX/IAAs (Dharmasiri *et al.*, 2005a; Kepinski and Leyser, 2005), triggering their oligo-ubiquitination (Gray *et al.*, 2001). After ubiquitination, these proteins are rapidly targeted for proteolysis (Thrower *et al.*, 2000; Ramos *et al.*, 2001). This may well be one of the most crucial steps in translating auxin signal into transcriptional information, as AUX/IAAs are known inhibitors of a specific family of transcription factors, called auxin response factors (ARFs).

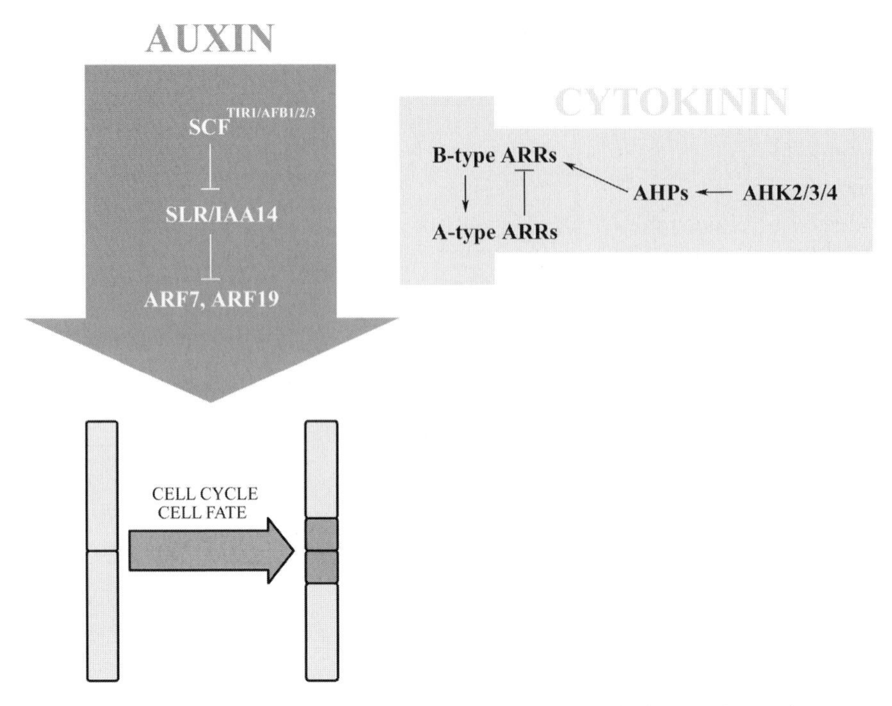

Figure 8.3 Scheme of auxin–cytokinin antagonism during lateral root initiation. When auxin concentrations are high, auxin binds directly to $SCF^{TIR1/AFB1/2/3}$ E3 ligases and dramatically increases their affinity for AUX/IAA proteins, such as SLR/IAA14. Upon interaction, SLR/IAA14 is ubiquitinated and targeted for proteolysis. SLR/IAA14 degradation derepresses the auxin-responsive transcription factors ARF7 and ARF19, which stimulate cell cycle activation and cell fate respecification during lateral root initiation. Cytokinins activate their signalling cascade through binding to membrane-bound AHK2/3/4, which results in the phosphorylation of AHPs that transmit the signal to the nucleus. In the nucleus, AHPs phosphorylate B-type ARRs that subsequently activate transcription of A-type ARRs among other genes. In a negative feedback loop, the A-type ARRs will on their turn repress B-type ARR activity. B-type ARR activity represses ARF7/ARF19-induced lateral root initiation at an as yet unknown level.

Despite their importance in signal transduction, to date no phenotypes have been observed in loss-of-function *aux/iaa* mutants, suggesting a high functional redundancy within the gene family (Overvoorde *et al.*, 2005). In the AUX/IAA protein structure four domains can be discerned (reviewed by Liscum and Reed, 2002). Conserved residues within domain II are essential for interaction with $SCF^{TIR1/AFB1/2/3}$ complexes (Ramos *et al.*, 2001; Dharmasiri *et al.*, 2005a; Kepinski and Leyser, 2005). Single amino acid changes within domain II impede such interactions, increasing the stability of the AUX/IAA protein. Several of such gain-of-function mutants have been identified, displaying altered root architectures, *axr2-1*, *axr3-1*, *bdl*, *shy2-2*, *iaa28-1*, *slr-1*, *msg2-1* and *axr5-1* (Timpte *et al.*, 1994; Rouse *et al.*, 1998; Hamann *et al.*, 1999; Tian and Reed, 1999; Rogg *et al.*, 2001; Fukaki *et al.*, 2002; Tatematsu *et al.*, 2004; Yang *et al.*, 2004). For *msg2-1*, *iaa28-1* and *axr5-1*, reductions in lateral root density can be observed, while in *axr3-1* and *slr-1* very little or no lateral roots are formed. Inhibition of AUX/IAA-mediated auxin signalling in xylem pole pericycle cells through tissue-specific misexpression of mIAA17 (*axr3*), mIAA12 (*bdl*) or mIAA14 (*slr*) abolishes lateral root initiation (De Smet *et al.*, 2007; Fukaki *et al.*, 2005), suggesting that auxin perception in xylem pole pericycle cells is required for lateral root initiation. In contrast to the high degree of functional redundancy, phenotypes ranging from opposing to similar can be obtained by overproducing different stabilized (distant or closely related) AUX/IAAs in the same tissue (Knox *et al.*, 2003; Weijers *et al.*, 2005), suggesting some level of functional divergence, even between closely related AUX/IAAs.

The ARF family counts 23 members, which are able to stimulate and/or inhibit auxin-mediated transcriptional changes. They contain a DNA-binding domain and can heterodimerize with AUX/IAA proteins. Because of the presence of a potent inhibitory domain in AUX/IAAs, AUX/IAA–ARF dimers are transcriptionally inactive. The phenotypes observed in plants producing stabilized AUX/IAA can be mimicked by knocking out the target ARF such as in *bdl/iaa12* and *mp/arf5* mutants, which both lack the embryonic root (Hardtke *et al.*, 2004). Similarly, *nph4 arf19* double mutants phenocopy the lack of lateral roots in *slr-1* (Okushima *et al.*, 2005; Wilmoth *et al.*, 2005). Furthermore, IAA14 interacts with NPH4/ARF7 and ARF19 in yeast two-hybrid assays (Fukaki *et al.*, 2005), suggesting that IAA14 is a repressor for NPH4/ARF7 and ARF19 both being essential for lateral root initiation.

Not all ARFs are believed to be activators of expression. Stabilizing ARF17 (Mallory *et al.*, 2005) or ARF16 (Wang *et al.*, 2005) transcripts strongly repressed lateral root development through modulation of auxin responses. The main task lying ahead is to identify the downstream targets of this complex web of stimulatory and inhibitory factors.

8.5 Post-transcriptional feedback mechanisms on auxin signalling

Besides the rigorous transcriptional feedback mechanisms, an additional layer of post-transcriptional regulation through miRNA-targeted mRNA degradation is emerging. DICER plays an essential role in processing of miRNA precursors into mature miRNAs (Kurihara and Watanabe, 2004) that guide RNA-induced RNA silencing complexes (RISC) to complementary target mRNAs (Vaucheret *et al.*, 2004). Knocking out DICER-LIKE in *Arabidopsis* resulted in a strong increase in lateral root density (Guo *et al.*, 2005). In contrast, impairing RISC function through *ago1* mutation decreased adventitious rooting in *Arabidopsis* (Sorin *et al.*, 2005). These findings suggest that transcripts of auxin-signalling components may be targets of miRNA-mediated mRNA degradation. Indeed, several miRNA targets are clearly involved in lateral root development. The mRNAs of auxin-binding F-box proteins, identified as a family of auxin receptors (see above) are targeted for degradation through the pathogen-inducible miR393 (Navarro *et al.*, 2006). ARF8 and ARF6 mRNAs are targeted for degradation through miR167. Moreover, ARF10, ARF16 and ARF17 mRNAs are targets for miR160 (Mallory *et al.*, 2005; Wang *et al.*, 2005).

Overproduction of miR160 enhances lateral root development, whereas plants expressing miR160-resistent ARF16 (Wang *et al.*, 2005) or ARF17 (Mallory *et al.*, 2005) display strongly impaired lateral root densities. In addition, NAC1, known as a positive regulator of lateral root formation (Xie *et al.*, 2000, 2002), is targeted by miR164 (Guo *et al.*, 2005). As miRNAs are just beginning to be understood, it is apparent that miRNA-mediated mRNA stability is an important regulatory mechanism to control auxin action.

8.6 Polar auxin transport defines lateral root boundaries

Detailed analyses have shown that auxin is present throughout all stages of developing lateral roots. Particularly, in developing lateral root primordia an auxin gradient, similar to that found in primary root meristems, can be visualized around the stem cell niche (Figure 8.1; Benková *et al.*, 2003). Chemically interfering with polar auxin transport results in the misspecification of the stem cell niche in the primary root meristem, demonstrated by ectopic quiescent centre identity (Sabatini *et al.*, 1999). Similarly, cell fates are misspecified in lateral root primordia when polar auxin transport is disrupted (Benková *et al.*, 2003; Geldner *et al.*, 2004). Polar auxin transport is dependent on the activity of PIN proteins (Petrášek *et al.*, 2006). Indeed, in higher order *pin* mutants lateral root boundaries are not correctly specified, as demonstrated by a continuous sheet of proliferating pericycle cells (Benková *et al.*, 2003). These data suggest a simple organogenesis model, in which orchestrated polar auxin transport accumulates auxin at the site of lateral root initiation and as a consequence deprives the surrounding tissues of auxin. Subsequently, an auxin gradient is set up within the developing lateral root meristem with an optimum at the tip and auxin deprivation at the lateral root base. Such a model of auxin accumulation deprivation is becoming a more and more accepted model for auxin-driven organogenesis. In the shoot apex, polar auxin transport has already been extensively implicated in lateral organ formation and positioning (Reinhardt, 2005). Computer models based on real PIN1 dynamics in the shoot epidermis have shown that polar auxin transport is responsible for phyllotactic spacing of leaves. In this model, auxin is pumped to the cell with the highest auxin concentration and ensures a local auxin accumulation and a peripheral auxin deprivation, defining the site of leaf initiation (Jönsson *et al.*, 2006; Smith *et al.*, 2006). Indeed when polar auxin transport (PAT) function is abolished, ring-shaped organs could be induced by local auxin application to the shoot apex (Reinhardt *et al.*, 2003). In the root, an equivalent situation is found in the formation of a continuous sheet of proliferating pericycle cells (Benková *et al.*, 2003; Geldner *et al.*, 2004). Given the strong positive regulatory effect of auxin on cell cycle activity, it is apparent that a local auxin gradient results in differential cell cycle activity, shaping the organ.

Despite the high importance of polar auxin transport in the different stages of lateral root development it seems unlikely to be sufficient to fulfil the requirements for organogenesis. Because of similarities between the primary root meristem and lateral root meristems, we will briefly discuss cell fate determinants that act in primary root organogenesis as well as in lateral root development. One of the earliest induced genes determining root identity in embryogenesis is *PLETHORA1* (*PLT1*), as it is expressed at a position correlating with the embryonic root stem cell niche. Furthermore, strong overexpression of *PLT1* can induce homeotic transformation of shoot to root identity. Interestingly, *PLT1* expression is downstream of the AUX/IAA–ARF auxin-signalling cascade and thus correlates with sites of auxin accumulation, such as lateral root initiation. Yet, *plt1plt2* mutants produce many lateral roots, which rapidly terminally differentiate (Aida *et al.*, 2004). As discussed above, auxin accumulation in developing lateral root primordia is imposed by polar auxin transport. Furthermore, PLT activity is also required for *PIN* gene expression to stabilize auxin accumulation, sustaining its own expression (Blilou *et al.*, 2005). Several of the *PIN* genes have been shown to be rapidly auxin inducible (Vieten *et al.*, 2005), whereas *PLT* expression was much slower (Aida *et al.*, 2004). Nevertheless, it is not the level of expression of *PIN* genes, per se, that drives the auxin flow, but rather their subcellular localization. During quiescent centre regeneration, cell fate changes brought about by PLT precede PIN polarity changes (Xu *et al.*, 2006). Similarly, for different stages of lateral root development, PIN polar localizations rapidly change (Benková *et al.*, 2003), suggestive of ever-changing cell fates within the developing lateral root primordium.

8.7 Cytokinins inhibit lateral root development

The balance of auxins and cytokinins is believed to be a key determinant in many developmental processes. In tissue cultures, high auxin-to-cytokinin ratios are used to induce rooting, while low auxin-to-cytokinin ratios favour shoot development. In roots, cytokinins seem to have a negative effect on lateral root formation (Figure 8.3). This is supported by the observed reduction of lateral root formation in tobacco plants overproducing cytokinin (Li *et al.*, 1992). On the other hand, decreasing cytokinin content by overproduction of cytokinin oxidases dramatically promotes root expansion by more extensive branching and higher rates of root growth (Werner *et al.*, 2001, 2003). These experiments suggest that next to auxin, cytokinin content is a major determinant of lateral root formation.

Cytokinin is perceived through a small family of sensor histidine kinases, AHK2, AHK3 and CRE1/AHK4. Mutant analysis showed that AHK2 and AHK3 are the main contributors to the cytokinin-mediated lateral root repression (Riefler *et al.*, 2006). The cytokinin receptors are believed to transmit the signal through phosphotransfer proteins (AHPs) leading ultimately to an altered phosphorylation state of the *Arabidopsis* response regulators (ARRs). Furthermore, it was inferred that higher order mutants of AHP show decreased cytokinin sensitivity and consistent developmental defects (Hutchison *et al.*, 2006). Type-B ARRs have a receptor and a DNA-binding domain and may serve as transcription factors that translate the cytokinin signal into a primary transcriptional cytokinin response (Lohrmann *et al.*, 2001; Sakai *et al.*, 2001; Hosoda *et al.*, 2002). Overexpression of *ARR1* or *ARR2* increases sensitivity to cytokinin (Hwang and Sheen, 2001; Sakai *et al.*, 2001). Conversely, at cytokinin concentrations limiting for lateral root formation in wild type, still several lateral roots were present in higher order mutants of type-B ARRs (Mason *et al.*, 2005), consistent with a decreased cytokinin sensitivity. Type-A *ARR*s are primary cytokinin-responsive genes (Taniguchi *et al.*, 1998; D'Agostino *et al.*, 2000) that are believed to be directly activated through type-B ARRs (Rashotte *et al.*, 2003). In contrast to type-B ARRs, type-A ARRs are negative regulators of cytokinin response and thus repress their own production (Hwang and Sheen, 2001). In accordance, hexuple mutants in type-A ARRs display an increased cytokinin sensitivity coupled with a reduced lateral root density (To *et al.*, 2004). All these data point unanimously towards an inhibitory role for cytokinins in root development. However, abrogating cytokinin response in the root by knocking out all cytokinin receptors results in a complete loss of meristem function (Higuchi *et al.*, 2004), suggesting that a minimal cytokinin response is required for root development. Recently, it has been shown that cytokinins inhibit lateral root initiation specifically at the level of G2 to M transition. Expression of G2-to-M regulatory genes, such as A- and B-type cyclins, was shown to be repressed by cytokinin treatment, whereas expression of G1-to-S regulatory genes was not affected (Li *et al.*, 2006).

As reduction in cytokinin content or signalling results in higher lateral root densities, it is beyond doubt that cytokinins antagonize auxin activity. On the other hand, auxin readily stimulates oxidative breakdown of cytokinins (Zhang *et al.*, 1995) and represses cytokinin biosynthesis, whereas cytokinin only mildly affects auxin content (Nordström *et al.*, 2004). In accordance to these data, the primary cytokinin-responsive type-A ARR5 is found to be downregulated at sites of lateral root development (Lohar *et al.*, 2004). These data are consistent with the established idea that a balanced response to auxin and cytokinin is required for lateral root initiation.

8.8 Brassinosteroids regulate auxin transport

Brassinosteroids have recently been shown to promote lateral root development synergistically with auxin (Bao *et al.*, 2004). Consistently, the signalling pathways for auxin and brassinosteroids have been shown to converge at the level of transcription during hypocotyl elongation (Mockaitis and Estelle, 2004; Nemhauser *et al.*, 2004). However, as auxin-responsive genes respond much slower to brassinosteroids than to auxins (Goda *et al.*, 2004), it seems more likely that brassinosteroids affect auxin-responsive gene expression indirectly. Recently, Li *et al.* (2005) showed that brassinosteroids stimulate basipetal polar auxin transport in roots, which was correlated with an increased *PIN1* and *PIN2* expression. Furthermore, PIN2 appears to be regulated by brassinosteroids also at the post-transcriptional level. In plants overexpressing *PIN2–GPF*, only a small domain of expression can be observed. However, by applying brassinosteroids, the *PIN2–GFP* expression domain could be enlarged considerably (Li *et al.*, 2005). Furthermore, in *pin2* mutants the promotive effect of brassinosteroids on lateral root formation is nearly completely abolished (Li *et al.*, 2005), suggesting a link between brassinosteroids, PIN2 abundance and lateral root formation. In contrast to this, auxin stimulates PIN2 degradation (Abas *et al.*, 2006) probably with the involvement of the SCF regulatory protein AXR1 (Sieberer *et al.*, 2000). It will be of interest to identify the F-box protein that mediates PIN2 ubiquitination and at which level brassinosteroids interfere with PIN2 degradation.

8.9 Light alters auxin sensitivity

Light is one of the fundamental elements required for sustaining autotrophic plant growth. The most apparent example of the role of light in plant growth and development is photomorphogenesis. Although roots remain predominantly in the dark, light also appears to play a role in regulating root development, at least in part by regulating auxin sensitivity. Unexpectedly, light might reach roots, even when submerged in the soil, as it can be conducted via the vascular system (mainly xylem) of woody (Sun *et al.*, 2003) and herbaceous plants (Sun *et al.*, 2005).

The molecular mechanisms of light signalling are well studied and characterized. COP1 is an E3 ubiquitin ligase that represses light signalling by targeting signal transduction machinery for degradation (reviewed by Yi and Deng, 2005). One of the best characterized targets of COP1 is the bZIP transcription factor HY5 (Osterlund *et al.*, 2000). Mutants in *cop1* and *hy5* impose opposite effects on lateral root development, in which *cop1* exhibits a defect in lateral root density whereas the *hy5* mutation enhances lateral root initiation and lateral root elongation (Oyama *et al.*, 1997; Ang *et al.*, 1998). These phenotypes are most likely to be achieved through alterations in auxin sensitivity. Indeed, HY5 has recently been implicated in direct binding to the promoters of the auxin-signalling inhibitors AXR2/IAA7 and SLR/IAA14 (Cluis *et al.*, 2004). In light conditions, COP1 is inactive and HY5 is stable to induce expression of *AXR2/IAA7* and *SLR/IAA14*, thereby repressing auxin signalling and lateral root formation. Therefore, the COP1/HY5-dependent control on root development might represent the naturally occurring mechanism to avoid rooting in the above-ground parts of the plant.

8.10 Conclusions and perspectives

Little is known about the molecular mechanisms of auxin-driven cell cycle progression. During lateral root initiation, auxin signalling converges onto cell cycle activation. Nevertheless, cell cycle activation in xylem pole pericycle cells does not necessarily result in the formation of a lateral root. This implies that lateral root initiation requires additional processes, such as the specification of a differential cell fate between the daughter cells within a lateral root initiation site. It will be of interest to identify components involved in the interplay between cell cycle progression and the respecification of cell fates during lateral root initiation. Because of the recent development of a lateral-root-inducible system (Himanen *et al.*, 2002) it is possible to use high-throughput molecular tools, such as microarray analysis.

Carefully dissecting lateral root initiation will yield many novel and exciting insights into cell cycle regulation, auxin-signalling cascades as well as the complex phenomenon of hormonal crosstalk.

References

Abas, L., Benjamins, R., Malenica, N., *et al.* (2006) Intracellular trafficking and proteolysis of the *Arabidopsis* auxin-efflux facilitator PIN2 are involved in root gravitropism. *Nat Cell Biol* **8**, 249–256.

Aida, M., Beis, D., Heidstra, R., *et al.* (2004) The *PLETHORA* genes mediate patterning of the *Arabidopsis* root stem cell niche. *Cell* **119**, 109–120.

Ang, L.-H., Chattopadhyay, S., Wei, N., *et al.* (1998) Molecular interaction between COP1 and HY5 defines a regulatory switch for light control of *Arabidopsis* development. *Mol Cell* **1**, 213–222.

Bao, F., Shen, J., Brady, S.R., Muday, G.K., Asami, T. and Yang, Z. (2004) Brassinosteroids interact with auxin to promote lateral root development in Arabidopsis. *Plant Physiol* **134**, 1624–1631.

Beeckman, T., Burssens, S. and Inzé, D. (2001) The peri-cell-cycle in *Arabidopsis*. *J Exp Bot* **52**, 403–411.

Benfey, P.N. and Scheres, B. (2000) Root development. *Curr Biol* **10**, R813–R815.

Benková, E., Michniewicz, M., Sauer, M., *et al.* (2003) Local, efflux-dependent auxin gradients as a common module for plant organ formation. *Cell* **115**, 591–602.

Blilou, I., Frugier, F., Folmer, S., *et al.* (2002) The *Arabidopsis HOBBIT* gene encodes a CDC27 homolog that links the plant cell cycle to progression of cell differentiation. *Genes Dev* **16**, 2566–2575.

Blilou, I., Xu, J., Wildwater, M., *et al.* (2005) The PIN auxin efflux facilitator network controls growth and patterning in *Arabidopsis* roots. *Nature* **433**, 39–44.

Boudolf, V., Barrôco, R., de Almeida Engler, J., *et al.* (2004a) B1-type cyclin-dependent kinases are essential for the formation of stomatal complexes in *Arabidopsis thaliana*. *Plant Cell* **16**, 945–955.

Boudolf, V., Vlieghe, K., Beemster, G.T.S., *et al.* (2004b) The plant-specific cyclin-dependent kinase CDKB1;1 and transcription factor E2Fa–DPa control the balance of mitotically dividing and endoreduplicating cells in Arabidopsis. *Plant Cell* **16**, 2683–2692.

Cluis, C.P., Mouchel, C.F. and Hardtke, C.S. (2004) The *Arabidopsis* transcription factor HY5 integrates light and hormone signaling pathways. *Plant J* **38**, 332–347.

D'Agostino, I.B., Deruère, J. and Kieber, J.J. (2000) Characterization of the response of the Arabidopsis response regulator gene family to cytokinin. *Plant Physiol* **124**, 1706–1717.

del Pozo, J.C., Boniotti, M.B. and Gutierrez, C. (2002a) Arabidopsis E2Fc functions in cell division and is degraded by the ubiquitin-SCF^{AtSKP2} pathway in response to light. *Plant Cell* **14**, 3057–3071.

del Pozo, J.C., Dharmasiri, S., Hellmann, H., Walker, L., Gray, W.M. and Estelle, M. (2002b) AXR1-ECR1-dependent conjugation of RUB1 to the Arabidopsis cullin AtCUL1 is required for auxin response. *Plant Cell* **14**, 421–433.

De Smet, I., Tetsumura, T., De Rybel, B., *et al.* (2007) Auxin-dependent regulation of lateral root positioning in the basal meristem of *Arabidopsis*. *Development* **134**, 681–690.

De Veylder, L., Beeckman, T., Beemster, G.T.S., *et al.* (2001) Functional analysis of cyclin-dependent kinase inhibitors of Arabidopsis. *Plant Cell* **13**, 1653–1668.

De Veylder, L., Beeckman, T., Beemster, G.T.S., *et al.* (2002) Control of proliferation, endoreduplication and differentiation by the *Arabidopsis* E2Fa/DPa transcription factor. *EMBO J* **21**, 1360–1368.

Dewitte, W., Riou-Khamlichi, C., Scofield, S., *et al.* (2003) Altered cell cycle distribution, hyperplasia, and inhibited differentiation in Arabidopsis caused by the D-type cyclin CYCD3. *Plant Cell* **15**, 79–92.

Dharmasiri, N., Dharmasiri, S. and Estelle, M. (2005a) The F-box protein TIR1 is an auxin receptor. *Nature* **435**, 441–445.

Dharmasiri, N., Dharmasiri, S., Weijers, D., *et al.* (2005b) Plant development is regulated by a family of auxin receptor F box proteins. *Dev Cell* **9**, 109–119.

Dharmasiri, S., Dharmasiri, N., Hellmann, H. and Estelle, M. (2003) The RUB/Nedd8 conjugation pathway is required for early development in *Arabidopsis*. *EMBO J* **22**, 1762–1770.

Dittmer, H.J. (1937) A quanitative study of the roots and root root hairs of a winter rye plant (*Secale cereale*). *Am J Bot* **24**, 417–420.

Fukaki, H., Nakao, Y., Okushima, Y., Theologis, A. and Tasaka, M. (2005) Tissue-specific expression of stabilized SOLITARY-ROOT/IAA14 alters lateral root development in Arabidopsis. *Plant J* **44**, 382–395.

Fukaki, H., Tameda, S., Masuda, H. and Tasaka, M. (2002) Lateral root formation is blocked by a gain-of-function mutation in the *SOLITARY-ROOT/IAA14* gene of *Arabidopsis*. *Plant J* **29**, 153–168.

Geldner, N., Richter, S., Vieten, A., *et al.* (2004) Partial loss-of-function alleles reveal a role for *GNOM* in auxin transport-related, post-embryonic development of *Arabidopsis*. *Development* **131**, 389–400.

Goda, H., Sawa, S., Asami, T., Fujioka, S., Shimada, Y. and Yoshida, S. (2004) Comprehensive comparison of auxin-regulated and brassinosteroid-regulated genes in Arabidopsis. *Plant Physiol* **134**, 1555–1573.

Gray, W.M., del Pozo, J.C., Walker, L., *et al.* (1999) Identification of an SCF ubiquitin-ligase complex required for auxin response in *Arabidopsis thaliana*. *Genes Dev* **13**, 1678–1691.

Gray, W.M., Kepinski, S., Rouse, D., Leyser, O. and Estelle, M. (2001) Auxin regulates SCF^{TIR1}-dependent degradation of AUX/IAA proteins. *Nature* **414**, 271–276.

Gray, W.M., Muskett, P.R., Chuang, H.W. and Parker, J.E. (2003) Arabidopsis SGT1b is required for SCF^{TIR1}-mediated auxin response. *Plant Cell* **15**, 1310–1319.

Guo, H.S., Xie, Q., Fei, J.F. and Chua, N.-H. (2005) MicroRNA directs mRNA cleavage of the transcription factor *NAC1* to downregulate auxin signals for Arabidopsis lateral root development. *Plant Cell* **17**, 1376–1386.

Hamann, T., Mayer, U. and Jürgens, G. (1999) The auxin-insensitive *bodenlos* mutation affects primary root formation and apical-basal patterning in the *Arabidopsis* embryo. *Development* **126**, 1387–1395.

Hardtke, C.S., Ckurshumova, W., Vidaurre, D.P., *et al.* (2004) Overlapping and non-redundant functions of the *Arabidopsis* auxin response factors *MONOPTEROS* and *NONPHOTOTROPIC HYPOCOTYL 4*. *Development* **131**, 1089–1100.

Higuchi, M., Pischke, M.S., Mähönen, A.P., *et al.* (2004) *In planta* functions of the *Arabidopsis* cytokinin receptor family. *Proc Natl Acad Sci USA* **101**, 8821–8826.

Himanen, K., Boucheron, E., Vanneste, S., de Almeida Engler, J., Inzé, D. and Beeckman, T. (2002) Auxin-mediated cell cycle activation during early lateral root initiation. *Plant Cell* **14**, 2339–2351.

Hosoda, K., Imamura, A., Katoh, E., *et al.* (2002) Molecular structure of the GARP family of plant Myb-related DNA binding motifs of the Arabidopsis response regulators. *Plant Cell* **14**, 2015–2029.

Hutchison, C.E., Li, J., Argueso, C., *et al.* (2006) The *Arabidopsis* histidine phosphotransfer proteins are redundant positive regulators of cytokinin signaling. *Plant Cell* **18**, 3073–3087.

Hwang, I. and Sheen, J. (2001) Two-component circuitry in *Arabidopsis* cytokinin signal transduction. *Nature* **413**, 383–389.

Jönsson, H., Heisler, M.G., Shapiro, B.E., Meyerowitz, E.M. and Mjolsness, E. (2006) An auxin-driven polarized transport model for phyllotaxis. *Proc Natl Acad Sci USA* **103**, 1633–1638.

Kepinski, S. and Leyser, O. (2005) The *Arabidopsis* F-box protein TIR1 is an auxin receptor. *Nature* **435**, 446–451.

Knox, K., Grierson, C.S. and Leyser, O. (2003) *AXR3* and *SHY2* interact to regulate root hair development. *Development* **130**, 5769–5777.

Kurihara, Y. and Watanabe, Y. (2004) *Arabidopsis* micro-RNA biogenesis through Dicer-like 1 protein functions. *Proc Natl Acad Sci USA* **101**, 12753–12758.

Kurup, S., Runions, J., Köhler, U., Laplaze, L., Hodge, S. and Haseloff, J. (2005) Marking cell lineages in living tissues. *Plant J* **42**, 444–453.

Laskowski, M.J., Williams, M.E., Nusbaum, H.C. and Sussex, I.M. (1995) Formation of lateral root meristems is a two-stage process. *Development* **121**, 3303–3310.

Li, L., Xu, J., Xu, Z.-H. and Xue, H.-W. (2005). Brassinosteroids stimulate plant tropisms through modulation of polar auxin transport in *Brassica* and *Arabidopsis*. *Plant Cell* **17**, 2738–2753.

Li, X., Mo, X., Shou, H. and Wu, P. (2006) Cytokinin-mediated cell cycling arrest of pericycle founder cells in lateral root initiation of *Arabidopsis*. *Plant Cell Physiol* **47**, 1112–1123.

Li, Y., Hagen, G. and Guilfoyle, T.J. (1992) Altered morphology in transgenic tobacco plants that overproduce cytokinins in specific tissues and organs. *Dev Biol* **153**, 386–395.

Liscum, E. and Reed, J.W. (2002) Genetics of Aux/IAA and ARF action in plant growth and development. *Plant Mol Biol* **49**, 387–400.

Lohar, D.P., Schaff, J.E., Laskey, J.G., Kieber, J.J., Bilyeu, K.D. and Bird, D.M. (2004) Cytokinins play opposite roles in lateral root formation, and nematode and Rhizobial symbioses. *Plant J* **38**, 203–214.

Lohrmann, J., Sweere, U., Zabaleta, E., *et al.* (2001) The response regulator ARR2, a pollen-specific transcription factor involved in the expression of nuclear genes for components of mitochondrial complex I in *Arabidopsis*. *Mol Genet Genomics* **265**, 2–13.

Lui, H., Wang, H., DeLong, C., Fowke, L.C., Crosby, W.L. and Fobert, P.R. (2000) The *Arabidopsis* Cdc2a-interacting protein ICK2 is structurally related to ICK1 and is a potent inhibitor of cyclin-dependent kinase activity *in vitro*. *Plant J* **21**, 379–385.

Magyar, Z., De Veylder, L., Atanassova, A., Bako, L., Inzé, D. and Bögre, L. (2005) The role of the *Arabidopsis* E2FB transcription factor in regulating auxin-dependent cell division. *Plant Cell* **17**, 2527–2541.

Malamy, J.E. and Benfey, P.N. (1997) Organization and cell differentiation in lateral roots of *Arabidopsis thaliana*. *Development* **124**, 33–44.

Mallory, A.C., Bartel, D.P. and Bartel, B. (2005) MicroRNA-directed regulation of Arabidopsis *AUXIN RESPONSE FACTOR17* is essential for proper development and modulates expression of early auxin response genes. *Plant Cell* **17**, 1360–1375.

Mason, M.G., Mathews, D.E., Argyros, D.A., *et al.* (2005) Multiple type-B response regulators mediate cytokinin signal transduction in *Arabidopsis*. *Plant Cell* **17**, 3007–3018.

Mockaitis, K. and Estelle, M. (2004) Integrating transcriptional controls for plant cell expansion. *Genome Biol* **5**, 245.1–245.4.

Navarro, L., Dunoyer, P., Jay, F., *et al.* (2006) A plant miRNA contributes to antibacterial resistance by repressing auxin signaling. *Science* **312**, 436–439.

Nemhauser, J.L., Mockler, T.C. and Chory, J. (2004) Interdependency of brassinosteroid and auxin signaling in *Arabidopsis*. *PLoS Biol* **2**, e258.

Nordström, A., Tarkowski, P., Tarkowska, D., *et al.* (2004) Auxin regulation of cytokinin biosynthesis in *Arabidopsis thaliana*, a factor of potential importance for auxin – cytokinin-regulated development. *Proc Natl Acad Sci USA* **101**, 8039–8044.

Okushima, Y., Overvoorde, P.J., Arima, K., *et al.* (2005) Functional genomic analysis of the *AUXIN RESPONSE FACTOR* gene family members in *Arabidopsis thaliana*, unique and overlapping functions of *ARF7* and *ARF19*. *Plant Cell* **17**, 444–463.

Osterlund, M.T., Hardtke, C.S., Wei, N. and Deng, X.W. (2000) Targeted destabilization of HY5 during light-regulated development of *Arabidopsis*. *Nature* **405**, 462–466.

Overvoorde, P.J., Okushima, Y., Alonso, J.M., *et al.* (2005) Functional genomic analysis of the *AUXIN/INDOLE-3-ACETIC ACID* gene family members in *Arabidopsis thaliana*. *Plant Cell* **17**, 3282–3300.

Oyama, T., Shimura, Y. and Okada, K. (1997) The *Arabidopsis HY5* gene encodes a bZIP protein that regulates stimulus-induced development of root and hypocotyl. *Genes Dev* **11**, 2983–2995.

Parker, G.C., Anastassova-Kristeva, M., Eisenberg, L.M., *et al.* (2005) Stem cells: shibboleths of development, part II: toward a functional definition. *Stem Cells Dev* **14**, 463–469.

Petrášek, J., Mravec, J., Bouchard, R., *et al.* (2006) PIN proteins perform a rate-limiting function in cellular auxin efflux. *Science* **312**, 914–918.

Ramos, J.A., Zenser, N., Leyser, O. and Callis, J. (2001) Rapid degradation of auxin/indoleacetic acid proteins requires conserved amino acids of domain II and is proteasome dependent. *Plant Cell* **13**, 2349–2360.

Rashotte, A.M., Carson, S.D.B., To, J.P.C. and Kieber, J.J. (2003) Expression profiling of cytokinin action in Arabidopsis. *Plant Physiol* **132**, 1998–2011.

Reinhardt, D. (2005) Phyllotaxis – a new chapter in an old tale about beauty and magic numbers. *Curr Opin Plant Biol* **8**, 487–493.

Reinhardt, D., Pesce, E.-R., Stieger, P., *et al.* (2003) Regulation of phyllotaxis by polar auxin transport. *Nature* **426**, 255–260.

Riefler, M., Novak, O., Strnad, M. and Schmülling, T. (2006) *Arabidopsis* cytokinin receptor mutants reveal functions in shoot growth, leaf senescence, seed size, germination, root development, and cytokinin metabolism. *Plant Cell* **18**, 40–54.

Rogg, L.E., Lasswell, J. and Bartel, B. (2001) A gain-of-function mutation in *IAA28* suppresses lateral root development. *Plant Cell* **13**, 465–480.

Rouse, D., Mackay, P., Stirnberg, P., Estelle, M. and Leyser, O. (1998) Changes in auxin response from mutation in an *AUX/IAA* gene. *Science* **279**, 1371–1373.

Sabatini, S., Beis, D., Wolkenfelt, H., *et al.* (1999) An auxin-dependent distal organizer of pattern and polarity in the *Arabidopsis* root. *Cell* **99**, 463–472.

Sakai, H., Honma, T., Aoyama, T., *et al.* (2001) ARR1, a transcription factor for genes immediately responsive to cytokinins. *Science* **294**, 1519–1521.

Segers, G., Gadisseur, I., Bergounioux, C., *et al.* (1996) The *Arabidopsis* cyclin-dependent kinase gene *cdc2bAt* is preferentially expressed during S and G2 phases of the cell cycle. *Plant J* **10**, 601–612.

Shen, W.-H. (2002) The plant E2F–Rb pathway and epigenetic control. *Trends Plant Sci* **7**, 505–511.

Shostak, S. (2006) (Re)defining stem cells. *BioEssays* **28**, 301–308.

Sieberer, T., Seifert, G.J., Hauser, M.-T., Grisafi, P., Fink, G.R. and Luschnig, C. (2000) Post-transcriptional control of the *Arabidopsis* auxin efflux carrier EIR1 requires AXR1. *Curr Biol* **10**, 1595–1598.

Smith, R.S., Guyomarc'h, S., Mandel, T., Reinhardt, D., Kuhlemeier, C. and Prusinkiewicz, P. (2006) A plausible model of phyllotaxis. *Proc Natl Acad Sci USA* **103**, 1301–1306.

Sorin, C., Bussell, J.D., Camus, I., *et al.* (2005) Auxin and light control of adventitious rooting in Arabidopsis require ARGONAUTE1. *Plant Cell* **17**, 1343–1359.

Sun, Q., Yoda, K. and Suzuki, H. (2005) Internal axial light conduction in the stems and roots of herbaceous plants. *J Exp Bot* **56**, 191–203.

Sun, Q., Yoda, K., Suzuki, M. and Suzuki, H. (2003) Vascular tissue in the stem and roots of woody plants can conduct light. *J Exp Bot* **54**, 1627–1635.

Taniguchi, M., Kiba, T., Sakakibara, H., Ueguchi, C., Mizuno, T. and Sugiyama, T. (1998) Expression of *Arabidopsis* response regulator homologs is induced by cytokinins and nitrate. *FEBS Lett* **429**, 259–262.

Tatematsu, K., Kumagai, S., Muto, H., *et al.* (2004) *MASSUGU2* encodes Aux/IAA19, an auxin-regulated protein that functions together with the transcriptional activator NPH4/ARF7 to regulate differential growth responses of hypocotyl and formation of lateral roots in *Arabidopsis thaliana*. *Plant Cell* **16**, 379–393.

Thrower, J.S., Hoffman, L., Rechsteiner, M. and Pickart, C.M. (2000) Recognition of the polyubiquitin proteolytic signal. *EMBO J* **19**, 94–102.

Tian, Q. and Reed, J.W. (1999) Control of auxin-regulated root development by the *Arabidopsis thaliana SHY2/IAA3* gene. *Development* **126**, 711–721.

Timpte, C., Wilson, A.K. and Estelle, M. (1994) The *axr2-1* mutation of *Arabidopsis thaliana* is a gain-of-function mutation that disrupts an early step in auxin response. *Genetics* **138**, 1239–1249.

To, J.P.C., Haberer, G., Ferreira, F.J., *et al.* (2004) Type-A Arabidopsis response regulators are partially redundant negative regulators of cytokinin signaling. *Plant Cell* **16**, 658–671.

Vanneste, S., De Rybel, B., Beemster, G.T.S., *et al.* (2005) Cell cycle progression in the pericycle is not sufficient for SOLITARY ROOT/IAA14-mediated lateral root initiation in *Arabidopsis thaliana*. *Plant Cell* **17**, 3035–3050.

Vaucheret, H., Vazquez, F., Crete, P. and Bartel, D.P. (2004) The action of ARGONAUTE1 in the miRNA pathway and its regulation by the miRNA pathway are crucial for plant development. *Genes Dev* **18**, 1187–1197.

Verkest, A., de O. Manes, C.-L., Vercruysse, S., *et al.* (2005) The cyclin-dependent kinase inhibitor KRP2 controls the onset of the endoreduplication cycle during Arabidopsis leaf development through inhibition of mitotic CDKA;1 kinase complexes. *Plant Cell* **17**, 1723–1736.

Vieten, A., Vanneste, S., Wisniewska, J., *et al.* (2005) Functional redundancy of PIN proteins is accompanied by auxin-dependent cross-regulation of PIN expression. *Development* **132**, 4521–4531.

Wang, J.-W., Wang, L.-J., Mao, Y.-B., Cai, W.-J., Xue, H.-W. and Chen, X.-Y. (2005) Control of root cap formation by microRNA-targeted auxin response factors in Arabidopsis. *Plant Cell* **17**, 2204–2216.

Weijers, D., Benkova, E., Jäger, K.E., *et al.* (2005) Developmental specificity of auxin response by pairs of ARF and Aux/IAA transcriptional regulators. *EMBO J* **24**, 1874–1885.

Werner, T., Motyka, V., Laucou, V., Smets, R., Van Onckelen, H. and Schmülling, T. (2003) Cytokinin-deficient transgenic Arabidopsis plants show multiple developmental alterations indicating opposite functions of cytokinins in the regulation of shoot and root meristem activity. *Plant Cell* **15**, 2532–2550.

Werner, T., Motyka, V., Strnad, M. and Schmülling, T. (2001) Regulation of plant growth by cytokinin. *Proc Natl Acad Sci USA* **98**, 10487–10492.

Wildwater, M., Campilho, A., Perez-Perez, J.M., *et al.* (2005) The *RETINOBLASTOMA-RELATED* gene regulates stem cell maintenance in *Arabidopsis* roots. *Cell* **123**, 1337–1349.

Wilmoth, J.C., Wang, S., Tiwari, S.B., *et al.* (2005) NPH4/ARF7 and ARF19 promote leaf expansion and auxin-induced lateral root formation. *Plant J* **43**, 118–130.

Xie, Q., Frugis, G., Colgan, D. and Chua, N.-H. (2000) *Arabidopsis* NAC1 transduces auxin signal downstream of TIR1 to promote lateral root development. *Genes Dev* **14**, 3024–3036.

Xie, Q., Guo, H.S., Dallman, G., Fang, S., Weissman, A.M. and Chua, N.-H. (2002) SINAT5 promotes ubiquitin-related degradation of NAC1 to attenuate auxin signals. *Nature* **419**, 167–170.

Xu, J., Hofhuis, H., Heidstra, R., Sauer, M., Friml, J. and Scheres, B. (2006) A molecular framework for plant regeneration. *Science* **311**, 385–388.

Yang, X., Lee, S., So, J.-H., *et al.* (2004) The IAA1 protein is encoded by *AXR5* and is a substrate of SCF^{TIR1}. *Plant J* **40**, 772–782.

Yi, C. and Deng, X.W. (2005) COP1 – from plant photomorphogenesis to mammalian tumorigenesis. *Trends Cell Biol* **15**, 618–625.

Zhang, R., Zhang, X., Wang, J., Letham, D.S., McKinney, S.A. and Higgins, T.J.V. (1995) The effect of auxin on cytokinin levels and metabolism in transgenic tobacco tissue expressing an *ipt* gene. *Planta* **196**, 84–94.

9 Cell cycle control during leaf development

Andrew J. Fleming

9.1 Introduction

Leaves are made of cells. Moreover, plant histology reveals that leaves contain distinct tissues which, to a large extent, are distinguishable by their relative cell size and shape (e.g. epidermis, palisade mesophyll and spongy mesophyll). Therefore, it appears self-evident that the control of the cell cycle (and the associated process of cytokinesis) must be intimately linked with the form and histology of a leaf. As will be seen in the rest of this chapter, to some extent this apparently simplistic expectation is met by reality. However, there are many strands of data which indicate that an interpretation of a leaf as a body constructed from cells in the manner of building blocks stacked upon each other is not valid and that many aspects of leaf development are not dependent on a rigid control of the cell cycle. Indeed, as will be discussed, it is possible to seriously disrupt the 'normal' pattern of cell cycle progression and cell division during leaf development and still obtain a leaf that is fully functional. Thus, the role of the cell cycle in leaf development is in many ways surprisingly debatable. At the most extreme, it is possible to postulate that its role may simply be restricted to generating a compartmented volume upon which regulators can act over space and time to determine morphology, size and differentiation. Aspects of this debate will be discussed in this chapter.

At this point, it is probably necessary to define the different elements of leaf development, since the cell cycle may have a different significance for each of these facets. Firstly, there is morphogenesis, that is the successive change in three-dimensional shape that a leaf undergoes during development. Secondly, there is growth, which will determine the absolute size that a mature leaf can achieve under optimal conditions. Morphogenesis is dependent on differential growth. Thirdly, there is growth rate, that is the rapidity with which a leaf will achieve a specific size and shape. Finally, there is differentiation, that is the mechanism by which different parts of a leaf will take on specific biochemical and biophysical attributes to allow that tissue to perform a specific role as a part of a whole functional organ and plant. Since leaves are made of cells, it is clear that the cell cycle and its control has the potential to influence all these attributes of the leaf and, indeed, that in the normal progression of development all these attributes are coordinated. At the same time, it is also clear that (at least conceptually) the cell cycle may not be absolutely necessary for any of these attributes. For example, if there were no biophysical limit on the size of a cell one could imagine a single cell undergoing anisotropic growth to form a leaf of particular shape. Certainly, single-celled giant algae can form unique and complex forms in the absence of any division process. The form of this theoretical leaf-cell would be controlled by cytoskeletal/cell wall synthesis and extension processes, as would the rate of growth. The differentiation of such a super-cell would be controlled by differential gene expression, leading to specific biochemical properties that would not a priori depend on any particular phase of the cell cycle. The final size of such a theoretical leaf-cell would also not by necessity be dependent on any cell cycle property. The spatial differentiation within such a giant single cell would, of course, be rather limited, being dependent on the ability to target particular biochemical attributes to different parts of the cell (although it is clear that such subcellular targeting and organization occurs in 'normal' cells). In such a hypothetical scenario, one might envision that the formation of cells (compartments) within such a giant leaf-cell might primarily occur as a mechanism by which different spatial regions could be shut off from each other to allow more specific differentiation to occur. Such chopping up of a large space would also, of course, offer the advantage of providing better structural strength and integrity to the organ.

Although put here in very simplified form, the theoretical scenario envisioned above (often referred to as an organismal view of development) has been proposed as an alternative to the traditional idea of leaf development being driven by cell-cycle-associated processes that lead to an accumulation of cells in particular regions of the developing leaf, thus driving morphogenesis, determining size and defining differentiation (cellular theory of development). Although obviously put here in an extreme theoretical form, such thought experiments are useful in challenging accepted (but actually unproven) hypotheses on the role of the cell cycle and division in development. These alternative views have been considered in a number of recent papers and reviews (Kaplan, 2001; Fleming, 2002, 2005; Tsukaya, 2002, 2003; Beemster *et al.*, 2003). To what extent do the latest data support or disprove these alternate views?

In this chapter, the processes of cell division and the cell cycle that occur during leaf development will be described. In addition, and perhaps more importantly, an attempt will be made to assess the significance of these processes. To what extent are they causally involved in the various attributes of leaf development? To what extent can they be viewed as parallel or correlative facets? In this analysis, repeated reference to the organismal and cellular theories of leaf development described above will be made.

9.2 The cell cycle and cell division during leaf initiation

9.2.1 Patterns of the cell cycle and cell division during leaf initiation

Leaf initiation is a growth process. A portion of tissue in a specialized organ, the shoot apical meristem (SAM), grows out of the surface to form a visibly distinct organ. At this early point in our analysis it is vital to distinguish and define growth and cell division and, most importantly, to make the point clear that the two terms are not synonymous. Growth is an irreversible increase in size. Cell division is a process by which a single cell splits into two cells. Cell division may be (and, indeed, generally is) accompanied by an increase in cell size (i.e. cell growth). However, the two characteristics of cell division and growth are clearly separable (Figure 9.1). Cell division (by definition) leads to the formation of more cells. If there is a limit to

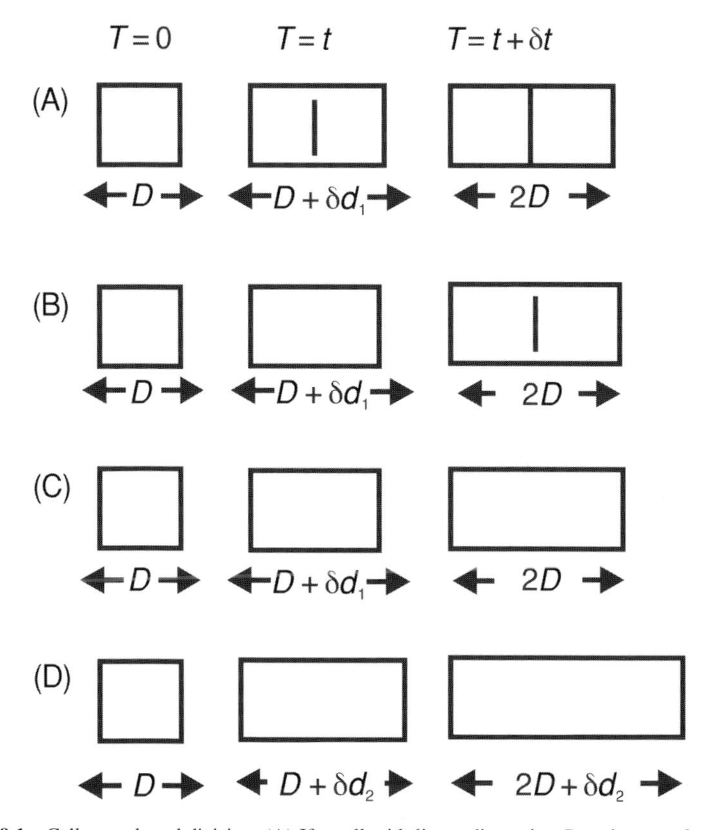

Figure 9.1 Cell growth and division. (A) If a cell with linear dimension D at time $t = 0$ undergoes anisotropic growth, then after a time t, the linear dimension will have attained a size $D + \delta d_1$, at which cell division is initiated (incomplete cross-wall). The daughter cells generated will continue to grow until at time $t + \delta t$ two daughter cells, each of size D, have been formed. (B) If the set size for division to occur is shifted to a higher value, then division occurs at a later time (e.g. $t + \delta t$). At this time point, the daughter cells have the same size as in (A) and the growth of the tissue is the same as in (A). However, provided the daughter cells fulfil their growth potential, final growth will be greater at later time points compared with (A). (C) If proliferation rate falls to zero, growth rate of the cell may be exactly the same as in (A) and (B). Growth occurs independent of cell division. If a limit is set on the final potential size of a cell, final growth in (C) will be less than in (A) and (B), again provided that the cells in (A) and (B) fulfil their growth potential. (D) If the switch from cell-division-associated growth (A and B) to non-cell-division-associated growth (C) involves an increase in growth rate, then in any time period $t + \delta t$ a larger cell will be produced ($2D + \delta d_2$). However, a limit on growth will be set by the maximum size that an individual cell can attain. The generation of more cells (as in A and B) will lead to a greater potential final organ size (at a slower growth rate), but only if this generation of cells is translated into final individual cell growth.

the size that any individual cell can achieve, then the more cells that are produced, the larger the potential final size of an organ. If more cells are produced in one part of an organ than in another, then the potential final growth of that organ part will be greater. Cell production can thus be viewed as an investment that may be realized as a final increase in overall organ growth. However, this will occur only if all the individual cells generated undergo appropriate growth. Again, increased cell division does not necessarily mean increased growth, and increased growth is not necessarily caused by increased cell division (since individual cells may grow more than others). Indeed, plants are characterized by a massive range of component cell sizes.

Although this point may seem laboured, it is an important distinction that many published papers and reviews fail to take into account. Thus, although the SAM can be referred to as a centre of cell proliferation (which it certainly is), it is more accurate to describe the SAM as a growing organ associated with a relatively high rate of cell division. This definition highlights the important points that the SAM continually generates new tissue upon which (as we will see later) growth factors can act to determine organogenesis, and that although cell division is relatively high in the SAM it is in no way absent in adjacent parts of the plant, including the leaves derived from it. So, how are growth, the cell cycle and cell division controlled within the SAM, and how are these related but separable processes related to leaf initiation?

A basic paradigm of the molecular function of the SAM is that a small group of cells at the heart of the organ express a homeodomain transcription factor (WUSCHEL) that is required for the maintenance of growth and cell division in the surrounding tissue (again, it must be stressed here that although most authors refer simply to cell division, they are actually using the term to mean both growth and cell division) (Schoof *et al.*, 2000). Loss of WUSCHEL activity leads to eventual loss of growth and cell division in the SAM and termination of plant growth, since the source of cellular building blocks required for growth ceases. The *WUSCHEL*-expressing domain lies proximal to a surface group of cells of the SAM that express a gene, *CLAVATA3*, which encodes an extracellular protein. WUSCHEL acts at a distance by an as yet unknown mechanism to promote the expression of *CLAVATA3*. The CLAVATA3 protein acts as an inhibitor of *WUSCHEL* expression, most probably via a receptor–kinase complex encoded by *CLAVATA1* and *CLAVATA2* (Clark *et al.*, 1997; Brand *et al.*, 2000). These two genes are expressed in a domain of cells encompassing the WUSCHEL-expressing cells. This WUSCHEL/CLAVATA paradigm provides a feedback loop whereby, once the system is established, any tendency for an increase in growth (and thus cell number) within the SAM is counteracted by an increase in CLAVATA3 expression (thus inhibiting WUSCHEL expression and decreasing the promotion of growth and cell number). Conversely, any tendency for a decrease in growth and cell number in the SAM leads to a decrease in CLAVATA3 expression and a derepression of WUSCHEL activity, thus promoting growth and cell proliferation. This system depends on a consistent relationship between growth and cell division within the SAM; yet in plants the regulatory mechanism linking these two key parameters is unknown. This regulatory link is key to plant development, since as tissue leaves the SAM domain, cell size generally increases despite cell proliferation being maintained. This implies that the set cell volume at which cell division is occurring must be changing in a controlled fashion or, alternatively, the increase in cell volume is entirely due to an increase in vacuolar volume and cell division is occurring at a set cytoplasmic/nuclear volume which is constant. Without accurate quantitation of these subcellular volumes, it is difficult to distinguish these possibilities; yet identifying which is correct is of fundamental importance for our understanding of the control of cell division in plants and how this is integrated with growth.

With respect to leaf initiation, not all regions of the SAM (and thus, by inference, not all cells within the SAM) are competent to make the transition to leaf differentiation. In particular, leaf initiation can occur only on the periphery of the SAM, the tissue in a central zone apparently being recalcitrant to the factors that promote leaf formation. Exactly how this specification between central and peripheral zone occurs is unclear, although recent data from the Meyerowitz group (using elegant visualization techniques coupled with inducible zone-specific gene expression constructs) show that alteration in CLAVATA3 activity can both re-specify the position of the central zone/peripheral zone boundary and influence the growth rate and cell division activity at a distance from the CLAVATA3-expressing domain (Reddy and Meyerowitz, 2005).

The earliest marker for leaf initiation is a switch in PIN protein distribution, which is predicted to lead to a local accumulation and flux of auxin at the presumptive site of leaf formation (Benková *et al.*, 2003; Reinhardt *et al.*, 2003b). Auxin has been closely linked with the promotion of cell cycle events and can lead to the transcriptional promotion or repression of a number of genes linked with the cell cycle (Menges *et al.*, 2002). However, at the same time auxin has long been linked with cell expansion events via, for example, the acid growth hypothesis (Hager, 2003). Thus, the deduction that auxin accumulates at the site of leaf formation does not automatically mean that the cell cycle and cell division are promoted there. Indeed, visualization of cell division dynamics at the site of leaf initiation has failed to reveal an increase of cell division frequency compared to equivalent regions around the SAM circumference that are not involved in leaf initiation (Grandjean *et al.*, 2004; Reddy *et al.*, 2004; Reddy and Meyerowitz, 2005). Although the primary transcriptional response to local accumulation of auxin in the SAM is unknown, a number of distinct transcriptional patterns are observed at the site of leaf formation, and these are presumably linked to auxin accumulation. For example, transcripts for the homeodomain transcriptional regulators STM and KNAT1 are found throughout the SAM but disappear at the site of leaf initiation (McConnell *et al.*, 2001; Byrne, 2005). At the same time, transcripts for MYB transcription factors accumulate in the tissue determined to form a leaf.A complex network of interactions occurs between these transcription factors to delineate presumptive leaf tissue from the SAM, but whether an altered pattern of cell cycle gene expression is required for this switch in fate is unclear. For example, the expression of a *CYCLIND3* gene is upregulated during primordium initiation in *Antirrhinum* (Gaudin *et al.*, 2000), but it is unclear whether this specific expression pattern is causally involved in the occurrence of morphogenic processes. Parallel to the altered patterns of transcription factors, genes encoding proteins involved in hormone biosynthesis or signal transduction show altered patterns of expression in tissue undergoing determination to form a leaf (Sakamoto *et al.*, 2001; Hay *et al.*, 2002). Thus, an enzyme required for gibberellic acid biosynthesis increases in the presumptive leaf tissue. However, although gibberellic acid (like auxin) could target the expression of genes involved in the cell cycle, there is as yet no evidence that this happens in the SAM. Moreover, like auxin, it is also easy to envisage that these hormones and their linked signalling pathways impinge on tissue growth rather than directly on cell division. These initial triggers might not actually require altered gene expression, but might, for example, influence cytoskeletal/cell wall complexes or vesicle trafficking, which restrain or enable expansion growth. Thus, although we are beginning to unravel the molecular processes underpinning leaf initiation, there are as yet no definitive data showing that an altered frequency of the cell cycle or cell division is required for this process. Indeed, one of the few genes characterized to show an increased expression at the presumptive site of leaf formation encodes the cell wall protein expansin (Cho and Kende, 1997; Reinhardt *et al.*, 1998). This protein is predicted to increase cell wall extensibility, thus promoting growth (Cosgrove, 2000). At least some genes encoding expansin are auxin induced (Caderas *et al.*, 2000), so these observations fit with a non-cell division dependent initial process of leaf initiation.

That cell cycle and cell division processes may not be primary targets for the hormonal system involved in leaf initiation comes from direct observation of the cellular dynamics in the SAM during leaf initiation. These data do not reveal any increase in cell division frequency at the site of presumptive leaf formation (Grandjean *et al.*, 2004; Reddy *et al.*, 2004). What has been observed, however, is that there is often a change in cell division patterning at the site of presumptive leaf formation (Reddy *et al.*, 2004) (Figure 9.2). In particular, in the tunica layers there is often a switch from purely anticlinal cell division to a pattern in which the orientation of cell division is more random (Lyndon and Robertson, 1976). The role of this conserved switching of cell division pattern in leaf initiation is discussed in the next section.

9.2.2 Manipulation of the cell cycle and cell division during leaf initiation

The above description has emphasized the patterns of cell division during leaf initiation. The consistency in cell division pattern observed in the SAM suggests that these patterns play a functional role.A number of experiments have been performed in which parameters of the cell cycle and division have been manipulated, either specifically within the SAM or, more commonly, throughout the developing plant including the SAM. These experiments have shed light on the significance of the observed commonality of cell division pattern within the SAM and its requirement during leaf initiation.

One of the earliest investigations on the potential linkage of cell division pattern and leaf initiation came from the studies of Foard (1971) who utilized radiation to inhibit cell cycle processes in young grass seedlings. The most surprising observation made was that despite evidence that cell division was indeed inhibited in these seedlings, the process of leaf development continued to occur for at least a limited

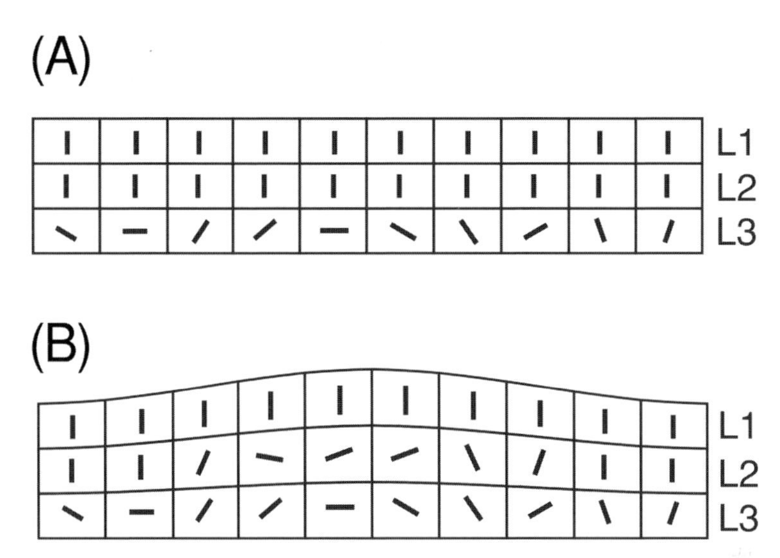

Figure 9.2 Cell division pattern and leaf initiation. (A) The shoot apical meristem consists of three generative layers, L1, L2 and L3. In most of the meristem, divisions (indicated by thick lines within each cell) are restricted to an anticlinal orientation in the outer two cell layers. In the inner L3 layer, cell division orientation is more random. (B) At the position where leaf initiation (outgrowth) is just starting, there is a loss of the tight regulation of anticlinal cell division orientation, particularly in the L2layer. Is this altered cell division pattern causal or correlative with leaf initiation?

period of time. Although it was somewhat unclear whether the development that was observed reflected the continued growth of leaves already initiated at the time of irradiation rather than actual *de novo* formation of leaves, these examples clearly demonstrated the potential separation of cell division processes from growth. More refined analysis became possible in the 1980s and 1990s with the advent of transgenic techniques, which allowed the manipulation of specific cell cycle genes, and the widespread use of mutational approaches to analyze plant development.

As documented elsewhere in this volume, a large number of plant cell cycle genes have now been identified and various approaches taken to investigate the outcome on plant development of either mutation of these genes or, more frequently, misexpression of the products of these genes. In the following section, a summary is made of the conclusions of these studies with respect to leaf initiation (Hemerly *et al.*, 1995; Cockcroft *et al.*, 2000; Wang *et al.*, 2000; De Veylder *et al.*, 2001, 2002; Autran *et al.*, 2002; Wyrzykowska *et al.*, 2002).

A general conclusion from these experiments has been that plants display a great plasticity with respect to their ability to absorb altered parameters of the cell cycle and cell division while maintaining a relatively normal morphology, differentiation and functionality. For example, although overexpression of cell cycle genes can lead to clear changes in cell division frequency and, as will be discussed in the next section, altered leaf size and shape, the process of leaf initiation remains relatively unperturbed. These observations suggest that, providing at least a basic level of competence for cell division is achieved within a plant, the actual process of leaf initiation is not dependent on a particular rate or pattern of cell division. Changes in these parameters are observed during the normal processes of leaf formation, but they are not absolutely required. This implies either that default mechanisms exist which can compensate for abnormalities in cell division during leaf initiation (i.e. that cell division is normally important during leaf initiation but that, *in extremis*, safety mechanisms can cut in to salvage the process of leaf formation) or that the process of leaf formation is independent of cell division and that the consistent patterns of cell division observed during leaf formation are a consequence of a consistent mechanism of leaf formation which, as one of its downstream targets, influences cell division. Recent data have shed light on what this consistent mechanism might be.

As mentioned in the previous section, the flux of the growth regulator auxin is controlled by a transport system closely associated with the cellular asymmetric distribution of a family of PIN proteins. Although definitive evidence that these membrane-localized proteins actually function to transport auxin is still lacking, the pattern of PIN protein distribution predicts auxin flux. Within the SAM, PIN protein distribution predicts an accumulation of auxin at the site of presumptive leaf formation, and localized ectopic application of auxin is sufficient to induce leaf formation. The open question is what is the immediate downstream target of localized auxin accumulation within the SAM. Although auxin has frequently been associated with changes in cell division frequency, it has also been long implicated in growth processes. It seems highly likely that cellular processes of both division and growth are promoted by auxin accumulation in the SAM. However, experiments in which cell division frequency was locally stimulated within the SAM (via a transgenic approach in which either a plant *CYCLINA* or a yeast *CDC25* was transiently overexpressed) did not lead to any promotion or disruption of leaf initiation (Wyrzykowska *et al.*, 2002). On the other hand, similar experiments in which a cell-wall-loosening protein (expansin) was locally induced or applied to the SAM led to leaf initiation (Pien *et al.*, 2001). The simplest interpretation of these data is that although under normal circumstances auxin leads to a coordinated promotion of growth and cell division at the presumptive site of leaf formation, the actual process of morphogenesis is dependent on altered tissue growth characteristics (most probably defined by the cell wall), and the observed cell division pattern is a consequence of this growth and does not drive morphogenesis. Similarly, when the orientation of cell division was disrupted in the SAM (via misexpression of the dynamin-like protein phragmoplastin) leaf formation was not disrupted (Wyrzykowska and Fleming, 2003). It therefore appears that the highly conserved pattern of cell division observed in the SAM is not causally involved in leaf initiation.

This raises the question of why such a conserved pattern exists. One simple explanation is that in a growing three-dimensional ball of cells, tissue towards the surface will automatically experience tangential forces that will stretch the surface. If, as proposed over a hundred years ago by Hofmeister (1863), the orientation of plant cell division normally occurs perpendicular to the main axis of tissue extension, then the fact that anticlinal cell divisions are observed in the outer tissue of a domed SAM may not be surprising (i.e. it simply reflects the biophysical stresses within the tissue and the cellular response to such stresses). Although the primary cause of such a conserved pattern of division may be geometric, it might have a significant outcome on the functioning of the SAM. Plant cells are connected to each other via complex pores termed *plasmodesmata*. These may either be primary plasmodesmata (formed within the new cell wall laid down at the phragmoplast, thus connecting daughter cells) or secondary plasmodesmata (formed *de novo* between cells that are not direct siblings). The layered structure of the outer SAM would predict that the anticlinal cell divisions lead to a distinct pattern of plasmodesmatal connection in the outer layers (termed *tunica*) of the SAM separate from the inner body (corpus) of the SAM. Numerous data now demonstrate the potential flux of information-carrying molecules (RNA, transcription factors) via plasmodesmata, leading to a proposed symplastic-based information superhighway within plants. The functional significance of such a plasmodesmata-controlled flux of information is still somewhat controversial, but it would allow for cell division pattern to influence developmental decisions (Lucas *et al.*, 1993; Haywood *et al.*, 2002; Kim *et al.*, 2002). Indeed, experiments described above in which the plane of cell division was altered in the SAM led to altered patterns of accumulation of transcription factors (KNOX and ARP) involved in the control of leaf development (Wyrzykowska and Fleming, 2003). Although it is unclear whether the pattern of protein distribution was altered in the SAM, these experiments demonstrate the potential influence of cell division pattern on gene expression pattern and, thus, development.

More recently, the potential importance of the conserved pattern of cell division in the tunica layer has been highlighted by the finding that the PIN proteins associated with auxin flux are generally restricted to the outer cell layer. Ablation of this layer leads to blockage of leaf initiation (Reinhardt *et al.*, 2003a). Thus, although disruption of the anticlinal pattern of division in the outer tunica does not disrupt leaf formation, the maintenance of some level of cell division in this layer (providing a conduit for auxin flux) is probably essential for leaf formation.

9.2.3 The role of the cell cycle and cell division during leaf initiation

The data indicate that provided a certain basal competence for division is maintained, which precludes lethality, leaf formation can occur. Despite the occurrence of conserved patterns of cell division within the SAM, it appears that these patterns are not required for leaf formation. Altered frequency and orientation of cell division in the SAM does not lead to abnormal leaf initiation. There are also no data to suggest that cells incorporated into a leaf must pass through a particular stage of the cell cycle or that initiation of an endocycle is required during leaf initiation.

With respect to differentiation, it is clear that at an early stage of determination of tissue within the SAM to form a leaf, distinct changes in gene expression occur, most notably in *KNOX* and *ARP*-type genes. There are, however, few data to indicate that these patterns of gene expression are cell cycle linked. It seems, therefore, that the main function of cell division within the SAM is to generate a volume (field) of growing cells upon which the growth factors regulating leaf initiation can act.

9.3 The cell cycle and cell division during leaf growth

9.3.1 Patterns of the cell cycle and cell division during leaf growth

On initiation, leaf primordia from different plants have very similar forms, i.e. they are relatively broad, finger-like projections arising on the periphery of the SAM. Subsequently, leaves of different plants take on very different final sizes and shapes, and even within one plant the leaves formed early in development are often distinct in form from those formed later. What is the role of cell division and cell cycle in the mechanism by which different leaves achieve distinct size and form?

To understand the potential role of cell division in leaf morphogenesis, it is first necessary to examine the pattern of cell division during leaf development post-initiation. This has been done in great detail via kinematic analysis coupled with techniques to assess the nuclear C value of cells within the leaf at different stages of development, as well by the use of reporter gene constructs designed to provide data on the patterning of mitosis during leaf development. The following description is based on a number of papers, particularly Poethig and Sussex (1985a, b), Pyke *et al.* (1991), Donnelly *et al.* (1999), Fiorani *et al.* (2000), Granier *et al.* (2000), Tardieu and Granier (2000), Beemster *et al.* (2006).

Immediately after initiation, leaf primordia undergo a high rate of relative expansion growth coupled with a high rate of cell division throughout. As development proceeds, the relative growth rate decreases and termination of cell division occurs in a gradient starting at the distal tip of the leaf and progressing down towards the proximal region of the petiole. Thus, in a developing leaf there are two processes occurring: growth accompanied by cell division and growth not accompanied by cell division (Figure 9.1). Tissue in which growth is occurring linked with cell division may actually be growing at a lower rate than the tissue in which cell division has terminated and tissue is growing without cell division. Indeed, the available data suggest that whereas proliferating cells grow in a linear fashion, non-proliferating but growing tissue grows in an exponential fashion. However, the catch is that once the non-dividing tissue has reached a maximum component cell size, further growth cannot occur (i.e. final tissue and component cell size has been achieved). Tissue in which cell division is maintained may initially grow relatively slowly compared to tissue that has terminated cell division, but since more cells are produced a greater final tissue size can be achieved if the final growth potential of each individual cell is realized. The attainment of this final growth potential requires, however, that the tissue eventually switches from a proliferative phase to a non-proliferative phase. The final size of a leaf, therefore, depends upon the integration in time and space of phases of cell-division-associated and non-cell-division-associated growth. In addition to this potentially highly complicated integration of cell division events, it must be borne in mind that leaves are three-dimensional objects normally consisting of several layers of tissue. Each of these layers consists of cells that may undergo distinctly different spatial and temporal patterns of cell division, leading to the observed distinct histology of a leaf. Thus, for example, cell division rates in palisade mesophyll may be different from those observed in the adjacent spongy mesophyll, which will themselves be distinct from the adjacent epidermis. Yet, despite these different patterns of cell division and, thus, cell cycle events, the overall growth of the three-dimensional leaf volume is coordinated to create an approximately planar structure. As has been pointed out, maintenance of such a planar solid form in a growing system is (conceptually) not a trivial problem and small abnormalities in the appropriate gradients of growth required for such morphogenesis could lead to potentially major outcomes in the final form of the leaf (Nath *et al.*, 2003). If cell division processes drive growth, then they must be precisely coordinated across many tissue layers showing differences both in pattern of division and in timing of termination of proliferation.

Kinematic analyses of cell division processes during leaf development have focused on the epidermal layers that lend themselves to non-invasive analysis of growth rate, division pattern and nuclear DNA content. These data (e.g. Granier and Tardieu, 1998; Tardieu and Granier, 2000; De Veylder *et al.*, 2001) provide a quantitative analysis of the growth processes over different parts of the leaf surface during different phases of development and provide data on the component processes of cell division rate and distribution. In dicotyledon leaves, cell division occurs throughout the leaf during and immediately after formation (Figure 9.3). Cell cycle duration is approximately constant. Cell cycle duration then increases as cells become blocked in G1, with the length of the G2–M being relatively constant. This cessation of proliferation occurs in a wave from the distal tip towards the base of the leaf and it occurs with different timing in different cell layers. At the same time, endoreduplication becomes more common. Thus, cells within young leaf primordia generally possess either a 2C or a 4C nuclear DNA content. As growth of the leaf progresses, cells with 8C nuclear content appear at the distal tip of the leaf, indicating that an extra round of DNA synthesis has occurred without cytokinesis. As the wave of termination in cell proliferation occurs from the distal to proximal region of the leaf, there is a following wave of endoreduplication. During this phase the cells undergo significant enlargement (non-cell-division-associated growth), raising the possibility that the observed endoreduplication is causally involved in the growth that occurs (Sugimoto-Shirasu and Roberts, 2003).

The above description has described the development in *Arabidopsis* which generates a simple dicotyledon leaf. Other dicotyledons form more complex leaves in which individual leaflets arise from a single petiole or rachis. Observational data show that at an early stage in development of these more complex (often termed *compound*) leaves, bulges of tissue occur along the flank of the main initial primordium, and the final leaflets arise from these bulges. The cells within these bulges maintain a meristematic potential, allowing the continued generation and subsequent growth of these cells to produce the observed leaflets of the complex leaf. Most work on the control of the formation of complex leaves has focused on transcriptional regulators, in particular the KNOX and ARP families of transcription factors (Sinha *et al.*, 1993; Chuck *et al.*, 1996; Schneeberger *et al.*, 1998; Bharathan *et al.*, 2002; Kim *et al.*, 2003; Tsiantis and Hay, 2003; Harrison *et al.*, 2005). However, how the altered expression of these transcriptional regulators feeds into the actual observed differential growth processes is unclear (Mele *et al.*, 2003) and, in particular,

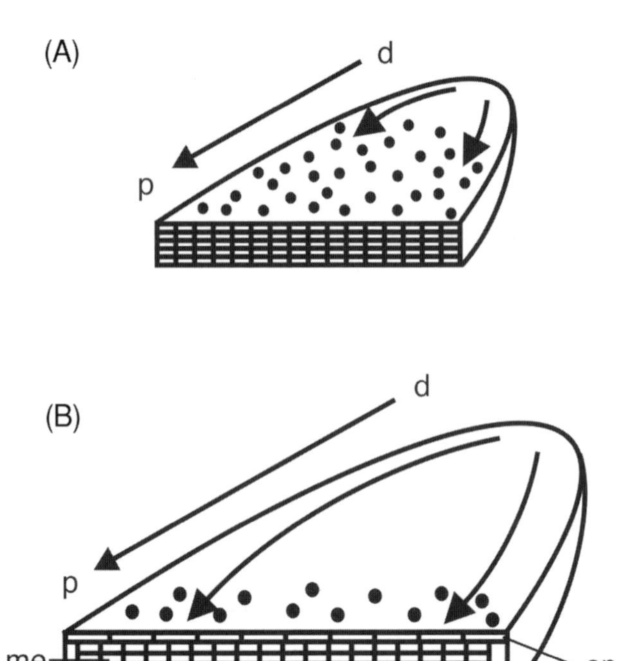

Figure 9.3 Cell division pattern during leaf growth. (A) In a young leaf primordium, cell division (circles) occurs throughout the leaf, but as development progresses a termination of cell division is initiated at the distal (d) tip of the leaf. (B) This wave of division termination proceeds from distal (d) to proximal (p) regions of the leaf (arrows). Moreover, this termination occurs with different dynamics in the different cell layers of the leaf, with termination occurring in the presumptive epidermis (ep) before the mesophyll (me).

how the cell division capability is maintained in certain regions of the leaf is unknown.

9.3.2 Manipulation of the cell cycle and cell division during leaf growth

In considering these experiments, a clear distinction must be made between those investigations in which the gene altered unambiguously encodes a protein directly involved in the cell cycle and those in which although changes in cell division have obviously occurred, the gene product manipulated is at least one step removed (or more) from processes directly involved in the cell cycle. The interpretation of the former is much simpler than the latter.

A number of experiments have been performed in which parameters of the cell cycle have been altered during leaf development. The outcome on leaf morphogenesis has been varied, depending on the exact nature of the manipulation performed. Generally, manipulation of gene products directly involved in the cell cycle has led to altered frequency of cell division, which has led to either no overt change in leaf morphogenesis or phenotypes in which leaf size has become smaller (despite an increase in cell number). Increase in leaf size or alteration in leaf shape has generally been achieved by manipulation of gene products that, besides perhaps influencing cell division, also probably influence many other factors within the developing leaf. Thus, in these instances although increases in cell number are observed, it is impossible to say whether this is the primary mechanism by which altered leaf growth is achieved.

Going through these published data, overexpression of *CYCLIND2* led to an increased rate of growth in tobacco plants, but the final size and shape of the leaves formed was approximately normal (Cockcroft *et al.*, 2000), whereas overexpression of *E2F* and *DP* led to smaller leaves consisting of a large number of relatively small cells (De Veylder *et al.*, 2002). Overexpression of CDK inhibitors led to the formation of smaller leaves consisting of relatively large cells (De Veylder *et al.*, 2001). These data are consistent with the idea outlined above that cell division alone is not driving leaf morphogenesis but that the attainment of an appropriate leaf size requires the appropriate timing and place of switching from a phase of cell-division-associated growth to a non-cell-division-associated growth and that the attributes of these two phases of plant tissue growth are different. An exception to this relative lack of influence of manipulation of cell cycle gene expression on leaf morphogenesis comes from experiments in which a specific cyclin (*CYCLIND3*) was overexpressed (Riou-Khamlichi *et al.*, 1999), resulting in twisted leaves. Although care must be taken in the interpretation of phenotypes resulting from ectopic expression of a gene product, these data indicate that some *CYCLIND3* genes might have a special role in influencing the growth characteristics of the tissue in which they are expressed. It is noticeable that the plants overexpressing *CYCLIND3* became independent of a requirement for exogenous cytokinin in tissue culture experiments, indicating that overexpression of this cyclin probably influences a phase of the cell cycle responsive to this growth regulator. The mechanism of this interaction is as yet unknown but elucidating the mechanisms (both feed-forward and feedback) by which cell cycle components and plant growth regulators interact is a major task for the future.

As outlined above, with the exception of the experiments with *CYCLIND3*, investigations in which cell cycle gene expression has been manipulated have shown relatively minor changes in leaf form. In contrast, genetic manipulation of many other gene products has resulted in leaves of distinct form. Although in many of these reports the altered form has been described as being causally related to changes in cell division, in all instances the final target processes that mediated the observed changes in final leaf form are unknown.

Altered expression of HD–ZIP transcription factors leads to leaves lacking dorsiventrality, i.e. they become more round in cross section, lacking a lamina (McConnell and Barton, 1998; McConnell *et al.*, 2001). This change in morphology is accompanied by a change in cell types associated with abaxial and adaxial tissue identity, supporting the hypothesis that the formation of a flattened lamina requires the juxtaposition of tissue of adaxial and abaxial identity, mediated by transcription factors of the *YABBY* and *KANADI* gene families (Waites *et al.*, 1998; Eshed *et al.*, 2001; Kerstetter *et al.*, 2001; Golz and Hudson, 2002; Eshed *et al.*, 2004). However, although these genetic changes clearly lead to a change in leaf morphology and histology, the mechanism by which they are achieved is unclear. Changes in cell division pattern certainly occur and, therefore, cell cycle genes are certainly included as a subgroup of the target genes. However, whether changes in cell cycle gene expression are primary targets, occur in parallel to altered expression of other genes whose products influence tissue growth or, indeed, are secondary indirect targets of other primary target genes is unclear.

Misexpression of other transcription factors (e.g. *KNOX* genes) also leads to altered leaf morphogenesis, but again the question remains open as to whether cell cycle gene products represent the mechanism by which the observed changes in growth are mediated (Sinha *et al.*, 1993; Schneeberger *et al.*, 1995; Kerstetter and Hake, 1997; Sinha, 1999), but see also (Mizukami and Fischer, 2000). The available evidence indicates that gene products involved in growth regulator synthesis are among the target genes affected by these transcription factors (Sakamoto *et al.*, 2001; Hay *et al.*, 2002). Such growth factors are likely to have wide-ranging effects on gene targets (nuclear gene expression) and non-gene targets (cytoskeletal and vesicle trafficking), which will significantly influence both the rate and vector of tissue growth, as well as cell cycle and cytokinesis events. One key output of *KNOX* overexpression is the suppression of *ARP*-like gene expression, and the evidence indicates that ARP-like transcription factors are intimately involved in the maintenance of cell division in the leaf (McHale and Koning, 2004). However, as with KNOX transcription factors, the precise downstream factors by which these transcriptional regulators might influence cell division pattern are unclear. *ARP*-like gene expression is also associated with the ability of complex leaves to form leaflets, again indicating some linkage with the maintenance of meristematic capability (Bharathan *et al.*, 2002). Untangling and deciphering these different target processes and relating them to the observed growth processes during leaf development is a major task that has yet to be undertaken.

Another group of transcription factors shown to be involved in influencing leaf form is the TCP family (Nath *et al.*, 2003; Palatnik *et al.*, 2003). Thus, misexpression of *TCP* leads to a crumpled leaf phenotype. Moreover, analysis reveals that this is linked with an altered pattern of cell division, with the normal pattern of termination of cell division along the leaf lamina being delayed, i.e. the pattern of cell division termination is altered and this correlates with an altered leaf form. However, again, whether cell cycle genes are the primary targets for such TCP transcription factors is unclear. Recent data indicate that both cell cycle genes and ribosomal protein genes (whose products can be related to the translational capacity of a cell and, thus, its growth potential) are likely targets for TCP factors (Li *et al.*, 2005). This raises the possibility that TCP factors act as integrators of the cell cycle and growth but does not answer the question whether one of these target processes is more important in the control of leaf form. Dissection of TCP target genes and the outcome of their independent misexpression should address this problem.

The finding that *TCP* genes might target ribosomal proteins relates to earlier findings in which a molecular genetic approach was taken to identify genes involved in controlling leaf form. This led to (at the time) surprising identification of a ribosomal protein (Van Lijsebettens *et al.*, 1994). However, there is now a growing realization that basic translational capacity might significantly influence developmental growth capability (Fingar *et al.*, 2002; Nijhout, 2003; Rudra and Warner, 2004). Whether such basic attributes of a cell relate only to cell-division-linked growth or also to non-cell division linked growth (normally attributed to vacuolar and cell-wall-based processes) is as yet unknown.

Molecular genetic analysis of leaf shape mutants has also revealed that gene products involved in brassinolide biosythesis have a role in altering leaf shape (Kim *et al.*, 1998, 1999, 2005). Although, as with the transcription factor based work described above, there are clear changes in the number of cells generated within these leaves, it is also clear that changes in growth factor level leads to a plethora of changes in gene expression. These include those related to cell cycle gene products, but assigning a special role to these genes awaits further analysis.

Mutations in *STRUWWELPETER* lead to putative disruption of RNA polymerase recruitment, i.e. have a potentially general influence on growth (Autran *et al.*, 2002). Leaf size is smaller and leaves are more elongated with a tendency to gain a radial symmetry. Cell size is slightly larger than normal and the leaves generated consist of fewer cells than normal. There is clearly an influence on the cell cycle as there is a precocious arrest of cell division during leaf development. The question remains, however, whether the clearly observed changes in cell division pattern are causally related to the phenotype observed. Altered RNA polymerase function clearly has the potential to influence a wide range of cellular activities beyond that of the cell cycle.

ELONGATA mutants display elongated leaves as a result of mutation in a gene encoding a putative histone acetyl transferase, an enzyme implicated in general regulation of gene expression. Again, as with the *STRUWWELPETER* and *PFL* mutants, it seems that abrogation or disruption of components of the general machinery of cell metabolism can lead to an apparently specific morphogenic outcome. In the case of the *ELONGATA* mutant the rate of cell proliferation is reduced, but again the question remains of whether this is a causal or a parallel outcome to altered morphogenesis (Nelissen *et al.*, 2005).

If, as postulated above, disruption of components of the general machinery of cell metabolism can lead to apparently specific (but distinct) morphogenic outcomes, how does this come about? As yet, we have little idea as to the answer to this basic question. What is the relationship between cellular and organ metabolism and growth? Does this relationship vary during development? How is it related to the transcriptional networks that define developmental progression? Answers to these fundamental questions await further research.

The above section has considered cell division frequency. What of the orientation of cell division? Does this play a role in leaf morphogenesis? As with leaf initiation, it appears that as long as a basal competence for cytokinesis is present, relatively normal leaf development can occur. For example, in the *TANGLED* mutant of maize all cross-walls are aberrantly orientated, yet the overall leaf morphology is approximately normal (Smith *et al.*, 1996, 2001). The *TONNEAU* mutant of *Arabidopsis* has a more extreme problem with cytokinesis with division plane being apparently randomized (Traas *et al.*, 1995). Although the early growth of the plant is highly abnormal, development still occurs to generate a plant with recognizable stem, leaves and flowers. Cytokinesis depends on the appropriate delivery of vesicles containing new cell wall components to the forming cell plate. This system must be coordinated with cell cycle gene products. Mutants have been identified in which components of the cytokinesis system cannot sustain the formation of intact, functional, new cell wall, and in these instances the outcome is lethality for the plant, even though rudimentary seedlings can sometimes be formed (Lukowitz *et al.*, 1996; Assaad *et al.*, 2001; Mayer and Jürgens, 2004). These data again fit with the idea that providing a basic competence for cell division and cytokinesis is achieved, plant development (and, thus, leaf morphogenesis) can occur. However, when this lower limiting capability is lost, further growth and development is abrogated and leaf growth cannot occur.

9.3.3 The role of the cell cycle and cell division during leaf growth

During leaf development an ordered pattern of cell division occurs, which defines the classical histology of the leaf. This terminal differentiation of the leaf involves the gradual exit of cells from the cell cycle. Once cells have ceased cycling, growth can occur only via cell expansion and, since there is a biophysical limit on the size that an individual cell can achieve, the final size of a leaf will be determined by the timing and spatial distribution of this termination of the cell cycle. Thus, the competence to undergo cell division is required for leaf growth. However, an increased ability to undergo cell division does not lead to increased growth per se. An increased competence for cell division may lead to more cells being generated, but this will lead to increased growth only if all of these cells undergo an appropriate expansion (growth) process. Data from experiments in which cell cycle elements have directly been promoted or inhibited point to a supracellular control of growth and setting of size. The nature of this supracellular control is still obscure (Day and Lawrence, 2000), but experiments that have led to increased leaf size have generally involved either transcription factors or growth regulators that are likely to target a combination of growth and cell division parameters. In these cases, an observation of increased cell number does not necessarily imply that the cell cycle was the prime causal agent in the phenotype observed.

9.4 The cell cycle and cell division during leaf differentiation

9.4.1 Patterns of the cell cycle and cell division during leaf differentiation

Classical leaf histology is dependent on (and, indeed, reflects) specific patterns of cell division. This leads to specific cell types characterized by specific cell shape and size. To what extent are the specific functions of these different cell types dependent on the patterns of cell division that give rise to them? Is differentiation dependent on progression through the cell cycle?

At the earliest stages following leaf initiation, the constituent cells of a leaf appear very similar to those observed in the SAM from which it is derived. However, gradually distinct cell types (as defined by size, shape and position) become apparent. For example, in the central core of the leaf primordium a chain of relatively thin, elongated cells becomes apparent, which presages the mid-rib vasculature of the leaf. At about the same time, cells along the edge of the leaf margin cease division and enter a phase of expansion to form the margin cells. In addition, selected cells of the epidermis start to undergo differentiation to form trichomes and, slightly later, a particular pattern of division and differentiation to form stomata. Within the leaf primordium blade, cell division frequency and plane of division are maintained so that a layered structure of roughly equivalent cells is produced. However, as growth of the leaf continues, distinctions in the rate and orientation of cell division occur to generate an adaxial tier(s) of cells with an elongated axis perpendicular to the plane of the lamina. These will form the palisade mesophyll, specialized for optimum photosynthesis, whereas the abaxial spongy mesophyll (more specialized for gas exchange) is characterized by a more rotund cell form and, via pulling apart of some cells, the formation of air spaces. As described in the previous section, the termination of cell division occurs at different times in the different histological layers, with, for example, cell division occurring for a longer period in the palisade mesophyll than the adjacent epidermis. Concomitant with this wave of termination of cell division, cell cycling continues in an endoreduplication mode to generate nuclei with C values greater than 4C. In some cells of the leaf, these C values can reach high levels (see Chapter 11) and a correlation can be made between the size of cell and the C value attained, raising the possibility that endoreduplication plays a causal role in this process (Sugimoto-Shirasu and Roberts, 2003). However, at least some data indicate that this correlation is not absolute, as discussed in the next section.

9.4.2 Manipulation of the cell cycle and cell division during leaf differentiation

Returning to the previously described experiments in which gene products closely associated with the cell cycle have been either overexpressed or suppressed during leaf development, the general outcome has been that although cell size and shape may have been altered, a functional leaf histology is still attained. These observations can be compared with mutational approaches in which cell division pattern has been massively disrupted, yet cell-type-specific marker genes are still appropriately expressed (Yadegari *et al.*, 1994). These data suggest that the patterning of different tissues within the leaf is dependent on supracellular regulators and that all cells within the field of a particular regulator become determined for a particular cell fate, irrespective of the particular pattern of cell division that occurs within those cells. Particular cell size and shape may be required for optimum function of a tissue, but basic differentiation does not depend on a particular pattern of cell division, with the exception of a few specialized cell types (e.g. stomata) where clearly functionality is dependent on the juxtaposition of cells of appropriate form.

As cells cease proliferation and enter a pathway of differentiation they frequently enter a phase of endoreduplication. The potential linkage of endoreduplication and cell differentiation is discussed elsewhere in this volume (Vlieghe *et al.*, 2006, this volume).

Does differentiation within a leaf depend upon a particular phase of the cell cycle? Although such experiments are difficult to perform in intact tissue, there are data from *in vitro* systems (notably *Xinnia*) suggesting that specific events of differentiation may be linked to passage via specific phases of the cell cycle (Milioni *et al.*, 2002; Groover *et al.*, 2003). These transitions are often linked with the presence of hormonal cues, such as auxin. How easily such observations on *in vitro* cultured cells can be transposed onto intact tissue (such as a leaf) is difficult to judge.

As with the previous two sections, the most dramatic influences on cell differentiation in the developing leaf have been achieved by the manipulation of transcriptional regulators and growth regulators which are liable to target altered expression of a whole suite of gene products, not just those involved in the cell cycle. It is highly likely that these regulatory factors influence genes involved in differentiation directly and do not depend initially on altered cell cycle gene expression to move the cell into a particular phase of the cycle. As has been described in previous sections, the altered expression of transcription factors (e.g. KNOX-like and ARP-like) leads to an altered leaf phenotype, which can be interpreted as an altered level of commitment to particular pathways of differentiation. This is observed as, for example, the maintenance of cells in a proliferative phase with lack of acquisition of appropriate tissue histology (McHale and Koning, 2004). However, the mechanism of this linkage is unknown. Some specific transcription factors involved in specific aspects of leaf differentiation have been identified. For example, the APL MYB transcription factor is required for vascular differentiation (Bonke *et al.*, 2003), and there is a large body of evidence describing the factors involved in trichome differentiation (see Chapter 11) and stomatal differentiation (Nadeau and Sack, 2003; Bergmann *et al.*, 2004). As a result of these analyses both transcription factors and signal transduction components involved in particular elements of leaf-cell differentiation have been uncovered. In the case of trichomes, a clear relationship with the cell cycle has been uncovered. This is described in more detail in Chapter 11.

9.4.3 The role of the cell cycle and cell division during leaf differentiation

For differentiation to occur, the volume of a leaf must be compartmented into cells. For optimum leaf function, the pattern of division in the different putative histological zones of a leaf must be tightly controlled. This involves a control on the rate of progress through the cell cycle, the orientation and position of the new cell wall laid down during cytokinesis, the timing of termination of cell-division-associated growth and the entry to non-cell-division-associated growth, and the entry into and termination of an endoreduplication phase of the cell cycle. Disruption of any of these different parameters will influence the final histology of a leaf and the potential function of the constituent cells. However, it is clear that despite changes in the above-listed parameters a functional leaf can still be formed, with appropriate gene expression patterns and biochemical specialization still occurring across a field of cells showing abnormal patterns of cell division and the cell cycle. These observations are consistent with a view of leaf development in which gradients of growth regulators (probably set up by spatial regulation of expression of key transcriptional regulators) lead to regions of differential growth and, at the same time via the differential expression of downstream transcription factors, the acquisition of different tissue fates. The differential growth underlying morphogenesis is due to a combination of target processes including cytoplasmic translational capacity, vacuolar expansion properties and cell wall extensibility. These processes are linked to and coordinated with events of the cell cycle. This linkage with the cell cycle is required not only to generate new cells that can act as the raw material for further differential growth, but also for the acquisition of an appropriate histology for the function of the different tissue types. Deviation from the normal pattern of linkage is likely to lead to abnormal tissue function, which under most conditions is likely to be detrimental for leaf function but which may also act as a source of variation upon which evolutionary pressures can function. For certain specific cell types an appropriate pattern of cell cycle progression and cell division pattern is likely to be essential for function, for example stomatal complex formation. For most other tissue types, abnormal patterns of cell cycle and division will detract from function but will not lead to total abrogation of function.

9.5 Conclusions

Clearly, leaf development normally requires cell division to provide the units for differentiation. Also, clearly, since there is a biophysical limit to the maximum size that a cell can achieve, any block or mutation that leads to a serious problem in cell proliferation will have a detrimental effect on leaf size and shape. However, taken overall, it seems that as long as a basic competence for cell division is present within a plant, and as long as the cell cycle can be exited so that cell expansion processes can occur, it seems that many aspects of leaf development are not dependent on either the cell cycle or cell division. This does not mean that appropriate control of the cell cycle and cell division is not important to allow the formation of an optimally functional leaf. Indeed, clearly, the functioning of some specific cell types (such as stomata) is dependent on an appropriate pattern of cell division. Loss of such cell types (again notably stomata) is likely to have a profound outcome on leaf physiology and growth. However, such outcomes can be viewed as an indirect effect of cell division and the cell cycle on leaf development. In many instances where mutational approaches have revealed an altered leaf development and an altered pattern of cell division, it is very difficult to infer an unambiguous causal relationship from cell division to morphogenesis, size and differentiation. Unless the gene product mutated or whose expression has been altered can be shown or inferred to have an affect primarily only on a component of the cell cycle or cytokinesis, there is always the possibility that other target processes are in fact the causal agent of the observed phenotype. Methods to tease out the different strands of target processes from pleiotropic regulators need to be applied before a final consensus on the role of cell division in leaf development can be deduced. The methods and tools for this analysis are now at hand.

Although at first consideration the role of the cell cycle and cell division in leaf development might seem obvious, it is hopefully clear from this chapter that it has proved surprisingly difficult to pin down the precise requirement and contribution of the cell cycle to leaf development. Normally, tissue growth and cell cycling and division are closely coordinated. At the same time, this relationship varies during development, leading to the formation of differently sized cells within an organ of consistent size and shape. Dissecting out these two distinct but linked processes (growth and division) has not proved trivial. Because of the perceived primacy of cell division and the significant advances in our understanding of the cell cycle, there has been a tendency to relate all observed changes in growth to the cell cycle. This interpretation may be correct, but it is also possible to see the same phenomena in a different light, i.e. a growth light. Our understanding of the mechanism and control of growth is still astonishingly limited. Clearly, the formation of a functioning leaf requires a coordination of the cell cycle and growth. A greater understanding of both these processes and how they are intertwined is required before we can make a full assessment of the role of the cell cycle in leaf development.

Acknowledgments

Thanks go to Dr C. Fleming for assistance in correcting the manuscript. Research from the author's laboratory was funded by the BBSRC, University of Sheffield and the Swiss National Science Foundation.

References

Assaad, F.F., Huet, Y., Mayer, U. and Jürgens, G. (2001) The cytokinesis gene KEULE encodes a Sec1 protein that binds the syntaxin KNOLLE. *J Cell Biol* **152**, 531–543.

Autran, D., Jonak, C., Belcram, K., *et al.* (2002) Cell numbers and leaf development in Arabidopsis: a functional analysis of the *STRUWWELPETER* gene. *EMBO J* **21**, 6036–6049.

Beemster, G.T., Fiorani, F. and Inzé, D. (2003) Cell cycle: the key to plant growth control? *Trends Plant Sci* **8**, 154–158.

Beemster, G.T.S., Vercruysse, S., De Veylder, L., Kuiper, M. and Inzé, D. (2006) The *Arabidopsis* leaf as a model system for investigating the role of cell cycle regulation in organ growth. *J Plant Res* **119**, 43–50.

Benková, E., Michniewicz, M., Sauer, M., *et al.* (2003) Local, efflux-dependent auxin gradients as a common module for plant organ formation. *Cell* **115**, 591–602.

Bergmann, D.C., Lukowitz, W. and Somerville, C.R. (2004) Stomatal development and pattern controlled by a MAPKK kinase. *Science* **304**, 1494–1497.

Bharathan, G., Goliber, T.E., Moore, C., Kessler, S., Pham, T. and Sinha, N.R. (2002) Homologies in leaf form inferred from *KNOXI* gene expression during development. *Science* **296**, 1858–1860.

Bonke, M., Thitamadee, S., Mähönen, A.P., Hauser, M.T. and Helariutta, Y. (2003) APL regulates vascular tissue identity in *Arabidopsis*. *Nature* **426**, 181–186.

Brand, U., Fletcher, J.C., Hobe, M., Meyerowitz, E.M. and Simon, R. (2000) Dependence of stem cell fate in Arabidopsis on a feedback loop regulated by CLV3activity. *Science* **289**, 617–619.

Byrne, M.E. (2005) Networks in leaf development. *Curr Opin Plant Biol* **8**, 59–66.

Caderas, D., Muster, M., Vogler, H., *et al.* (2000) Limited correlation between expansin gene expression and elongation growth rate. *Plant Physiol* **123**, 1399–1414.
Cho, H.T. and Kende, H. (1997) Expression of expansin genes is correlated with growth in deepwater rice. *Plant Cell* **9**, 1661–1671.
Chuck, G., Lincoln, C. and Hake, S. (1996) KNAT1 induces lobed leaves with ectopic meristems when overexpressed in Arabidopsis. *Plant Cell* **8**, 1277–1289.
Clark, S.E., Williams, R.W. and Meyerowitz, E.M. (1997) The *CLAVATA1* gene encodes a putative receptor kinase that controls shoot and floral meristem size in Arabidopsis. *Cell* **89**, 575–585.
Cockcroft, C.E., den Boer, B.G., Healy, J.M. and Murray, J.A. (2000) Cyclin D control of growth rate in plants. *Nature* **405**, 575–579.
Cosgrove, D.J. (2000) Loosening of plant cell walls by expansins. *Nature* **407**, 321–326.
Day, S.J. and Lawrence, P.A. (2000) Measuring dimensions, the regulation of size and shape. *Development* **127**, 2977–2987.
De Veylder, L., Beeckman, T., Beemster, G.T., *et al.* (2002) Control of proliferation, endoreduplication and differentiation by the *Arabidopsis* E2Fa-DPa transcription factor. *EMBO J* **21**, 1360–1368.
De Veylder, L., Beeckman, T., Beemster, G.T.S., *et al.* (2001) Functional analysis of cyclin-dependent kinase inhibitors of *Arabidopsis*. *Plant Cell* **13**, 1653–1668.
Donnelly, P.M., Bonetta, D., Tsukaya, H., Dengler, R.E. and Dengler, N.G. (1999) Cell cycling and cell enlargement in developing leaves of Arabidopsis. *Dev Biol* **215**, 407–419.
Eshed, Y., Baum, S.F., Perea, J.V. and Bowman, J.L. (2001) Establishment of polarity in lateral organs of plants. *Curr Biol* **11**, 1251–1260.
Eshed, Y., Izhaki, A., Baum, S.F., Floyd, S.K. and Bowman, J.L. (2004) Asymmetric leaf development and blade expansion in Arabidopsis are mediated by *KANADI* and *YABBY* activities. *Development* **131**, 2997–3006.
Fingar, D.C., Salama, S., Tsou, C., Harlow, E. and Blenis, J. (2002) Mammalian cell size is controlled by mTOR and its downstream targets S6K1and 4EBP1/eIF4E. *Genes Dev* **16**, 1472–1487.
Fiorani, F., Beemster, G.T.S., Bultynck, L. and Lambers, H. (2000) Can meristematic activity determine variation in leaf size and elongation rate among four Poa species? A kinematic study. *Plant Physiol* **124**, 845–856.
Fleming, A.J. (2002) The mechanism of leaf morphogenesis. *Planta* **216**, 17–22.
Fleming, A.J. (2005) The control of leaf development. *New Phytol* **166**, 9–20.
Foard, D.E. (1971) Initial protrusion of a leaf primordium can form without concurrent periclinal cell divisions. *Can J Bot* **49**, 1601–1603.
Gaudin, V., Lunness, P.A., Fobert, P.R., *et al.* (2000) The expression of *D-cyclin* genes defines distinct developmental zones in snapdragon apical meristems and is locally regulated by the *cycloidea* gene. *Plant Physiol* **122**, 1137–1148.
Golz, J.F. and Hudson, A. (2002) Signalling in plant lateral organ development. *Plant Cell* **14**, S277–S288.
Grandjean, O., Vernoux, T., Laufs, P., Belcram, K., Mizukami, Y. and Traas, J. (2004) In vivo analysis of cell division, cell growth, and differentiation at the shoot apical meristem in Arabidopsis. *Plant Cell* **16**, 74–87.
Granier, C. and Tardieu, F. (1998) Spatial and temporal analyses of expansion and cell cycle in sunflower leaves. A common pattern of development for all zones of a leaf and different leaves of a plant. *Plant Physiol* **116**, 991–1001.
Granier, C., Turc, O. and Tardieu, F. (2000) Co-ordination of cell division and tissue expansion in sunflower, tobacco, and pea leaves, dependence or independence of both processes? *J Plant Growth Regul* **19**, 45–54.
Groover, A.T., Pattishall, A. and Jones, A.M. (2003) *IAA8* expression during vascular cell differentiation. *Plant Mol Biol* **51**, 427–435.
Hager, A. (2003) Role of the plasma membrane H^+-ATPase in auxin-induced elongation growth: historical and new aspects. *J Plant Res* **116**, 483–505.

Harrison, C.J., Corley, S.B., Moylan, E.C., Alexander, D.L., Scotland, R.W. and Langdale, J.A. (2005) Independent recruitment of a conserved developmental mechanism during leaf evolution. *Nature* **434**, 509–514.

Hay, A., Kaur, H., Phillips, A., Hedden, P., Hake, S. and Tsiantis, M. (2002) The gibberellin pathway mediates KNOTTED1-type homeobox function in plants with different body plans. *Curr Biol* **12**, 1557–1565.

Haywood, V., Kragler, F. and Lucas, W.J. (2002) Plasmodesmata: pathways for protein and ribonucleoprotein signaling. *Plant Cell* **14**(Suppl.), S303–S325.

Hemerly, A., de Almeida Engler, J., Bergounioux, C., *et al.* (1995) Dominant negative mutants of the Cdc2 kinase uncouple cell division from iterative plant development. *EMBO J* **14**, 3925–3936.

Hofmeister, W. (1863) Zusütze und Berichtigungen zu den 1851verüffentlichen Untersuchungen der Entwicklung hüherer Kryptogamen. *Jahrb Wiss Bot* **3**, 259–293.

Kaplan, D.R. (2001) Fundamental concepts of leaf morphology and morphogenesis: a contribution to the interpretation of molecular genetic mutants. *Int J Plant Sci* **162**, 465–474.

Kerstetter, R.A., Bollman, K., Taylor, R.A., Bomblies, K. and Poethig, R.S. (2001) *KANADI* regulates organ polarity in *Arabidopsis*. *Nature* **411**, 706–709.

Kerstetter, R.A. and Hake, S. (1997) Shoot meristem formation in vegetative development. *Plant Cell* **9**, 1001–1010.

Kim, G.-T., Fujioka, S., Kozuka, T., *et al.* (2005) CYP90C1and CYP90D1are involved in different steps in the brassinosteroid biosynthesis pathway in *Arabidopsis thaliana*. *Plant J* **41**, 710–721.

Kim, G.-T., Tsukaya, H., Saito, Y. and Uchimiya, H. (1999) Changes in the shapes of leaves and flowers upon overexpression of cytochrome P450 in *Arabidopsis*. *Proc Natl Acad Sci USA* **96**, 9433–9437.

Kim, G.-T., Tsukaya, H. and Uchimiya, H. (1998) The *ROTUNDIFOLIA3* gene of *Arabidopsis thaliana* encodes a new member of the cytochrome P-450 family that is required for the regulated polar elongation of leaf cells. *Genes Dev* **12**, 2381–2391.

Kim, J.Y., Yuan, Z., Cilia, M., Khalfan-Jagani, Z. and Jackson, D. (2002) Intercellular trafficking of a *KNOTTED1* green fluorescent protein fusion in the leaf and shoot meristem of *Arabidopsis*. *Proc Natl Acad Sci USA* **99**, 4103–4108.

Kim, M., McCormick, S., Timmermans, M. and Sinha, N. (2003) The expression domain of PHANTASTICA determines leaflet placement in compound leaves. *Nature* **424**, 438–443.

Li, C., Potuschak, T., Colón-Carmona, A., Gutiérrez, R.A. and Doerner, P. (2005) *Arabidopsis* TCP20 links regulation of growth and cell division control pathways. *Proc Natl Acad Sci USA* **102**, 12978–12983.

Lucas, W.J., Ding, B. and van Der Schoot, C. (1993) Plasmodesmata and the supracellular nature of plants. *New Phytol* **125**, 435–476.

Lukowitz, W., Mayer, U. and Jürgens, G. (1996) Cytokinesis in the Arabidopsis embryo involves the syntaxin-related *KNOLLE* gene product. *Cell* **84**, 61–71.

Lyndon, R.F. and Robertson, E.S. (1976) The quantitative ultrastructure of the pea shoot apex in relation to leaf initiation. *Protoplasma* **87**, 387–402.

Mayer, U. and Jürgens, G. (2004) Cytokinesis: lines of division taking shape. *Curr Opin Plant Biol* **7**, 599–604.

McConnell, J.R. and Barton, M.K. (1998) Leaf polarity and meristem formation in Arabidopsis. *Development* **125**, 2935–2942.

McConnell, J.R., Emery, J., Eshed, Y., Bao, N., Bowman, J. and Barton, M.K. (2001) Role of *PHABULOSA* and *PHAVOLUTA* in determining radial patterning in shoots. *Nature* **411**, 709–713.

McHale, N.A. and Koning, R.E. (2004) *PHANTASTICA* regulates development of the adaxial mesophyll in Nicotiana leaves. *Plant Cell* **16**, 1251–1262.

Mele, G., Ori, N., Sato, Y. and Hake, S. (2003) The *knotted1*-like homeobox gene *BREVIPEDICELLUS* regulates cell differentiation by modulating metabolic pathways. *Genes Dev* **17**, 2088–2093.

Menges, M., Hennig, L., Gruissem, W. and Murray, J.A. (2002) Cell cycle-regulated gene expression in Arabidopsis. *J Biol Chem* **277**, 41987–42002.

Milioni, D., Sado, P.E., Stacey, N.J., Roberts, K. and McCann, M.C. (2002) Early gene expression associated with the commitment and differentiation of a plant tracheary element is revealed by cDNA-amplified fragment length polymorphism analysis. *Plant Cell* **14**, 2813–2824.

Mizukami, Y. and Fischer, R.L. (2000) Plant organ size control, *AINTEGUMENTA* regulates growth and cell numbers during organogenesis. *Proc Natl Acad Sci USA* **97**, 942–947.

Nadeau, J.A. and Sack, F.D. (2003) Stomatal development, cross talk puts mouths in place. *Trends Plant Sci* **8**, 294–299.

Nath, U., Crawford, B.C.W., Carpenter, R. and Coen, E. (2003) Genetic control of surface curvature. *Science* **299**, 1404–1407.

Nelissen, H., Fleury, D., Bruno, L., *et al.* (2005) The elongata mutants identify a functional Elongator complex in plants with a role in cell proliferation during organ growth. *Proc Natl Acad Sci USA* **102**, 7754–7759.

Nijhout, H.F. (2003) The control of growth. *Development* **130**, 5863–5867.

Palatnik, J.F., Allen, E., Wu, X., *et al.* (2003) Control of leaf morphogenesis by microRNAs. *Nature* **425**, 257–263.

Pien, S., Wyrzykowska, J., McQueen-Mason, S., Smart, C. and Fleming, A. (2001) Local expression of expansin induces the entire process of leaf development and modifies leaf shape. *Proc Natl Acad Sci USA* **98**, 11812–11817.

Poethig, R.S. and Sussex, I.M. (1985a) The developmental morphology and growth dynamics of the tobacco leaf. *Planta* **165**, 158–169.

Poethig, R.S. and Sussex, I.M. (1985b) The cellular parameters of leaf development in tobacco: a clonal analysis. *Planta* **165**, 170–184.

Pyke, K.A., Marrison, J.L. and Leech, R.M. (1991) Temporal and spatial development of the cells of the expanding first leaf of *Arabidopsis thaliana* (L.). *Heynh J Exp Bot* **42**, 1407–1416.

Reddy, G.V., Heisler, M.G., Ehrhardt, D.W. and Meyerowitz, E.M. (2004) Real-time lineage analysis reveals oriented cell divisions associated with morphogenesis at the shoot apex of *Arabidopsis thaliana*. *Development* **131**, 4225–4237.

Reddy, G.V. and Meyerowitz, E.M. (2005) Stem-cell homeostasis and growth dynamics can be uncoupled in the *Arabidopsis* shoot apex. *Science* **310**, 663–667.

Reinhardt, D., Frenz, M., Mandel, T. and Kuhlemeier, C. (2003a) Microsurgical and laser ablation analysis of interactions between the zones and layers of the tomato shoot apical meristem. *Development* **130**, 4073–4083.

Reinhardt, D., Pesce, E.-R., Stieger, P., *et al.* (2003b) Regulation of phyllotaxis by polar auxin transport. *Nature* **426**, 255–260.

Reinhardt, D., Wittwer, F., Mandel, T. and Kuhlemeier, C. (1998) Localized upregulation of a new expansin gene predicts the site of leaf formation in the tomato meristem. *Plant Cell* **10**, 1427–1437.

Riou-Khamlichi, C., Huntley, R., Jacqmard, A. and Murray, J.A. (1999) Cytokinin activation of *Arabidopsis* cell division through a D-type cyclin. *Science* **283**, 1541–1544.

Rudra, D. and Warner, J.R. (2004) What better measure than ribosome synthesis? *Genes Dev* **18**, 2431–2436.

Sakamoto, T., Kamiya, N., Ueguchi-Tanaka, M., Iwahori, S. and Matsuoka, M. (2001) *KNOX* homeodomain protein directly suppresses the expression of a gibberellin biosynthetic gene in the tobacco shoot apical meristem. *Genes Dev* **15**, 581–590.

Schneeberger, R., Tsiantis, M., Freeling, M. and Langdale, J.A. (1998) The rough sheath2 gene negatively regulates homeobox gene expression during maize leaf development. *Development* **125**, 2857–2865.

Schneeberger, R.G., Becraft, P.W., Hake, S. and Freeling, M. (1995) Ectopic expression of the *knox* homeo box gene *rough sheath1* alters cell fate in the maize leaf. *Genes Dev* **9**, 2292–2304.

Schoof, H., Lenhard, M., Haecker, A., Mayer, K.F.X., Jürgens, G. and Laux, T. (2000) The stem cell population of *Arabidopsis* shoot meristems in maintained by a regulatory loop between the *CLAVATA* and *WUSCHEL* genes. *Cell* **100**, 635–644.

Sinha, N. (1999) Leaf development in angiosperms. *Annu Rev Plant Physiol Plant Mol Biol* **50**, 419–446.

Sinha, N.R., Williams, R.E. and Hake, S. (1993) Overexpression of the maize homeo box gene, *KNOTTED-1*, causes a switch from determinate to indeterminate cell fates. *Genes Dev* **7**, 787–795.

Smith, L.G., Gerttula, S.M., Han, S. and Levy, J. (2001) Tangled1: a microtubule binding protein required for the spatial control of cytokinesis in maize. *J Cell Biol* **152**, 231–236.

Smith, L.G., Hake, S. and Sylvester, A.W. (1996) The *tangled-1*mutation alters cell division orientations throughout maize leaf development without altering leaf shape. *Development* **122**, 481–489.

Sugimoto-Shirasu, K. and Roberts, K. (2003) "Big it up": endoreduplication and cell-size control in plants. *Curr Opin Plant Biol* **6**, 544–553.

Tardieu, F. and Granier, C. (2000) Quantitative analysis of cell division in leaves: methods, developmental patterns and effects of environmental conditions. *Plant Mol Biol* **43**, 555–567.

Traas, J., Bellini, C., Nacry, P., Kronenberger, J., Bouchez, D. and Caboche, M. (1995) Normal differentiation patterns in plants lacking microtubular preprophase bands. *Nature* **375**, 676–677.

Tsiantis, M. and Hay, A. (2003) Comparative plant development: the time of the leaf? *Nat Rev Genet* **4**, 169–180.

Tsukaya, H. (2002) Interpretation of mutants in leaf morphology: genetic evidence for a compensatory system in leaf morphogenesis that provides a new link between cell and organismal theories. *Int Rev Cytol* **217**, 1–39.

Tsukaya, H. (2003) Organ shape and size: a lesson from studies of leaf morphogenesis. *Curr Opin Plant Biol* **6**, 57–62.

Van Lijsebettens, M., Vanderhaeghen, R., De Block, M., Bauw, G., Villarroel, R. and Van Montagu, M. (1994) An S18 ribosomal protein gene copy at the Arabidopsis *PFL* locus affects plant development by its specific expression in meristems. *EMBO J* **13**, 3378–3388.

Waites, R., Selvadurai, H.R.N., Oliver, I.R. and Hudson, A. (1998) The PHANTASTICA gene encodes a MYB transcription factor involved in growth and dorsoventrality of lateral organs in *Antirrhinum*. *Cell* **93**, 779–789.

Wang, H., Zhou, Y., Gilmer, S., Whitwill, S. and Fowke, L.C. (2000) Expression of the plant cyclin-dependent kinase inhibitor ICK1 affects cell division, plant growth and morphology. *Plant J* **24**, 613–623.

Wyrzykowska, J. and Fleming, A. (2003) Cell division pattern influences gene expression in the shoot apical meristem. *Proc Natl Acad Sci USA* **100**, 5561–5566.

Wyrzykowska, J., Pien, S., Shen, W.H. and Fleming, A.J. (2002) Manipulation of leaf shape by modulation of cell division. *Development* **129**, 957–964.

Yadegari, R., Paiva, G., La, J., *et al.* (1994) Cell differentiation and morphogenesis are uncoupled in Arabidopsis raspberry embryos. *Plant Cell* **6**, 1713–1729.

10 Physiological relevance and molecular control of the endocycle in plants

Kobe Vlieghe, Dirk Inzé and Lieven De Veylder

10.1 Introduction

During the classical mitotic cell cycle, DNA that is duplicated during the S phase is equally divided during the M phase, so that each daughter cell produced after cytokinesis possesses a genomic DNA content that is identical to that of its parents, being 2C (C equals the haploid DNA content). In this cycle, the S phase and the M phase are separated from each other by two intervening gap phases, G1 and G2. In contrast to the traditional mitotic cell cycle, during the endoreduplication cycle, no cytokinesis occurs between successive rounds of DNA replication. In this manner, the DNA content of the cell is doubled with every new round of DNA replication, resulting in the formation of cells with a DNA ploidy level of 2C, 4C, 8C, 16C, 32C and so on.

Endoreduplication can happen in different ways. When DNA replication is accompanied with mitosis, but without cytokinesis, polynucleated cells are formed; however, when repeated replication of DNA takes place without formation of new nuclei during the telophase, endoreduplication can result in polyploidy or polyteny (Figure 10.1). In polyploid cells, the sister chromatids separate and return to the interphase state as in the mitotic cell cycle, resulting in an increase in the number of chromosomes that retain their individual identity within the original nuclear envelope (Geitler, 1939). On the contrary, polyteny is characterized by the occurrence of chromosome duplication without any DNA condensation/decondensation steps and without sister chromatid segregation, forming 'giant' multistranded chromosomes with 2^n chromatids. As a consequence, the number of chromosomes within the nuclear envelope remains the same (Lorz, 1947; Levan and Hauschka, 1953). In addition, any variation between polyteny and polyploidy is possible.

10.2 Occurrence and physiological role of endoreduplication in nature

Endoreduplication occurs in a wide variety of cell types in arthropods, mammals and plants. Various biological processes, including cell differentiation, cell expansion, metabolic activity and resistance against irradiation stress, have been proposed to be involved in the endoreduplication process. Nevertheless, the evidence for most of these claims remains circumstantial and the physiological role of endoreduplication is still poorly characterized. Below, we will give an overview on the data that link endoreduplication to specific physiological functions.

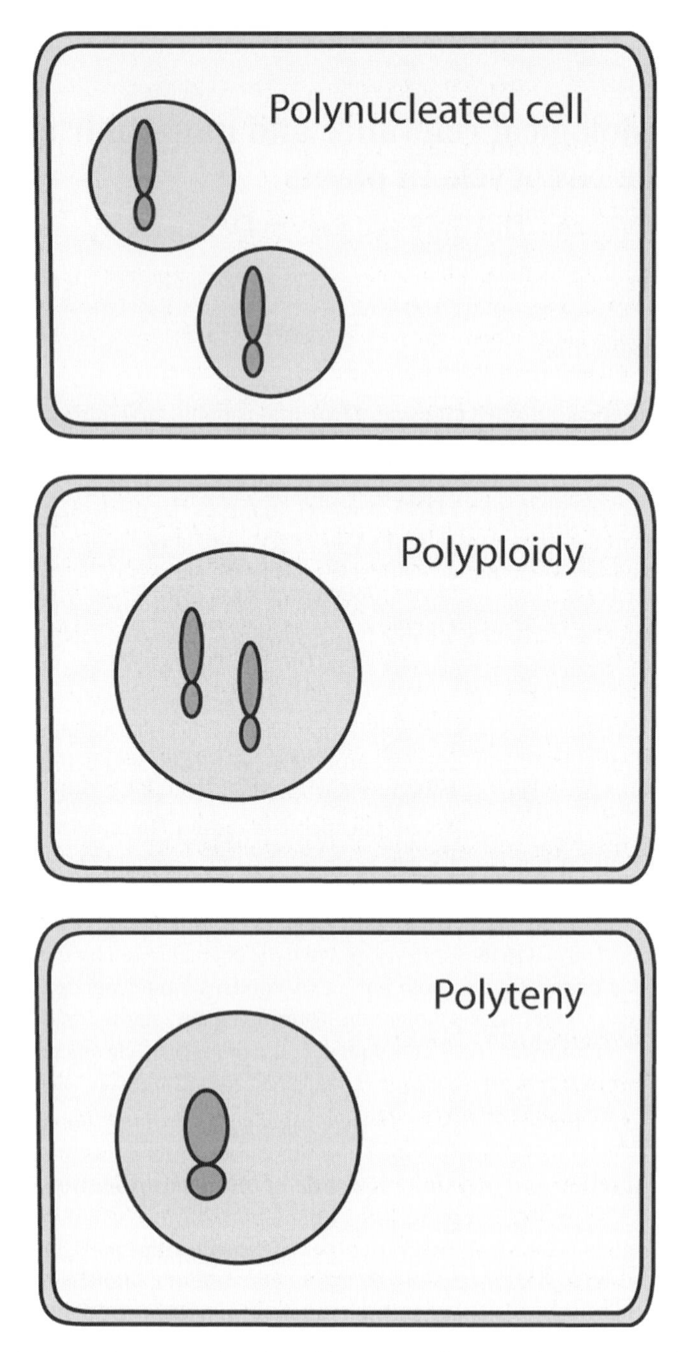

Figure 10.1 Schematic representation of the different forms of endoreduplication: polynucleated cells, polyploidy and polyteny.

10.2.1 Endoreduplication in nonplant species

For nonplant species, the endoreduplication process is probably best characterized in *Drosophila melanogaster*, in which the endoreduplication cycle has been demonstrated to play a prominent role during many different aspects of life. In this organism, endoreduplication is mainly correlated with growth and increased metabolic activity. During oogenesis, specialized germline cells, called nurse cells, grow enormously while undergoing 10–12 endocycles, thereby synthesizing the future egg contents at extremely high rates and enabling oogenesis to proceed rapidly (Spradling, 1993). Later, at the larval stage, endoreduplication is thought to play an important role in the growth of the larva, whose mass expands 200-fold primarily by an increase in cell size rather than in cell number (Edgar and Orr-Weaver, 2001). This growth is accompanied by the development of a large range of endopolyploid larval cells, with a DNA content ranging from 16C up to 1024C. Only the cells that will contribute to the adult body, the imaginal cells, undergo mitosis. Later, at the adult stage, endoreduplication is observed in the salivary glands, where it is supposed to be responsible for the high metabolic activity of these cells. Similarly, a study of genome sizes of spiders revealed the presence of endopolyploid nuclei in the silk and poison glands of *Pholcus phalangioides*, suggesting that the strongly amplified genomes in these organs are required for their high protein output (Gregory and Shorthouse, 2003).

Endoreduplication is also found among mammalian cell types. In the liver, acytokinetic mitosis gives rise to cells with multiple nuclei, although some polyploid cells with single nuclei are also observed (Brodsky and Uryvaeva, 1977; Kudryavtsev *et al.*, 1993). A significant number of myocytes are polyploid (Sandritter and Scomazzoni, 1964). Also megakaryocytes, producing the blood platelets, undergo endoreduplication (Datta *et al.*, 1996; Zhang *et al.*, 1996b), and in the placenta, polyploid giant trophoblasts can be found (Soares *et al.*, 1998). Here again, in all cases, endoreduplication is thought to correlate with the metabolic activity of the cells.

10.2.2 Endoreduplication in plants

10.2.2.1 Endoreduplication and growth

In plants, endoreduplication is observed more frequently than in mammals, although more often in species with small genomes, such as *Arabidopsis thaliana* (Galbraith *et al.*, 1991), *Lycopersicon esculentum* (tomato) (Smulders *et al.*, 1994), *Cucumis sativus* (cucumber) (Gilissen *et al.*, 1993), *Medicago sativa* (alfalfa) and *Medicago truncatula* (barrel medic) (Kondorosi *et al.*, 2000), *Brassica oleracea* (wild cabbage) (Kudo and Kimura, 2001a, b) and *Mesembryanthemum crystallinum* (ice plant) (De Rocher *et al.*, 1990). Accordingly, Nagl (1976) claimed a negative correlation between genome size and extent of endopolyploidization, suggesting that polyploidy represents an evolutionary strategy to compensate for a lack of phylogenetic increase in nuclear DNA. Because in many cases the DNA ploidy level of cells and their size are correlated, nuclear enlargement might be required to support voluminous cells in endoreduplicating species. Such a correlation can, for instance, be seen for the epidermal pavement cells of the developing leaf of *Arabidopsis*. By combining DNA ploidy measurements and growth kinematic analyses, *Arabidopsis* leaf development can be divided into different stages (Boudolf *et al.*, 2004; Beemster *et al.*, 2005; Vlieghe *et al.*, 2005). Initially, cellular divisions are the primary contributors to leaf growth and, at that stage, the cell number increases exponentially. For the first leaf pair of plants grown *in vitro*, all cells participate in the division process until 9 days after sowing. Then, from day 9 until day 12, cell division rates decrease rapidly and cells start to expand until approximately day 19, when the leaves reach their mature size. It is only during the cell expansion phase that cells undergo endoreduplication. Moreover, in the mature leaves, the DNA ploidy level of the individual epidermal cells is correlated with their size (Melaragno *et al.*, 1993).

Analogously, a correlation is found between the DNA ploidy level and the growth of the trichomes, which are specialized epidermal cells that project from the leaf surface and have been proposed to provide a range of benefits, including protection against insect attack, evaporation and excess of light (Hülskamp *et al.*, 1994, 1999). *Arabidopsis* leaf trichome cells initially undergo three rounds of endoreduplication, concomitant with the outgrowth of a cell with two branches. The formation of the third branch is accompanied by a fourth endocycle, resulting in a nuclear DNA level up to 32C. Mutants that increase or reduce the ploidy level in trichomes invariably result in increased or reduced final cell size (Folkers *et al.*, 1997).

Similarly, a relationship between endoreduplication and cell growth is observed during hypocotyl development in *A. thaliana*. At the end of embryogenesis and just before germination, all cells in the embryo have a 2C DNA content. As soon as the embryos start to germinate, cells undergo one or two endoreduplication cycles, and finally up to 80% of the hypocotyl cells become polyploid. Remarkably, there are almost no cell divisions in the endodermis and in the cortex of the hypocotyl. Most of the initial growth of the seedling is therefore not obtained through cell divisions, but via the elongation of hypocotyl cells that have undergone endoreduplication (Gendreau *et al.*, 1997; Traas *et al.*, 1998).

A correlation between the DNA ploidy level and the size of cells has been observed as well for epidermal cells of *Zea mays* (maize), petals of wild cabbage, and petioles of alfalfa and *M. truncatula* (Cavallini *et al.*, 1997; Cebolla *et al.*, 1999; Kudo and Kimura, 2002), strongly suggesting that DNA ploidy not merely correlates with cell size, but controls it. However, several other observations refute this hypothesis. Beemster *et al.* (2002) did not find a link between endoreduplication and the size of mature root cells from different *Arabidopsis* ecotypes. Analogously, a decrease in the cell number, obtained by genetic interference with the cell cycle in *Arabidopsis* and *Nicotiana tabacum* (tobacco) leaves, results in large cells with a low DNA content (Hemerly *et al.*, 1995; Wang *et al.*, 2000; De Veylder *et al.*, 2001; Jasinski *et al.*, 2002).

Recent experiments in *Arabidopsis* that specifically inhibited the endoreduplication process in trichome cells has revealed growth at two levels: at the first level, growth depends on DNA, because trichomes with a reduced DNA ploidy level are clearly smaller; at the second level, a DNA-independent growth mechanism is operational, as illustrated by growth within a certain range without the need to increase the DNA content (Schnittger *et al.*, 2003). A similar conclusion can be made for the *Arabidopsis* leaf epidermal cells in which the endoreduplication cycle is specifically inhibited (own unpublished results). This ploidy-independent growth mechanism might explain the situations wherein cell size and the DNA content are not clearly correlated. In these cases, mechanisms that control cell growth in a ploidy-independent manner might become dominant, masking the casual relationship between DNA content and cell size.

Interestingly, despite the existing relationship between genome size and the endoreduplication level, Barow and Meister (2003) demonstrated that the negative correlation between genome size and endopolyploidization is not so universal and illustrated that endoreduplication is rather related to the taxonomic position and the life cycle of the plants. The observed faster development of endopolyploid species than that of species without endopolyploidy with the same genome size might be achieved by combining the advantages of small and large genomes. Because they have a shorter cell cycle time, plant species with a small genome are expected to grow faster than those with larger genomes. On the other hand, a large genomic content might support DNA content-driven cell expansion (Brandham and West, 1993). Thus, in endopolyploid species, growth of organs might be more rapid than that of nonendopolyploid species, because the initial growth by cell division is followed by endoreduplication-driven cell expansion (Barow and Meister, 2003).

10.2.2.2 Endoreduplication and metabolism

Like in mammals and arthropods, endoreduplication in plants is often correlated with the differentiation of specific cell types that are highly specialized and/or unusually large, such as raphide crystal idioblasts in *Vanilla planifolia* (vanilla orchid) (Kausch and Horner, 1984), suspensor cells in *Phaseolus coccineus* (scarlet runner bean) (Nagl, 1974), root hairs in *Elodea canadensis* (American waterweed) (Dosier and Riopel, 1978), basal cells of the anther hairs of *Bryonia dioica* (white bryony) (Barlow, 1975), parenchyma in seedlings of *Vanda* (Orchidaceae) (Alvarez and Sagawa, 1965; Alvarez, 1968), cotyledons of *Arachis hypogaea* (peanut) (Dhillon and Miksche, 1982), root parenchyma (Marciniak and Bilecka, 1985) and cortex cells from various species (Olszewska, 1976; Olszewska and Kononowicz, 1979). In these tissues, endoreduplication might serve as a mechanism to increase the metabolic output of the cells by increasing the number of housekeeping genes, such as ribosomes and, hence, their transcription levels. The necessity of endoreduplication in supporting the metabolism of cells is most clearly illustrated in studies on pathogenic and symbiotic plant interactions. DNA synthesis has been demonstrated to be essential for the establishment of the metabolically active feeding site upon nematode infection (de Almeida Engler *et al.*, 1999; Favery *et al.*, 2002). Analogously, in nodules of *Medicago* sp., the inefficient nitrogen fixation observed in correlation with a reduction in endopolyploidy indicates that endoreduplication cycles do not simply accompany but play a central role in nodule development (Foucher and Kondorosi, 2000; Vinardell *et al.*, 2003; Kondorosi and Kondorosi, 2004). Repeated endoreduplication cycles during symbiotic cell development might have dual functions; on the one hand, they might ensure extreme enlargement of cells to host the bacteroids and, on the other hand, provide energy and nutrient supply for the bacteroids by increased transcriptional and metabolic activities of the host cell (Kondorosi and Kondorosi, 2004).

In most angiosperms, fertilization of the egg cell by a male gamete is accompanied by a second 'fertilization', in a process called 'double fertilization'. Here, a second male gamete unites with two other female nuclei, called the polar nuclei, to form a triploid cell that later develops into the endosperm, a tissue that acts as a food reserve for the growing embryo. In maize, at the initial phase of development, the triploid endosperm nucleus undergoes synchronous divisions without cytokinesis, forming a syncytium composed of hundreds of nuclei (Kieselbach, 1949). Later on, cell walls surround these nuclei, so that by four days after pollination the endosperm is completely cellularized, followed by a period of growth characterized by numerous cell divisions and differentiation of several cell types. This mitotic phase of development is essentially completed 10–12 days after pollination. At this time, cells in the central starchy endosperm cease to divide and start to engage in multiple cycles of endoreduplication (Kowles and Phillips, 1985; Dilkes *et al.*, 2001). Coincidentally, these cells become enlarged and begin to accumulate starch and storage proteins (Lopes and Larkins, 1993). In some plant species, endoreduplication also occurs at other stages of fruit development; for example, in tomato development of the pericarp and the jelly-like locular tissue is associated with an increase in the nuclear DNA ploidy (Joubès *et al.*, 1999; Joubès and Chevalier, 2000). Here again, endoreduplication during endosperm and fruit development might provide a mechanism to boost the level of gene expression by increasing the number of DNA templates from which RNA transcripts are generated. In agreement with this hypothesis, endoreduplication in the maize endosperm is accompanied with a reduction in chromatin-condensing proteins and the accumulation of a high-mobility group (HMG) protein associated with an open chromatin conformation (Zhao and Grafi, 2000). HMG proteins bind enhancer-like elements in the promoters of endosperm storage protein genes and assist in the binding of transcription factors (Grasser *et al.*, 1990; Schultz *et al.*, 1996). These data indicate that endoreduplication might enhance the transcription of storage proteins. However, this role for the endocycle has been debated, because only slight differences in the expression level of maize endosperm storage protein genes have been observed in transgenic kernels in which the endocycle was partially inhibited by a transgenic approach, suggesting that endoreduplication might play only a minor role in the control of the transcription of genes associated with starch and storage protein synthesis (Leiva-Neto *et al.*, 2004).

10.2.2.3 Endoreduplication and stress

Another role played by endoreduplication might be growth support under stress conditions. Because many endopolyploid species grow in ecological niches that require a fast development, endoreduplication might sustain growth by promoting cell expansion under environmental conditions that inhibit mitosis, such as low temperature and drought (Barow and Meister, 2003). Alternatively, the extra copies of the genome that are produced during endoreduplication could also be regarded as back-up copies that provide the cells with extra wild-type genes in case a gene gets mutated as a consequence of, for example, UV damage. On the other hand, protection against stress could also be mediated in a metabolic way by the possibility of increased transcription of genes with a role in the production of metabolic compounds, such as flavonoids. There is good evidence that flavonoids that absorb strongly in the UV-B region of the solar spectrum protect plants from the effects of UV-B radiation by acting as simple sunscreens (Jordan, 1996; Reuber *et al.*, 1996).

10.3 Molecular control of the endocycle

The endoreduplication cycle can essentially be seen as a mitotic cell cycle in which the M phase is skipped. In this view, the endoreduplicating cells are expected to have simplified their cycle through the elimination of the molecular components that are required to progress through mitosis. By contrast, because cells that divide and endoreduplicate both need to replicate their DNA, they probably share the apparatus necessary to progress through the S phase. A shared component could be the A-type cyclin-dependent kinase (CDKA), demonstrated to be essential for both the mitotic cell cycle and the endocycle. Its role in the mitotic cell cycle is illustrated by the observed reduction in the cell number in plants that overexpress a dominant negative *CDKA* allele (Hemerly *et al.*, 1995), whereas its role in the endoreduplication cycle is demonstrated by the observed reduction in the mean C value upon inactivation of CDKA activity in the developing maize endosperm (Leiva-Neto *et al.*, 2004). In agreement with its dual role, CDKA activity has been detected in both mitotically dividing and endoreduplicating tissues (Grafi and Larkins, 1995; Coelho *et al.*, 2005; Verkest *et al.*, 2005). The involvement of CDKA;1 in these two cycles can additionally be deduced from the phenotypic analysis of plants that strongly overproduce CDKA inhibitory proteins, resulting into an inhibition of both the mitotic cell cycle and the endoreduplication cycle (see below).

CDK activity at the G1 to S transition is presumably required to activate the mechanisms that trigger DNA replication via the retinoblastoma (Rb)–E2F pathway. Genes in this pathway have been shown to essentially regulate the G1 to S transition. In *A. thaliana*, eight E2F transcription factors have been identified that can be divided into three distinct classes based on their gene structure and biochemical activity (Vandepoele *et al.*, 2002). E2Fa–E2Fc have a similar domain organization, characterized by a single highly conserved DNA-binding domain, followed by a DP heterodimerization domain. Dimerization of the E2Fs with the DP proteins, which contribute to a second DNA-binding domain, is a prerequisite for tight sequence-specific binding to the promoter regions of the E2F-responsive genes. Both E2Fa and E2Fb possess a transcriptional activation domain and, correspondingly, have been demonstrated to promote cell division through the induction of genes required for DNA synthesis (De Veylder *et al.*, 2002; Rossignol *et al.*, 2002; Kosugi and Ohashi, 2003; Sozzani *et al.*, 2006). By contrast, E2Fc lacks a clear activation domain and rather operates as a transcriptional suppressor, thereby forming a second class within the family of E2F transcription factors (del Pozo *et al.*, 2002). The third class of

E2F genes encodes proteins that are able to bind DNA as a monomer because of the presence of a duplicated DNA-binding domain and holds three members: E2Fd/DEL2, E2Fe/DEL1 and E2Ff/DEL3 (de Jager *et al.*, 2001; Kosugi and Ohashi, 2002; Mariconti *et al.*, 2002; Vandepoele *et al.*, 2002).

The co-overexpression of *E2Fa* with *DPa* triggers a dual phenotype in both *Arabidopsis* and tobacco. Some cell types have been seen to undergo ectopic cell divisions, while others are stimulated to endoreduplicate. As a result, leaf cells with enlarged nuclei coexist with a high number of smaller cells (De Veylder *et al.*, 2002; Kosugi and Ohashi, 2003). Similarly, microscopic analysis on *Arabidopsis* hypocotyls revealed cell files undergoing ectopic cell divisions, alternated with cell files displaying enlarged cells that have undergone excessive endoreduplication (De Veylder *et al.*, 2002). Cell files that divide ectopically are those in which stomata are formed (De Veylder *et al.*, 2002). Recently, similar, but less outspoken, phenotypes have been observed in plants overexpressing *E2Fb–DPa* (Sozzani *et al.*, 2006).

The coexistence of smaller cells that divide ectopically and larger cells that undergo extra rounds of endoreduplication suggests that E2F activity modulates the cell cycle in a cell-type-specific manner. Also the overproduction of downstream targets of E2F–DP, such as the DNA replication-licensing factors CDC6 and CDT1, triggers a similar dual phenotype (Castellano *et al.*, 2001, 2004). In all cases, cells competent to divide and with a limited stem cell potential are stimulated to undergo ectopic cell division, whereas endoreduplication is activated in the nondividing differentiating cells. Based on these observations, the differential response of cells toward increased E2F activity might depend on the presence of a mitosis-inducing factor (MIF). In this model, cells in which the S phase is stimulated by E2F activity will progress through mitosis in the presence of the MIF, resulting into cell division. By contrast, in the absence of MIF activity, increased expression of S-phase-specific genes might trigger cycles of DNA replication, resulting in endopolyploidy (De Veylder *et al.*, 2002).

The identity of the MIF is still obscure; however, the M-phase-specific CDKB1;1 is a very likely component. Plants overexpressing a dominant negative allele of *CDKB1;1* display a premature onset of their endoreduplication program, demonstrating that CDKB1;1 activity is required to inhibit the endocycle. Moreover, when the mutant allele of *CDKB1;1* is overexpressed in an E2Fa–DPa-overproducing background, it enhances the endoreduplication phenotype and the extra mitotic cell divisions normally induced by E2Fa–DPa are repressed (Boudolf *et al.*, 2004). In agreement with CDKB1;1 being part of the hypothetical MIF, the onset of endoreduplication has been correlated with the inhibition of M-phase-associated CDK activity in the maize endosperm (Grafi and Larkins, 1995).

However, CDKB1;1 cannot be the sole component of the MIF because the overexpression of the wild-type allele does not influence the balance between mitotically dividing and endoreduplicating cells, neither in wild type, nor in *E2Fa–DPa*-overexpressing plants (Boudolf *et al.*, 2004). The reason might be the need of cyclins to activate CDK activity. The A-type cyclins seem to be good candidates to work in collaboration with CDKB1;1. In tobacco, *CYCA3;2* is necessary during the normal mitotic cell cycle, as evidenced from experiments in which its antisense expression severely interferes with embryogenesis. *CYCA3;2* expression also seems to be downregulated to allow the initiation of the endoreduplication process, because overproduction of the tobacco CYCA3;2 in *Arabidopsis* results in reduced endoreduplication (Yu *et al.*, 2003). The same inhibitory effect on endoreduplication is seen upon the overexpression of the A-type cyclin *CYCA2;3*. Correspondingly, in plants lacking CYCA2;3 the onset of the endoreduplication cycle is premature and the DNA ploidy level increases (Imai *et al.*, 2006). Here again, the data indicate that the expression of *CYCA2;3* needs to be downregulated to suppress the mitotic cell cycle, triggering the endocycle to start. In agreement with this hypothesis, the transition from mitotic division to endoreduplication has been correlated with a decrease in expression of mitotic cyclins during *Arabidopsis* leaf and tomato fruit development (Joubès *et al.*, 1999; Beemster *et al.*, 2005).

Besides an elimination of MIF activity, CDKA activity must probably also be downregulated to enable the endocycle onset, as suggested by the inhibition of the endoreduplication cycle upon the overexpression of the D-type cyclin *CYCD3;1* that binds to A-type, but not B-type, CDKs (Healy *et al.*, 2001; Schnittger *et al.*, 2002; Dewitte *et al.*, 2003). In mitotically dividing cells, the genomic material is replicated only once per cell cycle through the inhibitory association of the CDKs with the origins of replication and the CDK activity has to drop to low levels before another round of replication is possible (Wuarin *et al.*, 2002). Thus, a decrease in CDK activity might be required to escape the mechanisms that prevent the rereplication of DNA during the cell cycle without the need to progress through mitosis. Recent data indicate that this decrease in CDK activity could be mediated through the action of CDK-inhibitory proteins (CKIs). CKIs have been proven to be important regulators of the endoreduplication cycle in several organisms. Overexpression of *Rum1* in *Schizosaccharomyces pombe* (fission yeast) promotes nucleus enlargement and polyploidy via the inhibition of M-phase CDKs (Moreno and Nurse, 1994). Ectopic production of $p57^{Kip2}$ in mammalian trophoblasts induces giant cell formation, whereas production of a stable form of the protein blocks endoreduplication (Hattori *et al.*, 2000). Recently, the CDK inhibitor Dacapo has been demonstrated to function in the mitosis to endocycle transition in *Drosophila* follicle cells and to play an important role in endocycling nurse cells (Hong *et al.*, 2003; Shcherbata *et al.*, 2004).

Proteins related to the class of mammalian Kip/Cip CKIs have been identified in plants, designated Kip-related proteins (KRPs) in *Arabidopsis*, although some members are also known as interactors of Cdc2 kinases (ICKs) (Wang *et al.*, 1997; Lui *et al.*, 2000; De Veylder *et al.*, 2001; Vandepoele *et al.*, 2002). The effect of ICKs/KRPs on the endoreduplication process appears to depend on the concentration: high ICK/KRP levels result in a decrease in the DNA ploidy level (De Veylder *et al.*, 2001; Jasinski *et al.*, 2002; Zhou *et al.*, 2002; Schnittger *et al.*, 2003), whereas at low levels, ICK1/KRP1 or ICK2/KRP2 both positively control the endocycle onset (Verkest *et al.*, 2005; Weinl *et al.*, 2005). The different phenotypes observed in the strongly versus the weakly overexpressing lines are a consequence of the binding preference of ICK2/KRP2 toward specific cyclin/CDK complexes. In the weakly overexpressing lines, solely CDKs active in the mitotic cell cycle are targeted and the specific inhibition of these complexes triggers a premature endoreduplication onset. By contrast, in the high *ICK2/KRP2* overexpressing lines, the binding preference toward the mitotic CDKs is lost, and the mitotic cell cycle and endocycle are both inhibited (Verkest *et al.*, 2005).

A role for ICKs/KRPs in the endocycle onset is also supported from research in other plant species. During tomato fruit development, maximal *LeKRP1* expression coincides with the disappearance of cell division activity and the appearance of endoreduplication in the gel tissue. On the other hand, *LeKRP2* is preferentially expressed at the onset of fruit maturation, suggesting a role for the ICK/KRPs during the endoreduplication process itself (Bisbis *et al.*, 2006). In maize, *Zeama;KRP1* transcripts and proteins are present during the period that encompasses the onset of endoreduplication and differentiation. Curiously, whereas *Zeama;KRP2* RNA can be detected in the developing endosperm between 7 and 21 days after pollination, the corresponding protein level decreases after 13 days (Coelho *et al.*, 2005). The discrepancy between the mRNA and its protein amounts suggests regulation of ICK/KRP activity at the posttranscriptional level. Also in developing leaves of *Arabidopsis*, the *ICK2/KRP2* mRNA and protein levels are not clearly correlated (Verkest *et al.*, 2005). Whereas dividing and endoreduplicating leaf tissues display an equal amount of mRNA transcripts, the ICK2/KRP2 protein can only be detected in the latter. Biochemical analysis indicated that the ICK2/KRP2 protein is phosphorylated by CDKs, and that this phosphorylation event triggers the destruction of the CKI by the proteasome. The observed accumulation of ICK2/KRP2 in plants whose CDKB1;1 activity is reduced strongly suggests that this CDK is in part responsible for the targeted destruction of ICK2/KRP2. Because CDKA is the main target of ICK/KRPs, the phosphorylation of the ICK/KRPs by CDKB1;1 hints at a mechanism in which a decrease in CDKB1;1 activity controls the inhibition of CDKA activity. As long as cells divide, CDKB1;1 activity remains high and ICK/KRP activity is inhibited through phosphorylation, with ICK/KRP breakdown and a high CDKA activity level as a consequence. However, as CDKB1;1 activity decreases, ICK/KRPs will become stabilized and will inhibit CDKA activity, with a drop in the latter's activity as a result (Figure 10.2). This mechanism couples reduced CDKB with decreased CDKA activities. Such a coordinated decrease in CDK activity might be required to ensure a correct endoreduplication onset (Verkest *et al.*, 2005).

A decrease in CDK activity at the onset of the endocycle could as well be achieved by changing the phosphorylation status of the CDK subunit (De Veylder *et al.*, 2003). Mammalian CDK activity is negatively regulated by phosphorylation on the Thr14/Tyr15 residues through the activity of the WEE1 kinase, which is counteracted by the CDC25 phosphatase. Whereas the existence of true CDC25 orthologs in plants is still debated (Landrieu *et al.*, 2004; Bleeker *et al.*, 2006), clear homologs of the *WEE1* gene have been described for plants (Sun *et al.*, 1999; Sorrell *et al.*, 2002; Gonzalez *et al.*, 2004). In maize, *Zeama;WEE1* transcripts accumulate in the endosperm at the onset of endoreduplication (Sun *et al.*, 1999). Similarly, a temporal link between expression and start of the endoreduplication cycle can be observed in tomato, where besides accumulation of transcripts in mitotically active organs, *LeWEE1* is also expressed in the jelly-like locular tissue concomitant with

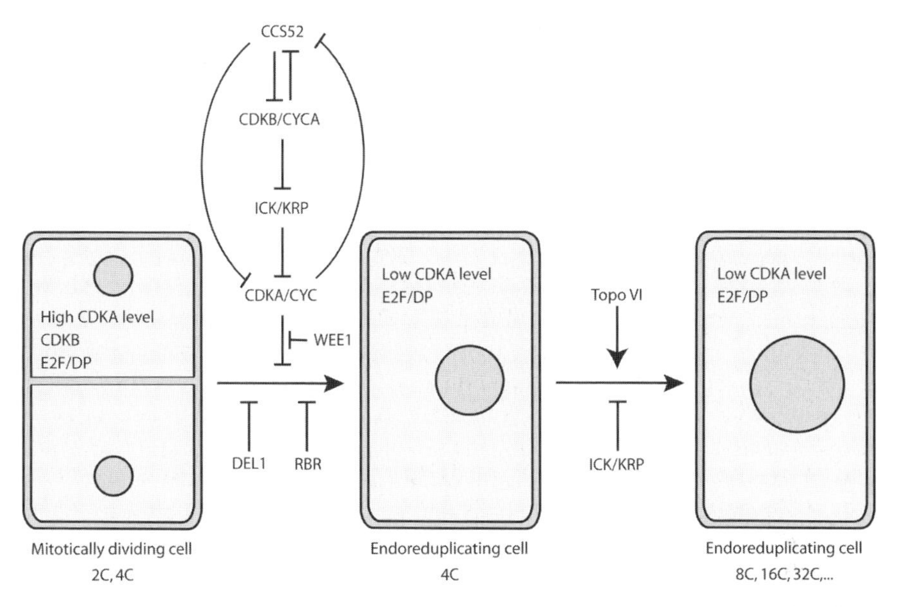

Figure 10.2 Model illustrating the molecular interactions that regulate transition to the endocycle.

endoreduplication during fruit development (Gonzalez *et al.*, 2004). As such, based on their expression profiles in maize and tomato, plant WEE1 plays probably a significant role in the onset to the endocycle (Figure 10.2), although this still has to be proven functionally.

A third way to achieve a decrease in CDK activity is through the activation of the ubiquitin–proteasome pathway. Ubiquitin-dependent proteolysis ensures that specific protein functions are turned off in a unidirectional fashion. Polyubiquitylation of proteins involves at least three enzyme activities. The ubiquitin-activating enzyme (E1) forms a high-energy bond with ubiquitin, that is then transesterified to an ubiquitin-conjugating enzyme (E2). The transfer of ubiquitin to the target protein substrate requires an ubiquitin–protein ligase (E3). Polyubiquitylation of a protein is sufficient to target its degradation by a large ATP-dependent multicatalytic protease, the 26S proteasome. The selection and specific timing of polyubiquitination of the target proteins are conferred by different E3 ubiquitin ligases. One of these E3 complexes, designated the anaphase-promoting complex (APC) is intimately connected with the onset of the endocycle by controlling the proteolysis of mitotic cyclins, as demonstrated by the observation that the overproduction of the APC-activating protein CCS52A (also known in nonplant species as Hct1, Fizzy-related, Srw1, or Ste9) of *Medicago* sp. in yeast cells is sufficient to trigger the proteolysis of its mitotic cyclins, resulting into endoreduplication. Moreover, downregulation of the *CCS52A* expression in *Medicago* sp. plants significantly reduces the DNA ploidy level (Cebolla *et al.*, 1999; Vinardell *et al.*, 2003; Tarayre *et al.*, 2004). Corroborating its anticipated role in the endoreduplication onset, *CCS52A* expression is induced in differentiating cells undergoing endoreduplication during nodule development upon *Rhizobium* infection and during giant cell formation after nematode infection (Cebolla *et al.*, 1999; Favery *et al.*, 2002). Moreover, higher expression levels of *CCS52A* coincide with higher levels of endoreduplication in *Medicago* sp. nodules, suggesting that in plants the degree of endopolyploidy and the level of *CCS52A* expression are directly correlated.

In the plant genomes, two classes of *CCS52* genes are encoded: *CCS52A* and *CCS52B*. Whereas *Medicago* sp. has only one member in each class (Cebolla *et al.*, 1999; Vinardell *et al.*, 2003; Tarayre *et al.*, 2004), *Arabidopsis* has two *CCS52A* genes, *CCS52A1* and *CCS52A2*, and one *CCS52B* gene (Fülöp *et al.*, 2005). Although the specific role of these different classes remains unknown, expression analysis hints at distinctive activity for the A-type and B-type *CCS52* genes. Both in *Medicago* and *Arabidopsis*, expression of *CCS52B* is restricted to G2/M and M phase, whereas that of *CCS52A* is constitutive during the whole cell cycle in *Medicago*, and from late M phase until early G2 phase in *Arabidopsis* (Tarayre *et al.*, 2004; Fülöp *et al.*, 2005). Also, the differential binding preferences of the *Arabidopsis* CCS52 proteins to distinct sets of cyclins suggest that CCS52A proteins might be critical for APC activity in all cell cycle phases, whereas the CCS52B proteins could play specific roles in M-phase progression (Fülöp *et al.*, 2005). Similarly, their function during plant development could be distinct and specific. Because the APC^{CCS52A} is dispensable for nodule meristem formation in *Medicago* but necessary for endoreduplication cycles and differentiation of nodule cells, CCS52A might play a major role in postmitotic, differentiating cells, in which degradation of specific APC targets could contribute to the differentiation of given cell types, tissues or organs (Vinardell *et al.*, 2003).

The manner in which CCS52 regulates the activity of mitotic cyclin/CDK complexes is quite complicated: on the one hand, CCS52 proteins can target the mitotic cyclins for degradation by the APC, thereby inhibiting the mitotic CDK activity, but, on the other hand, they are phosphorylated by the mitotic cyclin/CDK complexes and block their APC-activating function (Tarayre *et al.*, 2004) (Figure 10.2). Similarly, the yeast homologs of CCS52, Cdh1/Hct1 in *Saccharomyces cerevisiae* and Ste9/Srw1 in *S. pombe* have also been shown to contain cyclin-dependent phosphorylation sites that vary in numbers and positions as well as in effects on activity. Cdh1 is inactivated by hyperphosphorylation, which prevents its association with the APC (Zachariae *et al.*, 1998; Jaspersen *et al.*, 1999; Kramer *et al.*, 2000), triggers its proteolysis (Blanco *et al.*, 2000), and leads to its translocation from the nucleus to the cytoplasm (Jaquenoud *et al.*, 2002; Zhou *et al.*, 2003). Moreover, binding of inhibitory proteins such as the spindle checkpoint components BubR1 and Mad2 has been shown to repress the Cdh1 function (reviewed in Harper *et al.*, 2002; Peters, 2002; Castro *et al.*, 2005). These mechanisms are probably operational in plants as well.

Yet another factor that might inactivate MIF activity and/or downregulate the CDKA activity is E2Fe/DEL1, an atypical member of the E2F family of transcription factors that inhibits specifically endoreduplication in *Arabidopsis*. Ectopic expression of *E2Fe/DEL1* reduces endoreduplication, whereas loss of E2Fe/DEL1 function causes an increase in the DNA ploidy level. Likewise, E2Fe/DEL1 inhibits the endoreduplication phenotype, but not the ectopic cell divisions that result from the overexpression of both *E2Fa* and *DPa*, illustrating that E2Fe/DEL1 specifically represses the endocycle rather than being an antagonist of the E2F pathway. Because *E2Fe/DEL1* transcripts are detected exclusively in mitotically dividing cells, E2Fe/DEL1 might preserve the mitotic state of proliferating cells by suppressing transcription of genes that are required for cells to enter the DNA endoreduplication cycle (Vlieghe *et al.*, 2005). However, currently the molecular mechanism by which E2Fe/DEL1 suppresses the endocycle remains unknown (Figure 10.2).

During the endocycle itself, successive rounds of DNA replication in endoreduplicating cells need S-phase inhibitors to be downregulated. Neutralization of one of these inhibitors, Rb, can be mediated by the activity of cyclin/CDK complexes (Hunter and Pines, 1994; Sherr, 1994). *In vitro* phosphorylation of a plant Rb-related (RBR) protein by an S-phase kinase has been observed in endoreduplicating cells from the maize endosperm, thereby probably releasing E2F activity and transcription of S-phase genes (Grafi *et al.*, 1996). Moreover, during maize endosperm development, downregulation of *RBR3* expression is concomitant with the occurrence of endoreduplication (Sabelli *et al.*, 2005). Nevertheless, to trigger endoreduplication, downregulation of RBR activity must be coordinated with the mechanisms that control the decrease in M-phase CDK activity, as clearly demonstrated by experiments in which inactivation of tobacco RBR activity causes extra rounds of DNA replication in endoreduplicating tissues, as well as prolonged cell proliferation in tissues that are normally differentiated (Park *et al.*, 2005). Similarly, induced inactivation of RBR in young leaves of *Arabidopsis* stimulates cell division and leads to epidermal hyperplasia, while in older leaves it triggers the endoreduplication process (Desvoyes *et al.*, 2006). These data again stress the importance of inhibiting the MIF action when proceeding to the endocycle.

The topoisomerase VI homologs from *Arabidopsis*, HYP7/RHL1, HYP6/RHL3/AtTOP6B and RHL2/AtSPO11-3 form a functional complex that is responsible for chromosome decatenation at the end of the DNA replication and that is required for successive rounds of endoreduplication. Whereas wild-type leaf cells from *Arabidopsis* endoreduplicate up to four times, resulting in a DNA ploidy level of 32C, mutants in either of the three topoisomerase genes can only complete the first two rounds of endoreduplication to 8C, correlated with an extreme dwarf growth phenotype (Hartung *et al.*, 2002; Sugimoto-Shirasu *et al.*, 2002; Yin *et al.*, 2002). A reduction in the number of endocycles is also observed in dark-grown hypocotyls and trichome cells. In these specific mutants, incomplete decatenation of replicated DNA might result in progressively entangled chromosomes that either physically block further DNA replication or initiate a checkpoint mechanism to block further endocycles (Sugimoto-Shirasu *et al.*, 2002, 2005) (Figure 10.2).

Besides the examples outlined above, other genes and mutants have been described that affect the endoreduplication process, specifically during trichome development, such as *SIM*, *KAK*, *CPR5* and others. These genes will be discussed in more detail in Chapter 11.

10.4 Environmental and hormonal control of the endocycle

Endoreduplication can be triggered by a number of stimuli, including plant hormones and environmental signals. Phytohormones directly affect the expression and activity of regulatory key cell cycle genes (Stals and Inzé, 2001). For example, auxin has been linked to the regulation of protein turnover via the ubiquitin–proteasome pathway (Leyser *et al.*, 1993; del Pozo and Estelle, 1999; Parry and Estelle, 2006), which controls the stability of certain cell cycle regulatory proteins (King *et al.*, 1996). Cytokinins have been implicated in the activation of M phase through tyrosine phosphorylation of CDKA (Zhang *et al.*, 1996a, 2005), as well as in the activation of S phase CDKs (Riou-Khamlichi *et al.*, 1999). By contrast, the hormonal regulation of endoreduplication remains relatively poorly characterized. In the case of auxin, it is difficult to draw any conclusions on its importance, because different results have been obtained for distinct model organisms. For cultured tobacco cells, an auxin-alone signal induces elongation and DNA endoreduplication, whereas addition of auxin and cytokinin causes the cells to divide actively (Valente *et al.*, 1998). On the contrary, no change in DNA ploidy levels is observed when auxin is applied to cultured *Pisum sativum* (pea) root cortex cells, but endoreduplication is triggered in the presence of both auxin and cytokinin (Libbenga and Torrey, 1973). When auxin is applied to *Prunus armeniaca* (apricot) trees, it provokes an increase in fruit size because of the endoreduplication-driven enlargement of mesocarp cell volume (Bradley and Crane, 1955). Similarly, in cultured haploid *Petunia hybrida* leaf tissues, auxin treatment induces endopolyploidy by doubling the chromosome number (Liscum and Hangarter, 1991).

Whereas treatments with brassinolides, cytokinin and abscisic acid only slightly affect endoreduplication, gibberellins (GAs) have been shown to control DNA synthesis during *Arabidopsis* hypocotyl development. Mutants with a perturbed GA biosynthesis or sensitivity display an important reduction in their endoreduplication level. Although it can be debated whether the effects observed might be secondary to those of the hormone on cell expansion, addition of GA to the mutants restores the wild-type DNA ploidy levels at concentration levels that are 100-fold lower than those required to rescue the cell growth phenotype (Gendreau *et al.*, 1999). An enhancing effect of GAs on endoreduplication has also been described in pea (Mohamed and Bopp, 1980) and on DNA synthesis in *Triticum aestivum* (wheat) leaves, although this effect appeared to depend on the cultivar (Cavallini *et al.*, 1995).

A second phytohormone with a positive effect on the endoreduplication cycle is ethylene. In wild-type *Arabidopsis* seedlings grown in the presence of 1-aminocyclopropane-1-carboxylic acid, the direct precursor of ethylene biosynthesis, an extra round of endoreduplication can be observed (Gendreau *et al.*, 1999). Similarly, in hypocotyls of *Cucumis sativus* (cucumber), transient ethylene exposure stimulates DNA synthesis, while it suppresses cytokinesis with an eightfold increase in DNA content as a result (Dan *et al.*, 2003).

Recently, specific sterols have been proposed to control the endocycle as well, as evidenced from the *frill1* (*frl1*) mutant of *Arabidopsis* in which the sterol biosynthesis pathway is disturbed. The resulting alteration of the normal sterol profile in this mutant leads to ectopic endoreduplication in petals and enhanced endoreduplication in the rosette leaves (Hase *et al.*, 2005).

Besides the hormonal regulation of endoreduplication, plant growth conditions seem to influence ploidy patterns as well. Leaves of tomato plants grown in a greenhouse reach higher endoreduplication levels than those of plants grown *in vitro* (Smulders *et al.*, 1994), and the same tendency is observed for *Solanum tuberosum* (potato) plants (Uijtewaal, 1987). One of the environmental factors with an impact on endoreduplication seems to be the light regime. In the *Arabidopsis* hypocotyl, the third endocycle is inhibited by light through the action of the red/far-red light photoreceptor phytochrome (Gendreau *et al.*, 1998). During photomorphogenesis in wild-type plants, up to two rounds of endoreduplication are observed in hypocotyl cells, whereas a third round takes place only during skotomorphogenesis. In contrast to wild-type plants, phytochrome A mutants also have 16C nuclei in hypocotyl cells when grown under continuous far-red light, stressing the importance of phytochrome in the repression of endoreduplication. Similarly, during epicotyl elongation of pea in the dark, an endocycle occurs leading to an 8C DNA content, while the nuclear level in the light is only 4C (Van Oostveldt and Van Parijs, 1975).

Water availability seems to be another environmental factor that influences endocycles. Whereas normally the number of endoreduplicating cells in the maize endosperm steadily increases from 9 to 13 days after pollination, in plants that are exposed to a water deficit the proportion of endoreduplicating nuclei initially increases, but is halted after a prolonged deficit, so that at 13 days after pollination it is less than in the control plants. The increase in endoreduplicating cells 9 days after pollination is probably due to the higher sensitivity of mitotic cycles toward drought, allowing endoreduplication to augment the existing pool of nuclei to higher DNA contents 9 days after pollination. However, as the duration of stress stretches on, both cell proliferation and endoreduplication are inhibited (Setter and Flannigan, 2001).

Similarly, short-term high-temperature treatment does not affect endoreduplication in maize endosperm, but prolonged high-temperature treatment results in reduced ploidy levels (Engelen-Eigles *et al.*, 2000). Nevertheless, a mild increase in growth temperature had a positive effect on endoreduplication cycles in tomato pericarp cells (Bertin, 2005). The lower amount of endoreduplicating nuclei at high temperature might be associated with disruption of cellular and nuclear integrity (Commuri and Jones, 1999).

10.5 Outlook

Although much progress has been made in recent years, more extensive studies are necessary to expand our present knowledge of endoreduplication in plants. Recently, several mutants with altered DNA ploidy levels have become available and these mutants will help us to unravel the remaining mysteries about endoreduplication. Characterization of these mutants under various environmental conditions will shed more light on the physiological role of endoreduplication. Moreover, extensive molecular analyses on these mutants will broaden the insight into the regulatory mechanisms behind endoreduplication. In the future, this knowledge may help us to interfere with endocycling in a useful manner.

Acknowledgments

The authors thank Martine De Cock for help in preparing the manuscript. This work was supported by the Interuniversity Poles of Attraction Programme-Belgian Science Policy (P5/13). K.V. is indebted to the Institute for the Promotion of Innovation by Science and Technology in Flanders for a predoctoral fellowship. L.D.V. is a Postdoctoral Fellow of the Research Foundation-Flanders.

References

Alvarez, M.R. (1968) Quantitative changes in nuclear DNA accompanying postgermination embryonic development in *Vanda* (Orchidaceae). *Am J Bot* **55**, 1036–1041.

Alvarez, M.R. and Sagawa, Y. (1965) A histochemical study of embryo development in *Vanda* (Orchidaceae). *Caryologia* **18**, 251–261.

Barlow, P.W. (1975) The polytene nucleus of the giant hair cell of *Bryonia* anthers. *Protoplasma* **83**, 339–349.

Barow, M. and Meister, A. (2003) Endopolyploidy in seed plants is differently correlated to systematics, organ, life strategy and genome size. *Plant Cell Environ* **26**, 571–584.

Beemster, G.T.S., De Veylder, L., Vercruysse, S., *et al.* (2005) Genome-wide analysis of gene expression profiles associated with cell cycle transitions in growing organs of Arabidopsis. *Plant Physiol* **138**, 734–743.

Beemster, G.T.S., De Vusser, K., De Tavernier, E., De Bock, K. and Inzé, D. (2002) Variation in growth rate between Arabidopsis ecotypes is correlated with cell division and A-type cyclin-dependent kinase activity. *Plant Physiol* **129**, 854–864.

Bertin, N. (2005) Analysis of the tomato fruit growth response to temperature and plant fruit load in relation to cell division, cell expansion and DNA endoreduplication. *Ann Bot* **95**, 439–447.

Bisbis, B., Delmas, F., Joubès, J., *et al.* (2006) Cyclin-dependent kinase (CDK) inhibitors regulate the CDK-cyclin complex activities in endoreduplicating cells of developing tomato fruit. *J Biol Chem* **281**, 7374–7383.

Blanco, M.A., Sánchez-Díaz, A., de Prada, J.M. and Moreno, S. (2000) $APC^{ste9/srw1}$ promotes degradation of mitotic cyclins in G1 and is inhibited by cdc2 phosphorylation. *EMBO J* **19**, 3945–3955.

Bleeker, P.M., Hakvoort, H.W.J., Bliek, M., Souer, E. and Schat, H. (2006) Enhanced arsenate reduction by a CDC25-like tyrosine phosphatase explains increased phytochelatin accumulation in arsenate-tolerant *Holcus lanatus*. *Plant J* **45**, 917–929.

Boudolf, V., Vlieghe, K., Beemster, G.T.S., *et al.* (2004) The plant-specific cyclin-dependent kinase CDKB1;1 and transcription factor E2Fa-DPa control the balance of mitotically dividing and endoreduplicating cells in Arabidopsis. *Plant Cell* **16**, 2683–2692.

Bradley, M.V. and Crane, J.C. (1955) The effect of 2,4,5-trichlorophenoxyacetic acid on cell and nuclear size and endopolyploidy in parenchyma of apricot fruits. *Am J Bot* **42**, 273–281.

Brandham, P.E. and West, J.P. (1993) Correlation between nuclear DNA values and differing optimal ploidy levels in *Narcissus*, *Hyacinthus* and *Tulipa* cultivars. *Genetica* **90**, 1–8.

Brodsky, V.Y. and Uryvaeva, I.V. (1977) Cell polyploidy: its relation to tissue growth and function. *Int Rev Cytol* **50**, 275–332.

Castellano, M.M., Boniotti, M.B., Caro, E., Schnittger, A. and Gutierrez, C. (2004) DNA replication licensing affects cell proliferation or endoreplication in a cell type-specific manner. *Plant Cell* **16**, 2380–2393.

Castellano, M.M., del Pozo, J.C., Ramirez-Parra, E., Brown, S. and Gutierrez, C. (2001) Expression and stability of Arabidopsis CDC6 are associated with endoreplication. *Plant Cell* **13**, 2671–2686.

Castro, A., Bernis, C., Vigneron, S., Labbé, J.-C. and Lorca, T. (2005) The anaphase-promoting complex: a key factor in the regulation of cell cycle. *Oncogene* **24**, 314–325.

Cavallini, A., Baroncelli, S., Lercari, B., Cionini, G., Rocca, M. and D'Amato, F. (1995) Effect of light and gibberellic acid on chromosome endoreduplication in leaf epidermis of *Triticum durum* Desf. *Protoplasma* **186**, 57–62.

Cavallini, A., Natali, L., Cionini, G., Balconi, C. and D'Amato, F. (1997) Inheritance of nuclear DNA content in leaf epidermal cells of *Zea mays* L. *Theor Appl Genet* **94**, 782–787.

Cebolla, A., Vinardell, J.M., Kiss, E., *et al.* (1999) The mitotic inhibitor *ccs52* is required for endoreduplication and ploidy-dependent cell enlargement in plants. *EMBO J* **18**, 4476–4484.

Coelho, C.M., Dante, R.A., Sabelli, P.A., *et al.* (2005) Cyclin-dependent kinase inhibitors in maize endosperm and their potential role in endoreduplication. *Plant Physiol* **138**, 2323–2336.

Commuri, P.D. and Jones, R.J. (1999) Ultrastructural characterization of maize (*Zea mays* L.) kernels exposed to high temperature during endosperm cell division. *Plant Cell Environ* **22**, 375–385.

Dan, H., Imaseki, H., Wasteneys, G.O. and Kazama, H. (2003) Ethylene stimulates endoreduplication but inhibits cytokinesis in cucumber hypocotyl epidermis. *Plant Physiol* **133**, 1726–1731.

Datta, N.S., Williams, J.L., Caldwell, J., Curry, A.M., Ashcraft, E.K. and Long, M.W. (1996) Novel alterations in CDK1/cyclin B1 kinase complex formation occur during the acquisition of a polyploid DNA content. *Mol Biol Cell* **7**, 209–223.

de Almeida Engler, J., De Vleesschauwer, V., Burssens, S., *et al.* (1999) Molecular markers and cell cycle inhibitors show the importance of cell cycle progression in nematode-induced galls and syncytia. *Plant Cell* **11**, 793–807.

de Jager, S.M., Menges, M., Bauer, U.-M. and Murray, J.A.H. (2001) *Arabidopsis* E2F1 binds a sequence present in the promoter of S-phase-regulated gene *AtCDC6* and is a member of a multigene family with differential activities. *Plant Mol Biol* **47**, 555–568.

del Pozo, J.C., Boniotti, M.B. and Gutierrez, C. (2002) Arabidopsis E2Fc functions in cell division and is degraded by the ubiquitin–SCF^{AtSKP2} pathway in response to light. *Plant Cell* **14**, 3057–3071.

del Pozo, J.C. and Estelle, M. (1999) F-box proteins and protein degradation: an emerging theme in cellular regulation. *Plant Mol Biol* **44**, 123–128.

De Rocher, E.J., Harkins, K.R., Galbraith, D.W. and Bohnert, H.J. (1990) Developmentally regulated systemic endopolyploidy in succulents with small genomes. *Science* **250**, 99–101.

Desvoyes, B., Ramirez-Parra, E., Xie, Q., Chua, N.-H. and Gutierrez, C. (2006) Cell type-specific role of the retinoblastoma/E2F pathway during Arabidopsis leaf development. *Plant Physiol* **140**, 67–80.

De Veylder, L., Beeckman, T., Beemster, G.T.S., *et al.* (2001) Functional analysis of cyclin-dependent kinase inhibitors of Arabidopsis. *Plant Cell* **13**, 1653–1667.

De Veylder, L., Beeckman, T., Beemster, G.T.S., *et al.* (2002) Control of proliferation, endoreduplication and differentiation by the *Arabidopsis* E2Fa/DPa transcription factor. *EMBO J* **21**, 1360–1368.

De Veylder, L., Joubès, J. and Inzé, D. (2003) Plant cell cycle transitions. *Curr Opin Plant Biol* **6**, 536–543.

Dewitte, W., Riou-Khamlichi, C., Scofield, S., *et al.* (2003) Altered cell cycle distribution, hyperplasia and inhibited differentiation in Arabidopsis caused by the D-type cyclin CYCD3. *Plant Cell* **15**, 79–92.

Dhillon, S.S. and Miksche, J.P. (1982) DNA content and heterochromatin variations in various tissues of peanut (*Arachis hypogaea*). *Am J Bot* **69**, 219–226.

Dilkes, B.P., Dante, R.A., Coelho, C. and Larkins, B.A. (2001) Genetic analyses of endoreduplication in *Zea mays* endosperm: evidence of sporophytic and zygotic maternal control. *Genetics* **160**, 1163–1177.

Dosier, L.W. and Riopel, J.L. (1978) Origin, development and growth of differentiating trichoblasts in *Elodea canadensis*. *Am J Bot* **65**, 813–822.

Edgar, B.A. and Orr-Weaver, T.L. (2001) Endoreplication cell cycles: more for less. *Cell* **105**, 297–306.

Engelen-Eigles, G., Jones, R.J. and Phillips, R.L. (2000) DNA endoreduplication in maize endosperm cells: the effect of exposure to short-term high temperature. *Plant Cell Environ* **23**, 657–663.

Favery, B., Complainville, A., Vinardell, J.M., *et al.* (2002) The endosymbiosis-induced genes *ENOD40* and *CCS52a* are involved in endoparasitic-nematode interactions in *Medicago truncatula*. *Mol Plant-Microbe Interact* **15**, 1008–1013.

Folkers, U., Berger, J. and Hülskamp, M. (1997) Cell morphogenesis of trichomes in *Arabidopsis*: differential control of primary and secondary branching by branch initiation regulators and cell growth. *Development* **124**, 3779–3786.

Foucher, F. and Kondorosi, E. (2000) Cell cycle regulation in the course of nodule organogenesis in *Medicago*. *Plant Mol Biol* **43**, 773–786.

Fülöp, K., Tarayre, S., Kelemen, Z., *et al.* (2005) *Arabidopsis* anaphase-promoting complexes: multiple activators and wide range of substrates might keep APC perpetually busy. *Cell Cycle* **4**, 1084–1092.

Galbraith, D.W., Harkins, K.R. and Knapp, S. (1991) Systemic endopolyploidy in *Arabidopsis thaliana*. *Plant Physiol* **96**, 985–989.

Geitler, L. (1939) Die Entstehung der polyploiden Somakerne der Heteropteren durch Chromosomenteilung ohne Kernteilung. *Chromosoma* **1**, 1–22.

Gendreau, E., Höfte, H., Grandjean, O., Brown, S. and Traas, J. (1998) Phytochrome controls the number of endoreduplication cycles in the *Arabidopsis thaliana* hypocotyl. *Plant J* **13**, 221–230.

Gendreau, E., Orbovic, V., Höfte, H. and Traas, J. (1999) Gibberellin and ethylene control endoreduplication levels in the *Arabidopsis thaliana* hypocotyl. *Planta* **209**, 513–516.

Gendreau, E., Traas, J., Desnos, T., Grandjean, O., Caboche, M. and Höfte, H. (1997) Cellular basis of hypocotyl growth in *Arabidopsis thaliana*. *Plant Physiol* **114**, 295–305.

Gilissen, L.J.W., van Staveren, M.J., Creemers-Molenaar, J. and Verhoeven, H.A. (1993) Development of polysomaty in seedlings and plants of *Cucumis sativus* L. *Plant Sci* **91**, 171–179.

Gonzalez, N., Hernould, M., Delmas, F., *et al.* (2004) Molecular characterization of a *WEE1* gene homologue in tomato (*Lycopersicon esculentum* Mill.). *Plant Mol Biol* **56**, 849–861.

Grafi, G., Burnett, R.J., Helentjaris, T., *et al.* (1996) A maize cDNA encoding a member of the retinoblastoma protein family: involvement in endoreduplication. *Proc Natl Acad Sci USA* **93**, 8962–8967.

Grafi, G. and Larkins, B.A. (1995) Endoreduplication in maize endosperm: involvement of M phase-promoting factor inhibition and induction of S phase-related kinases. *Science* **269**, 1262–1264.

Grasser, K.D., Maier, U.-G., Haass, M.M. and Feix, G. (1990) Maize high mobility group proteins bind to CCAAT and TATA boxes of a zein gene promoter. *J Biol Chem* **265**, 4185–4188.

Gregory, T.R. and Shorthouse, D.P. (2003) Genome sizes of spiders. *J Hered* **94**, 285–290.

Harper, J.W., Burton, J.L. and Solomon, M.J. (2002) The anaphase-promoting complex: it's not just for mitosis any more. *Genes Dev* **16**, 2179–2206.

Hartung, F., Angelis, K.J., Meister, A., Schubert, I., Melzer, M. and Puchta, H. (2002) An archaebacterial topoisomerase homolog not present in other eukaryotes is indispensable for cell proliferation of plants. *Curr Biol* **12**, 1787–1791.

Hase, Y., Fujioka, S., Yoshida, S., Sun, G., Umeda, M. and Tanaka, A. (2005) Ectopic endoreduplication caused by sterol alteration results in serrated petals in *Arabidopsis*. *J Exp Bot* **56**, 1263–1268.

Hattori, N., Davies, T.C., Anson-Cartwright, L. and Cross, J.C. (2000) Periodic expression of the cyclin-dependent kinase inhibitor $p57^{Kip2}$ in trophoblast giant cells defines a G2-like gap phase of the endocycle. *Mol Biol Cell* **11**, 1037–1045.

Healy, J.M.S., Menges, M., Doonan, J.H. and Murray, J.A.H. (2001) The *Arabidopsis* D-type cyclins CycD2 and CycD3 both interact *in vivo* with the PSTAIRE cyclin-dependent kinase Cdc2a but are differentially controlled. *J Biol Chem* **276**, 7041–7047.

Hemerly, A., de Almeida Engler, J., Bergounioux, C., *et al.* (1995) Dominant negative mutants of the Cdc2 kinase uncouple cell division from iterative plant development. *EMBO J* **14**, 3925–3936.

Hong, A., Lee-Kong, S., Iida, T., Sugimura, I. and Lilly, M.A. (2003) The $p27^{cip/kip}$ ortholog *dacapo* maintains the *Drosophila* oocyte in prophase of meiosis I. *Development* **130**, 1235–1242.

Hülskamp, M., Miséra, S. and Jürgens, G. (1994) Genetic dissection of trichome cell development in Arabidopsis. *Cell* **76**, 555–566.

Hülskamp, M., Schnittger, A. and Folkers, U. (1999) Pattern formation and cell differentiation: trichomes in *Arabidopsis* as a genetic model system. *Int Rev Cytol* **186**, 147–178.

Hunter, T. and Pines, J. (1994) Cyclins and cancer II: cyclin D and CDK inhibitors come of age. *Cell* **79**, 573–582.

Imai, K.K., Ohashi, Y., Tsuge, T., *et al.* (2006) The A-type cyclin CYCA2;3 is a key regulator of ploidy levels in *Arabidopsis* endoreduplication. *Plant Cell* **18**, 382–396.

Jaquenoud, M., van Drogen, F. and Peter, M. (2002) Cell cycle-dependent nuclear export of Cdh1p may contribute to the inactivation of APC/C^{Cdh1}. *EMBO J* **21**, 6515–6526.

Jasinski, S., Riou-Khamlichi, C., Roche, O., Perennes, C., Bergounioux, C. and Glab, N. (2002) The CDK inhibitor NtKIS1a is involved in plant development, endoreduplication and restores normal development of cyclin D3;1-overexpressing plants. *J Cell Sci* **115**, 973–982.

Jaspersen, S.L., Charles, J.F. and Morgan, D.O. (1999) Inhibitory phosphorylation of the APC regulator Hct1 is controlled by the kinase Cdc28 and the phosphatase Cdc14. *Curr Biol* **9**, 227–236.

Jordan, B.R. (1996) The effects of ultraviolet-B radiation on plants: a molecular perspective. *Adv Bot Res* **22**, 98–162.

Joubès, J. and Chevalier, C. (2000) Endoreduplication in higher plants. *Plant Mol Biol* **43**, 735–745.

Joubès, J., Phan, T.-H., Just, D., *et al.* (1999) Molecular and biochemical characterization of the involvement of cyclin-dependent kinase A during the early development of tomato fruit. *Plant Physiol* **121**, 857–869.

Kausch, A.P. and Horner, H.T. (1984) Increased nuclear DNA content in raphide crystal idioblasts during development in *Vanilla planifolia* L. (Orchidaceae). *Eur J Cell Biol* **33**, 7–12.

Kieselbach, T.A. (1949) The structure and reproduction of corn. *Research Bulletin* 161. Agricultural Experimental Station, Lincoln, NE.

King, R.W., Deshaies, R.J., Peters, J.-M. and Kirschner, M.W. (1996) How proteolysis drives the cell cycle. *Science* **274**, 1652–1659.

Kondorosi, E. and Kondorosi, A. (2004) Endoreduplication and activation of the anaphase-promoting complex during symbiotic cell development. *FEBS Lett* **567**, 152–157.

Kondorosi, E., Roudier, F. and Gendreau, E. (2000) Plant cell-size control: growing by ploidy? *Curr Opin Plant Biol* **3**, 488–492.

Kosugi, S. and Ohashi, Y. (2002) E2Ls, E2F-like repressors of *Arabidopsis* that bind to E2F sites in a monomeric form. *J Biol Chem* **277**, 16553–16558.

Kosugi, S. and Ohashi, Y. (2003) Constitutive E2F expression in tobacco plants exhibits altered cell cycle control and morphological change in a cell type-specific manner. *Plant Physiol* **132**, 2012–2022.

Kowles, R.V. and Phillips, R.L. (1985) DNA amplification patterns in maize endosperm nuclei during kernel development. *Proc Natl Acad Sci USA* **82**, 7010–7014.

Kramer, E.R., Scheuringer, N., Podtelejnikov, A.V., Mann, M. and Peters, J.-M. (2000) Mitotic regulation of the APC activator proteins CDC20 and CDH1. *Mol Biol Cell* **11**, 1555–1569.

Kudo, N. and Kimura, Y. (2001a) Flow cytometric evidence for endopolyploidization in cabbage (*Brassica oleraceae L.*) flowers. *Sex Plant Reprod* **13**, 279–283.

Kudo, N. and Kimura, Y. (2001b) Patterns of endopolyploidy during seedling development in cabbage (*Brassica oleraceae* L.). *Ann Bot* **87**, 275–281.

Kudo, N. and Kimura, Y. (2002) Nuclear DNA endoreduplication during petal development in cabbage: relationship between ploidy levels and cell size. *J Exp Bot* **53**, 1017–1023.

Kudryavtsev, B.N., Kudryavtseva, M.V., Sakuta, G.A. and Stein, G.I. (1993) Human hepatocyte polyploidization kinetics in the course of life cycle. *Virchows Arch B Cell Pathol Incl Mol Pathol* **64**, 387–393.

Landrieu, I., da Costa, M., De Veylder, L., *et al.* (2004) A small CDC25 dual-specificity tyrosine-phosphatase isoform in *Arabidopsis thaliana*. *Proc Natl Acad Sci USA* **101**, 13380–13385. (Erratum: *Proc Natl Acad Sci USA* **101**, 16391).

Leiva-Neto, J.T., Grafi, G., Sabelli, P.A., *et al.* (2004) A dominant negative mutant of cyclin-dependent kinase A reduces endoreduplication but not cell size or gene expression in maize endosperm. *Plant Cell* **16**, 1854–1869.

Levan, A. and Hauschka, T.S. (1953) Endomitotic reduplication mechanisms in ascites tumors of the mouse. *J Natl Cancer Inst* **14**, 1–43.

Leyser, H.M.O., Lincoln, C.A., Timpte, C., *et al.* (1993) *Arabidopsis* auxin-resistance gene *AXR1* encodes a protein related to ubiquitin-activating enzyme E1. *Nature* **364**, 161–164.

Libbenga, K.R. and Torrey, J.G. (1973) Hormone induced endoreplication prior to mitosis in cultured pea root cells. *Am J Bot* **60**, 293–299.

Liscum, E. and Hangarter, R.P. (1991) Manipulation of ploidy level in cultured haploid *Petunia* tissue by phytohormone treatments. *J Plant Physiol* **138**, 33–38.

Lopes, M.A. and Larkins, B.A. (1993) Endosperm origin, development, and function. *Plant Cell* **5**, 1383–1399.

Lorz, A.P. (1947) Supernumerary chromosomal reproduction: polytene chromosomes, endomitosis, multiple chromosome complexes, polysomaty. *Bot Rev* **13**, 597–624.

Lui, H., Wang, H., DeLong, C., Fowke, L.C., Crosby, W.L. and Fobert, P.R. (2000) The *Arabidopsis* Cdc2a-interacting protein ICK2 is structurally related to ICK1 and is a potent inhibitor of cyclin-dependent kinase activity *in vitro*. *Plant J* **21**, 379–385.

Marciniak, K. and Bilecka, A. (1985) Changes in nuclear, nucleolar and cytoplasmic RNA content during growth and differentiation of root parenchyma cells in plant species with different dynamics of DNA endoreplication. *Folia Histochem Cytobiol* **23**, 231–245.

Mariconti, L., Pellegrini, B., Cantoni, R., *et al.* (2002) The E2F family of transcription factors from *Arabidopsis thaliana*. Novel and conserved components of the retinoblastoma/E2F pathway in plants. *J Biol Chem* **277**, 9911–9919.

Melaragno, J.E., Mehrotra, B. and Coleman, A.W. (1993) Relationship between endopolyploidy and cell size in epidermal tissue of Arabidopsis. *Plant Cell* **5**, 1661–1668.

Mohamed, Y. and Bopp, M. (1980) Distribution of polyploidy in elongating and non-elongating shoot axis of *Pisum sativum*. *Z Pflanzenphysiol* **98**, 25–33.

Moreno, S. and Nurse, P. (1994) Regulation of progression through the G1 phase of the cell cycle by the *rum1*$^{+}$ gene. *Nature* **367**, 236–242.

Nagl, W. (1974) The *Phaseolus* suspensor and its polytene chromosomes. *Z Pflanzenphysiol* **73**, 1–44.

Nagl, W. (1976) DNA endoreduplication and polyteny understood as evolutionary strategies. *Science* **261**, 614–615.

Olszewska, M.J. (1976) Autoradiographic and ultrastructural study of *Cucurbita pepo* root cells during their growth and differentiation. *Histochemistry* **49**, 157–175.

Olszewska, M.J. and Kononowicz, A.K. (1979) Activities of DNA polymerases and RNA polymerases detected in situ in growing and differentiating cells of root cortex. *Histochemistry* **59**, 311–323.

Park, J.-A., Ahn, J.-W., Kim, Y.-K., *et al.* (2005) Retinoblastoma protein regulates cell proliferation, differentiation, and endoreduplication in plants. *Plant J* **42**, 153–163.

Parry, G. and Estelle, M. (2006) Auxin receptors: a new role for F-box proteins. *Curr Opin Cell Biol* **18**, 152–156.

Peters, J.-M. (2002) The anaphase-promoting complex: proteolysis in mitosis and beyond. *Mol Cell* **9**, 931–943.

Reuber, S., Bornman, J.F. and Weissenböck, G. (1996) A flavonoid mutant of barley (*Hordeum vulgare* L.) exhibits increased sensitivity to UV-B radiation in the primary leaf. *Plant Cell Environ* **19**, 593–601.

Riou-Khamlichi, C., Huntley, R., Jacqmard, A. and Murray, J.A.H. (1999) Cytokinin activation of *Arabidopsis* cell division through a D-type cyclin. *Science* **283**, 1541–1544.

Rossignol, P., Stevens, R., Perennes, C., *et al.* (2002) AtE2F-a and AtDP-a, members of the E2F family of transcription factors, induce *Arabidopsis* leaf cells to re-enter S phase. *Mol Genet Genomics* **266**, 995–1003.

Sabelli, P.A., Dante, R.A., Leiva-Neto, J.T., Jung, R., Gordon-Kamm, W.J. and Larkins, B.A. (2005) RBR3, a member of the retinoblastoma-related family from maize, is regulated by the RBR1/E2F pathway. *Proc Natl Acad Sci USA* **102**, 13005–13012.

Sandritter, W. and Scomazzoni, G. (1964) Deoxyribonucleic acid content (Feulgen photometry) and dry weight (interference microscopy) of normal and hypertrophic heart muscle fibers. *Nature* **202**, 100–101.

Schnittger, A., Schöbinger, U., Bouyer, D., Weinl, C., Stierhof, Y.-D. and Hülskamp, M. (2002) Ectopic D-type cyclin expression induces not only DNA replication but also cell division in *Arabidopsis* trichomes. *Proc Natl Acad Sci USA* **99**, 6410–6415.

Schnittger, A., Weinl, C., Bouyer, D., Schöbinger, U. and Hülskamp, M. (2003) Misexpression of the cyclin-dependent kinase inhibitor *ICK1/KRP1* in single-celled Arabidopsis trichomes reduces endoreduplication and cell size and induces cell death. *Plant Cell* **15**, 303–315.

Schultz, T.F., Spiker, S. and Quatrano, R.S. (1996) Histone H1 enhances the DNA binding activity of the transcription factor EmBP-1. *J Biol Chem* **271**, 25742–25745.

Setter, T.L. and Flannigan, B.A. (2001) Water deficit inhibits cell division and expression of transcripts involved in cell proliferation and endoreduplication in maize endosperm. *J Exp Bot* **52**, 1401–1408.

Shcherbata, H.R., Althauser, C., Findley, S.D. and Ruohola-Baker, H. (2004) The mitotic-to-endocycle switch in *Drosophila* follicle cells is executed by Notch-dependent regulation of G1/S, G2/M and M/G1 cell-cycle transitions. *Development* **131**, 3169–3181.

Sherr, C.J. (1994) G1 phase progression: cycling on cue. *Cell* **79**, 551–555.

Smulders, M.J.M., Rus-Kortekaas, W. and Gilissen, L.J.W. (1994) Development of polysomaty during differentiation in diploid and tetraploid tomato (*Lycopersicon esculentum*) plants. *Plant Sci* **97**, 53–60.

Soares, M.J., Müller, H., Orwig, K.E., Peters, T.J. and Dai, G. (1998) The uteroplacental prolactin family and pregnancy. *Biol Reprod* **58**, 273–284.

Sorrell, D.A., Marchbank, A., McMahon, K., Dickinson, J.R., Rogers, H.J. and Francis, D. (2002) A *WEE1* homologue from *Arabidopsis thaliana*. *Planta* **215**, 518–522.

Sozzani, R., Maggio, C., Varotto, S., *et al.* (2006) Interplay between Arabidopsis activating factors E2Fb and E2Fa in cell cycle progression and development. *Plant Physiol* **140**, 1355–1366.

Spradling, A.C. (1993) Developmental genetics of oogenesis. In: *The Development of Drosophila melanogaster* (eds Bate, M. and Martínez-Arias, A.). Cold Spring Harbor Press, Cold Spring Harbor, NY, pp. 1–70.

Stals, H. and Inzé, D. (2001) When plant cells decide to divide. *Trends Plant Sci* **6**, 359–364.

Sugimoto-Shirasu, K., Roberts, G.R., Stacey, N.J., McCann, M.C., Maxwell, A. and Roberts, K. (2005) RHL1 is an essential component of the plant DNA topoisomerase VI complex and is required for ploidy-dependent cell growth. *Proc Natl Acad Sci USA* **102**, 18736–18741.

Sugimoto-Shirasu, K., Stacey, N.J., Corsar, J., Roberts, K. and McCann, M.C. (2002) DNA topoisomerase VI is essential for endoreduplication in *Arabidopsis*. *Curr Biol* **12**, 1782–1786.

Sun, Y., Dilkes, B.P., Zhang, C., *et al.* (1999) Characterization of maize (*Zea mays* L.) Wee1 and its activity in developing endosperm. *Proc Natl Acad Sci USA* **96**, 4180–4185.

Tarayre, S., Vinardell, J.M., Cebolla, A., Kondorosi, A. and Kondorosi, E. (2004) Two classes of the Cdh1-type activators of the anaphase-promoting complex in plants: novel functional domains and distinct regulation. *Plant Cell* **16**, 422–434.

Traas, J., Hülskamp, M., Gendreau, E. and Höfte, H. (1998) Endoreduplication and development: rule without dividing? *Curr Opin Plant Biol* **1**, 498–503.

Uijtewaal, B.A. (1987) Ploidy variability in greenhouse cultured and in vitro propagated potato (*Solanum tuberosum*) monohaploids (2n=x=12) as determined by flow cytometry. *Plant Cell Rep* **6**, 252–255.

Valente, P., Tao, W. and Verbelen, J.-P. (1998) Auxins and cytokinins control DNA endoreduplication and deduplication in single cells of tobacco. *Plant Sci* **134**, 207–215.

Vandepoele, K., Raes, J., De Veylder, L., Rouzé, P., Rombauts, S. and Inzé, D. (2002) Genome-wide analysis of core cell cycle genes in Arabidopsis. *Plant Cell* **14**, 903–916.

Van Oostveldt, P. and Van Parijs, R. (1975) Effect of light on nucleic-acid synthesis and polyploidy level in elongating epicotyl cells of *Pisum sativum*. *Planta* **124**, 287–295.

Verkest, A., de O. Manes, C.-L., Maes, S., *et al.* (2005) The cyclin-dependent kinase inhibitor KRP2 controls the mitosis-to-endocycle transition during Arabidopsis leaf development through a specific inhibition of the mitotic CDKA;1 kinase complexes. *Plant Cell* **17**, 1723–1736.

Vinardell, J.M., Fedorova, E., Cebolla, A., *et al.* (2003) Endoreduplication mediated by the anaphase-promoting complex activator CCS52A is required for symbiotic cell differentiation in *Medicago truncatula* nodules. *Plant Cell* **15**, 2093–2105.

Vlieghe, K., Boudolf, V., Beemster, G.T.S., *et al.* (2005) The DP-E2F-like *DEL1* gene controls the endocycle in *Arabidopsis thaliana*. *Curr Biol* **15**, 59–63.

Wang, H., Fowke, L.C. and Crosby, W.L. (1997) A plant cyclin-dependent kinase inhibitor gene. *Nature* **386**, 451–452.

Wang, H., Zhou, Y., Gilmer, S., Whitwill, S. and Fowke, L.C. (2000) Expression of the plant cyclin-dependent kinase inhibitor ICK1 affects cell division, plant growth and morphology. *Plant J* **24**, 613–623.

Weinl, C., Marquardt, S., Kuijt, S.J.H., *et al.* (2005) Novel functions of plant cyclin-dependent kinase inhibitors, ICK1/KRP1, can act non-cell-autonomously and inhibit entry into mitosis. *Plant Cell* **17**, 1704–1722.

Wuarin, J., Buck, V., Nurse, P. and Millar, J.B.A. (2002) Stable association of mitotic cyclin B/Cdc2 to replication origins prevents endoreduplication. *Cell* **111**, 419–431.

Yin, Y., Cheong, H., Friedrichsen, D., *et al.* (2002) A crucial role for the putative *Arabidopsis* topoisomerase VI in plant growth and development. *Proc Natl Acad Sci USA* **99**, 10191–10196.

Yu, Y., Steinmetz, A., Meyer, D., Brown, S. and Shen, W.-H. (2003) The tobacco A-type cyclin, *Nicta;CYCA3;2*, at the nexus of cell division and differentiation. *Plant Cell* **15**, 2763–2777.

Zachariae, W., Schwab, M., Nasmyth, K. and Seufert, W. (1998) Control of cyclin ubiquitination by CDK-regulated binding of Hct1 to the anaphase promoting complex. *Science* **282**, 1721–1724.

Zhang, K., Diederich, L. and John, P.C.I. (2005) The cytokinin requirement for cell division in cultured *Nicotiana plumbaginifolia* cells can be satisfied by yeast Cdc25 protein tyrosine phosphatase. Implications for mechanisms of cytokinin response and plant development. *Plant Physiol* **137**, 308–316.

Zhang, K., Letham, D.S. and John, P.C.L. (1996a) Cytokinin controls the cell cycle at mitosis by stimulating the tyrosine dephosphorylation and activation of $p34^{cdc2}$-like H1 histone kinase. *Planta* **200**, 2–12.

Zhang, Y., Wang, Z. and Ravid, K. (1996b) The cell cycle in polyploid megakaryocytes is associated with reduced activity of cyclin B1-dependent Cdc2 kinase. *J Biol Chem* **271**, 4266–4272.

Zhao, J. and Grafi, G. (2000) The high mobility group I/Y protein is hypophosphorylated in endoreduplicating maize endosperm cells and is involved in alleviating histone H1-mediated transcriptional repression. *J Biol Chem* **275**, 27494–27499.

Zhou, Y., Ching, Y.-P., Chun, A.C.S. and Jin, D.-Y. (2003) Nuclear localization of the cell cycle regulator CDH1 and its regulation by phosphorylation. *J Biol Chem* **278**, 12530–12536.

Zhou, Y., Fowke, L.C. and Wang, H. (2002) Plant CDK inhibitors: studies of interactions with cell cycle regulators in the yeast two-hybrid system and functional comparisons in transgenic *Arabidopsis* plants. *Plant Cell Rep* **20**, 967–975.

11 Insights into the endocycle from trichome development

John C. Larkin, Matthew L. Brown and Michelle L. Churchman

11.1 Introduction

The trichomes of *Arabidopsis thaliana* are large branched single-celled hairs that extend from the epidermis of various shoot organs (Figure 11.1A). Their ready accessibility on the surface of the plant, combined with the ease of genetic analysis in Arabidopsis, has made trichome differentiation one of the most thoroughly studied models of plant cell differentiation. A great deal is known about the molecular genetics of the trichome cell fate decision, and the study of trichome development has given insights into the role of the cytoskeleton in cell expansion as well (Larkin *et al.*, 2003; Szymanski, 2005). During differentiation, developing trichomes replicate their nuclear DNA without concomitant mitosis or cytokinesis, a variant of the cell cycle called endoreduplication or endoreplication, reviewed by Vlieghe and colleagues elsewhere in this volume (see Chapter 10). Trichome nuclei typically undergo three to four rounds of endoreduplication, resulting in a final DNA content of approximately 32C in mature trichomes, and a number of mutants affecting endoreduplication in trichomes have been isolated (Hülskamp *et al.*, 1994). Because of our detailed understanding of many aspects of trichome differentiation, trichomes are an excellent model for studying the coordination of cell differentiation and the cell cycle, particularly with regard to the initiation of endoreduplication.

Both the mitotic cell cycle and the endoreduplication cell cycle (endocycle) of plants are reviewed in detail elsewhere in this volume; a good overall review is that of Dewitte and Murray (2003). Briefly, cell cycle transitions in plants, as in other organisms, are driven by a class of serine–threonine protein kinases called cyclin-dependent kinases (CDKs). The kinase activity of these CDKs is triggered by binding to specific cyclin (CYC) partners; expressions of many of the cyclin proteins are expressed in a cell cycle stage-specific manner, and thus serve to regulate the activity of the various CYC/CDK complexes. There are three major classes of cyclins involved in cell cycle regulation, A-type cyclins (CYCA), B-type cyclins (CYCB), and D-type cyclins (CYCD). CYCD/CDK complexes control the G1/S transition by inhibiting the retinoblastoma-related protein (RBR), resulting in the transcription of genes involved in DNA replication. CYCB/CDK complexes control the G2/M transition. The role of CYCA/CDK complexes is less clear; they appear to play various roles in maintaining S phase and in the G2/M transition.

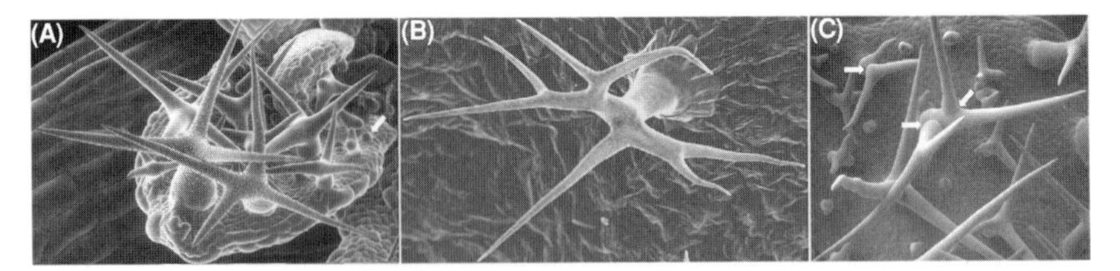

Figure 11.1 Trichome phenotypes and endoreduplication. (A) Scanning electron micrograph showing wild-type trichome development. Note that wild-type leaf trichomes have three or four branches and are unicellular. A newly initiated trichome is indicated by the arrow. (B) Scanning electron micrograph of a trichome of a plant overexpressing *GL3* (M. Brown and J.C. Larkin, unpublished observations). Trichomes of these plants have an increased nuclear DNA content. Note the increased branching typical of trichomes with a greater than wild-type DNA content relative to the wild-type trichomes in (A). (C) Scanning electron micrograph of *sim-1* mutant trichomes. Note cell junctions of multicellular trichomes (arrows).

The above summary paints the main functions of the cyclins with a broad brush, while in reality, this standard view is mostly based on generalization from experiments performed with one or two family members. Much evidence suggests that individual CYC/CDK complexes have diverse roles. Plants have larger gene families than other eukaryotes for most cell cycle regulators, suggesting that the plant cell cycle may have greater complexity than that of other organisms, perhaps because altered growth patterns represent one of the few options for sessile plants to respond to environmental changes. The specific role of many of these individual cyclins remains to be tested. If the plant cell cycle has evolved extra flexibility through functional diversity, then some individual CYC/CDK complexes may turn out to have specific developmental functions.

It is obvious that development and the cell cycle must be closely coordinated; cell division is usually coordinated with growth, and in most cases differentiating cells no longer divide. However, closer inspection of developmental patterns reveals a myriad of modified cell cycles tailored to meet the needs of particular developmental contexts. For example, in many organisms the egg is large, and after fertilization undergoes rapid divisions with minimal G1 or G2 phases, with little or no growth for the first few cell cycles. Early Xenopus development is such a case, where 12 rounds of division occur without gap phases or zygotic transcription, indicating that the necessary cell cycle components are stored as protein or mRNA (Newport and Kirschner, 1982). The end result is the conversion of a large zygote into many small cells that will begin the process of developmental specialization. Another example is found in the rapid divisions that constitute the Drosophila bristle cell lineage, where the first two daughter cells resulting from the initial division both enter S phase during mitotic telophase, indicating that they have skipped G1 entirely (Audibert *et al.*, 2005). Stem cells represent another example of the control of the cell cycle by developmental context (Watt and Hogan, 2000). Typically, stem cells divide infrequently and remain undifferentiated with the potential for unlimited division, while producing a population of daughter cells, termed *transient amplifying cells*, that divide rapidly a few times before differentiating. The rapid divisions of the transient amplifying cells serve to amplify the population of differentiated cells ultimately produced. The divisions that produce transient amplifying cells are thus asymmetric with respect to cell division rate, producing one slowly dividing stem cell daughter, and one rapidly dividing daughter that is fated to differentiate after a limited number of divisions.

Endoreduplication is another variant of the standard cell cycle that often occurs in terminally differentiating cells. Mechanistically, endocycles appear to require oscillations of G1/S CDK activity (i.e. CYCD/CDK complexes) to allow relicensing of replication origins between each round of DNA replication, coupled with inhibition of mitosis due to the absence of G2/M phase CDK activity (i.e. CYCB/CDK complexes) (Edgar and Orr-Weaver, 2001; Larkins *et al.*, 2001). Thus, endoreduplication in plants typically requires CYCD/CDK complexes, but not CYCB/CDK complexes. The nuclear DNA content of endoreduplicated cells is often correlated with cell size, and endoreduplication is commonly associated with very large or metabolically active cells. Although the function of endoreduplication is not clearly understood, it is generally assumed that the extra genome copies are necessary to support the greater volume of cytoplasm in larger cells, or that the suppression of mitosis, in combination with the coupling of the cell cycle to cell growth, provides a developmental route to the production of large cells when they are favored by natural selection (Edgar and Orr-Weaver, 2001; Larkins *et al.*, 2001). While the canonical cell cycle produces daughter cells of a constant size with each round of the cycle, repeated endocycles result in a single very large cell. In this sense, endocycling as a developmental strategy is the developmental inverse of the common embryonic strategy of rapid divisions without growth that convert the zygote to many smaller cells. The well-studied trichome system provides an ideal framework for investigating the integration of endoreduplication with development.

11.2 The regulation and cell cycle context of trichome development

The first detectable sign of trichome initiation is an enlarged nucleus, indicative of the onset of endoreduplication (Hülskamp *et al.*, 1994). The newly initiated trichome cell (Figure 11.1A, arrow) expands out of the plane of the leaf surface and produces three or four branches (Figure 11.1A). During this period, the cells undergo about four rounds of endoreduplication, resulting in a final average nuclear DNA content of 16–32C, the highest level of endoreduplication in leaf tissue, although individual trichomes show a wide range of DNA contents (Melaragno *et al.*, 1993; Hülskamp *et al.*, 1994; Mauricio and Rausher, 1997). After a phase of rapid expansion, trichomes develop a thick papillate secondary cell wall. At maturity, this heavily armored cell wall, along with the pointed ends of the branches, is thought to provide mechanical protection against insects (Mauricio and Rausher, 1997), although other possible roles include scattering of UV light, insulation, and control of evaporation (Johnson, 1975; Larkin *et al.*, 1996).

Trichomes are the first differentiated cell type to develop within the adaxial epidermis of *Arabidopsis* leaves (Figure 11.1A), and this places their development in an unusual developmental context relative to the cell cycle. While cell differentiation is typically pictured as a choice between alternative differentiated states, in the case of trichome differentiation, cells choose between either differentiating into a nondividing trichome or continuing to proliferate. The adaxial epidermis of the *Arabidopsis* first leaf is descended from approximately six founder cells that generate the mature epidermis via approximately 11 rounds of division (Larkin *et al.*, 1996). Trichome development is first detectable after the fourth epidermal cell cycle, when the adaxial epidermis consists of about 100 cells, beginning near the tip of the leaf and proceeding basipetally. Initiation of trichomes continues for approximately the next four cell cycles.

Ultimately, the appearance of asymmetrically dividing meristemoid mother cells heralds the onset of stomatal development. These meristemoid mother cells give rise to a population of transient amplifying cells that ultimately produce both stomata and the bulk of the ordinary epidermal pavement cells through a highly regulated series of divisions (Larkin *et al.*, 2003). These divisions occupy approximately the last three epidermal cell cycles (Larkin *et al.*, 1996), by which time the initiation of new trichomes has largely ceased, although the degree of overlap between the onset of stomatal development and the end of the trichome development phase remains to be explored. Epidermal pavement cells themselves have endoreduplicated nuclei with a DNA content of approximately 4–8C, while stomatal guard cells have a 2C DNA content (Melaragno *et al.*, 1993).

The molecular genetics of trichome development has been extensively reviewed (Larkin *et al.*, 2003; Schellmann and Hülskamp, 2005). Trichome initiation is promoted by a transcription factor complex containing an R2R3 MYB protein encoded by the *GLABRA1* (*GL1*) gene and a bHLH protein encoded by the *GLABRA3* (*GL3*) gene, in conjunction with the WD-repeat protein encoded by the *TRANSPARENT TESTA GLABRA* (*TTG*) gene. Mutations in any of these genes severely compromise the formation of trichomes, and the protein products of these three genes have been shown to interact. Only a small fraction of leaf epidermal cells become trichomes, and these cells are nonrandomly spaced; trichomes occur adjacent to one another much less frequently than would be expected by chance (Larkin *et al.*, 1996). This spacing pattern is maintained by the action of a small family of R3MYB proteins that lack a transcriptional activation domain, the most well studied of which is *TRYPTCHON* (*TRY*) (Schellmann *et al.*, 2002). The TRY protein is thought to move between cells via plasmodesmata to inhibit trichome development in cells adjacent to a developing trichome in a process called lateral inhibition (Larkin *et al.*, 2003). Some of the best evidence that TRY may act as a diffusible inhibitor comes from recent work on CPC, a homolog of TRY that is involved in patterning the root epidermis (Wada *et al.*, 2002; Kurata *et al.*, 2005). TRY has been shown to compete with GL1for binding to GL3 (Esch *et al.*, 2003), and is thought to block trichome initiation by preventing formation or function of the GL1/GL3 transcription complex. The existence of both an activating complex and a diffusible inhibitor for trichome development provides the basic components of a classical lateral inhibition patterning mechanism in which the interplay of local activation and inhibition at a distance spontaneously generates a stable spacing pattern (Turing, 1952), similar to patterning mechanisms found in a wide variety of organisms (Larkin *et al.*, 2003).

Trichome development is also under the control of hormonal and environmental signals, and is positively regulated both by long days and by gibberellins (GAs) (Chien and Sussex, 1996; Telfer *et al.*, 1997). Mutants lacking GA cease production of trichomes when maternal GA is exhausted, and exogenous GA application can restore trichome development. GA was shown to regulate *GL1* expression levels, and constitutive expression of *GL1* in conjunction with constitutive expression of a maize *GL3* homolog was shown to bypass the GA requirement for trichome development (Perazza *et al.*, 1998). These results suggest that GA regulates trichome development by regulating expression of components of the trichome development transcription complex. GA regulation of trichome development may be relevant to endoreduplication of trichome nuclei, because GA has been shown to regulate endoreduplication during hypocotyl development (Gendreau *et al.*, 1999). Plant hormones involved in plant defense also influence trichome density; jasmonic acid appears to act synergistically with GA to increase trichome density in *Arabidopsis*, while salicylic acid application reduces trichome density (Traw and Bergelson, 2003).

11.3 Regulation of endoreduplication during trichome development

11.3.1 Control of trichome endoreduplication by developmental regulators

Classical genetic screens have revealed several mutants that affect trichome endoreduplication levels. Some of these contain mutations in genes involved in trichome development, and are likely to act upstream of the cell cycle itself to control the degree of endoreduplication (Table 11.1). There is often, though not always, a correlation between endoreduplication level and cell size in Arabidopsis, and mutant trichomes with increased nuclear DNA content virtually always are larger than the wild type and have extra branches (Figure 11.1B), while trichomes with reduced DNA content usually are smaller with reduced branching. Thus, mutations affecting the level of endoreduplication in trichomes are relatively easy to identify. Several of these mutations are in genes encoding the transcription factors controlling the initiation of trichome development. Loss-of-function *gl3* mutants have a decreased trichome nuclear DNA content and decreased branching (Hülskamp *et al.*, 1994). Conversely, the *gl3-sst* allele, a novel gain-of-function allele, results in trichomes with large nuclei that are reported to have increased DNA content (Esch *et al.*, 2003), and overexpression of the cloned *GL3* gene results in an increased nuclear DNA content as well as increased size and branching of the trichome (Payne *et al.*, 2000; M. Brown and J. Larkin, unpublished observations; Figure 11.1B). Loss-of-function *try* mutants have increased trichome size and branching, and increased trichome nuclear DNA content (Schellmann *et al.*, 2002). Although the main role of the TRY protein is to inhibit neighboring cells from developing as trichomes, the gene is actually transcribed in developing trichomes, and the observation that *try* mutants affect trichome size, branching and DNA content in addition to disrupting the distribution of trichomes on the leaf indicates that TRY acts to inhibit the

Table 11.1 Genes affecting the cell cycle during trichome development

Gene	Function	Manipulation	Cell cycle	Reference
Developmental regulators of endoreduplication in trichomes				
GL3	BHLH TF	LOF/GOF	Reduced/ increased DNA	1,2
TRY	Myb TF	LOF	Increased DNA	1
SPY	GA signaling	LOF	Increased DNA	3
KAK	HECT E3-ligase	LOF	Increased DNA	3
Genes affecting trichome endoreduplication levels whose roles are unclear				
CPR5	Unknown	LOF	Decreased DNA	4
AtFIP37	*p35S* OE	Increased DNA	5	
Immunophilin				
RFI	Not cloned, *GL3*?	GOF?	Increased DNA	3
PYM	Not cloned	Unknown	Increased DNA	3
Regulators of the G1/S transition and S-phase components				
RBR	Rb-related	RepA inhibition	Increased DNA	6
E2Fa/Dpa	TF	*p35S* OE	Increased DNA	7
CDC6	Replic. Licensing	*p35S* OE	Increased DNA	8
CDT1	Replic. Licensing	*p35S* OE	Increased DNA	8
RHL1	Topo IV	LOF	Decreased DNA	9
RHL2	Topo IV	LOF	Decreased DNA	10
HYP6	Topo IV	LOF	Decreased DNA	11
Negative cell cycle regulators of trichome endoreduplication levels				
ICK1/KRP1	CDK inhibitor	*pGL2* OE	Decreased DNA,	12
				Cell death
CYCA2;3	A-cyclin	LOF, OE	Increased/ decreased DNA	13
Genes affecting division potential of developing trichomes				
SIM	Unknown	LOF	Trich. cell division	14
CYCB1;2	B-cyclin	*pGL2* OE	Trich. cell division	15
CYCD3;1	D-cyclin	*pGL2* OE	Trich. cell division	16

Abbreviations: Trich., trichome; TF, transcription factor; LOF, loss of function; GOF, gain of function; Topo IV, topoisomerase IV; *p35S* OE, overexpression from the *CAMV 35S* promoter; Replic., replication; *pGL2* OE, overexpression from the *GLABRA2* promoter. *References*: 1, Hülskamp *et al.*, 1994; 2, Esch *et al.*, 2003; 3, Perazza *et al.*, 1999; 4, Kirik *et al.*, 2001; 5, Vespa *et al.*, 2004; 6, Desvoyes *et al.*, 2006; 7, De Veylder *et al.*, 2002; 8, Castellano *et al.*, 2004; 9, Sugimoto-Shirasu *et al.*, 2005; 10, Hartung *et al.*, 2002; 11, Sugimoto-Shirasu *et al.*, 2002; 12, Schnittger *et al.*, 2003; 13, Imai *et al.*, 2006; 14, Walker *et al.*, 2000; 15, Schnittger *et al.*, 2002a; 16, Schnittger *et al.*, 2002b.

trichome development to some extent even within the developing trichomes, consistent with its site of transcription. *GL3* and *TRY* encode transcription factors involved in initiating the trichome developmental program, and one of the first events in this program is the switch from the mitotic to the endocycle. It is likely that the products of these genes act upstream of the cell cycle proper, and regulate transcription of cell cycle components involved in the switch to endocycling.

Mutations in *SPINDLY* (*SPY*) define another gene that is likely to act upstream of the cell cycle to limit the degree of endoreduplication. Loss-of-function *spy* mutants have large, extra-branched trichomes with an increased level of nuclear DNA (Perazza *et al.*, 1999), and the gene encodes a putative O-linked *N*-acetylglucosamine transferase that plays an inhibitory role in GA signal transduction, presumably by posttranslationally modifying an unknown target protein (Swain *et al.*, 2002). Given the role of GA in trichome development outlined above, it is possible that the effect of *SPY* on endoreduplication is mediated by *GL3* and *TRY*. Another gene with a similar mutant phenotype of increased trichome size, branching and endoreduplication is *KAKTUS* (*KAK*) (Perazza *et al.*, 1999). The *KAK* gene encodes a HECK-class ubiquitin E3-ligase (Downes *et al.*, 2003; El Refy *et al.*, 2003). Mutations in *kak* also affect endoreduplication in other cell types, such as those in hypocotyl and cotyledons. Although the role of *KAK* is unclear, some circumstantial evidence suggests that the gene product may regulate aspects of GA signaling related to endoreduplication. Light-grown *kak* mutants have longer hypocotyls than wild-type controls, are hypersensitive to GA-induced hypocotyl elongation, and have increased endoreduplication levels in hypocotyls (Downes *et al.*, 2003; El Refy *et al.*, 2003). Because endoreduplication in hypocotyls is known to be affected by GA and light, these results suggest that *KAK* modulates a GA signaling pathway leading to endoreduplication in trichomes and hypocotyl.

Several additional genes affect the endoreduplication level when mutant or overexpressed, and are likely to act upstream of the cell cycle proper, their role in the process is not understood (Table 11.1). Mutations in the *CONSTITUTIVE PATHOGEN RESPONSE5* (*CPR5*) gene were originally identified based on their constitutive expression of pathogen response genes, but *cpr5* mutations also result in small plants with reduced cell number and reduced endoreduplication (Kirik *et al.*, 2001). The *cpr5* trichomes are small, reduced in branching, and have a reduced DNA content. The gene encodes a protein of unknown function with putative transmembrane domains. CPR5 has variously been suggested to function as a component of pathogen response signaling (Bowling *et al.*, 1997), as a regulator of cell proliferation and cell death (Kirik *et al.*, 2001), and as a repressor of senescence responses (Yoshida *et al.*, 2002); the role of the CPR5 protein in endoreduplication remains unknown. This extreme pleiotropy and suggested role in signaling pathways is consistent with an upstream regulatory role.

Another protein that affects endoreduplication levels in trichomes in a poorly understood fashion is AtFIP37, an immunophilin-interacting protein (Vespa *et al.*, 2004). Overexpression of *AtFIP37* results in large, extra-branched trichomes that have increased levels of endoreduplication. In animals, immunophilins, which are a family of proteins having peptidylprolyl *cis–trans* isomerase activity, can act in signaling pathways upstream of the cell cycle. ATFIP37 interacts with the plant FKBP12 immunophilin, and it has some similarity to mammalian proteins involved in mRNA splicing, but its role in endoreduplication remains unclear.

Mutations in two additional genes, *POLYCOME* (*PYM*) and *RASTAFARI* (*RFI*) result in large, extra-branched trichomes that have increased levels of endoreduplication (Perazza *et al.*, 1999). Neither of these genes has been isolated to date, so their functions remain unknown. However, *rfi* maps very close to the position of *gl3* and its phenotype resembles the *gl3* GOF phenotype, and thus it is possible that *rfi* is a *gl3* GOF mutation. Both of these genes show genetic interactions with regulators of trichome development, and may act in upstream developmental pathways, but their exact role is unknown.

11.3.2 Regulators of the G1/S transition and S-phase progression affect endoreduplication levels in trichomes

Investigation of known cell cycle genes in trichomes has also contributed to our understanding of endoreduplication in trichomes, and as might be expected, regulators of the G1/S transition play a key role in this process (Table 11.1). Co-overexpression of *E2Fa* and *DPa*, which encode two key transcription factors that function as a heterodimer under the control of RBR to regulate entry into S phase, increases the DNA content of endoreduplicating cells, including trichomes (De Veylder *et al.*, 2002). In contrast, in mitotically dividing tissues the number of cells increases, suggesting that increased levels of the E2Fa–DPa heterodimer promote the type of cell cycle that has already been programmed by other factors (De Veylder *et al.*, 2002). An alternative, but not mutually exclusive, possible role for E2F/DP heterodimers in endoreduplication is suggested by work in Drosophila, where loss of active repression functions of dE2F2 results in greatly reduced endoreduplication due to failure to repress cyclin E during gap phases in endoreduplicating cells, while mitotically dividing tissues remain unaffected (Weng *et al.*, 2003). It is possible that a repressive function of E2Fa–DPa heterodimers plays a role in endoreduplication in Arabidopsis as well, although this hypothesis has not been tested.

Co-overexpression of *E2Fa* and *DPa* from the *CAMV35S* promoter also promotes increased expression of many factors associated with DNA replication, including DNA polymerase α, *ORC*, *MCM* and *CDC6* (De Veylder *et al.*, 2002). It is thus not surprising that overexpression of *CDC6* or *CDT1* results in a similar phenotype: increased division in cell lineages that divide mitotically, and increased endoreduplication in lineages that normally endoreduplicate, including trichomes (Castellano *et al.*, 2004). In this work, it was shown that in addition to having an increased DNA content, the trichomes also had increased branching, typical of the effect of increased endoreduplication on trichomes. The observation that overexpression of E2Fa/DPa target genes can lead to increased endoreduplication indicates that in Arabidopsis, E2Fa and DPa can regulate endoreduplication at least in part by their role as positive regulators of S-phase genes.

This conclusion is supported by the recent observation that disruption of the function of the Arabidopsis retinoblastoma homolog RBR by inducible expression of the viral RepA protein results in increased endoreduplication and branching of trichomes, increased endoreplication throughout the leaf, and increased proliferation of cells that have not entered the endoreduplication pathway (Desvoyes *et al.*, 2006). E2F/DP targets such as *CDC6*, *CDC* and *ORC* subunits are activated in these plants. Thus, inactivation of RBR appears to function via activating E2F/DP heterodimers and activates E2F/DP target genes. As a result, cells that are already endoreduplicating, including trichomes, exhibit increased levels of endoreduplication, while cells that are still in the mitotic cycle undergo more divisions. Taken together, these observations indicate that regulation of the RBR/E2F/DP switch at the G1/S transition controls the degree of mitotic division or endoreduplication, but as proposed by De Veylder and coworkers (De Veylder *et al.*, 2002), the switch between these two alternate versions of the cell cycle must lie elsewhere.

One class of functionally defined mutants that reduce the endoreduplication level and have a role related to DNA replication affects genes encoding subunits of topoisomerase IV. Thus, LOF mutations in the topoisomerase IV subunit genes *ROOTHAIRLESS2* (*RHL2*), *HYPOCOTYL6* (*HYP6*) and *ROOTHAIRLESS1* (*RHL1*) result in dwarf plants with small trichomes that have reduced endoreduplication levels throughout the plant, including in trichomes (Hartung *et al.*, 2002; Sugimoto-Shirasu *et al.*, 2002; Yin *et al.*, 2002). Topoisomerase IV is apparently dispensable for DNA replication in the mitotic cell cycle, which presumably requires a different topoisomerase. Topoisomerase IV has been hypothesized to be required for resolving entangled chromatids during endoreduplication (Sugimoto-Shirasu *et al.*, 2002, 2005).

11.3.3 Inhibitors of trichome endoreduplication levels

In addition to RBR, several other proteins have been implicated as negative regulators of endoreduplication levels in trichomes (Table 11.1). One class of such inhibitors is the ICK/KRP family of CDK inhibitors (Wang *et al.*, 1997; Lui *et al.*, 2000; De Veylder *et al.*, 2001). Overexpression of the CDK inhibitor *ICK1/KRP1* from the relatively trichome-specific *GL2* promoter (*pGL2*) reduced endoreduplication levels to approximately 8C, eliminating two rounds of endoreduplication, and the trichomes are reduced in size and have fewer branches (Schnittger *et al.*, 2003). Overexpression of *ICK1/KRP1*^{109}, a deleted form of the gene lacking an N-terminal-encoded negative regulatory domain, increases the severity of these phenotypes. Trichomes on *ICK1/KRP1*^{109} plants also undergo cell death shortly after maturation, as indicated by degeneration of the heterochromatin chromocenters, disappearance of nucleoli and vital dye staining. Another interesting observation is that while trichomes of *pGL2:ICK1/KRP1* plants have undergone one less round of endoreduplication than trichomes of *gl3* mutant plants, the trichomes of *pGL2:ICK1/KRP1* plants are larger than the small trichomes of *gl3* mutants. This demonstrates that nuclear DNA content and cell size are not strictly coupled in Arabidopsis trichomes, and suggests that the regulation of cell size by *GL3* may not be based solely on control of endoreduplication. One reasonable explanation for this difference is that *GL3* regulates endoreduplication in trichomes by controlling the expression of appropriate cell cycle regulators, and at the same time independently regulates the expression of various genes involved more directly in the control of cell expansion that are essential for the extremely rapid cell growth exhibited by trichomes during early development.

This study has also given some insight into the *in vivo* interaction partners of ICK1/KRP1that are involved in this inhibition of endoreduplication. Previous work indicated that ICK1/KRP1 interacts in a yeast two-hybrid assay with CDKA;1and CYCD3;1 (Wang *et al.*, 1998). Expression of *CDKA;1* and *CYCD3;1* from the *pGL2* promoter could rescue the reduced-branching phenotype of *pGL2:ICK1/KRP1* plants, but expression of *CDKB;1, CYCB1;2* or *CYCD4;1* could not rescue the phenotype (Schnittger *et al.*, 2003). These results are consistent with the yeast two-hybrid results, and indicate that ICK1/KRP1 is likely to regulate endoreduplication levels by interacting with CDKA;1 and specific D-cyclins, but not by interaction with CDKB or mitotic B-cyclins.

The trichome-specific overexpression of *ICK1/KRP1* has given another unexpected insight about the factors controlling endoreduplication. It was observed that in these plants the size of the epidermal cells immediately neighboring the trichomes increased by more than tenfold, as estimated from their surface area (Weinl *et al.*, 2005). These cells also exhibit greatly increased endoreduplication, indicating that while overexpression of *ICK1/KRP1* in developing trichomes results in reduced endoreduplication, endoreduplication in the cells neighboring an overexpressing trichome is increased. In wild-type plants, the trichome-neighboring cells are differentiated into a specialized 'socket cell' phenotype. The cells are more rectangular in shape and their long axes are radially arranged around the trichome that they neighbor. This socket cell fate develops late relative to trichome development, with the socket cell morphology not becoming distinct until the trichomes are almost mature. Further evidence for specialization of these cells comes from several enhancer-trap lines that show socket cell-specific reporter gene expression and thus act as markers of socket cell fate. These reporters are expressed in the enlarged socket cells surrounding *ICK1/KRP1* overexpressing trichomes, demonstrating that these enlarged cells still retain the socket cell fate, and do not have higher endoreduplication due to switching to a different, more highly endoreduplicated cell fate such as the trichome fate. Once the trichome dies, neighboring socket cells can reenter the mitotic cell cycle in spite of being endoreduplicated, as indicated by the expression of a *CYCB* reporter, and in at least some cases they completed cell division (Weinl *et al.*, 2005).

YFP:ICK1/KRP1 fusions expressed from the *pGL2* promoter move to neighboring socket cells, and a pGL2:ICK1/KRP1^{109}:YFP fusion can also trigger increased cell size and endoreduplication (Weinl *et al.*, 2005). Controls indicate that there is no expression of *pGL2* in socket cells of these plants, and thus it appears that movement of the *ICK1/KRP1* protein to neighboring cells triggers the increased endoreduplication. Trichome-neighboring cells contain lower levels of the nuclear-localized YFP fusion than do the trichome nuclei that express the transgene and are the source of the fusion protein. The authors suggest that the most likely explanation for the differential effect of *ICK1/KRP1* on cell cycle patterns in trichomes and socket cells is that different levels of the protein result in different cell cycle responses, with very high levels in trichomes causing cell cycle arrest, and lower levels diffusing into neighboring socket cells causing endoreduplication. These results are consistent with recent results showing that moderate levels of ICK2/KRP2 expression can trigger endoreduplication onset in mitotically dividing cells (Verkest *et al.*, 2005).

It should be noted that while *CDKA;1* appears to be expressed in trichomes, *CYCD3;1* is not expressed in trichomes (Schnittger *et al.*, 2002a), and *ICK1/KRP1* has never been shown to be expressed in trichomes. Also, the most dramatic results in these studies were obtained from overexpression of the *ICK1/KRP1*^{109} deletion allele, although results from overexpression of the wild-type *ICK1/KRP1* protein were generally similar. Thus, while these results give insight into how endoreduplication could be regulated, they do not demonstrate the regulatory mechanism that actually occurs in developing trichomes. However, there are multiple ICK/KRP proteins, and multiple D-type cyclins, that may be expressed in trichomes, and this work provides some of our strongest hints regarding the regulation of endoreduplication.

A particularly exciting recent finding is the demonstration that CYCA2;3 plays a role in endocycling during trichome development. Aoyama and colleagues noticed that a *CYCA2;3* promoter-*GUS* reporter gene fusion was expressed in trichomes toward the end of the branch initiation phase, but not in newly initiated or mature trichomes (Imai *et al.*, 2006). On the basis of this expression pattern, they hypothesized that CYC2;3might function in the termination of endoreduplication. This was confirmed by showing that insertion mutants in the gene causing loss of function resulted in increased DNA contents and branching in trichomes, and increased DNA contents in endoreduplicated leaf cells. Additionally, expression of a *CYCA2;3* gene with a mutation in the destruction box repressed endoreduplication, and CYCA2;3 was shown to interact with CDKA in an *in vivo* pull-down experiment. A particularly important point is that of the known cell cycle regulators affecting trichome endoreduplication levels; CYCA2;3 is the only one for which there is good evidence that it is expressed in trichomes in a functional way with respect to its role in the process. However, the evidence for transcription of the gene in developing trichomes is based solely on a reporter gene construct; demonstration of transcripts in trichomes by *in situ* hybridization would provide stronger evidence for the timing of expression relative to the end of endoreduplication.

11.3.4 Genes affecting division potential of developing trichomes

Wild-type trichomes are single cells that replicate their DNA approximately four times without dividing, while neighboring protodermal cells continue to divide mitotically (Figure 11.1A). The contrast between these two neighboring types of cells in the developing epidermis – mitotically dividing protodermal cells and endoreduplicating trichomes – emphasizes the fact that the transition to the endocycle requires suppression of mitosis. Both classical genetic approaches and overexpression of known cell cycle regulators have given insights into this aspect of trichome endoreduplication (Table 11.1). Mutations in the *SIAMESE* (*SIM*) gene have a much different phenotype than that of other genes involved in endoreduplication during trichome development; *sim* loss-of-function mutants produce multicellular trichomes as well as clusters of adjacent trichomes (Walker *et al.*, 2000; Figure 11.1C). These multicellular trichomes have nuclei in each cell, fully formed walls between cells, and are very similar to wild-type trichomes in morphology. These cell divisions occur early during trichome development, during the time when endoreduplication would normally be occurring. A surprising aspect revealed by *sim* mutations is that continued division is remarkably compatible with relatively normal differentiation of trichomes, although this is consistent with other work suggesting that in plants differentiation proceeds at least somewhat independently of division.

Several lines of evidence indicate that *SIM* encodes a repressor of mitosis needed to establish endoreduplication during trichome development. First, the individual nuclei of multicellular *sim* trichomes have a lower DNA content averaging approximately 8C, compared to the approximately 32C of wild-type trichome nuclei. Second, while *try* mutant trichomes have increased levels of nuclear DNA, corresponding to an extra round of endoreduplication, *sim try* double mutants have a nuclear DNA content similar to that of *sim* single mutant nuclei; the double mutants produce more cells than the *sim* single mutant. This result suggests that the additional endoreduplication cycle that occurs in the *try* mutant proceeded as a mitotic cycle in the absence of an inhibitor of mitosis encoded by *SIM*. Finally, while endoreduplication in many tissues such as leaf epidermal pavement cells and leaf mesophyll is not altered in *sim* mutants, *SIM* is required for one round of endoreduplication that occurs in dark-grown hypocotyls, demonstrating that its role is not exclusively limited to trichomes. The *SIM* gene has recently been isolated and found to encode a small nuclear protein of previously unknown function (Churchman *et al.*, 2006). Overexpression of *SIM* in leaves results in increased endoreduplication levels, particularly in epidermal cells, and preliminary evidence from *in vivo* fluorescence resonant energy transfer (FRET) experiments suggests that SIM may interact with CYCD/CDKA complexes. One published result consistent with *SIM* encoding an inhibitor of CYCD/CDK activity is the observation that expression of *ICK1/KRP1* from the *GL2* promoter can complement the multicellular trichome phenotype of *sim* mutants at moderate expression levels, and can partially restore endoreduplication to *sim* mutant trichomes (Weinl *et al.*, 2005). However, in this study high levels of *ICK1/KRP1* expression still result in cell death of *sim* trichomes, as it does in wild-type trichomes.

As in the case of control of endoreduplication levels, manipulation of the expression of known cell cycle regulators has also contributed to our understanding of the repression of mitosis during endoreduplication in trichomes. An obvious candidate for controlling entry into mitosis is CYCB, and B-cyclins are not expressed in developing trichomes (Schnittger *et al.*, 2002b). Expression in trichomes of a truncated version of CYCB1;2l acking the N-terminal 133amino acids results in multicellular trichomes and clusters of adjacent trichomes similar to those seen in *sim* mutants (Schnittger *et al.*, 2002b, 2005). This effect is only observed with the deleted version of the protein, and although the initial report indicated that this effect was specific for

CYCB1;2 expression, more recent information indicates that this may not be the case (Schnittger *et al.*, 2005). The deleted N-terminal region contains the cyclin destruction box, suggesting that the inability of the full length protein to activate mitosis is prevented by ubiquitin-mediated proteolysis. *KNOLLE*, a gene whose product is involved in cytokinesis, is expressed in the developing multicellular trichomes induced by expression of truncated CYCB1;2, indicating that the full cell division program has been activated in these trichomes. Within a multicellular trichome, the total DNA content of all nuclei does not exceed that expected for the single nucleus of an endoreduplicated wild-type trichome, indicating that CYCB1;2 is acting only to switch on mitosis, and does not affect the overall number of cell cycles traversed. In contrast to wild-type trichomes, *sim* mutant trichomes express the endogenous *CYCB1;2* gene, and it thus seems likely that *SIM* acts at least in part to prevent expression of B-cyclins at the transcriptional level. When the truncated CYCB1;2 was expressed in a *sim* mutant background, the number of cells per trichome was increased in an additive fashion over that seen for either the construct or the mutant alone. This suggests that *SIM* is involved in additional regulatory steps that restrict the ability of CYCB to promote mitosis in addition to controlling *CYCB* transcription.

More surprising is the observation that expression of *CYCD3;1* in trichomes from the *GL2* promoter can also trigger formation of multicellular trichomes (Schnittger *et al.*, 2002a). Developing trichomes of plants expressing this construct express transcripts for both *CYCB1;1* and *CYCB1;2*, indicating that mitosis in these trichomes is likely to act through B-cyclins. No multicellular trichomes were seen in plants containing a *pGL2:CYCD2;2* construct, demonstrating that not all D-cyclins can trigger division in trichomes. In contrast to the results seen for multicellular trichomes triggered by B-cyclin expression, the total DNA content per multicellular trichome in these plants was on average greater than that of unicellular wild-type trichomes (Schnittger *et al.*, 2002a). This suggests that overexpression of CYCD3;1 in trichomes promotes additional S-phases as well as triggering mitosis. Combining *pGL2:CYCD3;1* with the *sim* mutation resulted in a synergistic increase in the number of cells per trichome; a similar result was seen when *pGL2:CYCD3;1* and *pGL2:N-term*Δ*CYCB1;2* were combined. Thus, SIM must limit the ability of CYCD3;1 to trigger mitosis. One caution in interpreting these results is that *CYCD3;1* is not expressed in wild-type trichomes, and it is not known what D-cyclin is involved in endoreduplication during trichome development.

11.4 Conclusions and outlook

11.4.1 Basic mechanism of endoreduplication in trichomes resembles that of other cell types

As reviewed by Larkins *et al.* (2001), and discussed in more detail by Vlieghe *et al.* elsewhere in this volume (Chapter 10), the simplest model of the endocycle requires a single CDK activity that fluctuates in level, triggering S phase when the level is high, then dropping to a low level to allow licensing of replication origins, in conjunction with the suppression of mitosis-promoting factors. Not surprisingly, virtually all of the factors implicated in endoreduplication in other tissues affect the DNA content of endoreduplicated trichomes (Table 11.1). These include downregulation of RBR function and a concomitant increase in E2F/DP-regulated transcription of genes involved in replication licensing, as well as involvement of topoisomerase IV, thought to be required for untangling chromatids during endoreduplication. The multicellular trichome phenotype of mutations in the *sim* gene provides direct genetic evidence that a suppressor of mitosis-promoting factors is involved in establishing endoreduplication. This similarity to endoreduplication in other tissues, along with the ease of detecting mutations affecting endoreduplication in trichomes and the ready accessibility of trichomes for microscopic observation and cell biological manipulations, make Arabidopsis trichomes an excellent model for the study of endoreduplication in plants.

11.4.2 The role of D-cyclins in trichome endoreduplication

The classical role of CYCD/CDK complexes in mitotic cells is to phosphorylate and inactivate RBR proteins, leading to the activation of E2F/DP heterodimers and the transcription of S-phase-specific genes (Oakenfull *et al.*, 2002). Consistent with this role in plants, overexpression of *CYCD3;1* from the *CAMV 35S* promoter resulted in hyperproliferation of cells, inhibited endoreduplication and inhibited terminal differentiation (Dewitte *et al.*, 2003). In addition, a reduction in the proportion of cells in G1 relative to cells in G2 was observed in these plants, and an increased number of cells were found to express a *CYCB* reporter construct, indicative of an increased number of cells in G2. Similar results were obtained when *CYCD3;1* was expressed in synchronized Arabidopsis tissue culture cells (Menges *et al.*, 2006). This study showed that while the doubling time of the cultures was unchanged by expression of *CYCD3;1*, the G1 phase was shorter and G2 phase was longer than in control cells, and activation of G2/M-specific gene expression was delayed. These results argue strongly that the primary role of *CYCD3;1* is to drive the G1/S transition.

In contrast, the observation that multicellular trichomes are produced when *CYCD3;1* is expressed in trichomes from *pGL2* suggests that some D-cyclins can trigger entry into mitosis, at least in the context of trichome development (Schnittger *et al.*, 2002a). Consistent with this view, the available evidence to date suggests that SIM likely suppresses mitosis during endoreduplication in trichomes by functioning as an inhibitor of CYCD/CDK complexes, as discussed in detail above. This view might be reconciled with the more orthodox view of CYCD function if developing trichomes contain specific CYCD/CDK complexes not found in mitotic cells, and these complexes can trigger mitosis if not properly regulated, or if endoreduplicating trichomes lack a checkpoint that prevents CYCD/CDK complexes from triggering mitosis during the mitotic cell cycle. In any case, these observations suggest that D-cyclins have roles in the endocycle distinct from their role in the mitotic cycle.

11.4.3 A speculative model of endoreduplication during trichome development

Figure 11.2 shows a model for the establishment of endoreduplication in developing trichomes including both clearly demonstrated cell cycle interactions (solid arrows) and speculative interactions suggested by the work described in this review (dotted arrows); the phenotypes resulting from genetic manipulation of the relevant genes are listed in Table 11.1. One key contribution of the trichome system to our understanding of endoreduplication is to give a glimpse into the developmental control of the endocycle. GA and other hormones are known to regulate trichome development, and are thought to act upstream of the transcription factors regulating trichome development. The transcription factor GL3 plays a central role in the initiation of

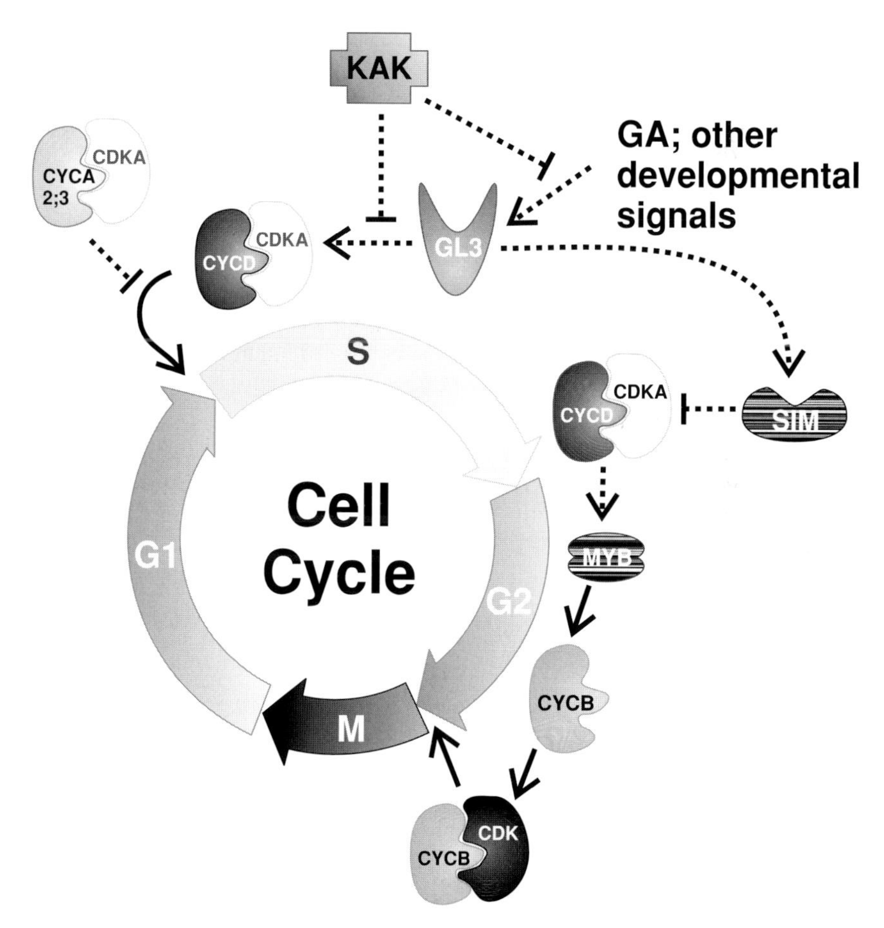

Figure 11.2 A speculative model for the establishment of the endocycle in developing trichomes. Gene names are given in the text, and phenotypes are summarized in Table 11.1. Arrows with pointed heads indicate positive regulation, arrows with flat heads indicate negative regulation, solid arrows indicate interactions with strong experimental support. Dotted arrows indicate hypothesized interactions.

trichome development (Larkin *et al.*, 2003), and based on the effect of altered *GL3* function on endoreduplication, GL3 is presumed to play a key role in establishing the endocycle in trichomes. The most straightforward way for this to occur would be by upregulation of D-cyclin expression, or of a regulator of D-cyclin function, under the direct control of GL3. An increase in functional D-cyclin would result in inactivation of RBR and entry into S phase. One protein whose role is unclear is that of KAK, a negative regulator of endoreduplication that encodes a HECT ubiquitin E3-ligase (Downes *et al.*, 2003; El Refy *et al.*, 2003). KAK presumably targets specific proteins for degradation, but which proteins? In principle, it could be a negative regulator of GA signaling, it could directly target GL3for degradation, or it could promote the degradation of inhibitors of CYCD/CDK complexes such as members of the ICK/KRP family. It could also target RBR, or the inhibitory E2F protein E2Fc for degradation (del Pozo *et al.*, 2002).

Establishment of the endocycle also requires inhibition of mitosis; the best candidate for a gene encoding an inhibitor of mitosis in endoreduplicating trichomes is *SIM* (Walker *et al.*, 2000), and more recent results support this view (M. Churchman, M. Brown, and J.C. Larkin, submitted). If GL3 regulates the switch to the endocycle, then *SIM* might also be a target of GL3 regulation, as indicated in Figure 11.2. SIM is depicted as an inhibitor of a CYCD/CDK complex, a reasonable hypothesis based on current evidence. The CYCD/CDK complex regulated by SIM is illustrated as a complex distinct from that regulating the G1/S transition. As indicated in the discussion above, this may be a complex containing a different D-cyclin, but it could also be a different posttranslational state of the same complex, or there may only be one CYCD/CDK complex, but the absence of a checkpoint normally present in mitotic cells may give the complex a second function that must be inhibited by SIM for endoreduplication to proceed.

How might SIM be preventing mitosis? In wild-type trichomes, B-cyclin expression is absent, but in *sim* mutants, dividing trichomes express B-cyclins (Schnittger *et al.*, 2002b), indicating that SIM acts upstream of B-cyclin transcription to inhibit mitosis. In mitotic cells, B-cyclin transcription begins in G2 and peaks at the G2/M transition (Ito *et al.*, 1997; Menges *et al.*, 2002). Transcription of B-cyclins in plants is controlled by 3R Myb transcription factors (Ito *et al.*, 2001). These Myb proteins are themselves controlled at the transcriptional and posttranscriptional levels. While expression of CYCD3;1 in mitotic cells results in a delay in B-cyclin transcription (Menges *et al.*, 2006), it is possible that in the context of trichome development, one or more D-cyclins can activate transcription or function of the Mybs controlling B-cyclin expression, and SIM is needed to inhibit this aberrant function of this CYCD/CDK complex for endoreduplication to proceed (Figure 11.2).

CYCA2;3 has been described as an inhibitor of endoreduplication that terminates the endocycle when it is induced late in trichome development (Imai *et al.*, 2006). The target of CYCA2;3/CDK complex kinase activity is unclear; in Figure 11.2, it is presumed to inhibit the G1/S transition in some fashion. CYC2;3 function is negatively regulated by the APC, which appears to be active in developing trichomes. The exact role of the APC in establishing the endocycle is unclear, and it has not been indicated in Figure 11.2.

11.4.4 Open questions and future prospects

Many important questions remain. One of the most crucial is what cell cycle proteins are actually expressed during trichome development? Many of the conclusions described here are the result of overexpression of cell cycle proteins that may not be expressed in developing trichomes, or are known not to be expressed there. An accurate cast list of the true players would be a big improvement! Also, overexpression is a powerful technique, but can give misleading results if higher concentrations of a protein cause it to bind to incorrect partners. Correct levels of CDK activity are obviously crucial, and there are many instances where a CDK complex triggers different outcomes at different levels of CDK activity. Loss-of-function phenotypes and domain-specific mutations in proteins that are known to function in wild-type trichomes would be expected to help greatly in untangling the true interactions controlling endocycling in trichomes.

Other important questions include the following: what are the roles of D-cyclin-containing CDK complexes in the endocycle? What are the targets of CYCA2;3/CDK complexes? What is the role of the APC in establishing and maintaining the endocycle? In animals, a variety of endocycle variants exist; are all plant endocycles the same? The ease of identifying trichome mutant phenotypes related to the endocycle, and the accessibility of trichomes to microscopic and other cell biological techniques, means that trichomes will be at the heart of answering many of these questions.

Acknowledgments

The authors wish to acknowledge the technical assistance of Margaret C. Henk for technical assistance with electron microscopy, and Kirsten Prüfer for critical reading of the manuscript. This work was supported by grants from the National Science Foundation (IOB 0444560), the Louisiana Governor's Biotechnology Initiative, and the University of Ghent to J.C.L.

References

Audibert, A., Simon, F. and Gho, M. (2005) Cell cycle diversity involves differential regulation of Cyclin E activity in the Drosophila bristle cell lineage. *Development* **132**, 2287–2297.

Bowling, S.A., Clarke, J.D., Liu, Y., Klessig, D.F. and Dong, X. (1997) The *cpr5* mutant of Arabidopsis expresses both NPR1-dependant and NPR1-independent resistance. *Plant Cell* **9**, 1573–1584.

Castellano, M.M., Boniotti, M.B., Caro, E., Schnittger, A. and Gutierrez, C. (2004) DNA replication licensing affects cell proliferation or endoreplication in a cell type-specific manner. *Plant Cell* **16**, 2380–2393.

Chien, J.C. and Sussex, I.M. (1996) Differential regulation of trichome formation on the adaxial and abaxial leaf surfaces by gibberellins and photoperiod in *Arabidopsis thaliana* (L.) Heynh. *Plant Physiol* **111**, 1321–1328.

Churchman, M.L., Brown, M.L., Kato, N., *et al.* (2006) SIAMESE, a plant-specific cell cycle regulator, controls endoreplication onset in *Arabidopsis thaliana*. *Plant Cell* **18**, 3145–3157.

De Veylder, L., Beeckman, T., Beemster, G.T.S., *et al.* (2001) Functional analysis of cyclin-dependent kinase inhibitors of Arabidopsis. *Plant Cell* **13**, 1653–1667.

De Veylder, L., Beeckman, T., Beemster, G.T.S., *et al.* (2002) Control of proliferation, endoreduplication and differentiation by the *Arabidopsis* E2Fa/DPa transcription factor. *EMBO J* **21**, 1360–1368.

del Pozo, J.C., Boniotti, M.B. and Gutierrez, C. (2002) Arabidopsis E2Fc functions in cell division and is degraded by the ubiquitinSCF^{AtSKP2} pathway in response to light. *Plant Cell* **14**, 3057–3071.

Desvoyes, B., Ramirez-Parra, E., Xie, Q., Chua, N.-H. and Gutierrez, C. (2006) Cell type-specific role of the retinoblastoma/E2F pathway during Arabidopsis leaf development. *Plant Physiol* **140**, 67–80.

Dewitte, W. and Murray, J.A. (2003) The plant cell cycle. *Annu Rev Plant Biol* **54**, 235–264.

Dewitte, W., Riou-Khamlichi, C., Scofield, S., *et al.* (2003) Altered cell cycle distribution, hyperplasia, and inhibited differentiation in Arabidopsis caused by the D-type cyclin CYCD3. *Plant Cell* **15**, 79–92.

Downes, B.P., Stupar, R.M., Gingerich, D.J. and Vierstra, R.D. (2003) The HECT ubiquitin-protein ligase (UPL) family in Arabidopsis: UPL3 has a specific role in trichome development. *Plant J* **35**, 729–742.

Edgar, B.A. and Orr-Weaver, T.L. (2001) Endoreplication cell cycles: more for less. *Cell* **105**, 297–306.

El Refy, A., Perazza, D., Zekraoui, L., *et al.* (2003) The Arabidopsis KAKTUS gene encodes a HECT protein and controls the number of endoreduplication cycles. *Mol Genet Genomics* **270**, 403–414.

Esch, J.J., Chen, M., Sanders, M., *et al.* (2003) A contradictory GLABRA3 allele helps define gene interactions controlling trichome development in Arabidopsis. *Development* **130**, 5885–5894.

Gendreau, E., Orbovic, V., Höfte, H. and Traas, J. (1999) Gibberellin and ethylene control endoreduplication levels in the *Arabidopsis thaliana* hypocotyl. *Planta* **209**, 513–516.

Hartung, F., Angelis, K.J., Meister, A., Schubert, I., Melzer, M. and Puchta, H. (2002) An archaebacterial topoisomerase homolog not present in other eukaryotes is indispensable for cell proliferation of plants. *Curr Biol* **12**, 1787–1791.

Hülskamp, M., Miséra, S. and Jürgens, G. (1994) Genetic dissection of trichome cell development in Arabidopsis. *Cell* **76**, 555–566.

Imai, K.K., Ohashi, Y., Tsuge, T., *et al.* (2006) The A-type cyclin CYCA2;3 is a key regulator of ploidy levels in *Arabidopsis* endoreduplication. *Plant Cell* **18**, 382–396.

Ito, M., Araki, S., Matsunaga, S., *et al.* (2001) G2/M-phase-specific transcription during the plant cell cycle is mediated by c-Myb-like transcription factors. *Plant Cell* **13**, 1891–1905.

Ito, M., Criqui, M.-C., Sakabe, M., *et al.* (1997) Cell-cycle-regulated transcription of A- and B-type plant cyclin genes in synchronous cultures. *Plant J* **11**, 983–992.

Johnson, H.B. (1975) Plant pubescence: an ecological perspective. *Bot Rev* **41**, 233–258.

Kirik, V., Bouyer, D., Schöbinger, U., *et al.* (2001) CPR5 is involved in cell proliferation and cell death control and encodes a novel transmembrane protein. *Curr Biol* **11**, 1891–1895.

Kurata, T., Ishida, T., Kawabata-Awai, C., *et al.* (2005) Cell-to-cell movement of the CAPRICE protein in Arabidopsis root epidermal cell differentiation. *Development* **132**, 5387–5398.

Larkin, J.C., Brown, M.L. and Schiefelbein, J. (2003) How do cells know what they want to be when they grow up? Lessons from epidermal patterning in Arabidopsis. *Annu Rev Plant Physiol Plant Mol Biol* **54**, 403–430.

Larkin, J.C., Young, N., Prigge, M. and Marks, M.D. (1996) The control of trichome spacing and number in *Arabidopsis*. *Development* **122**, 997–1005.

Larkins, B.A., Dilkes, B.P., Dante, R.A., Coelho, C.M., Woo, Y.M. and Liu, Y. (2001) Investigating the hows and whys of DNA endoreduplication. *J Exp Bot* **52**, 183–192.

Lui, H., Wang, H., DeLong, C., Fowke, L.C., Crosby, W.L. and Fobert, P.R. (2000) The *Arabidopsis* Cdc2a-interacting protein ICK2 is structurally related to ICK1and is a potent inhibitor of cyclin-dependent kinase activity *in vitro*. *Plant J* **21**, 379–385.

Mauricio, R. and Rauscher, M.D. (1997) Experimental manipulation of putative selective agents provides evidence for the role of natural enemies in the evolution of plant defense. *Evolution* **51**, 1435–1444.

Melaragno, J.E., Mehrotra, B. and Coleman, A.W. (1993) Relationship between endopolyploidy and cell size in epidermal tissue of Arabidopsis. *Plant Cell* **5**, 1661–1668.

Menges, M., Hennig, L., Gruissem, W. and Murray, J.A. (2002) Cell cycle-regulated gene expression in Arabidopsis. *J Biol Chem* **277**, 41987–42002.

Menges, M., Samland, A.K., Planchais, S. and Murray, J.A.H. (2006) The D-type cyclin CYCD3;1 is limiting for the G1-to-S-phase transition in Arabidopsis. *Plant Cell* **18**, 893–906.

Newport, J. and Kirschner, M. (1982) A major developmental transition in early Xenopus embryos: I. characterization and timing of cellular changes at the midblastula stage. *Cell* **30**, 675–686.

Oakenfull, E.A., Riou-Khamlichi, C. and Murray, J.A. (2002) Plant D-type cyclins and the control of G1 progression. *Philos Trans R Soc Lond B Biol Sci* **357**, 749–760.

Payne, C.T., Zhang, F. and Lloyd, A.M. (2000) GL3 encodes a bHLH protein that regulates trichome development in Arabidopsis through interaction with GL1 and TTG1. *Genetics* **156**, 1349–1362.

Perazza, D., Herzog, M., Hülskamp, M., Brown, S., Dorne, A.M. and Bonneville, J.M. (1999) Trichome cell growth in *Arabidopsis thaliana* can be derepressed by mutations in at least five genes. *Genetics* **152**, 461–476.

Perazza, D., Vachon, G. and Herzog, M. (1998) Gibberellins promote trichome formation by Up-regulating GLABROUS1 in arabidopsis. *Plant Physiol* **117**, 375–383.

Schellmann, S. and Hülskamp, M. (2005) Epidermal differentiation: trichomes in Arabidopsis as a model system. *Int J Dev Biol* **49**, 579–584.

Schellmann, S., Schnittger, A., Kirik, V., *et al.* (2002) TRIPTYCHON and CAPRICE mediate lateral inhibition during trichome and root hair patterning in Arabidopsis. *EMBO J* **21**, 5036–5046.

Schnittger, A., Schöbinger, U., Bouyer, D., Weinl, C., Stierhof, Y.-D. and Hülskamp, M. (2002a) Ectopic D-type cyclin expression induces not only DNA replication but also cell division in *Arabidopsis* trichomes. *Proc Natl Acad Sci USA* **99**, 6410–6415.

Schnittger, A., Schöbinger, U., Stierhof, Y.D. and Hülskamp, M. (2002b) Ectopic B-type cyclin expression induces mitotic cycles in endoreduplicating Arabidopsis trichomes. *Curr Biol* **12**, 415–420. (Erratum: *Curr Biol* **15**, 980).

Schnittger, A., Schöbinger, U., Stierhof, Y.D. and Hülskamp, M. (2005) Erratum: Ectopic B-type cyclin expression induces mitotic cycles in endoreduplicating Arabidopsis trichomes. *Curr Biol* **15**, 980.

Schnittger, A., Weinl, C., Bouyer, D., Schöbinger, U. and Hülskamp, M. (2003) Misexpression of the cyclin-dependent kinase inhibitor *ICK1KRP1* in single-celled Arabidopsis trichomes reduces endoreduplication and cell size and induces cell death. *Plant Cell* **15**, 303–315.

Sugimoto-Shirasu, K., Roberts, G.R., Stacey, N.J., McCann, M.C., Maxwell, A. and Roberts, K. (2005) RHL1 is an essential component of the plant DNA topoisomerase VI complex and is required for ploidy-dependent cell growth. *Proc Natl Acad Sci USA* **102**, 18736–18741.

Sugimoto-Shirasu, K., Stacey, N.J., Corsar, J., Roberts, K. and McCann, M.C. (2002) DNA topoisomerase VI is essential for endoreduplication in *Arabidopsis*. *Curr Biol* **12**, 1782–1786.

Swain, S.M., Tseng, T.S., Thornton, T.M., Gopalraj, M. and Olszewski, N.E. (2002) SPINDLY is a nuclear-localized repressor of gibberellin signal transduction expressed throughout the plant. *Plant Physiol* **129**, 605–615.

Szymanski, D.B. (2005) Breaking the WAVE complex: the point of Arabidopsis trichomes. *Curr Opin Plant Biol* **8**, 103–112.

Telfer, A., Bollman, K.M. and Poethig, R.S. (1997) Phase change and the regulation of trichome distribution in *Arabidopsis thaliana*. *Development* **124**, 645–654.

Traw, M.B. and Bergelson, J. (2003) Interactive effects of jasmonic acid, salicylic acid, and gibberellin on induction of trichomes in Arabidopsis. *Plant Physiol* **133**, 1367–1375.

Turing, A.M. (1952) The chemical basis of morphogenesis. *Phil Trans R Soc B* **641**, 37–72.

Verkest, A., de O. Manes, C.-L., Maes, S., *et al.* (2005) The cyclin-dependent kinase inhibitor KRP2 controls the mitosis-to-endocycle transition during Arabidopsis leaf development through a specific inhibition of the mitotic CDKA;1 kinase complexes. *Plant Cell* **17**, 1723–1736.

Vespa, L., Vachon, G., Berger, F., Perazza, D., Faure, J.-D. and Herzog, M. (2004) The immunophilin-interacting protein AtFIP37from Arabidopsis is essential for plant development and is involved in trichome endoreduplication. *Plant Physiol* **134**, 1283–1292.

Wada, T., Kurata, T., Tominaga, R., *et al.* (2002) Role of a positive regulator of root hair development, CAPRICE, in Arabidopsis root epidermal cell differentiation. *Development* **129**, 5409–5419.

Walker, J.D., Oppenheimer, D.G., Concienne, J. and Larkin, J.C. (2000) SIAMESE, a gene controlling the endoreduplication cell cycle in Arabidopsis thaliana trichomes. *Development* **127**, 3931–3940.

Wang, H., Fowke, L.C. and Crosby, W.L. (1997) A plant cyclin-dependent kinase inhibitor gene. *Nature* **386**, 451–452.

Wang, H., Qi, Q., Schorr, P., Cutler, A.J., Crosby, W.L. and Fowke, L.C. (1998) ICK1, a cyclin-dependent protein kinase inhibitor from *Arabidopsis thaliana* interacts with both Cdc2a and CYCD3, and its expression is induced by abscisic acid. *Plant J* **15**, 501–510.

Watt, F.M. and Hogan, B.L. (2000) Out of Eden: stem cells and their niches. *Science* **287**, 1427–1430.

Weinl, C., Marquardt, S., Kuijt, S.J.H., *et al.* (2005) Novel functions of plant cyclin-dependent kinase inhibitors, ICK1/KRP1, can act non-cell-autonomously and inhibit entry into mitosis. *Plant Cell* **17**, 1704–1722.

Weng, L., Zhu, C., Xu, J. and Du, W. (2003) Critical role of active repression by E2F and Rb proteins in endoreplication during Drosophila development. *EMBO J* **22**, 3865–3875.

Yin, Y., Cheong, H., Friedrichsen, D., *et al.* (2002) A crucial role for the putative Arabidopsis topoisomerase VI in plant growth and development. *Proc Natl Acad Sci USA* **99**, 10191–10196.

Yoshida, S., Ito, M., Nishida, I. and Watanabe, A. (2002) Identification of a novel gene HYS1/CPR5that has a repressive role in the induction of leaf senescence and pathogen-defense responses in Arabidopsis thaliana. *Plant J* **29**, 427–437.

12 Cell cycle control and fruit development

Christian Chevalier

12.1 Introduction

The fruit is a specialized organ which results from the development of the ovary after successful flower pollination and fertilization, and provides a suitable environment for seed maturation and seed dispersal mechanisms. It also represents an important source of nutrition for animals and humans, especially providing important components in human diet such as vitamins.

Owing to their importance in human nutrition and their economic inference, fleshy fruit species have been mostly the subjects of fruit developmental studies. Tomato (*Solanum lycopersicum* Mill.) has asserted as the model species for all fleshy fruits, since it presents a highly favourable biology with short life cycle, high multiplication rate, easy crosses and self-pollination. Molecular genetic analyses are supported with a wide range of genetic resources (cultivars, mutants, segregating populations) and several genomic tools (saturated genetic maps, BAC libraries, more than 239 500 expressed sequence tags [ESTs], available microarrays) (Van der Hoeven *et al.*, 2002; Alba *et al.*, 2004, 2005). Furthermore, an international program now aims at sequencing the genome of tomato, which should be completed in the coming years.

The formation of the fruit organ results from the relationship between cell division and cell expansion. These two developmental phenomena, which are under the control of complex interactions between internal signals (due to hormones) and external factors (carbon partitioning, environmental influences), represent crucial determinants of essential criteria for fruit quality such as the final size, weight and shape of fruits. In tomato, the cell number and consequently the extent of cell division activity determine the final fruit size, but how the cell number is regulated during development is still an important matter of investigation. Besides the setting of morphological quality traits, the organoleptic and nutritional quality traits of ripe tomato fruit relevant to composition in primary and secondary metabolites are also determined during the early stages of fruit development.

This chapter aims at describing the recent advances in the knowledge of cell cycle control and fruit organogenesis, focussing on tomato as a model, and will address the developmental, molecular, genetic and metabolic controls governing fruit growth.

12.2 Fruit development: a matter of cell number and cell size

Flower and fruit developmental programs are continuously linked, as the developmental patterns are determined very early in the floral initiation. The development of the ovary thus follows that of the other flower organs (Brukhin *et al.*, 2003). Following fertilization, the ovary wall develops into a fleshy pericarp, encompassing the placental tissue and the seeds, and gives rise to the tomato fruit berry.

12.2.1 Brief description of tomato fruit development

Tomato fruit organogenesis was classically described as proceeding from four distinct phases: fruit set (I), a phase of cell division (II) and a phase of cell expansion (III) both contributing to fruit growth, and ripening (IV) (Gillaspy *et al.*, 1993). As described earlier, this fruit developmental scheme is not that clear-cut, since the distribution of mitotic activities (Joubès *et al.*, 1999) and occurrence of cell expansion (Cheniclet *et al.*, 2005) are spatially and temporally regulated according to the various fruit tissues.

12.2.1.1 Fruit set

The ovary within the flower stops its growth as cell divisions cease shortly before anthesis (Ho and Hewitt, 1986). The development into fruit or fruit set is then dependent upon successful completion of pollination and fertilization. If failure of these processes occurs following anthesis, the flower aborts and abscises. If fertilization is successfully completed, the presence of fertilized ovules triggers very rapidly (within 2 days) the development of the ovary, which enters into a phase of rapid growth due to high mitotic activity.

Fruit set thus implies the generation of positive signals that drive the reactivation of cell-cycle-related genes involved in mitotic activity, but also of genes associated to the fruit-specific developmental programme. MADS-box genes encoding putative transcription factors were shown to be immediately induced after anthesis within the ovary wall (Busi *et al.*, 2003), suggesting that they could function as integrators of the ovary development signal triggered by fertilization. Barg *et al.* (2005) isolated a novel plant-specific short MYB-like gene (called *Lefsm1* for *fruit SANT/MYB-like 1*) whose expression starts to increase in post-anthesis ovaries. The ectopic expression of *Lefsm1* has a deep impact on the early stages of plant vegetative development, leading to reduced apical dominance (Barg *et al.*, 2005), which indicates that LeFSM1 may act as a more general regulator of plant developmental programmes. Whether fruit development has been impaired in these plants was neither described nor discussed.

12.2.1.2 Fruit growth

On fertilization, the fruit grows rapidly between 2 and 30 days post-anthesis (DPA); it then stops its growth as the maturation process starts (Breaker stage) (see Plate 12.1). Fruit growth is first obtained by cell divisions which resume in the ovary wall for up to 7–10 DPA (Varga and Bruinsma, 1986; Bohner and Bangerth,

1988). In the cherry tomato line Wva106, Cheniclet *et al.* (2005) showed that the increase in pericarp thickness in this early period of growth results from the generation of new cell layers, increasing in number from 8–9 to 14–15 layers (see Plate 12.1). These new cell layers arise predominantly from the outer subepidermis layer according to periclinal cell divisions, which are in fact completed within 5 DPA. Besides these so-called histogenic cell divisions, randomly oriented cell divisions occur in the pericarp up to 20 DPA, which follows the growth force imposed by the fruit-enlarging tissues (Cheniclet *et al.*, 2005). Similarly, the fruit epidermis also maintains growth-related cell divisions up to the onset of ripening (Joubès *et al.*, 1999). Hence, the simplistic description of fruit growth proceeding from two successive phases of cell division and cell expansion (Gillaspy *et al.*, 1993) has to be moderated since the pattern of cell division varies among the various fruit tissues. In cherry tomato fruits, this developmental scheme only applies to the placental locular (jelly-like) tissue composed of large and hyper-vacuolarized cells, where cell divisions cease after 10–15 DPA (Joubès *et al.*, 1999).

The very precise kinetic analysis of pericarp development recently published (Cheniclet *et al.*, 2005) emphasized the strong quantitative contribution of cell expansion in tomato fruit growth. In fertilized ovary, cell expansion also occurred very rapidly as it could be detected at 3–4 DPA, concomitantly to mitotic activity, and then lasts for 4 weeks up to the Mature Green (MG) stage, leading to full-sized fruit (see Plate 12.1). A very singular and so far unreported result from this kinetic analysis revealed that a rapid and transient increase in cell size occurs at the transition between MG and Breaker stages (Cheniclet *et al.*, 2005). This peak of cell enlargement drives a parallel increase in pericarp thickness without affecting the fruit size itself (see Plate 12.1) (Cheniclet *et al.*, 2005), suggesting it could be related to ripening-induced mechanisms such as cell wall modifications (Giovannoni, 2004).

Cell size in tomato fruit can reach spectacular levels such as hundreds of times the initial one (e. g., >0.4 mm in mean cell diameter inside the pericarp of some varieties; Cheniclet *et al.*, 2005). This variation in cell size is accompanied by an increase in the ploidy level, designated as endoreduplication, characterizing the development of the pericarp and jelly-like locular tissue from 10 DPA to the onset of ripening (Bergervoet *et al.*, 1996; Joubès *et al.*, 1999). Endoreduplication results from the ability of cells to modify their classical cell cycle in which DNA synthesis occurs independently from mitosis (cf. Chapter 'Endoreduplication' from Vlieghe *et al.*, this book). In the locular gel tissue, only endoreduplication occurs after 10–15 DPA as cell divisions cease (Joubès *et al.*, 1999), which makes this particular tissue a good model for spatial and temporal regulation of endoreduplication. Cell division, endoreduplication, and cell expansion occur at the same time in the pericarp at the early stages (Cheniclet *et al.*, 2005). At anthesis, the ovary contains roughly 50% of 2C (where C is the DNA content of a haploid genome) nuclei, 47% of 4C nuclei, and 3% of 8C nuclei, indicating that the first endocycles that give rise to 4C and 8C nuclei have already started before anthesis. Immediately after anthesis, the number of 2C nuclei decreases rapidly in the pericarp, that of 4C nuclei increases transiently at 10 DPA, and all other ploidy classes increase or appear successively up to 256C (i.e., seven rounds of endoreduplication cycle). The calculated mean C-value and mean cell area increase in parallel up to the Breaker stage in the pericarp, and both parameters are intimately correlated, suggesting that endoreduplication through cell size control contributes to setting the final fruit size (Cheniclet *et al.*, 2005).

The fruit biochemical composition is mainly determined during the growth period sustained by cell expansion: starch, organic acids, secondary metabolites, and aroma precursors accumulate and thus contribute to the organoleptic and nutritional quality of tomato fruit. These metabolic modifications concomitant to endoreduplication originate from a profound change in the cellular behaviour and gene expression program of fruit cells. Lemaire-Chamley *et al.* (2005) performed a large-scale analysis of gene expression using dedicated tomato cDNA microarrays. This analysis aimed at identifying genes linked to the differentiation of specialized tissues in early-developing tomato fruit, with a special emphasis on the exocarp (outer part of the pericarp) and the locular gel, two tissues differentially engaged in cell division and cell expansion, respectively. The different tissues from the expanding fruit display characteristic gene expression programs associated with their specialization. Many new genes of yet unknown function were highlighted, and gene expression was shown to be tightly regulated at the level of timing, intensity, and spatial localization. To summarize, gene expression in the exocarp relates to protective function and high metabolic activity, while gene expression in locular tissue is consistent with cell-expansion-related functions, such as water flow, organic acid synthesis, sugar storage, under the putative control of hormonal signalling (Lemaire-Chamley *et al.*, 2005).

12.2.1.3 Ripening

The final size of the fruit is reached at the end of the growth period mainly sustained by cell expansion, as fruit enlargement ceases when ripening processes are initiated (Gillaspy *et al.*, 1993). Ripening is characterized by a burst of ethylene production, deep chemical changes determining numerous fruit traits such as aroma, colour and composition and structural changes affecting the fruit texture, such as softening (Giovannoni, 2001, 2004). Within the frame of this book, we shall not refer any longer to ripening especially because cell division and cell expansion hardly occur when fruit is maturing.

12.2.2 Hormonal signalling in fruit set and development

Compelling evidence indicates that ovary fertilization and fruit set are under the influence of gibberellins and auxin produced by the germinating pollen (Gillaspy *et al.*, 1993). Following pollination, two peaks of auxin accumulation occur in developing tomato fruits. The first peak relates to the rapid fruit growth by intense mitotic activity, and is likely to originate from auxin import within the pollen and/or from auxin synthesized within the ovary wall and/or ovule. The second peak of auxin has been attributed to the embryo-derived auxin synthesis, which then promotes fruit growth by cell enlargement (Hocher *et al.*, 1992). As a consequence, fruit growth is coordinated with the embryo and seed development, and the number of seeds in the fruit is a key determinant of its final size (Varga and Bruinsma, 1986) because of hormone signals produced by seeds.

Fruit set can occur independently of pollination/fertilization, leading to the development of seedless parthenocarpic fruits. Parthenocarpy in tomato is either genetically (naturally) determined or artificially obtained on induction by exogenous application of auxin or to a lesser extent by gibberellins (for a review, see Gorguet *et al.*, 2005). Parthenocarpic fruit development implies that the arrest in ovary growth shortly before anthesis does not occur, resulting in larger ovaries most likely due to an extended period of cell divisions. Furthermore, parthenocarpic fruits are often of a smaller size than their normal seeded counterparts and deformed as the development of the jelly-like tissue filling the locular cavities is greatly impaired. The ovaries of parthenocarpic tomato lines display higher endogenous levels of both gibberellins and auxin than wild-type ones (for a review, see Gillaspy *et al.*, 1993), suggesting that parthenocarpy results from an impairment in the hormonal balance between gibberellins and auxin (Gorguet *et al.*, 2005). These increased levels of gibberellins and auxin can act as a substitute to the pollination signal, which activates cell division and subsequent fruit development. Therefore, the signal transduction pathway leading to fruit set involves the tight cooperation of (at least) these two hormones.

It has been known for long that auxin controls not only cell division, but also cell expansion. Therefore, during the later fruit development, it is thought that auxin plays a role in contributing to fruit growth by influencing cell enlargement. Interestingly, the comparative microarray analysis for exocarp and locular tissue gene expression – further supported by confirmatory RT-PCR analyses using mRNAs prepared from dissected fruit tissues (exocarp, mesocarp + endocarp, columella, locular tissue, and seed) – revealed that candidate genes in the auxin and gibberellin signalling pathway are preferentially expressed in locular tissue (Lemaire-Chamley *et al.*, 2005). These data suggested that auxin synthesis, transport, and responses play an important role during fruit development, by triggering or accelerating fruit growth through cell enlargement.

Transgenic lines altered for auxin biosynthesis, signalling, or perception (Balbi and Lomax, 2003; Gorguet *et al.*, 2005, and references therein; Wang *et al.*, 2005) have been useful to address the role of auxin in tomato fruit development, especially dealing with its regulatory function on gene expression. For instance, the auxin-resistant *diageotropica* mutant shows an altered phenotype for fruit development: fruit weight is reduced, as well as the number of locules and seeds, and the expression of a subset of *Aux/IAA* genes is altered (Balbi and Lomax, 2003). The *DIAGEOTROPICA* gene product has been recently identified as being a cyclophilin (Oh *et al.*, 2006), but how it interacts with the auxin signal transduction pathway is unknown. Similarly, the direct effects of the mutation on cell division and/or cell expansion driving fruit size needs to be documented.

In conclusion, an accumulating body of data establishes the roles of hormones during fruit development, but the interplay between hormonal signalling and cell cycle regulation that controls fruit growth through cell division and cell expansion remains to be fully deciphered.

12.3 Cell cycle gene expression and fruit development

12.3.1 Core cell cycle genes in tomato

The tomato genome is a modest-sized diploid genome with 12 chromosomes ($2n = 24$), accounting for approximately 950 Mb of DNA, whose sequencing is now in progress. In the last years, the Tomato Genomics Project funded by the National Science Foundation Plant Genome Research Program resulted in the development of an integrated set of experimental tools for use in tomato. A tomato EST database was developed with a special emphasis on sequences expressed during fruit development and maturation. BLAST searches against the database from the SOL Genomics Network (SGN; http://sgn.cornell.edu) allowed the identification of new cell cycle genes expressed from the tomato genome. The various genes encoding CDKs, cyclins, and other cell cycle regulators already identified from tomato (Joubès *et al.*, 1999, 2000b, 2001; Gonzalez *et al.*, 2004; Bisbis *et al.*, 2006) or from the Arabidopsis genome (Vandepoele *et al.*, 2002; Wang *et al.*, 2004; Menges *et al.*, 2005) were used as query sequences to search for homologues amongst the 239 500 ESTs deposited in the database, accounting for more than 41 000 unigenes. Only sequences giving E-values greater than 1e-005 were selected. The data related to this preliminary identification of expressed cell cycle genes in tomato are summarized in Table 12.1. The names of the identified genes were preceded with the species acronym Solly for *Solanum lycopersicum*, which now describes the common tomato and fits to the recently revised phylogenetic classification of the Solanaceae, where the genus *Lycopersicon* was reintegrated into the *Solanum* genus.

12.3.1.1 Cyclin-Dependent Kinases

The classification of CDK-related protein kinases in plants were divided into five conserved classes CDKA to CDKE (Joubès *et al.*, 2000 a). It was further completed by the creation of a new class: CDKF, corresponding to the distantly related CDK-activating kinase CAK1 (renamed CDKF;1; Vandepoele *et al.*, 2002). Recently, a global analysis of the core cell cycle regulators of Arabidopsis identified 17 new CDK-related sequences (Menges *et al.*, 2005) in addition to the previous 12 reported CDKs (Joubès *et al.*, 2000 a; Vandepoele *et al.*, 2002). Among these new genes two of them are closely related to the previously defined CDK classes, and display a conserved PLTSLRE motif. They were subsequently classified as a distinct CDK class named CDKG. The remaining 15 sequences showed significant conservation with CDK proteins, which led to their classification as CDK-like proteins (CKL).

BLAST searches against the tomato EST database using the *CDKA;*1, *CDKA;*2, *CDKB1*;1 and *CDKB2*;1 genes already identified in tomato did not reveal any other members of the CDKA and CDKB classes. A second member of the CDKC family, Solly;*CDKC*;2, was identified as corresponding to the EST309505, which displayed 79% of sequence identity with Solly;*CDKC*;1 (Joubès *et al.*, 2001). On the basis of protein sequence homologies using Arabidopsis CDKD to CDKG as query proteins, new genes from tomato were identified for the CDKD, CDKF and CDKG classes which were thus tentatively named Solly;*CDKD*;1, Solly;*CDKF*;1 and Solly;*CDKG*;1 respectively.

Table 12.1 Characteristics of core cell cycle genes identified so far in tomato

Gene	Accession no.	TC/EST no.	Features	Reference
Solly;*CDKA*;1	Y17225		PSTAIRE	Joubès *et al.* (1999)
Solly;*CDKA*;2	Y17226		PSTAIRE	Joubès *et al.* (1999)
Solly;*CDKB1*;1	AJ297916		PPTALRE	Joubès *et al.* (2002)
Solly;*CDKB2*;1	AJ297917		PPTTLRE	Joubès *et al.* (2002)
Solly;*CDKC*;1	AJ294903		PITAIRE	Joubès *et al.* (2002)
Solly;*CDKC*;2		EST309505	?	
Solly;*CDKD*;1		TC164384	NFTALRE	
Solly;*CDKF*;1		TC168616	?	
Solly;*CDKG*;1		TC165383	PLTSLRE	
Solly;*CKL1-L*		TC158570	?	
Solly;*CKL9-L*		EST544073	?	
Solly;*CKL10-L*		EST550664	?	
Solly;*CKL15-L*		EST338240	?	
Solly;*CYCA1*;1	AJ243451		LVEVxEEY	Joubès *et al.* (2000)
Solly;*CYCA2*;1	AJ243452		LVEVxEEY	Joubès *et al.* (2000)
Solly;*CYCA3*;1	AJ243453		LVEVxEEY	Joubès *et al.* (2000)
Solly;*CYCA3*;2		BG127151	LVEVxEEY	
Solly;*CYCA3*;3		BI928124	LVEVxEEY	
Solly;*CYCB1*;1	AJ243454		HxRF	Joubès *et al.* (2000)
Solly;*CYCB1*;2		TC158723	HxKF	
Solly;*CYCB1*;3		TC157942	HxKF	
Solly;*CYCB1*;4		TC158418	HxKF	
Solly;*CYCB2*;1	AJ243455		HxKF	Joubès *et al.*, 2000
Solly;*CYCB2*;2		TC159360	?	
Solly;*CYCB2*;3		TC166535	?	
Solly;*CYCD1*;1		TC159341	LxCxE	
Solly;*CYCD2*;1		BG129531	?	
Solly;*CYCD3*;1	AJ245415		LxCxE	Joubès *et al.*, 2000 Kvarnheden *et al.*, 2000–
Solly;*CYCD3*;2	AJ002589		LxCxE	Kvarnheden *et al.*, 2000–
Solly;*CYCD3*;3	AJ002590		LxCxE	Kvarnheden *et al.*, 2000–
Solly;*CYCH*;1		EST548942	none	
Solly;*CYCH*;2		EST398599	none	
Solly;*CYCC1*;1		TC165732		
Solly;*CYCL1*;1		TC156276		
Solly;*KRP1*	AJ441249		None	Bisbis *et al.*, 2006
Solly;*KRP2*	AJ441250		None	Bisbis *et al.*, 2006
Solly;*KRP3*		TC164327	None	
Solly;*CKS*		TC163745	None	
Solly;*DEL2*		TC165096	None	
Solly;*DPa*		TC157794	None	
Solly;*E2Fb*		EST289547	None	
Solly;*RBR*		TC168820	None	
Solly;*CDC25-L*		TC160603	None	
Solly;*WEE1*	AM180939		None	Gonzalez *et al.*, 2004

Four distinct tomato sequences corresponding to the Arabidopsis CDK-like (CKL) proteins could be identified in the tomato EST database. The translation product of the Tentative Consensus (TC) sequence TC158570 displayed 72% of identity to Arabidopsis CKL1. This TC was assembled from three EST sequences all originating from the same cDNA library, prepared with mRNAs of tomato flower buds at anthesis. The three other tomato sequences displayed significant identity percentages with Arabidopsis CKL9 (EST544073, 85%), CKL10 (EST550664, 73%) and CKL15 (EST338240, 94%). Since the number of members in this protein family is still unknown, these new identified proteins were assigned the designation CKL-like proteins relative to the Arabidopsis proteins: Solly;*CKL1-L*, Solly;*CKL9-L*, Solly;*CKL10-L* and Solly;*CKL15-L* respectively.

12.3.1.2 Cyclins

In addition to the classical CYCA, CYCB and CYCD family (Renaudin *et al.*, 1996), new cyclin families have been defined: CYCC and CYCH (Yamaguchi *et al.*, 2000), CYCL (Wang *et al.*, 2004), CYCP (Torres-Acosta *et al.*, 2004) and CYCT (Barrôco *et al.*, 2003). A genome-wide analysis of the cyclin family in Arabidopsis, including the distantly related Cyclins CYCJ18 (Abrahams *et al.*, 2001) and Solo Dancers (SDS; Azumi *et al.*, 2002) resulted in the identification of at least 50 distinct cyclins, putatively falling into 10 classes (Wang *et al.*, 2004). A new candidate was recently added to the list of cyclin-like genes (CYCLIN-LIKE, CYL1) (Menges *et al.*, 2005).

Only eight genes for tomato cyclins have been reported in the literature, representing single members of the CYCA1, CYCA2, CYCA3, CYCB1 and CYCB2 classes (Joubès *et al.*, 2000b) and three members of the CYCD3 class (Joubès *et al.*, 2000b; Kvarnheden *et al.*, 2000). BLAST searches against the tomato EST database using these eight tomato sequences and the various Arabidopsis cyclins and cyclin-like proteins allowed the identification of 13 new cyclins in tomato. Additional members of the CYCA3 (Solly;*CYCA3*;2 and Solly;*CYCA3*;3), CYCB1 (Solly;*CYCB1*;2, Solly;*CYCB1*;3 and Solly;*CYCB1*;4) and CYCB2 (Solly;*CYCB2*;2 and Solly;*CYCB2*;3) class were identified. Homologous sequences to CYCD1, CYCD2, CYCH;1, CYCH;2, CYCC1 and CYCL1 were also picked up from the interrogation of the database.

12.3.1.3 Other regulators

Recently, the existence of two cDNAs for CDK-specific inhibitors of the plant $p27^{Kip1}$-related protein (ICK/KRP) family (De Veylder *et al.*, 2001) was reported in tomato (Solly;*KRP1* and Solly;*KRP2*, formerly named Le*KRP1* and Le*KRP2* in Bisbis *et al.*, 2006). A third member could be identified in the EST database: Solly;*KRP3*. The high variability in length and primary structure within ICK/KRPs of a single plant species hampers a clear nomenclature of tomato ICK/KRPs relative to that of Arabidopsis. For instance, Solly;KRP1 and Solly;KRP2 displayed the highest sequence homology with Arath;KRP4 and Arath;KRP3 respectively, while Solly;KRP3, though closer to Solly;KRP1, shares a higher identity percentage with Arath;KRP3.

The identification of other cell cycle regulatory proteins through the tomato EST database interrogation was restricted to a limited number of candidate genes. In tomato, it seems that the CDK subunit (CKS) protein is encoded by a single copy gene: Solly;*CKS*. Only one cDNA representative of DEL2, DPa, E2Fb and retinoblastoma-related (RBR) proteins could be identified so far. It is likely that the narrow window cell cycle regulation of the genes involved in the RBR-E2 F pathway (Menges *et al.*, 2005) may explain the lack of ESTs for these genes in the database.

The WEE1 kinase which negatively regulates by phosphorylation the CDK–cyclin complex activity is encoded by a single gene in tomato mapped on chromosome 9 (Gonzalez *et al.*, 2004). Like in Arabidopsis, no clear CDC25 phosphatase could be identified. However, a homologous protein of the proposed CDC25-related protein (Arath;CDC25; Landrieu *et al.*, 2004) was identified from TC160603. This protein displayed 77% of identical residues with Arath;CDC25, and was thus named Solly;CDC25-L for CDC25-like protein.

12.3.2 Expression of cell cycle genes during fruit development

As described earlier, the early development of tomato fruit starts with a period of intense cell divisions occurring from ovary fertilization which is concomitant to anthesis, up to 8–10 DPA, and followed by a growth period associated to cell expansion up to the onset of ripening (MG stage). According to this developmental frame, the accumulation of cell cycle gene transcripts displays roughly two different patterns (see Figure 12.1A). First, the tomato genes encoding the various CDK as well as Solly;*CYCA3*;1, Solly;*CYCB2*;1 and Solly;*CYCD3*;1 are expressed as early as anthesis and up to 5 DPA, with maximum transcript accumulation at 3 DPA. Between 8 DPA and MG stage, these transcripts decrease gradually in abundance and become almost undetectable at the onset of ripening. Their expression decreased abruptly at 8 DPA and almost disappeared at the MG stage. Solly;*H4* used as a control in the RT-PCR assay displays quite a similar expression profile. Second, the transcripts for Solly;*CYCA1*;1, Solly;*CYCA2*;1 and Solly;*CYCB1*;1 appeared to be expressed later in fruit development. Transcripts of these genes were barely detected at anthesis, then accumulated substantially between 3 and 8 DPA, with a maximum at 5 DPA, and then decreased until the MG stage. Similar data were reported by Kvarnheden *et al.* (2000) for all the three CYCD3 subfamily members, namely Solly;*CYCD3*;1, Solly;*CYCD3*;2 and Solly;*CYCD3*;3, whose maximal expression peaks at 3 DPA.

In the course of fruit development, Solly;*KRP1*, Solly;*KRP2* and Solly;*WEE1* display a slightly different pattern of expression. Two peaks of expression are detected: first, transcripts are highly expressed as early as anthesis, with a maximum at 3 DPA; second, the expression increases again to reach a second peak at 15 DPA for Solly;*WEE1*, at 20 DPA for Solly;*KRP1* and at the MG stage for Solly;*KRP2*. This second peak of transcript expression thus occurs during the cell-expansion-dependent differentiation of tomato fruit cells (Gillaspy *et al.*, 1993), which is concomitant with endoreduplication (Joubès *et al.*, 1999; Cheniclet *et al.*, 2005).

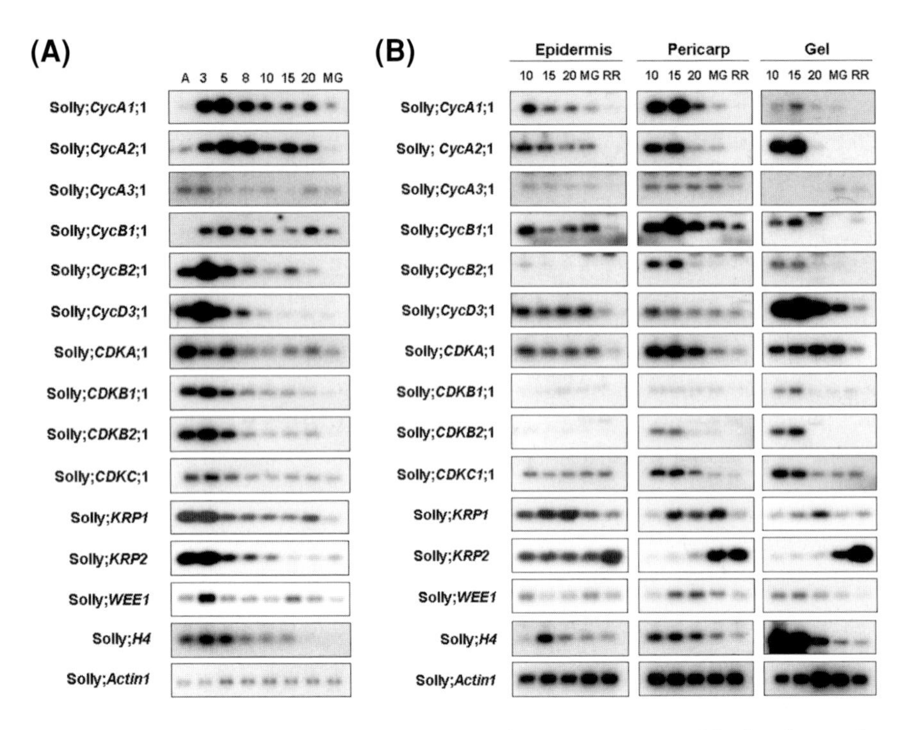

Figure 12.1 Cell cycle gene expression during tomato fruit development. (A) Kinetics of expression in whole tomato fruits, measured from anthesis up to Mature Green (MG) stage. Time points refer to days post-anthesis (DPA); (B) Kinetics of expression in dissected tomato fruit tissues: epidermis, pericarp and gel (locular) tissue, measured from 10 DPA up to Red Ripe (RR) stage.

12.3.3 Temporal expression of cell cycle genes in the different fruit tissues

The distribution of mitotic activity is not only temporally but also spatially regulated in the tomato fruit (Joubès *et al.*, 1999; Cheniclet *et al.*, 2005). Cell division occurs in the epidermis until the end of development. The fleshy pericarp is made of a heterogeneous tissue showing a gradient in cell size from the outer part (exocarp) to the inner part (endocarp) with important variations in cell ploidy level. Cells of the jelly-like (gel) locular tissues are characterized by an arrest of mitosis after 10–15 DPA and the concomitant endoreduplication of the nuclear DNA content.

To analyze their temporal and spatial expression patterns of some of these cell cycle genes, *in situ* hybridization of the mRNA has been performed using longitudinal sections of developing floral organs and fruits. In 6-DPA fruits, low levels of Solly;*CYCD3*;3 transcripts were found in the placenta proximal to seeds, while Solly;*CYCD3*;3 was highly expressed in the vascular tissue of young fruit and flower pedicule (Kvarnheden *et al.*, 2000), thus suggesting a specific role for Solly;*CYCD3*;3 in the development of vascular tissue. *In situ* hybridization with a Solly;*CDKC*;1-specific probe resulted in a uniformly distributed signal, which was observed in dividing tissues of flower buds such as the primordia of petals and stamens, in the L1 cell layer, and in ovules of fruits harvested at anthesis (Joubès *et al.*, 2001). Clearly, Solly;*CDKC*;1 does not display a patchy pattern of expression in agreement with a weak cell cycle regulation profile observed in *Arabidopsis* (Menges *et al.*, 2005). Both Solly;*KRP1* and Solly;*KRP2* are expressed within dividing tissues, and strong signals were obtained in the developing ovules and the carpel wall of floral buds (Bisbis *et al.*, 2006). However the expression of Solly;*KRP1* is restricted to a limited interval of the cell cycle as witnessed by a patchy expression pattern while the uniform distribution of Solly;*KRP2* transcripts suggested that the regulation of its expression was independent of the cell cycle. In flower buds and young developing fruits harvested during the early period of intense cell divisions, the accumulation of Solly;*WEE1* transcripts was strongly associated with the ovary wall, developing ovules and microspores, and systematically detected in the fruit territories where cell divisions occur predominantly (Gonzalez *et al.*, 2004). According to the observed patchy pattern, coupled *in situ* hybridization with DAPI staining and use of aphidicolin-synchronized tobacco BY-2 cultured cells clearly demonstrated that the expression of Solly;*WEE1* transcripts was restricted to the S phase (Gonzalez *et al.*, 2004), which is in complete agreement with the Arabidopsis microarray analysis by Menges *et al.* (2003).

The expression of cell cycle genes in the different fruit tissues have been investigated during fruit development to witness the temporal and spatial regulation of cell division (see Figure 12.1B). In the epidermis, most of the cell cycle genes appeared to be constitutively expressed between 10 DPA and the MG stage, and their expression decreased thereafter to become barely detectable at the Red Ripe (RR) stage. The determination of CDKA protein level and kinase activity correlates with the estimated mitotic index and Solly;*CDKA*;1 transcript level measured in the epidermis (Joubès *et al.*, 1999). This pattern of expression is in good agreement with the developmental scheme of fruit epidermis. Indeed, divisions still occur in the epidermis till very late in the developing fruit (Gillaspy *et al.*, 1993; Joubès *et al.*, 1999). As the fruit grows mainly by cell expansion from 10 DPA to the onset of maturation, the epidermis has to maintain a significant mitotic activity in order to follow the growth force imposed by the inner expanding tissues such as the pericarp and the placental tissue.

After anthesis and up to 10 DPA, cell divisions largely contribute to tomato fruit growth, especially in the fleshy part of the fruit, namely the pericarp. The production of new cell layers arises very early after anthesis and varies according to the genetic variability of tomato fruit size (Cheniclet *et al.*, 2005). This mitotic activity generating new cell layers within the pericarp is nicely reflected by the pattern of expression of mitotic cyclin genes Solly;*CYCA1*;1, Solly;*CYCA2*;1, Solly;*CYCB1*;1, Solly;*CYCB2*;1 and Solly;*CDKB2*;1, whose transcripts accumulate up to 15 DPA. mRNAs corresponding to CDKA, CDKC and WEE1 accumulate up to 20 DPA, and then decrease during ripening. The transcript accumulation profiles of Solly;*CYCA3*;1, Solly;*CYCD3;*1 and Solly;*KRP1* are rather constitutive between 10 DPA and the MG stage. Solly;*KRP2* displayed a remarkable expression profile as it became highly induced during ripening.

In the gel, the genes encoding the tomato CDKB1 and CDKB2 and mitotic cyclins CYCA1, CYCA2, CYCB1 and CYCB2 were mainly expressed up to 15 DPA. From 20 DPA to the ripening stage of the fruit, the transcripts of these genes disappeared and were virtually undetectable. Since these proteins are involved in the control of S/G2 and G2/M transition, their repression is in full agreement with the impairment of both mitosis and CDKA Histone H1 kinase activity in the gel after 15 DPA (Joubès *et al.*, 1999). Indeed, from 15 DPA to the onset of maturation, only endoreduplication occurs in the gel. Therefore, the sustained expression of Solly;*CDKA*;1, Solly;*CYCD3*;1, Solly;*WEE1* and tomato *KRP* genes in the gel up to the MG stage suggest their involvement in endoreduplication during tomato fruit development (Joubès *et al.*, 1999; Gonzalez *et al.*, 2004; Bisbis *et al.*, 2006). In the gel tissue, the CDKA protein level though diminishing could be still detected up to the RR stage, while the CDKA Histone H1 kinase activity was dramatically affected after 15 DPA (Joubès *et al.*, 1999). This coincides with the maximum of Solly;*KRP1* expression and the appearance of endoreduplication in the gel tissue. Interestingly, Solly;*KRP2* showed a different pattern of accumulation in the course of fruit development compared to that of Solly;*KRP1*. Solly;*KRP2* was preferentially expressed at the onset of fruit maturation, which is unexpected as mitotic activities are absent in the gel at this time and endoreduplication is retarded (our unpublished data). This raised the hypothesis that Solly;KRP2 could be involved in the inhibition of endoreduplication at the end of fruit development (Bisbis *et al.*, 2006).

12.4 Altering the cell cycle towards endoreduplication: a key feature for fruit growth

It has long been described that the variation in plant cell size could be accompanied by an increase in cell ploidy level as a result of endoreduplication (Joubès and Chevalier, 2000; Kondorosi and Kondorosi, 2004). The successive rounds of DNA synthesis during endoreduplication induce a spectacular increase in DNA content and a consequent hypertrophy of the nucleus, which can be observed in the course of tomato fruit development (Bergervoet *et al.*, 1996; Joubès *et al.*, 1999; Cheniclet *et al.*, 2005). This can influence the final size of the cell, which therefore may adjust its cytoplasmic volume with respect to the DNA content of the nucleus (according to the 'nuclear–cytoplasmic ratio' theory; Sugimoto-Shirasu and Roberts, 2003). In the course of fruit development, a clear correlation exists between the mean cell size in the pericarp of various tomato genotypes and the mean ploidy level (Cheniclet *et al.*, 2005), and therefore endoreduplication is thought to be an important developmental process contributing to the cell size control and consequently a determinant of fruit weight in tomato. However little is known about the molecular basis of cell size control in plants.

The endoreduplication cycle (endocycle) is made of the succession of S and G phases without mitosis, thus accounting for the cessation of cell division and the increase in ploidy level (Joubès and Chevalier, 2000; Edgar and Orr-Weaver, 2001;

Vlieghe *et al.*, this book). The progression within the endocycle may require the fluctuation in the activity of S-phase CDK activity between DNA synthesis and the gap phase. Among the potential mechanisms regulating the CDK activities in endoreduplicating tissues were proposed the involvements of the WEE1 kinase (Sun *et al.*, 1999) and the CDK-specific ICK/KRP (Coelho *et al.*, 2005; Verkest *et al.*, 2005 a; Weinl *et al.*, 2005).

12.4.1 Role of WEE1 in endoreduplication during tomato fruit development

So far in plants WEE1 homologues have been isolated from maize (Sun *et al.*, 1999), Arabidopsis (Sorrell *et al.*, 2002), tomato (Gonzalez *et al.*, 2004) and rice (accession number BAD10095). A functional analysis aimed at overexpressing the maize and Arabidopsis genes in *Schizosaccharomyces pombe* resulted in the inhibition of cell division and led to significant cell enlargement. Since the *WEE1* gene is significantly expressed in endoreduplicating cells of maize endosperm (Sun *et al.*, 1999) and tomato fruit tissues (Gonzalez *et al.*, 2004), the function of the WEE1 kinase in the control of cell size *in planta* is an intriguing matter of investigation.

Preliminary investigations have been performed to study the role of WEE1 during tomato fruit growth, its putative involvement in the negative regulation of CDK activity and in the consequent endoreduplication process and cell size control (Gonzalez *et al.*, submitted). Transgenic plants were generated with the aim to down-regulate the expression of the endogenous tomato *WEE1* gene. These plants (referred to as *35 S::Slwee1*^{AS}) overexpressed a fragment of the Solly;*WEE1* cDNA in an antisense orientation under the control of the constitutive Cauliflower Mosaic Virus (CaMV) 35 S promoter (Figure 12.2).

Primary transformants (T0 *35 S::Slwee1*^{AS} plants) displayed a gradation in mature fruit size: some plants produced fruits that were similar to wild-type ones; others produced fruits of a slightly smaller size. The observed fruit phenotypes were nicely correlated with the degree to which the endogenous Solly;*WEE1* gene was down-regulated: the lowest expression of endogenous Solly;*WEE1* was observed in *35 S::Slwee1*^{AS} lines, which produced fruits with the most altered size. A clear reduction in the level of endoreduplication was found in the T0 *35 S::Slwee1*^{AS} 25-DPA fruits when compared to wild type: the number of nuclei at the 4C and 8C DNA levels was increased, while that at the 16C, 32C and 64C DNA levels was decreased (see Figure 12.2A). In the following progeny (T1 *35 S::Slwee1*^{AS} plants) an important gradation in whole plant size was observed as illustrated in Figure 12.2B. The relative transcript levels corresponding to endogenous Solly;*WEE1* in the different T1 *35 S::Slwee1*^{AS} lines was determined by RT-PCR analysis, and a good correlation was found between the reduction of the plant size and the relative mRNA abundance. The molecular characterization of the *35 S::Slwee1*^{AS} plant revealed that it displayed only 27% of the wild-type Solly;*WEE1* transcript level (see Figure 12.2C). Furthermore, the CDK–cyclin histone H1 kinase activity measured in young leaves of the *35 S::Slwee1*^{AS} plant was enhanced as a result of the Solly;*WEE1* down-regulation. Most interestingly, a spectacular decrease in the amount of Tyr15-phosphorylated

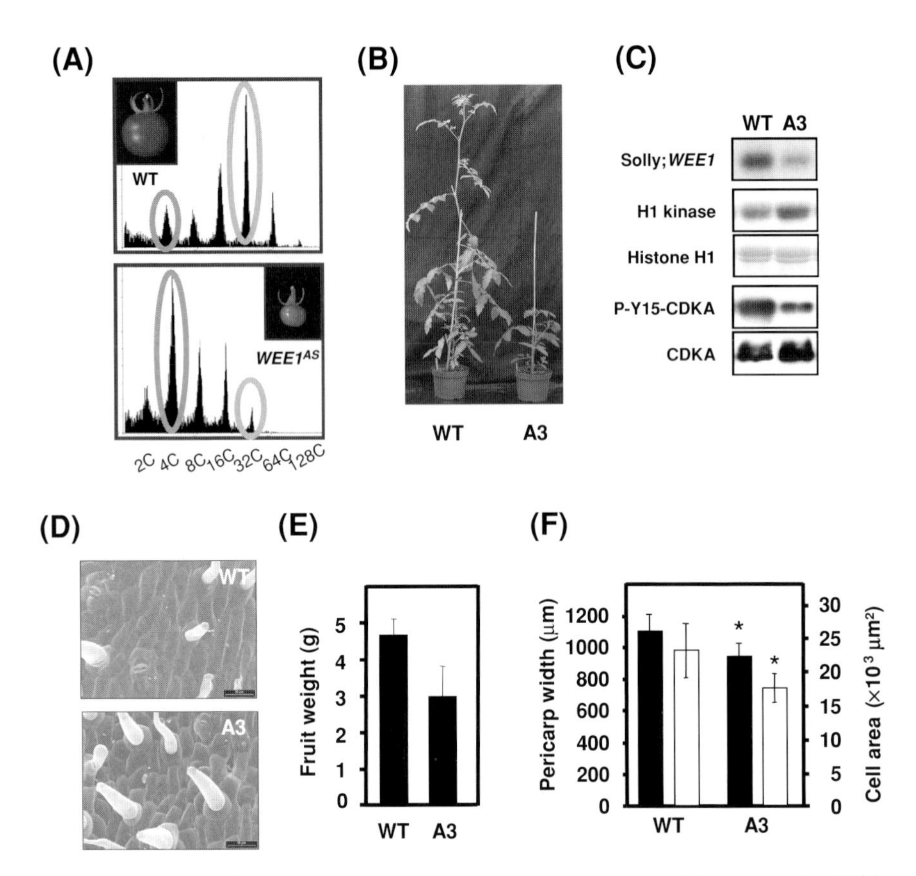

Figure 12.2 Phenotypic and molecular analysis of down-regulated tomato *WEE1* plants. (A) Ploidy level distribution of pericarp at 25 DPA in *35 S::Slwee1*^{AS} and wild-type (WT) fruits; (B) Growth phenotype of line A3 (*35 S::Slwee1*^{AS}) plant. Three months old A3 plant was compared with untransformed control plant (WT) of the same age; (C) Molecular analysis of line A3 (*35 S::Slwee1*^{AS}) plant compared to untransformed control plant (WT). Semi-quantitative RT-PCR analysis of the Solly;*WEE1* expression, kinase activity of CDK–cyclin complexes present in young leaves, and western blot analysis of young leaf protein extracts using anti-Tyr15 and anti-PSTAIR antisera were performed; (D) Cell size alteration of stem epidermal cells of line A3 (*35 S::Slwee1*^{AS}) plant compared with untransformed control plants (WT), viewed by electron scanning micrograph; (E) Fruit weight measurements of ripe fruits harvested from Line A3 compared to untransformed control fruits (WT) of the same age. Histograms represent the compiled data from all fruits harvested from the various plants; (F) Determination of pericarp width and mean cell area in *35 S::Slwee1*^{AS} Immature Green fruits (25 DPA) compared with untransformed control fruits (WT) of the same age. Star symbols above bars indicate that the means of A3 plants are significantly different ($P \leq 0.01$) in Student's tests.

CDKA could be observed when compared to control wild-type plants (see Figure 12.2C). Since Tyr15 is the natural target of the WEE1 kinase activity, thus leading to the inactivation of the CDKA activity, the down-regulation of Solly;*WEE1* in tomato transgenic plants resulted in an increased quantity of dephosphorylated CDKA, and thus increased CDKA activity.

In all the vegetative plant organs that were tested (cotyledons, young leaves and stems) we have been able to show that the effect of a Solly;*WEE1* down-regulation resulted in a short-cell phenotype (see Figure 12.2D).

In the *35S::Slwee1*^{AS} plants the down-regulation of Solly;*WEE1* had a deep impact on fruit size (see Figure 12.2E). A correlated 'scale-of-size' was observed for fruits: the smaller the plant, the smaller the fruit. The fruit size reduction results from a reduction in pericarp width (see Figure 12.2F), but not from a decreased number of cell layers, as the periclinal histogenic cell divisions giving rise to new cell layers that account for pericarp width (Cheniclet *et al.*, 2005) were not affected by the Solly;*WEE1* down-regulation. The reduction in pericarp width itself originates from a significant reduction of cell size (see Figure 12.2F), especially within cell layers 5 to 9 in the central pericarp, which is characterized by the presence of the largest cells in wild-type fruits (Cheniclet *et al.*, 2005). Finally, this observed reduction of cell size was also correlated with a modification in the ploidy profiles of pericarp cells towards an increase in low DNA levels (2C to 8C) and a decrease in high DNA levels (64C and 128C).

From all we know about yeast, the WEE1 kinase has a well-defined role in delaying entry into mitosis until the cell reaches a critical cell size (Fantes and Nurse, 1981; Russell and Nurse, 1987; Harvey and Kellog, 2003). In maize endosperm and tomato fruit development, the WEE1 kinase is associated to the endoreduplication process (Sun *et al.*, 1999; Gonzalez *et al.*, 2004), that is to say when mitosis is impaired. WEE1 is thus likely to target other CDK–cyclin complexes than mitotis-specific ones. It is thus an intriguing matter of investigation to decipher the subunit composition of the CDK–cyclin complex that is targeted by WEE1 during endoreduplication in relation to cell size control.

12.4.2 Role of ICK/KRP in endoreduplication during tomato fruit development

The misexpression of ICK/KRPs in *Arabidopsis thaliana* induces reduced endoreduplication levels and decreased numbers of cells, thus leading to altered vegetative and reproductive developments (Wang *et al.*, 2000; De Veylder *et al.*, 2001; Schnittger *et al.*, 2003). These phenotypes were attributed to the presumed function of ICK/KRPs in the control of the G1/S transition. Two recent reports revealed that ICK/KRPs can also function at the G2/M transition. Using various overexpressing lines in Arabidopsis, Weinl *et al.* (2005) and Verkest *et al.* (2005 a), respectively, demonstrated that ICK1/KRP1 and KRP2 act in a concentration-dependent manner. When ICK/KRPs are constitutively expressed slightly above their endogenous level, only mitotic cell cycle specific CDKA;1 complexes are affected, thus blocking the G2/M transition, while the endoreduplication cycle specific CDKA;1 complexes are unaffected. At high concentrations, both types of complexes are inhibited, resulting in a block of G1/S and G2/M transitions. Verkest *et al.* (2005 a) nicely demonstrated that CDKB1;1 activity in dividing cells controls the level of CDKA;1 activity through the phosphorylation of KRP2, thus mediating its degradation by the proteasome. As a result, when KRP2 protein level increases when CDKB1;1 activity decreases, cells enter the endoreduplication cycle. Fine-tuning of the ICK/KRP protein abundance is thus a key feature for cell cycle control, and especially to trigger the onset of the endoreduplication cycle.

In tomato fruit, the jelly-like (gel) locular tissue is made of large and hyper-vacuolarized cells which undergo multiple rounds of endoreduplication. Within this particular tissue, mitosis is arrested after 15 DPA and only endoreduplication occurs until then, concomitantly to a strong post-translational inhibitory regulation of the CDKA activity (Joubès *et al.*, 1999). The origin of this post-translational regulatory mechanism resides in the accumulation of a proteinaceous factor identified as a ICK/KRP CDK inhibitor named Solly;KRP1 (formerly LeKRP1; Bisbis *et al.*, 2006). The maximum of Solly;KRP1 accumulation parallels that of the corresponding mRNA and occurs at 20 DPA in the gel tissue. The presence of Solly;KRP1 was demonstrated to account for the inhibition of CDK–cyclin kinase activity during the development of the gel tissue in tomato fruit. Furthermore, it was shown that Solly;KRP1 was able to inhibit a putative kinase activity present in the endoreduplicating gel tissue of the fruit that targets RBR. Compelling evidence for the involvement of RBR pathway components in the control of endoreduplication arose from *in planta* functional analyses (De Veylder *et al.*, 2002; Park *et al.*, 2005). The CDK–cyclin complexes that drive the G1/S transition of the cell cycle (and most probably the G/S transition of the endocycle) by phosphorylating RBR are formed from the association of a D-type cyclin with CDKA (Nakagami *et al.*, 1999). It is tempting to speculate that the CDK–cyclin complex whose kinase activity is inhibited in fruit endoreduplicating cells is composed of CDKA and CYCD3;1, with which Solly;KRP1 strongly interacts (Bisbis *et al.*, 2006). These data obtained in tomato fruit are consistent with those presented by Coelho *et al.* (2005), who characterized two KRP proteins from the endoreduplicating maize endosperm tissue. The authors of this study showed the participation of Zeama;KRP;1 and Zeama;KRP;2 in the endoreduplication phase of endosperm development, and provided new data relative to the CDK–cyclin complex composition that both Zeama;KRP;1 and Zeama;KRP;2 may interact with and inhibit. The two maize KRP proteins were shown to specifically inhibit the presumed S-phase cyclin A1;3- and cyclin D5;1-associated CDK activities but not the M-phase CDK-CYCB1;3 complex activity (Coelho *et al.*, 2005). Furthermore, overexpression of Zeama;KRP;1 in embryonic calli ectopically expressing the wheat dwarf virus RepA protein, a counteractor of RBR function, resulted in an additional round of endoreduplication. These overall data obtained from the two most spectacular endoreduplicating plant systems, namely tomato fruit and maize endosperm, support the role for KRPs in the control of endoreduplication, especially at the G/S transition of the endocycle.

The molecular mechanisms governing the exit from the cell cycle and the commitment towards endoreduplication in plant cells now become more and more documented (Verkest *et al.*, 2005b; and references therein). Compelling evidence relates the onset of endoreduplication in various plant organs and tissues with a decrease in CDK activity which may be both triggered by the action of WEE1 kinase or ICK/KRPs. However, how the endocycle is controlled, that is the successive rounds of DNA replication implying the passage from a gap (G) phase to a DNA synthesis (S) phase and so on, is still poorly understood. During the development of the gel tissue in tomato fruits, WEE1 and ICK/KRPs are not only involved in the onset of endoreduplication but, more importantly, in the maintenance of the endocycle to give rise to polyploidization. From our observations, it is proposed as a working hypothesis that (i) WEE1 could be involved in the control of the endocycle G phase length to allow sufficient cell growth in response to nuclear DNA amplification and nucleus size increase and (ii) ICK/KRPs could regulate the phase transitions of the endocycle by modulating the CDKA-CYCD kinase activity on RBR in the late G phase of the endocycle to prevent premature passage into the S phase. Whether plant WEE1 and ICK/KRPs participate in these different mechanisms is an intriguing matter of investigation to understand the plant endocycle control.

12.5 Genetic control of fruit size

As a result of plant breeding, fruit from domesticated species became enlarged and of a much complicated shape than their ancestor wild species (Tanksley, 2004). Close wild relatives of the common tomato (*Solanum lycopersicum*) bear fruit less than 1 cm in diameter and weighing a few grams while cultivated tomato varieties can weigh up to 1000 g, with a tremendous increase in diameter. Hence, tomato represents the model of choice for investigating the genetic basis of fruit size and shape determination, because of the large morphological diversity that is encountered among tomato domesticated varieties and the genetic and genomic tools that have been developed these past 15 years. Among these tools the availability of marker-assisted mapping techniques using crosses between small and round wild tomatoes and domesticated tomatoes of various sizes and shapes contributed to enrich our current knowledge of the genetic control of fruit development (Grandillo *et al.*, 1999). In this part we shall focus on the fruit size trait, which is quantitatively determined and obviously directly linked to fruit growth by increasing cell number and/or cell size, but also importantly linked to fruit composition.

Nearly 30 quantitative trait loci (QTLs) for fruit size/weight have been detected in the tomato genome (Grandillo *et al.*, 1999). Among these QTLs, *fw2.2* accounts for as much as a 30% difference in fruit fresh weight between the domesticated tomato and its wild relatives (Alpert *et al.*, 1995). This major QTL effect is caused by a single gene (*fw2.2*) that to date stands for the only fruit-size-controlling locus to be cloned and characterized at the molecular level (Frary *et al.*, 2000). Most wild – small fruited – tomatoes (if not all) possess 'small fruit' alleles. Conversely, all cultivated – large fruited – tomatoes possess 'large fruit' alleles. Two nearly isogenic lines (NILs) containing, respectively, the small-fruit and the large-fruit allele were primarily used to address the *fw2.2* gene expression profile (Frary *et al.*, 2000). This expression analysis revealed that *fw2.2* was expressed in all pre-anthesis floral organs, with higher levels in carpels. Interestingly, the expression found in carpels of the small-fruited NIL was significantly more important than in the large-fruited NIL. A detailed transcript analysis of *fw2.2* developmental expression indicated that the differences in mRNA levels between the small- and large-fruit alleles result in both the timing of *fw2.2* transcription (heterochronic changes) and the overall quantity of transcripts (Cong *et al.*, 2002). The large-fruit allele is rapidly transcribed to reach a peak of expression around 6 to 8 DPA, whereas the small-fruit allele is transcribed more slowly and displays its maximum of expression nearly a week later (around 12–14 DPA). The *fw2.2* transcript levels in the small-fruited NIL are twice as much that in the large-fruited NIL all along fruit development. Interestingly, this discrepancy in the timing of *fw2.2* expression is inversely correlated to the mitotic activity of the developing fruit (Cong *et al.*, 2002). On the basis of these expression analyses, it was thus postulated that *fw2.2* functions as a negative regulator of cell division in pre-anthesis ovary and developing fruit (Frary *et al.*, 2000; Cong *et al.*, 2002). Liu *et al.* (2003) constructed an artificial *fw2.2* gene dosage series, which was useful to demonstrate clearly that fruit mass is negatively correlated with the *fw2.2* transcript level. Hence, the primary effect of *fw2.2* is to control the mitotic activity of the developing fruit and thus modulates final fruit size. The allele-specific effect on fruit size is exerted through heterochronic regulatory mutations within the gene promoter since comparative sequence analysis of *fw2.2* from the large- and small-fruit alleles showed that the sequence and structure of the protein were identical (Cong *et al.*, 2002).

FW2.2 is homologous to other plant proteins, but it does not share sequence similarities with any known cell cycle regulatory protein. The three-dimensional protein structure predicted some similarities with the oncogene RAS (Frary *et al.*, 2000). Furthermore, two transmembrane-spanning domains were predicted from the FW2.2 sequence analysis, suggesting that it could interact with the cell membrane (Tanksley, 2004). Unfortunately, the precise biological function of FW2.2 in controlling cell division remains unknown.

As mentioned earlier, tomato displays a wide diversity in fruit size. Interestingly, this diversity is determined after ovary fertilization, since a remarkable conservation of pericarp pattern, including cell layer number and cell size, is observed at anthesis in various tomato lines displaying a large range in fruit weight (from 3 to 600 g) (Cheniclet *et al.*, 2005). Afterwards, large variations of growth occur due to cell division, which increases the number of cell layers and endoreduplication-associated cell expansion, both driving the increase in pericarp thickness and final fruit size. It is likely that the *FW2.2* allele strength may contribute to the observed variations developing after anthesis. Obviously, many intrinsic and environmental parameters may influence fruit growth and consequently fruit size: the number of seeds, the number of carpellar locules and water and carbon nutrition.

12.6 Metabolic control of fruit development and growth

The fruit acquires its organoleptic quality traits in the early stages of development. During the growth phase, mainly associated with cell expansion, water, organic acids (primarily citric and malic acids) and minerals accumulate inside the vacuole of fruit cells (Coombe, 1976), while starch accumulates transiently and is converted later on to reducing sugars (Wang *et al.*, 1993). Fruit softening, colouring, and sweetening then occur during the ripening phase (Giovannoni, 2001, 2004). As a result fruit weight and fruit composition in primary and secondary metabolites are intimately linked during fruit development. However, the quantitative variations of these traits are antagonistically controlled, which then hampered breeding strategies. It is thus of prime importance to identify genes putatively involved in the control of these processes.

Several genes may be responsible for the variation in fruit size and composition. Dealing with the former trait we have seen above the importance of *fw2.2* as a major fruit weight QTL, putatively linked with the control of cell cycle genes. Genes involved in carbon metabolism or partitioning, or any gene specifically expressed during synthesis and accumulation of reserves may contribute to the elaboration of the fruit composition trait. Genetic analyses were performed to identify putative candidate genes linked to fruit weight and composition with the aim to dissect the molecular and physiological bases of the antagonism between size and composition in fruit. The map localization of QTLs controlling fruit size and composition and that of candidate genes were compared using a population of introgression lines harbouring one chromosome segment of a wild species of tomato in the same cultivated tomato background (Causse *et al.*, 2004). A few candidate genes showing co-localization with fruit size and composition QTLs were thus highlighted. As an example, the *Lin5* QTL (encoding the cell wall invertase) that controls fruit sugar content (Fridman *et al.*, 2000) was recovered. Interestingly, several co-localizations with fruit weight QTLs and cell cycle genes were also observed, such as *CycA1*;1, *CycA2*;1 and *CycD3*;1 (Causse *et al.*, 2004). Although the assessment of these genes as being responsible for the observed QTLs is still lacking, these results were consistent with their expected role in tomato fruit organogenesis and growth driven by mitotic activities.

Cell division is dependent upon environmental conditions, especially those influencing the nutritional status of cells. Mitogenic stimuli such as sugars or hormones regulate the expression of D-type cyclins, which integrate these signals and then drive the entry into the cell cycle (Dewitte and Murray, 2003). Under unfavourable growth conditions such as carbon limitation, cell cycle genes such as *CycD3;*1, *CYCB2*;1 and *CDKB2*;1 were deeply repressed in tomato cell suspension cultures and excised roots (Joubès *et al.*, 2000b, 2001; Devaux *et al.*, 2003). Similarly, the reduction in tomato fruit growth observed as a consequence of the decline in photo-assimilate supply in tomato plants submitted to extended darkness was related to a strong repression of these genes inside fruit tissues (Baldet *et al.*, 2002). As a sink organ, the development of fruit is highly dependent upon the partitioning of photo-assimilates, which, when modified, significantly affect fruit development and size through the modulation of cell number and cell size (Bohner and Bangerth, 1988; Bertin *et al.*, 2002).

As expected, modifying the carbohydrate partitioning between source and sink organs modulates the expression of cell cycle genes. When compared to tomato plants grown under a standard fruit load (five fruits per truss), tomato plants grown under a low fruit load (one fruit per truss) displayed an increased photo-assimilate availability in the plant and an increased growth rate in all plant organs analyzed, including flower and fruit (Baldet *et al.*, 2006). The larger size of flower and fruit was correlated with higher cell number in the pre-anthesis ovary owing to the acceleration of the flower growth rate and enhancement of mitotic activities inside the carpel as witnessed by the strong induction of both *CDKB2*;1 and *CYCB2*;1 genes in young flower buds. Remarkably, the transcript abundance of *CYCD3;*1 also increased more than fivefold at very early stages of flower development in low-fruit-load plants, which strongly suggests that the effects on cell number observed in the ovary are the result of changes in *CYCD3*;1 levels in the young flower buds (Baldet *et al.*, 2006). Conversely, the expression of *fw2.2* and Solly;*KRP1* are both reduced in the early stage of developing flowers on the reduction of fruit load, which is consistent with their expected function of negative regulators of mitosis.

Changes in carbohydrate partitioning may thus control fruit size through the regulation of cell-proliferation-related genes at very early stages of flower development. However, the connection of *fw2.2* and cell cycle genes is far from being clearly understood. *fw2.2* was proposed to directly control fruit size through the regulation of mitotic activities, which in turn has a consequence in sink–source relationships at the whole-plant level (Nesbitt and Tanksley, 2001). According to this model, the timing and intensity of *fw2.2* expression determine the ovary size in early-developing flowers, which subsequently result in increased fruit size and cell number in plants carrying the large-fruit (weak) *fw2.2* allele or decreased fruit size and cell number in plants expressing the small-fruit (strong) allele. As a result of larger fruit production, plants face increased sink strength, that is the capacity to attract photo-assimilates, and intra-inflorescence competition. It may then lead to flower abortion and subsequent decrease of the number of produced inflorescences. In accordance with this model, the work from Baldet *et al.* (2006) suggests that the primary role of *fw2.2* in tomato is to adjust flower/fruit growth to the prevailing plant conditions, for example, the level of photo-assimilate supply available for the development of the reproductive organs. According to the cell cycle gene expression analysis, *fw2.2* may function upstream of *CYCD3*;1 and *ICK/KRP* in a signalling pathway linking the modulation of cell cycle machinery with hormonal or sugar signals, which control flower/fruit development (Baldet *et al.*, 2006). The elucidation of this signalling pathway is a very exciting challenge for the plant development community, as FW2.2 is a plant-specific protein present in both monocotyledonous and dicotyledonous species, in important agronomic crops (e. g., maize and rice) and Arabidopsis (Frary *et al.*, 2000).

12.7 Conclusion

Tomato fruit organogenesis represents an accurate and interesting plant system to study at the molecular and functional level the regulations of the processes of cell division and cell expansion, especially the interplay between the classical cell cycle and the endoreduplication cycle. Among the different plant models, tomato is especially well suited to address this question. Indeed, the final size of the fruit depends upon both the cell number and cell size (strongly associated with endoreduplication). The originality of the tomato fruit model resides in the cellular and structural diversity of tissues that compose the fruit, and very importantly in the cell size attained (cell diameter higher than 0.5 mm at the end of fruit development) and ploidy levels (up to 512C in some genotypes) (Cheniclet *et al.*, 2005). Such levels of ploidy represent unmatched values in other model plants such as Arabidopsis or maize, where the maximum ploidy is limited, respectively, to 32C (epidermal cells of the hypocotyls; Melaragno *et al.*, 1993) or 96C (seed endosperm; Vilhar *et al.*, 2002).

The elucidation of the functional role of endoreduplication during tomato fruit growth remains far from being unravelled. Similarly, why such an extent in the phenomenon is observed in tomato represents a crucial interrogation. Several physiological significances have been attributed to endoreduplication in plants. Since polyploidization results from the amplification of chromatid DNA, it may provide a means to protect the genome from DNA-damaging conditions caused by environmental factors or uneven chromosome segregation. The amplification of the genomic DNA obviously multiplies the gene copy number, which in turn increases the availability of DNA templates to be used for increased gene expression. Endoreduplication was thus proposed to control transcriptional- and subsequent translational- and metabolic activities in cells. Although ploidy-dependent gene expression is correlated to specific cell functions in yeasts (Galitski *et al.*, 1999), this assertion has never been demonstrated in plant cells. Leiva-Neto *et al.* (2004) intended to investigate the putative relationship between endoreduplication-driven cell expansion and high metabolic activity in maize endosperm. This study showed that there was little difference in the level of gene expression with high and low levels of endoreduplication, resulting in only slight effects on the level of starch and storage protein accumulation. This work supported the idea that endoreduplication in seed storage tissues could provide a mechanism for storing nucleotides or nitrogen during embryogenesis and/or germination.

Endoreduplication is involved in the cell differentiation of trichomes in Arabidopsis, which undergo four endoreduplication rounds during morphogenesis to form a large, branched single cell with a DNA content of 32C (Hülskamp *et al.*, 1999). This endoreduplication-associated cell differentiation is genetically determined, since a direct correlation exists between the number of branches and the ploidy level in mutant trichomes (Perazza *et al.*, 1999). Endoreduplication was also shown as being involved in cell differentiation resulting from plant pathogenic and symbiotic interactions (for a review see Kondorosi and Kondorosi, 2004).

Together with cell differentiation, the increase in cell size is largely correlated to endoreduplication in the plant kingdom. The work from Cheniclet *et al.* (2005) nicely demonstrated that a positive correlation exists between cell size and ploidy, as well as between fruit size and ploidy, in tomato. In this chapter we intended to demonstrate that altering the cell cycle towards endoreduplication is an important mechanism for tomato fruit growth. Consequently, it was proposed that endoreduplication is likely to be a driving regulator for cell expansion, the ultimate driving force for fruit growth, and evenly an enhancer of cell growth rate leading to an accelerated fruit growth or bigger fruit size (Cheniclet *et al.*, 2005).

Acknowledgments

I would like to express my deepest thanks to the members of the 'Fruit Biology' Laboratory (UMR-619), and especially to my closest collaborators of the research group 'Fruit Organogenesis and Endoreduplication' who contributed to this chapter by the excellence of their work.

References

Abrahams, S., Cavet, G., Oakenfull, E.A., *et al.* (2001) A novel and highly divergent Arabidopsis cyclin isolated by complementation in budding yeast. *Biochim Biophys Acta* **1539**, 1–6.

Alba, R., Fei, Z., Payton, P., *et al.* (2004) ESTs cDNA microarrays, and gene expression profiling: tools for dissecting plant physiology and development. *Plant J* **39**, 697–714.

Alba, R., Payton, P., Fei, Z., *et al.* (2005) Transcriptome and selected metabolite analyses reveal multiple points of ethylene control during tomato fruit development. *Plant Cell* **17**, 2954–2965.

Alpert, K.B., Grandillo, S. and Tanksley, S.D. (1995) *fw2.2*: a major QTL controlling fruit weight is common to both red- and green-fruited tomato species. Theor. *Appl Genet* **91**, 994–1000.

Azumi, Y., Liu, D., Zhao, D., Li, W., Wang, G. and Ma, H.Y (2002) Homolog interaction during meiotic prophase I in Arabidopsis requires SOLO DANCERS genes encoding a novel cyclin-like protein. *EMBO J* **21**, 3081–3095.

Balbi, V. and Lomax, T.L. (2003) Regulation of early tomato fruit development by the *Diageotropica* gene. *Plant Physiol* **131**, 186–197.

Baldet, P., Devaux, C., Chevalier, C., Brouquisse, R., Just, D. and Raymond, P. (2002) Contrasted responses to carbohydrate limitation in tomato fruit at two stages of development. *Plant Cell Environ* **25**, 1639–1649.

Baldet, P., Hernould, M., Laporte, F., *et al.* (2006) The expression of cell proliferation-related genes in early developing flower is affected by fruit load reduction in tomato plants. *J Exp Bot* **57**, 961–970.

Barg, R., Sobolev, I., Eilon, T., *et al.* (2005) The tomato early fruit specific gene *Lefsm1* defines a novel class of plant-specific SANT/MYB domain proteins. *Planta* **221**, 197–211.

Barrôco, R.M., De Veylder, L., Magyar, Z., Engler, G., Inzé, D. and Mironov, V. (2003) Novel complexes of cyclin-dependent kinases and a cyclin-like protein from *Arabidopsis thaliana* with a function unrelated to cell division. *Cell Mol Life Sci* **60**, 401–412.

Bergervoet, J.H.W., Verhoeven, H.A., Gilissen, L.J.W. and Bino, R.J. (1996) High amounts of nuclear DNA in tomato (*Lycopersicon esculentum* Mill.) pericarp. *Plant Sci* **116**, 141–145.

Bertin, N., Gautier, H. and Roche C. (2002) Number of cells in tomato fruit depending on fruit position and source-sink balance during plant development. *Plant Growth Reg* **36**, 105–112.

Bisbis, B., Delmas, F., Joubès, J., *et al.* (2006) Cyclin-Dependent Kinase Inhibitors are involved in endoreduplication during tomato fruit development. *J Biol Chem* **281**, 7374–7383.

Bohner, J. and Bangerth, F. (1988) Cell number, cell size and hormone levels in semi-isogenic mutants of *Lycopersicon pimpinellifolium* differing in frut size. *Physiol Plant* **72**, 316–320.

Brukhin, V., Hernould, M., Gonzalez, N., Chevalier, C. and Mouras, A. (2003) Flower development schedule in tomato *Lycopersicon esculentum* cv. sweet cherry. *Sex Plant Reprod* **15**, 311–320.

Busi, M.V., Bustamante, C., D'Angelo, C., *et al.* (2003) MADS-box genes expressed during tomato seed and fruit development. *Plant Mol Biol* **52**, 801–815.

Causse, M., Duffe, P., Gomez, M.C., *et al.* (2004) A Genetic map of candidate genes and QTLs involved in tomato fruit size and composition. *J Exp Bot* **55**, 1671–1685.

Cheniclet, C., Rong, W.Y., Causse, M., *et al.* (2005) Cell expansion and endoreduplication show a large genetic variability in pericarp and contribute strongly to tomato fruit growth. *Plant Physiol* **139**, 1984–1994.

Coelho, C.M., Dante, R.A., Sabelli, P.A., *et al.* (2005) Cyclin-dependent kinase inhibitors in maize endosperm and their potential role in endoreduplication. *Plant Physiol* **138**, 2323–2336.

Cong, B., Liu, J. and Tanksley, S.D. (2002) Natural alleles of a tomato QTL modulate fruit size through heterochronic regulatory mutations. *Proc Natl Acad Sci USA* **99**, 13606–13611.

Coombe, B. (1976) The development of fleshy fruits. *Ann Rev Plant Physiol* **27**, 507–528.

Devaux, C., Baldet, P., Joubès, J., *et al.* (2003) Physiological, biochemical and molecular analysis of sugar-starvation responses in tomato roots. *J Exp Bot* **54**, 1–9.

De Veylder, L., Beeckman, T., Beemster, G.T.S., *et al.* (2001) Functional analysis of cyclin-dependent kinase inhibitors of Arabidopsis. *Plant Cell* **13**, 1–15.

De Veylder, L., Beeckman, T., Beemster, G.T.S., *et al.* (2002) Control of proliferation, endoreduplication and differentiation by the *Arabidopsis* E2Fa-DPa transcription factor. *EMBO J* **21**, 1360–1368.

Dewitte, W. and Murray, J.A.H. (2003) The plant cell cycle. *Annu Rev Plant Mol Biol* **54**, 235–264.

Edgar, B.A. and Orr-Weaver, T.L. (2001) Endoreplication cell cycles: more for less. *Cell* **105**, 297–306.

Fantes, P. and Nurse, P. (1981) Division timing: controls, models and mechanisms. In: *The Cell Cycle* (ed. John, P.C.L.). Cambridge University Press, Cambridge, pp. 11–33.

Frary, A., Nesbitt, T.C., Frary A., *et al.* (2000) *fw2.2*: a quantitative trait locus key to the evolution of tomato fruit size. *Science* **289**, 85–88.

Fridman, E., Pleban, T. and Zamir, D. (2000) A recombination hotspot delimits a wild-species quantitative trait locus for tomato sugar content to 484 bp within an invertase gene. *Proc Natl Acad Sci USA* **97**, 4718–4723.

Galitski ,T., Saldanha, A.J., Styles, C.A., Lander, E.S. and Fink, G.R. (1999) Ploidy regulation of gene expression *Science* **285**, 251–254.

Gillaspy, G., Ben-David, H. and Gruissem, W. (1993) Fruits: a developmental perspective. *Plant Cell* **5**, 1439–1451.

Giovannoni, J. (2001) Molecular biology of fruit maturation and ripening. *Annu Rev Plant Physiol Plant Mol Biol* **52**, 725–749.

Giovannoni, J. (2004) Genetic regulation of fruit development and ripening. *Plant Cell* 16, S170–S180.

Gonzalez, N., Hernould, M., Delmas, F., *et al.* (2004) Molecular characterization of a *WEE1* gene homologue in tomato (*Lycopersicon esculentum* Mill.) *Plant Mol Biol* **56**, 849–861.

Gorguet, B., van Huesden, A.W. and Lindhout, P. (2005) Parthenocarpic fruit development in tomato. *Plant Biol* **7**, 131–139.

Grandillo, S., Ku, H.-M. and Tanksley, S.D. (1999) Identifying loci responsible for natural variation in fruit size and shape in tomato. *Theor Appl Genet* **99**, 978–987.

Harvey, S.L. and Kellog, D.R. (2003) Conservation of mechanisms controlling entry into mitosis: budding yeast Wee1 delays entry into mitosis and is required for cell size control. *Curr Biol* **13**, 264–275.

Ho, L.C. and Hewitt, J.D. (1986) Fruit development. In: *The Tomato Crop. A Scientific Basis for Improvement* (eds Athernon J.R. and Rudich J.). Chapman and Hall, London, New York, pp. 201–239.

Hocher, V., Sotta, B., Maldiney, R., Bonnet, M. and Miginiac, E. (1992) Changes in indole-3-acetic acid levels during tomato (*Lycopersicon esculentum* Mill.) seed development. *Plant Cell Reprod* **11**, 253–256.

Hülskamp, M., Schnittger, A. and Folkers, U. (1999) Pattern formation and cell differentiation: trichomes in Arabidopsis as a genetic model system. *Int Rev Cytol* **186**, 147–178.

Joubès, J., Chevalier, C. 2000. Endoreduplication in higher plants. *Plant Mol Biol* **43**, 737–747.

Joubès, J., Chevalier, C., Dudits, D., *et al.* (2000a) Cyclin-dependent kinases related protein kinases in plants. *Plant Mol Biol* **43**, 607–621.

Joubès, J., Lemaire-Chamley, M., Delmas, F., *et al.* (2001) A new C-type cyclin-dependent kinase from tomato expressed in dividing tissues does not interact with mitotic and G1 cyclins. *Plant Physiol* **126**, 1403–1415.

Joubès, J., Phan, T.-H., Just, D., *et al.* (1999) Molecular and biochemical characterization of the involvement of Cyclin-Dependent Kinase CDKA during the early development of tomato fruit. *Plant Physiol* **121**, 857–869.

Joubès, J., Walsh, D., Raymond, P. and Chevalier, C. (2000b) Molecular characterization of the expression of distinct classes of cyclins during the early development of tomato fruit. *Planta* **211**, 430–439.

Kondorosi, E. and Kondorosi, A. (2004) Endoreduplication and activation of the anaphase-promoting complex during symbiotic cell development. *FEBS Lett* **567**, 152–157.

Kvarnheden, A., Jao, J.L., Zhan, X., O'Brien, I. and Moris, B.A.M. (2000) Isolation of three disctinct CycD3 genes expressed during fruit development in tomato. *J Exp Bot* **51**, 1789–1797.

Landrieu, I., da Costa, M., De Veylder, L., *et al.* (2004) A small CDC25 dual-specificity tyrosine-phosphatase isoform in *Arabidopsis thaliana. Proc Natl Acad Sci USA* **101**, 13380–13385.

Leiva-Neto, J.T., Grafi, G., Sabelli, P.A., *et al.* (2004) A dominant negative mutant of cyclin-dependent kinase A reduces endoreduplication but not cell size or gene expression in maize endosperm. *Plant Cell* **16**, 1854–1869.

Lemaire-Chamley, M., Petit, J., Garcia, V., *et al.* (2005) Changes in transcriptional profiles are associated with early fruit tissue specialization in tomato. *Plant Physiol* **139**, 292–299.

Liu, J., Cong, B. and Tanksley, S.D. (2003) Generation and analysis of an artificial gene dosage series in tomato to study the mechanisms by which the cloned quantitative trait locus *fw2.2* controls fruit size. *Plant Physiol* **132**, 292–299.

Melaragno, J.E., Mehrotra, B. and Coleman, A. W. (1993) Relationship between endopolyploidy and cell size in epidermal tissue of *Arabidopsis*. *Plant Cell* **5**, 1661–1668.

Menges, M., de Jager, S.M., Gruissem, W. and Murray, J.A.H. (2005) Global analysis of the core cell cycle regulators of Arabidopsis identifies novel genes, reveals multiple and highly specific profiles of expression and provides a coherent model for plant cell cycle control. *Plant J* **41**, 546–566.

Menges, M, Hennig, L., Gruissem, W. and Murray, J.A. (2003) Genome-wide gene expression in an Arabidopsis cell suspension. *Plant Mol Biol* **53**, 423–442.

Nakagami, H., Sekine, M., Murakami, H. and Shinmyo, A. (1999) Tobacco retinoblastoma-related protein phosphorylated by a distinct cyclin-dependent kinase complex with Cdc2/cyclin D *in vitro*. *Plant J* **18**, 243–252.

Nesbitt, T.C. and Tanksley, S.D. (2001) *fw2.2* directly affects the size of developing fruit, with secondary effects on fruit number and photosynthate partitioning. *Plant Physiol* **127**, 575–583.

Oh, K.C., Ivanchenko, M.C., White, T.J. and Lomax, T.L. (2006) The *diageotropica* gene of tomato encodes a cyclophilin: a novel player in auxin signalling. *Planta* **224**, 133–144.

Park, J.A., Ahn, J.W., Kim, Y.K., *et al.* (2005) Retinoblastoma protein regulates cell proliferation, differentiation, and endoreduplication in plants. *Plant J.* **42**, 153–163.

Perazza, D., Herzog, M., Hülskamp, M., Brown, S., Dorne, A.M. and Bonneville, J.M. (1999) Trichome cell growth in *Arabidopsis thaliana* can be derepressed by mutations in at least five genes. *Genetics* **152**, 461–476.

Renaudin, J.-P., Doonan, J.H., Freeman, D., *et al.* (1996) Plant cyclins: a unified nomenclature for plant A-, B- and D-type cyclins based on sequence organization. *Plant Mol. Biol* **32**, 1003–1018.

Russell P. and Nurse, P. (1987) Negative regulation of mitosis by wee1+, a gene encoding a protein kinase homolog. *Cell* **49**, 559–67.

Schnittger, A., Weinl, C., Bouyer, D., Schöbinger, U. and Hülskamp, M. (2003) Misexpression of the cyclin-dependent kinase inhibitor *ICK1/KRP1* in single-celled Arabidopsis trichomes reduces endoreduplication and cell size and induces cell death. *Plant Cell* **15**, 303–315.

Sorrell, D.A., Marchbank, A., McMahon, K., Dickinson, J.R., Rogers, H.J. and Francis, D. (2002) A *WEE1* homologue from *Arabidopsis thaliana*. *Planta* **215**, 518–522.

Sugimoto-Shirasu, K. and Roberts, K. (2003) "Big it up": endoreduplication and cell-size control in plants. *Curr Opin Plant Biol* **6**, 544–553.

Sun, Y., Dilkes, B.P., Zhang, C., *et al.* (1999) Characterization of maize (*Zea mays* L.) Wee1 and its activity in developig endosperm. *Proc Natl Acad Sci USA* **96**, 4180–4185.

Tanksley, S.D. (2004) The genetic, developmental, and molecular bases of fruit size and shape variation in tomato. *Plant Cell* **16**, S181–S189.

Torres-Acosta, J.A., de Almeida Engler, J., Raes, J., *et al.* (2004) Molecular characterization of Arabidopsis PHO80-like proteins, a novel class of CDKA;1-interacting cyclins. *Cell Mol Life Sci* **61**, 1485–1497.

Vandepoele, K., Raes, J., De Veylder, L., Rouzé, P., Rombauts, S. and Inzé, D. (2002) Genome-wide analysis of core cell-cycle genes in Arabidopsis. *Plant Cell* **14**, 903–916.

Van Der Hoeven, R., Ronning, C., Martin, G., Giovannoni, J. and Tanksley, S. (2002) Deductions about the number, organization, and evolution of genes in tomato genome based on analysis of a large expressed sequence tag collection and selective genomic sequencing. *Plant Cell* **14**, 1441–1456.

Varga, A. and Bruinsma, J. (1986) Tomato. In: *CRC Handbook of Fruit Set and Development* (ed. Monselise, S.P..) CRC Press, Boca Raton, FL, pp. 461–480.

Verkest, A., de O. Manes, C.L., Vercruysse, S., *et al.* (2005a) The cyclin-dependent kinase inhibitor KRP2 controls the onset of endoreduplication cycle during Arabidopsis leaf development through inhibition of mitotic CDKA;1 kinase complexes. *Plant Cell* 17, 1723–1736.

Verkest, A., Weinl, C., Inzé, D., De Veylder, L. and Schnittger, A. (2005b) Switching the cell cycle. Kip-related proteins in plant cell cycle control. *Plant Physiol* **139**, 1099–1106.

Vilhar, B., Kladnik, A., Blejec, A., Chourey, P.S. and Dermastia, M. (2002) Cytometrical evidence that the loss of seed weight in the *miniature1* seed mutant of maize is associated with reduced mitotic activity in the developing endosperm. *Plant Physiol* **129**, 23–30.

Wang, F., Sanz, A., Brenner, M.L. and Smith, A.G. (1993) Sucrose synthase, starch accumulation, and tomato fruit sink strength. *Plant Physiol* **101**, 321–327.

Wang, G., Kong, H., Sun, Y., *et al.* (2004) Genome-wide analysis of the cyclin family in Arabidopsis and comparative phylogenetic analysis of plant cyclin-like proteins. *Plant Physiol* **135**, 1084–1099.

Wang, H., Jones, B., Li, Z., *et al.* (2005) The tomato *Aux/IAA* transcription factor *IAA9* is involved in fruit development and leaf morphogenesis. *Plant Cell* **17**, 2676–2692.

Wang, H., Zhou, Y., Gilmer, S., Whitwill, S. and Fowke, L.C. (2000) Expression of the cyclin-dependent protein kinase inhibitor ICK1 affects cell division, plant growth and morphology. *Plant J* **24**, 613–623.

Weinl, C., Marquardt, S., Kuijt, S.J.H., *et al.* (2005) Novel functions of plant cyclin-dependent kinase inhibitors, ICK1/KRP1, can act non-cell-autonomously and inhibit entry into mitosis. *Plant Cell* **17**, 1704–1722.

Yamaguchi, M., Fabian, T., Sauter, M., *et al.* (2000) Activation of CDK-activating kinase is dependent on interaction with H-type cyclins in plants. *Plant J* **24**, 11–20.

13 Cell cycle and endosperm development

Paolo A. Sabelli, Hong Nguyen and Brian A. Larkins

13.1 Introduction

The seed endosperm results from the fertilization of the central cell of the embryo sac by one pollen sperm cell, while the embryo originates from the fertilization of the egg cell by the other pollen sperm cell. The foremost function of the endosperm is to provide nutrients to the developing and germinating embryo, and in many cases to the seedling as well. In many species, including *Arabidopsis*, the endosperm has an ephemeral existence, and it is largely digested and displaced by the developing embryo, so that the mature seed is almost devoid of endosperm tissue. In other species, such as cereals, the endosperm is a persistent seed structure. In addition to providing nutrients for the embryo, it is the most important source of calories for human and livestock nutrition and is the basis for countless manufactured foods, goods and biofuel.

Cell proliferation is the driving force behind the growth and development of the endosperm. Although several distinct patterns of endosperm development exist, the role of cell proliferation and its contribution to endosperm development have been best studied in cereals and the model dicot *Arabidopsis thaliana*, both of which have the most common type of endosperm development in angiosperms and share remarkably similar developmental patterns. Thus, in this chapter we focus on the role of the cell cycle in the development of maize and *Arabidopsis* endosperm. The reader interested in a broader discussion of endosperm development is referred to several excellent reviews (Friedman, 1998; Brown *et al.*, 1999; Berger, 2003; Costa *et al.*, 2004; Lersten, 2004; Olsen, 2004).

13.2 Endosperm development: a cell cycle perspective

In most angiosperms, the endosperm is the product of the fusion of a haploid sperm nucleus with two haploid polar nuclei (which, depending on the species, may have fused to give a diploid nucleus) in the central cell of the female gametophyte. As a result, the primary endosperm nucleus has a triploid DNA content (3C). Almost invariably, the endosperm begins mitosis earlier and at a faster rate that the zygote (Lersten, 2004). For example, by the time the zygote divides for the first time the endosperm contains 4–8 nuclei in maize (Randolph, 1936), and at least 16 nuclei in *Arabidopsis* (Brown *et al.*, 2003). In the most common type of early endosperm development, called 'coenocytic' (known also as nuclear) (Olsen, 2004), up to several thousand nuclei (e.g. 256–512 in maize, ~200 in *Arabidopsis*) are generated within the embryo sac, often through synchronous mitotic waves, in the absence of cytokinesis. These nuclei become localized at the periphery of the primary endosperm cell as a result of enlargement of the central vacuole. Following this coenocytic phase, the cellularization phase ensues, with the initiation of a network of internuclear radial microtubules that subdivides the endosperm in specific nuclear cytoplasmic domains (Figures 13.1A and 13.1B). Cell walls are then formed by adventitious phragmoplasts at sites of microtubule intersection, giving rise to tubular structures termed alveoli that surround each nucleus. Once the first endosperm cell layer is formed, the nuclei divide synchronously and periclinally, generating an adjacent, inner layer of alveoli (Figures 13.1C and 13.1D). This process proceeds centripetally and occurs without formation of the pre-prophase band, a cytoskeletal structure that typically marks the site of cell wall formation during division of somatic plant cells. Endosperm cellularization is normally completed by 4 days after pollination (DAP) in maize. In *Arabidopsis*, the endosperm is cellularized by the torpedo stage of embryo development, except for a delay in the chalazal region (Brown *et al.*, 1999), where nuclei perhaps first undergo endoreduplication (Boisnard-Lorig *et al.*, 2001). Following cellularization, a period of mitotic divisions takes place, which in maize lasts until 8–12 DAP in the central endosperm and continues until several days later in the aleurone and subaleurone layers (Kowles and Phillips, 1988). In this species, an overall peak in mitotic activity of 10% occurs about 8–10 DAP, which thereafter declines sharply.

The number of cells provides an indication of the extent of the mitotic phase of endosperm development. Estimates of cell number range over fourfold in various maize and wheat genotypes (e.g. 176 000–880 000 cells in maize endosperm), and approximately twofold in barley (Kvaale and Olsen, 1986), but it is not clear whether such intraspecific variation depends on differences in the experimental procedures used to count cells or the genotypes, environmental effects, or the developmental stages.

From around 8 to 10 DAP, maize endosperm cells switch from a mitotic to an endoreduplication cell cycle, where seemingly complete and reiterated rounds of DNA replication occur in the absence of chromatin condensation, sister chromatid segregation, or cytokinesis (reviewed in Larkins *et al.*, 2001; Sabelli *et al.*, 2005b). Because this is an asynchronous process, middevelopment (i.e. 18 DAP) maize endosperm contains a heterogeneous population of cells with nuclei varying in DNA content as a multiplier of the basic 3C value, and ploidy values up to 384C or more, indicating up to seven to eight rounds of endoreduplication (Kowles and Phillips, 1988). Flow-cytometric profiles clearly illustrate the dynamics of endoreduplication during the process of endosperm development (Figure 13.2). Cells toward the base of the endosperm and at its center are the first to engage in endoreduplication, and consequently they contain the largest amount of nuclear DNA at any stage. As a result, a clear gradient in nuclear size is observed in tissue sections, with the smallest nuclei (3C and 6C) located at the periphery of the endosperm and increasingly larger nuclei located toward the center. Cell size appears to be correlated with nuclear size and the extent of endoreduplication, as central endosperm cells

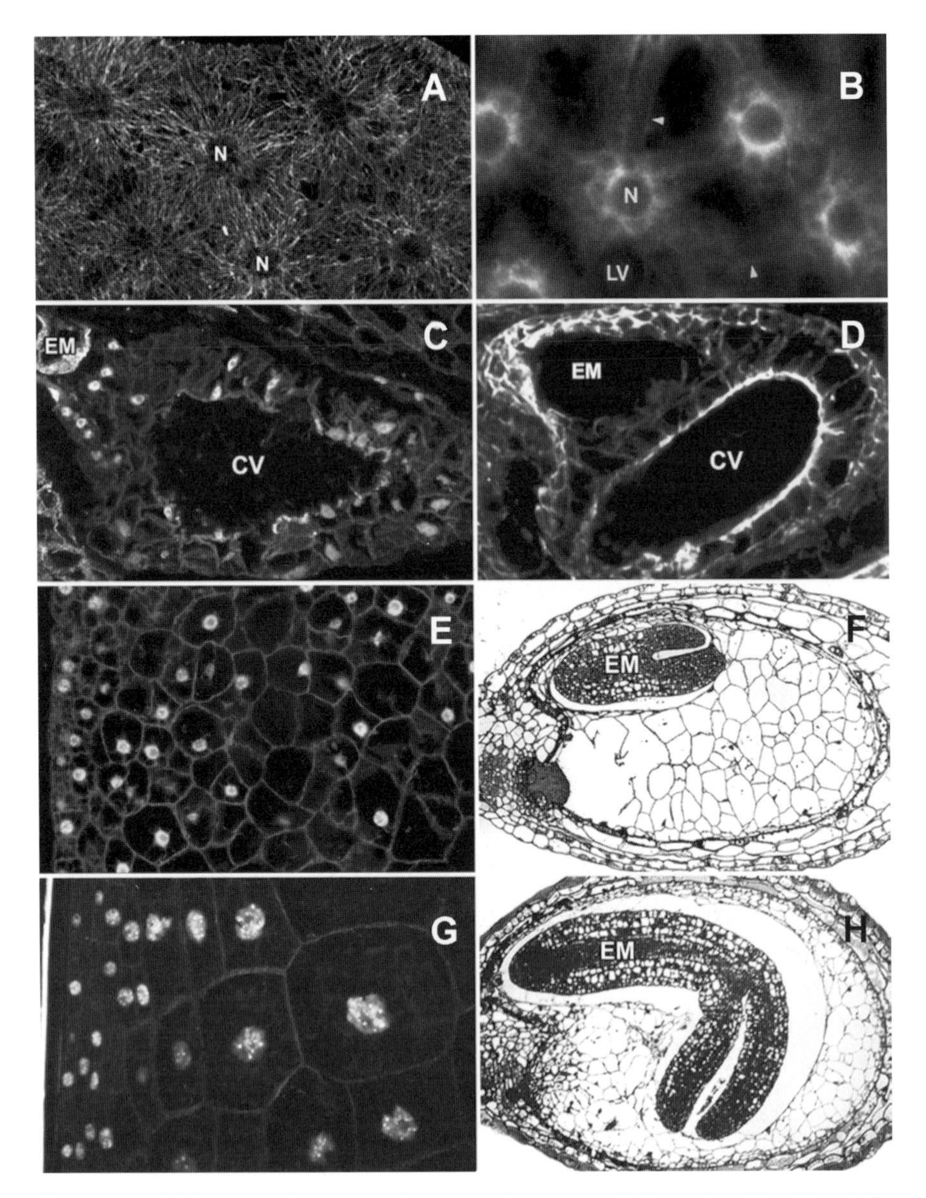

Figure 13.1 Major stages of endosperm development in maize (A, C, E and G) and *Arabidopsis* (B, D, F and H). (A and B) Syncytial stage characterized by acytokinetic mitosis; (C and D) Cellularization stage; (E and F) Mitotic stage; (G) Endoreduplication stage in maize and (H) Embryo growth and endosperm absorption in *Arabidopsis*. Tissue sections stained for tubulin (A and B) or DNA (C, D, E and F). Symbols: CV, central vacuole; EM, embryo; LV, lateral vacuole; N, nucleus. Arrowheads indicate radial microtubules. (Panels B, D and F are reproduced, with kind permission of the author and Springer Science and Business Media, from Brown *et al.*, 1999.)

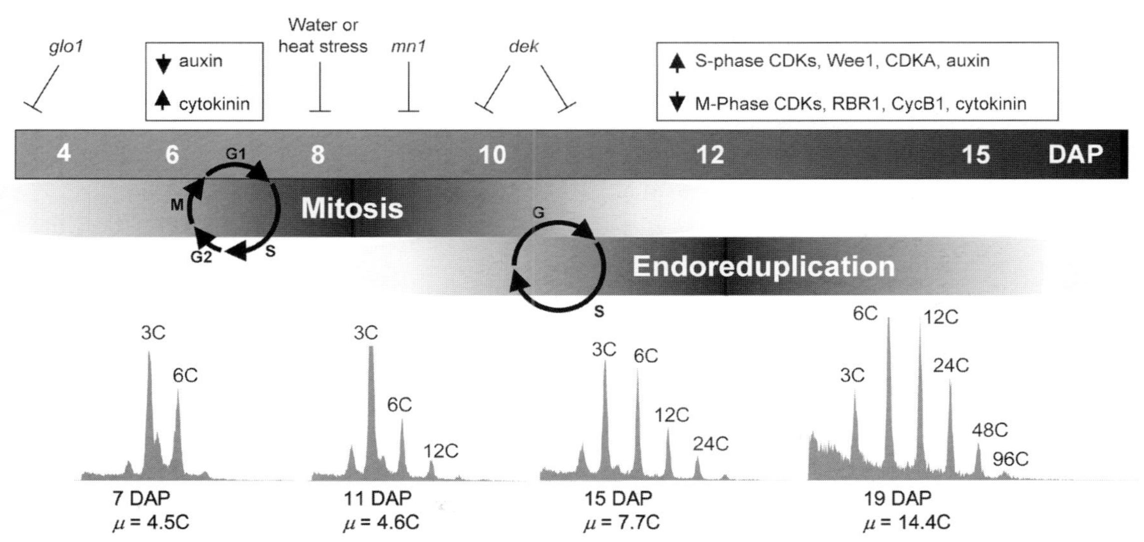

Figure 13.2 Factors affecting the mitotic (~4–12 DAP) and endoreduplication (~8–25 DAP) stages of maize endosperm development. Flow-cytometric profiles illustrate the change in nuclear ploidy during the transition from the mitotic stage (G1 and G2 nuclear DNA content peaks with mean ploidy = 4.5C at 7 DAP) to the endoreduplication stage (multiple peaks with increasing DNA content and mean ploidy). The top of the figure shows mutations or other factors that affect the mitotic and endoreduplication phases of endosperm development. (Adapted, with permission of the publisher, from Sabelli *et al.*, 2005b.)

are massive compared to peripheral and aleurone cells (Figure 13.1G) (Vilhar *et al.*, 2002). Coincident with the onset of endoreduplication, maize endosperm manifests a dramatic change in gene expression, leading to the rapid synthesis and accumulation of storage compounds, such as starch and storage proteins (zein). This relationship is the basis of the long-held view that the increased gene copy number resulting from endoreduplication of DNA supports high levels of gene expression and metabolic output.

Endosperm development in sorghum follows a pattern similar to that of maize, with endoreduplication initiating at an earlier stage (5 DAP) and generating nuclei up to 96C. Ploidy levels in sorghum endosperm nuclei are also correlated with cell size (Kladnik *et al.*, 2006). Studies in wheat have shown a pattern of endosperm development roughly similar to that in maize and sorghum, with an initial syncytium, a period of intense cell division generating at least 100 000 cells, and an endoreduplication phase resulting in ploidy values up to 24C (Chojecki *et al.*, 1986). Since bread wheat is a hexaploid species with three distinct genomes, the actual gene copy number (equivalent to that of a 72C DNA content in a diploid species) may be regarded as approaching that in maize and sorghum.

In *Arabidopsis* and many other angiosperms, mid-seed development is characterized by the absorption of the endosperm and expansion of the cotyledons of the developing embryo (Figure 13.1H), which become the site of storage metabolite (starch, storage protein, oil) accumulation. Even though the mature *Arabidopsis* endosperm consists of only one aleurone-like layer of cells, control of cell proliferation during its early development is crucially important for the size of the mature seed (Brown *et al.*, 1999).

13.3 Genetic control of endosperm cell proliferation

In maize, a relatively large collection of mutants has been described with defective development of the endosperm (*de* mutants), embryo (*emb* mutants), or both (most *dek* mutants) (Sheridan and Neuffer, 1980; Neuffer *et al.*, 1997) (Figure 13.2). Analysis of a group of 35 *dek* mutants revealed that all but one manifested defects in mitotic activity and endoreduplication (Kowles *et al.*, 1992), suggesting that to a large extent, but not always, the mutations affected regulatory factors common to the mitotic and endoreduplication cell cycles. However, the identification of one *dek* mutant with reduced endoreduplication but an unperturbed mitotic cell cycle suggested that certain genes may control endoreduplication, specifically, and that they might be identified through genetic screens. Although these mutants were identified several decades ago, little or nothing is known about how they impact cell cycle regulation. Because cell proliferation and development are tightly coupled (Gutierrez, 2005), it is generally quite difficult to attribute a developmental mutation to strictly a cell cycle defect. For example, cell proliferation requires a high-energy input, and therefore mutations in genes controlling house-keeping functions or carbohydrate metabolism can adversely affect the growth and development of the endosperm in ways that are difficult to distinguish from a cell cycle defect. *Miniature1* (*mn1*) and *dek1* are examples of such mutations. In *mn1*, loss of cell wall invertase INCW2 activity results in decreased mitotic activity, leading to a smaller endosperm (Vilhar *et al.*, 2002), whereas *dek1*, a mutation in a membrane-anchored, calpain-like cysteine proteinase, results in endosperm lacking the aleurone cell layer (Wang *et al.*, 2003) (Figure 13.2). Transposon-induced mutations are a valuable source of novel endosperm mutants that facilitate isolation of the affected gene responsible for the mutant phenotype. For example, the *globby1* (*glo1*) mutant results from a mutation that interferes with the syncytial and cellularization stages of early maize endosperm development (Costa *et al.*, 2003) (Figure 13.2). Also, the mitotic division plane is incorrectly formed in the *disorgal1* (*dil1*) and *disorgal2* (*dil2*) mutants, and this results in a disorganized aleurone (Lid *et al.*, 2004). Although these transposon-based mutations may soon allow the isolation of the affected genes, the identities of the *glo1*, *dil1* and *dil2* mutations are not currently known.

Screening of large *Arabidopsis* mutant collections has resulted in the isolation of a considerable number of mutations/genes affecting cell proliferation during endosperm development in this species (Table 13.1) (Lepiniec *et al.*, 2005). Often, *Arabidopsis* endosperm cell cycle mutants have a defect during nuclear divisions in the syncytium, such as in the *tor* (Menand *et al.*, 2002), *s5* (Weijers *et al.*, 2001), *atcul1* (Shen *et al.*, 2002), *atubp14* (Doelling *et al.*, 2001) and *orc2* (Collinge *et al.*, 2004) mutants, or during the cellularization stage, such as in the *ttn* group (Liu and Meike, 1998; Liu *et al.*, 2002; Tzafrir *et al.*, 2002), *atfh5* (Ingouff *et al.*, 2005a), *pilz* group (Mayer *et al.*, 1999; Steinborn *et al.*, 2002), *haiku/iku* (Garcia *et al.*, 2003, 2005; Luo *et al.*, 2005), *spätzle* (Sorensen *et al.*, 2002), *mini3* (Luo *et al.*, 2005), *prolifera* (Springer *et al.*, 2000; Holding and Springer, 2002) and *exs* (Canales *et al.*, 2002) mutants. The phenotype of these mutants is often small seeds, indicating that the proper sequence of events and extent/timing of nuclear division in the syncytium and subsequent cellularization are critical to the attainment of the normal size of the seed, which is made up almost exclusively of the embryo and the seed coat but not the endosperm. Thus, regulation of the cell cycle in the endosperm of *Arabidopsis* is important for seed size at maturity. Many of the mutations identified so far are pleiotropic, in that they also affect the development of the embryo to different extents. This is not surprising as many aspects of cell cycle regulation in the embryo and the endosperm may be regulated by the same sets of genes. However, the *haiku/iku* and *spätzle* mutations seem to affect exclusively cell proliferation in the endosperm, making them particularly interesting to study the development of this tissue. These mutants also indicate that although somatic cytokinesis and endosperm cellularization share many components, the latter requires the function of specific genes that are dispensable in the former.

Mutation of the *AtDPB2* gene, which encodes the large subunit of DNA polymerase ϵ, has recently revealed an essential role of this DNA polymerase in early nuclear endosperm division following fertilization (Ronceret *et al.*, 2005). Since *AtDPB2* (along with *AtPOL2a*, which encodes the other subunit of the enzyme) is most likely an E2F target, and because loss of function of RBR1 results in hyperproliferation of the central cell of the embryo sac (and in ovule abortion) (Ebel *et al.*, 2004), with a phenotype resembling that of the *fis* mutation (see below), DNA

Table 13.1 A list of mutants affecting cell cycle and the development of the endosperm in *Arabidopsis*

Mutant	Identity	Affected stage/phenotype	Reference
atdpb2	DNA pol ε subunit	Nuclear cell division	Ronceret *et al.*, 2005
cdc2a	CDKA;1	Endosperm development resulting from exclusive fertilization of egg	Nowack *et al.*, 2006
titan(ttn) group	SMC proteins	Giant and polyploid syncytial nuclei; defective cellularization	Liu and Meike, 1998; Liu *et al.*, 2002; Tzafrir *et al.*, 2002
atfh5	Formin	Delayed cellularization	Ingouff *et al.*, 2005a
orc2	ORC subunit	Premature arrest of nuclear divisions; few enlarged nuclei	Collinge *et al.*, 2004
pilz group	Tubulin folding complex subunits	Few enlarged nuclei; impaired cytokinesis and cellularization due to lack of microtubules	Mayer *et al.*, 1999; Steinborn *et al.*, 2002
haiku/iku	LRR kinase (HAIKU2)	Early cellularization; reduced growth; small seed	Garcia *et al.*, 2003, 2005; Luo *et al.*, 2005
spätzle	n.a.	Absence of cellularization/small seed	Sorensen *et al.*, 2002
knolle	Sintaxin	Absence of cellularization	Sorensen *et al.*, 2002
hinkel	Kinesin-related	Absence of cellularization	Sorensen *et al.*, 2002
runkel	n.a.	Partial cellularization	Nacry *et al.*, 2000
fis	PcG protein	Absence of cellularization	Ingouff *et al.*, 2005b
pleiade	MPAP65-3	Mild cellularization defect	Sorensen *et al.*, 2002
fass/ton2	Phosphatase 2A subunit	Defect in periclinal division	Sorensen *et al.*, 2002
mini3	WRKY10	Early cellularization; small seed	Luo *et al.*, 2005
prolifera	MCM7	Enlarged nuclei; defective cytokinesis	Springer *et al.*, 2000; Holding and Springer, 2002
tor	Target of rapamycin	Arrest of syncytial development; small seed	Menand *et al.*, 2002
s5	Ribosomal protein S5	Syncytial arrest of division and growth; small seed	Weijers *et al.*, 2001
exs	Receptor kinase	Delayed cellularization; small seed	Canales *et al.*, 2002
atcul1	SCF complex subunit	Arrest of central nuclei division	Shen *et al.*, 2002
atubp14	Allelic to *ttn6*	Defective syncytial nuclear division	Doelling *et al.*, 2001

Note: n.a.; identity not available.

polymerase ϵ may be the main target of a FIS–RBR1–E2F pathway that represses premature nuclear division within the central cell of the embryo sac but reactivates DNA replication upon double fertilization (Ronceret *et al.*, 2005). Interestingly, activation of proliferation of the central cell also appears to depend on an additional positive signal originating from the fertilization of the egg cell, as indicated by the development of unfertilized endosperm on exclusive fertilization of the egg cell by *cdc2a* pollen, even in a *fis1* mutant background (Nowack *et al.*, 2006).

13.4 The cell cycle molecular engine during endosperm development

The relatively large size of the maize endosperm, coupled with the ability to induce transgene expression within it beginning about 10 DAP using regulatory sequences of 27-kDa γ-zein gene, makes it possible to study cell cycle regulation during the endoreduplication phase of endosperm development (Larkins *et al.*, 2001). Work by Grafi and Larkins (1995) identified a peak in CDK activity at 10–12 DAP that did not simply depend on changes in the amount of the CDK protein. Experiments involving a mixture of 10 DAP and 15 DAP endosperm extracts revealed that the latter contained a specific inhibitory activity of CDK. In addition, mitotic endosperm cells at 8 DAP contained high levels of M-phase kinase activity, whereas endoreduplicated cells at 16 DAP displayed a peak in S-phase-like kinase activity. Thus, the onset of endoreduplication seems to involve the simultaneous inhibition of M-phase CDKs and the activation of S-phase CDKs, which is in agreement with data from other systems. Downregulation of M-phase CDK activity during endosperm endoreduplication correlates with decreased accumulation of mRNA encoding mitotic cyclin B1 (Sun *et al.*, 1999b). It also correlates with a peak (at 15 DAP) in transcript accumulation of a maize Wee1 kinase homolog, which was shown to inhibit cell division in *Schizosaccharomyces pombe* and mitotic CDK activity *in vitro* (Sun *et al.*, 1999a). Characterization of two CKIs from maize, Zeama;KRP;1 and Zeama;KRP;2, that inhibit cyclin A1;3 and D5;1 but not cyclin B1;3-associated CDK activities, suggested the Zeama;KRP;1 may contribute to the inhibition of CDK observed in endoreduplicating endosperm cells (Coelho *et al.*, 2005). Thus, it appears that modulation of CDK activity is an important aspect of the transition from a mitotic to an endoreduplication cell cycle.

Mounting evidence suggests a key role for the RBR-E2F pathway in coordinating cell cycle regulation with development and differentiation (Gutierrez, 2005) and it clearly is important during maize endosperm development. Three RBR genes have been isolated from maize: *RBR1* (Grafi *et al.*, 1996; Xie *et al.*, 1996), *RBR2* (Ach *et al.*, 1997a) and *RBR3* (Sabelli *et al.*, 2005a; Sabelli and Larkins, 2006). (See the Chapter ‘Function of the Retinoblastoma-Related Protein In Plants’ for a review on plant RBRs.) RBR1 and RBR2 proteins are about 90% identical in sequence and are likely paralogs with essentially redundant functions. The properties of RBR1 are consistent with its having an inhibitory role during the G1/S-phase transition. Its interaction with several proteins capable of inducing cell cycle progression (such as SV40 T-antigen, adenovirus E1A and papillomavirus E7, wheat dwarf virus RepA, tomato golden mosaic virus AL1, and plant cyclin Ds) indicates that, similarly to oncogenic transformation in mammals, relief from an RBR-block induces cell cycle entry and progression. RBR1 has also been implicated in suppressing gene expression through interaction with RPD3-type histone deacetylase (Rossi *et al.*, 2003) and members of the PcG protein family, such as FIE (Mosquna *et al.*, 2004). Analysis of RBR1 expression during maize endosperm development revealed that both the transcript and protein accumulate in endoreduplicating endosperm (Sabelli *et al.*, 2005a) – an observation that contrasts with the presumed upregulation of the G/S-phase transition in endoreduplicating cells. Although it is assumed that some S-phase CDK(s) phosphorylate and inactivate RBR1 during endoreduplication (Grafi *et al.*, 1996), the role of increased accumulation of RBR1 protein and its inactivation by phosphorylation during the transition from a mitotic to an endoreduplication cell cycle is unknown. Interestingly, RBR3, whose sequence is only about 50% identical to RBR1, appears to be an E2F target and is repressed by RBR1 (Sabelli *et al.*, 2005a; Sabelli and Larkins, 2006). RBR3, in contrast to RBR1, is preferentially expressed during the mitotic phase of endosperm development, and its expression sharply declines at the onset of endoreduplication and remains very low throughout this phase of development. This downregulation of RBR3 suggests that RBR1 is present in an active form that may be important in bringing endoreduplication to a halt and, perhaps, in targeting senescent endosperm cells for programmed cell death. The precise function of RBR3 in cell cycle regulation is unknown, though it is expected, by analogy with other systems, to have a negative role. It is puzzling that for a predicted cell cycle inhibitor, RBR3 is expressed at its highest level in mitotically active cells. Because of this and other similarities with the mammalian RBR, p107, it was proposed that RBR3 is the maize equivalent of p107 (Sabelli *et al.*, 2005a). This raises the intriguing possibility that a compensatory loop involving RBR1 and RBR3 might provide robust control over pocket protein activity. The complexity of the RBR family in maize, which resembles that found in mammals, seems to be a feature common and unique to the Gramineae, and perhaps monocots, among flowering plants (Sabelli *et al.*, 2005a; Sabelli and Larkins, 2006). The implications of a complex RBR family influencing endosperm development are not obvious. Animal RBR genes have distinct but overlapping roles, with the final outcome often depending on the physiological context in which they are studied. Forward and reverse genetic experiments aimed at dissecting the roles of RBR1 and RBR3 in maize are underway, and the results could provide insight regarding the mechanism of cell cycle regulation during endosperm development in all cereal species.

13.5 Role of CDKA in the endoreduplication cell cycle

We have identified many of the genes encoding key cell cycle regulators in maize and characterized their regulation and function during endosperm development. Recently, we generated transgenic plants overexpressing a wild-type CDKA (CDKA-WT) or its dominant-negative (CDKA-DN) allele, in which an Asp-146 to Asn-146 mutation interferes with ATP binding, thereby abolishing kinase activity (Leiva-Neto *et al.*, 2004). Expression of these transgenes was restricted to the endosperm by the 27-kDa γ-zein regulatory sequences beginning about 10 DAP, roughly coincident with the onset of endoreduplication. Expression of CDKA-WT and CDKA-DN in independent transgenic lines resulted in a roughly 50% increase and decrease, respectively, of kinase activity, as measured by phosphorylation of histone H1. Whereas CDKA-WT endosperm did not display any abnormalities in terms of cell division rates, nuclear DNA content, or cell number and size, CDKA-DN endosperms showed an approximately 50% decrease in endoreduplication, with virtually no cells exceeding a nuclear DNA content of 24C at 18 DAP – whereas 96C nuclei were routinely detected in wild-type endosperms from the same segregating ear (Figure 13.3). Although the size of CDKA-DN nuclei was visibly reduced compared to control, non-transgenic nuclei, especially within the larger cells in the central area of the starchy endosperm, there was no obvious change in average cell size between non-transgenic and transgenic endosperms (Figure 13.3). Both the expression of transcripts encoding starch biosynthetic enzymes and zeins, and the accumulation of starch and 27 kD γ-zein protein were only slightly reduced in CDKA-DN endosperms. These experiments implicate CDKA in the regulation of the endoreduplication cell cycle, though they suggest that this kinase is not rate-limiting for the process. They also show, at least in this context, that endoreduplication and nuclear size can be uncoupled from cell size and gene expression levels, challenging the long-standing hypothesis that endoreduplication in the developing endosperm drives cell growth and supports higher levels of gene expression. As an alternative function of endoreduplication in the endosperm, since endosperm cells undergo programmed cell death from about 20 DAP, it is possible that endopolyploid nuclei provide a reservoir of phosphate and nucleotides for the developing and germinating embryo.

13.6 Environmental and hormonal control of the cell cycle

It is clear that endosperm development is susceptible to environmental stress. In maize, the mitotic phase of development is particularly sensitive to heat (Engelen-Eigles *et al.*, 2001), water stress (Artlip *et al.*, 1995; Setter and Flannigan, 2001) and to ABA applied exogenously (Mambelli and Setter, 1998), whereas the endoreduplication phase is clearly more resilient (Figure 13.2). A recent microarray analysis of gene expression found that several genes involved in cell cycle regulation (i.e. β-tubulin, CDKA, MCM5, histone H2B, RPA1 and RNR) were downregulated after a regime of whole-plant water stress from 5 to 9 DAP (Yu and Setter, 2003). Interestingly, if the plants were watered and allowed to recover, H2B, RPA1 and RNR were upregulated by 12 DAP, suggesting that distinct cell cycle regulons respond differently to regimes of water stress and recovery.

It is well established that the highest level of cytokinin accumulation correlates with the mitotic phase of maize endosperm development, whereas the IAA concentration peaks at later developmental stages, suggesting an inherent specificity in the

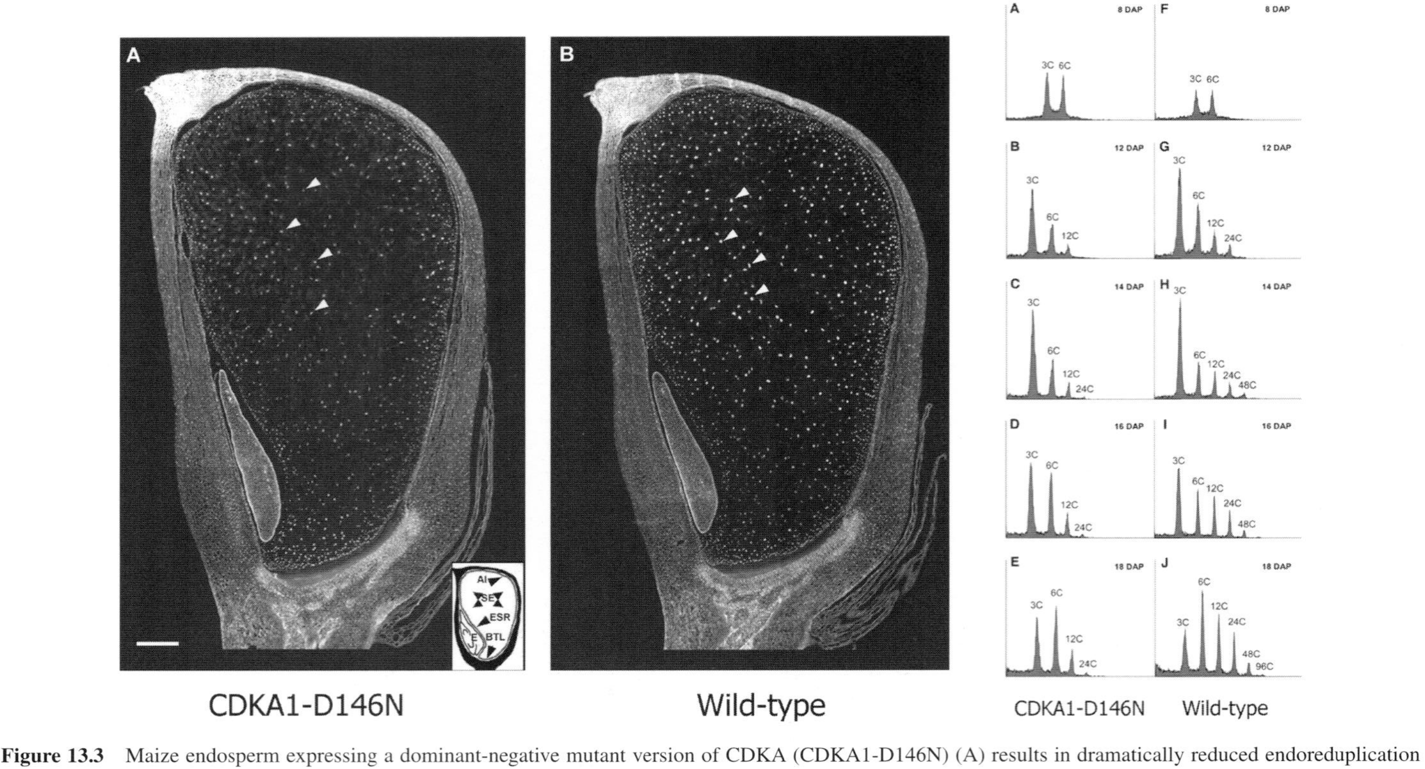

Figure 13.3 Maize endosperm expressing a dominant-negative mutant version of CDKA (CDKA1-D146N) (A) results in dramatically reduced endoreduplication profiles and ploidy levels (see right panel) as well as decreased nuclear sizes (arrows) compared to wild-type. (B) Nuclei were stained with DAPI. The panel on the right shows flow-cytometric profiles illustrating the residual level of endoreduplication in developing CDKA1-D146N (A–E) compared to wild-type (F–J) endosperm. The days after pollination (DAP) and nuclear ploidy levels are indicated. Abbreviations in the inset: Al, aleurone; SE, starchy endosperm; ESR, embryo surrounding region; BTL, basal transfer layer; E, embryo. (Reproduced from Leiva-Neto *et al.*, 2004, with permission from the American Society of Plant Biologists.)

functions of certain hormones. Indeed, different hormones appear to play opposing roles in endosperm development (Lur and Setter, 1993). Zeatin and zeatin riboside accumulation coincide with the peak in endosperm mitotic activity (9 DAP), whereas a sharp increase in auxin accumulation coincides with the onset of endoreduplication (11 DAP). Moreover, exogenous application of 2,4-D in early development (5–7 DAP) accelerates the onset of endoreduplication, nuclear enlargement and zein gene expression. Application of the anti-auxin 2-(*para*-chlorophenoxy)isobutyric acid has an opposing role and delays these processes. Thus, the transition from a mitotic to an endoreduplication cell cycle and the onset of zein gene expression coincide with a sharp decrease in the cytokinin-to-auxin ratio (Figure 13.2). In addition, disruption of polar auxin transport seems at least in part responsible for the pleiotropic effects of the *semaphore1* (*sem1*) mutant, which causes ectopic expression of *knox* genes in several tissues, including the endosperm, and a *dek*-like phenotype (Scanlon *et al.*, 2002). In spite of a large body of evidence for endosperm cell cycle control by both environmental and hormonal factors, information about the key elements linking them to the cell cycle machinery is largely missing, and this remains a challenge for future research.

13.7 Epigenetic control

Several aspects of endosperm development in maize are primarily under maternal control (Kowles *et al.*, 1997, and references therein), including cell number (Jones *et al.*, 1996) and endoreduplication (Cavallini *et al.*, 1995; Dilkes *et al.*, 2002). Maternal control of seed development appears to involve complex genetic and epigenetic interactions, and is only partially understood. However, among key factors that have been identified in *Arabidopsis* are the PcG-like proteins of the FIS class FIS1/MEDEA (Grossniklaus *et al.*, 1998), FIS2 (Luo *et al.*, 1999) and FIS3/FIE (Ohad *et al.*, 1999), which repress endosperm cell division in the absence of fertilization, and are expressed under maternal control. PcG proteins form multimeric complexes involved in chromatin remodeling and repression of gene expression, and play important roles in early endosperm development. In maize, two *fie* genes have been characterized and their expression in the endosperm was shown to be regulated by genomic imprinting (Danilevskaya *et al.*, 2003). Parental dosage effects, mediated by differential imprinting, have been proposed to impinge on distinct, though still unknown, cell cycle genes and regulate the transition from mitotic to endoreduplication cell cycles during maize endosperm development (Leblanc *et al.*, 2002). Increasing evidence points to a central role of the RBR-E2F pathway in regulating the interplay between cell cycle control and epigenetic mechanisms (Shen, 2002). Loss of RBR1 function results in uncontrolled cell proliferation in the *Arabidopsis* embryo sac, and a *fis*-like phenotype (Ebel *et al.*, 2004). In addition, RBR1 binds to specific polycomb subunits such as MSI (Ach *et al.*, 1997b; Kohler *et al.*, 2003) and FIE (Mosquna *et al.*, 2004), and represses gene expression through interaction with Rpd3-type histone deacetylases (Rossi *et al.*, 2003; Varotto *et al.*, 2003). Recently, an intriguing genetic connection between *MEDEA* and *ORC2* (a subunit of the origin replication complex required for DNA replication, which is conserved in eukaryotes) has been reported in *Arabidopsis* (Collinge *et al.*, 2004). *orc2* mutant endosperm aborts at an early stage with approximately four enlarged nuclei. Interestingly, *orc2* can be partly suppressed by mutation of the *MEDEA* gene, suggesting a potential role for *AtORC2* in regulating, and possibly coordinating DNA replication with the maintenance of epigenetic regulation of gene expression, which would be in agreement with similar roles of related ORC subunits in baker's yeast and fruit fly.

13.8 Perspectives

The seed endosperm provides an excellent system to study the relationship between cell cycle regulation and development, and their coordination in plants. Some outstanding questions for future research concern what controls the initial proliferation of the endosperm prior to zygotic cell division and whether, how, or to what extent, endosperm cell proliferation and development are coupled to these events in the embryo. Also, regulation of the mitotic synchrony observed during the coenocytic stage of endosperm development is an intriguing aspect about which virtually nothing is known. Many mutations affect the atypical cytokineses that take place during the endosperm cellularization phase, and they often have dramatic consequences for seed size and development. It will be both challenging and important to organize these mutations into genetic hierarchies and regulatory networks. Another important aspect of endosperm development concerns the regulation of endoreduplication and its function in cereals, as this process has the potential to provide a means to increase the yield of the most important crops worldwide.

The endosperm offers the possibility of dissecting, genetically and molecularly, how different types of cell cycles (i.e. acytokinetic mitosis, cytokinesis through formation of adventitious phragmoplast, mitosis and endoreduplication) are regulated in time and space, and how these cell cycles impact gene expression, the synthesis and accumulation of storage metabolites and cell death. A thorough understanding of the mechanisms that characterize endosperm development will have to integrate research on cell cycle regulation and cell differentiation with the notion that the endosperm is indeed composed of heterogeneous tissues comprising several functionally and developmentally distinct cell types and domains. In addition, although epigenetic control and parent-of-origin effects clearly influence endosperm development, knowledge of how the cell cycle is affected is far from complete. However, given the genetic resources and molecular tools currently available, there is little doubt that future research will soon make substantial progress in unraveling what lies at the interface between cell cycle regulation and epigenetic control of gene expression programs. Furthermore, studies on the regulation of key plant cell cycle genes during endosperm development, coupled with novel reverse genetic approaches, are providing important clues linking the cell cycle engine and the development of this tissue. Thus, knowledge gained from these studies will undoubtedly help pave the way to a coherent model that explains endosperm development and integrates cell cycle control with the action of genes, hormones, the environment and epigenetic factors.

Acknowledgments

We thank the US Department of Energy (Grant # DE-FG02–96ER20242) and Pioneer Hi-Bred Inc. for supporting the endosperm cell cycle project.

References

Ach, R.A, Durfee, T., Miller, A.B., *et al.* (1997a) RRB1 and RRB2 encode maize retinoblastoma-related proteins that interact with a plant D-type cyclin and geminivirus replication protein. *Mol Cell Biol* **17**, 5077–5086.

Ach, R.A., Taranto, P. and Gruissem, W. (1997b) A conserved family of WD-40 proteins binds to the retinoblastoma protein in both plants and animals. *Plant Cell* **9**, 1595–1606.

Artlip, T.S., Madison, J.T. and Setter, T.L. (1995) Water deficit in developing endosperm of maize: cell division and nuclear DNA endoreduplication. *Plant Cell Environ* **18**, 1034–1040.

Berger, F. (2003) Endosperm: the crossroad of seed development. *Curr Opin Plant Biol* **6**, 42–50.

Boisnard-Lorig, Colon-Carmona, A., Bauch, M., *et al.* (2001) Dynamic analyses of the expression of the HISTONE::YFP fusion protein in arabidopsis show that syncytial endosperm is divided in mitotic domains. *Plant Cell* **13**, 495–509.

Brown, R.C., Lemmon, B.E., Nguyen, H. and Olsen, O.-A. (1999) Development of endosperm in *Arabidopsis thaliana*. *Sex Plant Reprod* **12**, 32–42.

Brown, R.C., Lemmon, B.E. and Nguyen, H. (2003) Events during the first four rounds of mitosis establish three developmental domains in the syncytial endosperm of *Arabidopsis thaliana*. *Protoplasma* **222**, 167–174.

Canales, C., Bhatt, A.M., Scott, R. and Dickinson, H. (2002) EXS, a putative LRR receptor kinase, regulates male germline cell number and tapetal identity and promotes seed development in *Arabidopsis*. *Curr Biol* **12**, 1718–1727.

Cavallini, A., Natali, L., Balconi, C., *et al.* (1995) Chromosome endoreduplication in endosperm cells of two maize genotypes and their progenies. *Protoplasma* **189**, 156–162.

Chojecki, A.J.S, Bayliss, M.W. and Gale, M.D. (1986) Cell production and DNA accumulation in the wheat endosperm,and their association with grain weight. *Ann Bot* **58**, 809–817.

Coelho, C.M., Dante, R.A., Sabelli, P.A., *et al.* (2005) Cyclin-dependent kinase inhibitors in maize endosperms and their potential role in endoreduplication. *Plant Physiol* **138**, 2323–2336.

Collinge, M.A., Spillane, C., Köhler, C., Gheyselinck, J. and Grossniklaus, U. (2004) Genetic interaction of an origin recognition complex subunit and the *polycomb* group gene *MEDEA* during seed development. *Plant Cell* **16**, 1035–1046.

Costa, L.M., Gutierrez-Marcos, J.F. and Dickinson, H.G. (2004) More than a yolk: the short life and complex times of the plant endosperm. *Trends Plant Sci* **9**, 507–514.

Costa, L.M., Gutierrez-Marcos, J.F., Greenland, A.J., Brutnell, T.P. and Dickinson, H.G. (2003) The *globby1* (*glo1-1*) mutation disrupts nuclear and cell division in the developing maize seed causing aberrations in endosperm cell fate and tissue differentiation. *Development* **130**, 5009–5017.

Danilevskaya, O.N., Hermon, P., Hantke, S., Muszynski, M.G., Kollipara, K. and Ananiev, E.V. (2003) Duplicated fie genes in maize: expression pattern and imprinting suggest distinct functions. *Plant Cell* **15**, 425–438.

Dilkes, B.P., Dante, R.A., Coelho, C.M. and Larkins, B.A. (2002) Genetic variation in endoreduplication in maize endosperm is subject to parent-of-origin specific effects. *Genetics* **160**, 1163–1177.

Doelling, J.H., Yan, N., Kurepa, J., Walker, J. and Vierstra, R.D. (2001) The ubiquitin-specific protease UBP14 is essential for early embryo development in *Arabidopsis thaliana*. *Plant J* **27**, 393–405.

Ebel, C., Mariconti, L. and Gruissem, W. (2004) Plant retinoblastoma homologues control nuclear proliferation in the female gametophyte. *Nature* **429**, 776–780.

Engelen-Eigles, G., Jones, R.J. and Phillips, R.L. (2001) DNA endoreduplication in maize endosperm cells is reduced by high temperature during the mitotic phase. *Crop Sci* **41**, 1114–1121.

Friedman, W.E. (1998) The evolution of double fertilization and endosperm: an 'historical' perspective. *Sex Plant Reprod* **11**, 6–16.

Garcia, D., Fitz Gerald, J.N., and Berger, F. (2005) Maternal control of integument cell elongation and zygotic control of endosperm growth are coordinated to determine seed size in Arabidopsis. *Plant Cell* **17**, 52–60.

Garcia, D., Saingery, V., Chambrier, P., Mayer, U., Jürgens, G., and Berger, F. (2003) Arabidopsis *haiku* mutants reveal new controls of seed size by endosperm. *Plant Physiol* **131**, 1661–1670.

Grafi, G., Burnett, R.J., Helentjaris, T., *et al.* (1996) A maize cDNA encoding a member of the retinoblastoma protein family. *Proc Natl Acad Sci USA* **93**, 8962–8967.

Grafi, G. and Larkins, B.A. (1995) Endoreduplication in maize endosperm: involvement of M phase-promoting factor inhibition and induction of S phase-related kinases. *Science* **269**, 1262–1264.

Grossniklaus, U., Vielle-Calzada, J.P., Hoeppner, M.A. and Gagliano, W.B. (1998) Maternal control of embryogenesis by MEDEA, a polycomb group gene in *Arabidopsis*. *Science* **280**, 446–450.

Gutierrez, C. (2005) Coupling cell proliferation and development in plants. *Nat Cell Biol* **7**, 535–541.

Holding, D.R. and Springer, P.S. (2002) The *Arabidopsis* gene *PROLIFERA* is required for proper cytokinesis during seed development. *Planta* **214**, 373–382.

Ingouff, M., Fitz Gerald, J.N., Guerin, C., *et al.* (2005a) Plant formin AtFH5 is an evolutionarily conserved actin nucleator involved in cytokinesis. *Nat Cell Biol* **7**, 374–380.

Ingouff, M., Haseloff, J. and Berger, F. (2005b) Polycomb group genes control developmental timing of endosperm. *Plant J* **42**, 663–674.

Jones, R.J., Schreiber, B.M.N. and Roessler, J.A. (1996) Kernel sink capacity in maize: genotypic and maternal regulation. *Crop Sci* **36**, 301–306.

Kladnik, A., Chourey, P.S., Pring, D.R. and Dermastia, M. (2006) Development of the endosperm of *Sorghum bicolor* during the endoreduplication-associated growth phase. *J Cereal Sci* **43**, 209–215.

Kohler, C., Hennig, L., Bouveret, R., Gheyselinck, J., Grossniklaus, U. and Gruissem, W. (2003) *Arabidopsis* MSI1 is a component of the MEA/FIE Polycomb group complex and required for seed development. *EMBO J* **22**, 4804–4814.

Kowles, R.V., McMullen, M.D., Yerk, G., Phillips, R.L., Kramer, S. and Srienc, F. (1992) Endosperm mitotic activity and endoreduplication in maize affected by defective kernel mutations. *Genome* **35**, 68–77.

Kowles, R.V. and Phillips, R.L. (1988) Endosperm development in maize. *Int Rev Cytol* **112**, 97–136.

Kowles, R.V., Yerk, G.L., Haas, K.M. and Phillips, R.L. (1997) Maternal effects influencing DNA endoreduplication in developing endosperm of *Zea mays Genome* **40**, 798–805.

Kvaale, A. and Olsen, A. (1986) Rates of cell division in developing barley endosperms. *Ann Bot* **57**, 829–833.

Larkins, B.A., Dilkes, B.P., Dante, R.A., Coelho, C.M., Woo, Y-M. and Liu, Y. (2001) Investigating the hows and whys of endoreduplication. *J Exp Bot* **52**, 183–192.

Leblanc, O., Pointe, C. and Hernandez, M. (2002) Cell cycle progression during endosperm development in *Zea mays* depends on parental dosage effects. *Plant J* **32**, 1057–1066.

Leiva-Neto, J.T., Grafi, G., Sabelli, P.A., *et al.* (2004) A dominant negative mutant of CDKA reduces endoreduplication but not cell size or gene expression during maize endosperm development. *Plant Cell* **16**, 1854–1869.

Lepiniec, L., Devic, M. and Berger, F. (2005) Genetic and molecular control of seed development in *Arabidopsis*. In: *Plant Functional Genomics* (ed. Lester, D.). Food Products Press, Birghamton, NY, pp. 511–564.

Lersten, N.R. (2004) *Flowering Plant Embryology*. Blackwell Publishing, Ames, Iowa.

Lid, S.E., Al, R.H., Krekling, T., *et al.* (2004) The maize *disorganized aleurone layer 1 and 2* (*dil1, dil2*) mutants lack control of the mitotic division plane in the aleurone layer of developing endosperm. *Planta* **218**, 370–378.

Liu, C.M., McElver, J., Tzafrir, I., *et al.* (2002) Condensin and cohesion knockouts in *Arabidopsis* exhibit a titan seed phenotype. *Plant J* **29**, 405–415.

Liu, C.M. and Meinke, D.W. (1998) The *titan* mutants of *Arabidopsis* are disrupted in mitosis and cell cycle control during seed development. *Plant J* **16**, 21–31.

Luo, M., Bilodeau, P., Koltunow, A., Dennis, E.S., Peackock, W.J. and Chaudhury, A.M. (1999) Genes controlling fertilization-independent seed development in *Arabidopsis thaliana*. *Proc Natl Acad Sci USA* **96**, 296–301.

Luo, M., Dennis, E.S., Berger, F., Peacock, W.J. and Chaudhury, A. (2005) MINISEED3 (MINI3), a WRKY family gene, and HAIKU2 (IKU2), a leucine-rich repeat (LRR) KINASE gene, are regulators of seed size in *Arabidopsis*. *Proc Natl Acad Sci USA* **102**, 17531–17536.

Lur, H-S. and Setter, T.L. (1993) Role of auxin in maize endosperm development. *Plant Physiol* **103**, 273–280.

Mambelli, S. and Setter, T.L. (1998) Inhibition of maize endosperm cell division and endoreduplication by exogenously applied abscisic acid. *Physiol Plantarum* **104**, 266–277.

Mayer, U., Herzog, U., Berger, F., Inzé, D. and Jürgens, G. (1999) Mutations in the pilz group genes disrupt the microtubule cytoskeleton and uncouple cell cycle progression from cell division in *Arabidopsis* embryo and endosperm. *Eur J Cell Biol* **78**, 100–108.

Menand, B., Desnos, T., Nussaume, L., *et al.* (2002) Expression and disruption of the *Arabidopsis TOR* (*target of rapamycin*) gene. *Proc Natl Acad Sci USA* **99**, 6422–6427.

Mosquna, A., Katz, A., Shochat, S., Grafi, G. and Ohad, N. (2004) Interaction of FIE, a polycomb protein, with pRb: a possible mechanism regulating endosperm development. *Mol Gen Genomics* **271**, 651–657.

Nacry, P., Mayer, U. and Jürgens, G. (2000) Genetic dissection of cytokinesis. *Plant Mol Biol* **43**, 719–733.

Neuffer, M.G., Coe, E.H. and Wessler, S.R. (1997) *Mutants of Maize*. Cold Spring Harbor Laboratory Press, Cold Spring Harbor, New York.

Nowack, M.K., Grini, P.E., Jakoby, M.J., Lafos, M., Koncz, C. and Schnittger, A. (2006) A positive signal from the fertilization of the egg cell sets off endosperm proliferation in angiosperm embryogenesis. *Nat Genet* **38**, 63–67.

Ohad, N., Yadegari, R., Margossian, L., *et al.* (1999) Mutations in *FIE*, a WD polycomb group gene, allow endosperm development without fertilization. *Plant Cell* **11**, 407–416.

Olsen, O.A. (2004) Nuclear endosperm development in cereals and *Arabidopsis thaliana*. *Plant Cell* **16**, S214–S217.

Randolph, L.F. (1936) Developmental morphology of the caryopsis in maize. *J Agric Res* **53**, 881–916.

Ronceret, A., Guilleminot, J., Lincker, F., *et al.* (2005) Genetic analysis of two Arabidopsis DNA polymerase epsilon subunits during early embryogenesis. *Plant J* **44**, 223–236.

Rossi, V., Locatelli, S., Lanzanova, C., *et al.* (2003) A maize histone deacetylase and retinoblastoma-related protein physically interact and cooperate in repressing gene transcription. *Plant Mol Biol* **51**, 401–413.

Sabelli, P.A., Dante, R.A., Leiva-Neto, J.T., Jung, R., Gordon-Kamm, W.J. and Larkins, B.A. (2005a) RBR3, a member of the retinoblastoma-related family from maize, is regulated by the RBR1/E2F pathway. *Proc Natl Acad Sci USA* **102**, 13005–13012.

Sabelli, P.A. and Larkins, B.A. (2006) Grasses like mammals? Redundancy and compensatory regulation within the retinoblastoma protein family. *Cell Cycle* **5**, 352–355.

Sabelli, P.A., Leiva-Neto, J.T., Dante, R.A., Nguyen, H. and Larkins, B.A. (2005b) Cell cycle regulation during maize endosperm development. *Maydica* **50**, 485–496.

Scanlon, M.J., Henderson, D.C. and Bernstein, B. (2002) *SEMAPHORE1* functions during the regulation of ancestrally duplicated *knox* genes and polar auxin transport in maize. *Development* **129**, 2663–2673.

Setter, T.L. and Flannigan, B.A. (2001) Water deficit inhibits cell division and expression of transcripts involved in cell proliferation and endoreduplication in maize. *J Exp Bot* **52**, 1401–1404.

Shen, W-H. (2002) The plant E2F-Rb pathway and epigenetic control. *Trends Plant Sci* **7**, 505–511.

Shen, W.H., Parmentier, Y., Hellmann, H., *et al.* (2002) Null mutation of *AtCUL1* causes arrest in early embryogenesis in *Arabidopsis*. *Mol Biol Cell* **13**, 1916–1928.

Sheridan W.F. and Neuffer, M.G. (1980) Defective kernel mutants of maize. II. Morphological and embryo culture studies. *Genetics* **95**, 945–960.

Sorensen, M.B., Mayer, U., Lukowitz, W., *et al.* (2002) Cellularization in the endosperm of *Arabidopsis thaliana* is coupled to mitosis and shares multiple components with cytokinesis. *Development* **129**, 5567–5576.

Springer, P.S., Holding, D.R., Groover, A., Yordan, C. and Martienssen, R.A. (2000) The essential Mcm7 protein PROLIFERA is localized to the nucleus of dividing cells during the G_1 phase and is required maternally for early *Arabidopsis* development. *Development* **127**, 1815–1822.

Steinborn, K., Maulbetsch, C., Priester, B., *et al.* (2002) The *Arabidopsis PILZ* group genes encode tubulin-folding cofactor orthologs required for cell division but not cell growth. *Genes Dev* **16**, 959–971.

Sun, Y., Dilkes, B.P., Zhang, C-S., *et al.* (1999a) Characterization of maize (*Zea mays* L.) Wee1 and its activity in developing endosperm. *Proc Natl Acad Sci USA* **96**, 4180–4185.

Sun, Y., Flannigan, B.A. and Setter, T.L. (1999b) Regulation of endoreduplication in maize (*Zea mays* L.) endosperm. Isolation of a novel B1-type cyclin and its quantitative analysis. *Plant Mol Biol* **41**, 245–258.

Tzafrir, I., McElver, J.A., Liu, C.M., *et al.* (2002) Diversity of TITAN functions in *Arabidopsis* seed development. *Plant Physiol* **128**, 38–51.

Varotto, S., Locatelli, S., Canova, S., Pipal, A., Motto, M. and Rossi, V. (2003) Expression profile and cellular localization of maize Rpd3-type histone deacetylases during plant development. *Plant Physiol* **133**, 606–617.

Vilhar, B., Kladnik, A., Blejec, A., Chourey, P.S. and Dermastia, M. (2002) Cytometrical evidence that the loss of seed weight in the miniature1 seed mutant of maize is associated with reduced mitotic activity in the developing endosperm. *Plant Physiol* **129**, 23–30.

Wang, C., Barry, J.K., Min, Z., Tordsen, G. and Rao, A.G. (2003) The calpain domain of the maize DEK1 protein contains the conserved catalytic triad and functions as a cysteine proteinase. *J Biol Chem* **278**, 34467–34474.

Weijers, D., Franke-van Dijk, M., Vencken, R.J., Quint, A., Hooykaas, P. and Offringa, R. (2001) An *Arabidopsis* minute-like phenotype caused by a semi-dominant mutation in a *RIBOSOMAL PROTEIN S5* gene. *Development* **128**, 4289–4299.

Xie, Q., Sanz-Burgos, A.P., Hannon, G.H. and Gutierrez, C. (1996) Plant cells contain a novel member of the retinoblastoma family of growth regulatory proteins. *EMBO J* **15**, 4900–4908.

Yu, L-X. and Setter, T.L. (2003) Comparative transcriptional profiling of placenta and endosperm in developing maize kernels in response to water deficit. *Plant Physiol* **131**, 568–582.

14 Hormonal regulation of cell cycle progression and its role in development

Peter C.L. John

14.1 Introduction

It is becoming clear that auxin and cytokinin drive progression through the cell division cycle and also define the regions of cell division within apical meristems through the creation of regions of locally amplified hormone concentration. Mutations and transgenes that affect hormone concentration confirm the classic hypothesis that cytokinin is important in promoting shoot meristem identity and slowing root development, while conversely auxin preferentially promotes root development (Skoog and Miller, 1957). A necessity for cytokinin and auxin to initiate DNA replication and mitosis, represents control at the same transitions that are regulated in other eukaryotes but, although DNA replication has marked similarities with metazoa, the initiation of plant mitosis has unique features. At DNA replication, hormonally induced accumulation of D-cyclins directs phosphorylation to RB-related proteins, which then release S-phase transcription, as described in detail in accompanying chapters. More attention is paid here to control of mitosis, which requires increase in CDK enzyme activity, through removal of inhibitory phosphate from CDK as in other eukaryotes, but in plants it is dependent on cytokinin accumulation in the late G2 phase. Cytokinin abundance is essential for mitotic initiation but can be substituted by expression of yeast CDC25 tyrosine phosphatase that is specific for CDK.

Hormones that stimulate cell cycle progression also act at long distances in balancing growth of root and shoot and, additionally, within apical meristems they delimit cell division zones. In the shoot apex, abundance of cytokinin is a key regulatory element. Its induction of the cytokinin biosynthesizing enzyme IPT7 reinforces local cytokinin concentration and sustains cell cycle activity. Whereas in the root apex, auxin is the key hormone that specifies stem cell and meristem areas through local concentration of the hormone, which results from its recirculation under the direction of auxin efflux regulators of the PIN protein family. Hormones in multicellular plants therefore function to organize division zones as well as driving progression through the division cycle.

This review considers first the impacts of hormones on the division of an individual cell, and then presents recent molecular evidence that hormones arriving from long distance at the meristem regions of root and shoot are used to create zones of locally raised hormone concentration within the meristem. Finally, recent evidence is presented that long-distance transport of hormonal signals, and resulting changes in the ratio of hormones, determines the location of lateral shoot and root development and hence plant architecture.

14.2 Auxin and cytokinin have paramount roles in cell proliferation control

Auxin and cytokinin were initially pinpointed as the most potent regulators of plant cell division by classical studies of plant tissue culture. These revealed stimulation of division and capacity to induce development of root or shoot identity through change in hormone ratio. Appreciation of the key role of auxins and cytokinins is reinforced by the ability of *Agrobacterium tumefasciens* to induce tumours through genes for auxin and cytokinin synthesis, and is further underlined by the absence of any known mutants that are viable while unable to synthesize these hormones. This review narrows its focus to auxin and cytokinin, which are both absolutely required for division and meristem function.

14.3 Growth and cell cycle gene expression induced by auxin and cytokinin

Auxin and cytokinin are classically referred to as growth factors, and some of their effects on cell division derive from growth stimulation. Growth is certainly required for the maintenance of average cell size in proliferating cell populations and in the individual cell there is a widely observed and probably universal need to attain a critical minimum cell size for division (reviewed by Fantes *et al.*, 1975; Fantes and Nurse, 1981; Mitchison, 2003). This is clear in yeasts, where conditional division mutants are available and reveal that S phase is initiated at constant cell size even when growth rate changes (e.g. Carter and Jagardish, 1975). A cell size requirement is also detected in mammalian cells (e.g. Zetterberg and Killander, 1965; Dolznig *et al.*, 2004) and is implied in plant cells from the increasing probability of division as cells enlarge (Canovas *et al.*, 1990) which conforms to the eukaryote-wide observation that cell size is stabilized in proportion with nuclear DNA content by initiation of division at the appropriate cell size (Gregory, 2001). A further, probably universal, coupling to growth comes from requirement for a significant growth rate, seen in animal cells (Zetterberg and Larsson, 1985) and yeast cells, where faster growth allows more efficient translation of essential cell division proteins, in particular the G1 cyclin (CLN3) of budding yeast (Polymenis and Schmidt, 1997) and the mitotic phosphatase CDC25 and B-cyclin (CDC13) of fission yeast (Daga and Jiminez, 1999). These are essential for the key transitions G1/S and G2/M phase (see later) and accumulate only when cells are growing well, since a potential hairpin configuration in their mRNA is prevented from forming by a high density of active ribosomes on the messenger.

Growth is therefore a prerequisite for sustained proliferation of all cells, and one action of plant growth factors that drive the cell cycle is indeed to stimulate growth. Auxin in particular often stimulates biosynthesis and cell enlargement (Skoog and

Miller, 1957; Theologis *et al.*, 1985; Lincoln *et al.*, 1990). This is evident in culture, where cells without auxin cease growth and therefore do not divide, whereas cells with auxin but lacking cytokinin continue in growth but are blocked in division (e.g. Skoog and Miller, 1957). However, an additional direct influence of auxin and cytokinin on division is indicated by their induction of some cell cycle genes.

Hormone-induced expression of cell cycle genes can proceed rapidly in cells primed to respond to hormone, as in the elongation zone of pea roots, where auxin induces a rise in the transcript of *CDKA* within 10 min and a stable, threefold higher level by 3 h (John *et al.*, 1993). Cells in this zone have been recently formed and they have *CDKA* mRNA above the basal level found in fully differentiated cells. In normal development they respond to the tip having grown more distant by initiating local cell division and growth in new lateral root primordia. Such spacing of lateral roots, by initiation at a distance from the root tip, aids the efficient mining of soil for nutrients, and the signal for lateral root induction may be attainment of a sufficiently high ratio of auxin to cytokinin (as will be discussed further); therefore, exposure to auxin can be interpreted as the premature arrival of a signal to which the tissue is primed to respond. Induction of *CDKA* expression in response to auxin has also been detected in whole roots within 2 or 3 days (Martinez *et al.*, 1992; Hemerly *et al.*, 1993) and the CDK protein itself is detectable after 2 days in auxin-treated cotyledon explants (Gorst *et al.*, 1991). These slower induction responses, in cells not programmed to proliferate unless wounded, suggest that auxin often first stimulates cell growth before inducing cell cycle catalysts. A significant basal level of CDKA protein and transcript is found in all viable plant cells, even if not actively dividing (Gorst *et al.*, 1991; Martinez *et al.*, 1992), probably reflecting the unique general capacity of plant cells to return to division.

Auxin raises the capacity for division by inducing CDKA but additional proteins are required for division and auxin can induce some of these after a lag during which growth occurs or prior cycle events are completed. Two examples are, expression from the mitotic cyclin *CYCA2;1* promoter in *Arabidopsis* root within 12 h of exposure to auxin [synthetic NAA] (Burssens *et al.*, 2000) and from the promoter of mitosis-specific CDKB (*CDKB2;1*) of *Medicago*, in a synchronous culture stimulated with auxin (synthetic 2,4-D). The *CDKB* gene was expressed after a 9 h lag during which cells entered late G2 phase and came to require CDKB for mitosis (Zhiponova *et al.*, 2006). The particular transcription factors that are most significant in the expression of cell cycle genes are not yet known.

Cytokinins are particularly strong inducers of the D-cyclins, which have well-characterized effects on the induction of S phase in animals and plants through activation of S phase transcription (described later). Induction of D-cyclin is therefore a major mechanism by which cytokinin stimulates division. The question is, can cytokinin also induce accompanying growth? Although development of most animal cell types is remarkably unaffected by the absence of D-Cyclins (Kozar *et al.*, 2004), these cyclins are essential for growth and proliferation of the major lymphoid and myeloid blood cell lines, and furthermore, these become malignant if *D-cyclin* expression is upregulated (Bergsagel *et al.*, 2005). This has been interpreted as a possible indication that D-cyclins stimulate growth as well as division (e.g. Sherr,

1994), perhaps because the substrates of CDK/CYCLIN D could include transcription factors for genes that drive growth, or because CDK/CYCLIN D in animal cells could phosphorylate signal transduction proteins that are normally phosphorylated in response to growth factors. If such mechanisms operated in plants, then cytokinins that induce D-type cyclins and cell division may also be responsible for meristem growth.

Evidence for D-CYCLIN enhancement of growth is not conclusive. Expression of CYCLIN D in cultured mammalian cells accelerates entry into S phase and therefore reduces the number of cells in the G1 phase, but cells were smaller through having accelerated division without accompanying increase in growth (Ohtsubo and Roberts, 1993; Quelle *et al.*, 1993; Resnitzky *et al.*, 1994). Similar reduction in cell size from transgenic stimulation of division occurs in the *Drosophila* wing (Neufeld *et al.*, 1998). But growth stimulation by raised D-CYCLIN sometimes can occur, as in the *Drosophila* eye when CYCLIN D complexes with CDK4, although this CDK is not essential for growth and has no plant analogue (Frei and Edgar, 2004). Cell growth is therefore not an obligatory accompaniment of cell cycle progression or consequence of raised CYCLIN D. In plants also, it is not clear that D-cyclins regulate growth, since in cultured Arabidopsis cells CDK/CYCLIN D2 activity rose only after growth had been resumed for 2 h and CDK/CYCLIN D3 increased even later at 4 h (Riou-Khamlichi *et al.*, 2000). Induction of growth cannot confidently be attributed to CDK/ CYCLIN D2 since the *CYCD2;1* gene of Arabidopsis induces faster growth only when heterologously expressed in tobacco seedlings and without the change in abundance of G1 phase cells, which would be expected if it affected the cell cycle (Cockroft *et al.*, 2000). Similarly, overexpression of *CYCD3;1* in Arabidopsis prevents termination of cell division in the developing leaf but it does not increase growth, and therefore reduces cell size (Dewitte *et al.*, 2003). It is therefore difficult to conclude that cytokinin induction of CYCLIN D induces plant growth as well as cell division.

14.4 Does cell cycle progression affect growth?

The foregoing discussion opens the broader topic of whether the formation of new cells results in growth. Hormones that influence division will be of broader importance if control of division also regulates growth. Growth regulation by division might occur according to the hypothesis that growth is cell based, that is growth occurs because progress through the cell cycle itself stimulates growth or daughter cells are programmed to grow to a certain final size. Consistent with cell-based growth, the size of rodents increases when CDK/CYCLIN activity is increased by knockout of the CKI class inhibitor p27KIP1. The modified mice have more cells and, although cells are unable to quite reach full size, the animals do increase somewhat in size (Nakayama *et al.*, 1996).

A quite different interpretation of growth is that growth itself is regulated and forms an organ or organism of regulated final size, while cells divide as necessary to accommodate the increasing mass. In this case the general stability of cell size must result from an additional mechanism coupling division to growth. Such adjustment of division to growth is already in operation in the unicellular organisms from which multicells evolved. In unicellular organisms, a general requirement of adequate cell size for commitment to division has been noted in unicellular fungi (Fantes and Nurse, 1981; Mitchison, 2003) and plants (Donnan and John, 1983) and has been conserved through the transition to multicellular animals (e.g. Dolznig *et al.*, 2004) and, available evidence suggests, also to higher plants (Canovas *et al.*, 1990).

Current evidence is against cell-based growth in lateral organs, such as the leaf, where the determinate final size of the organ conflicts with any requirement that cells should reach a postulated programmed size. Cell size was not maintained when cell number in the Arabidopsis leaf was increased by overexpression of *CYCLIN D3* and markedly smaller cells resulted (Dewitte *et al.*, 2003); however, local induction of cell division was to some extent able to attract growth within the developing leaf to the affected region (Wyrzykowska *et al.*, 2002). Cell-based growth might therefore be thought more likely to occur in indeterminately growing terminal apices, where growth might extend to accommodate a programmed final cell size in spite of larger cell number. But our current analysis of ectopic cyclin expression in the root tip (Ruhu Qi and Peter CL John, unpublished observations) indicates that growth there also is not cell based.

Cell division may not therefore be the engine of plant growth, but it remains at least a vital process for the continuation of growth by maintaining cells at a physiologically effective size; small enough to allow adequate levels of gene transcripts and other nucleus-derived elements to permeate the cytoplasm. At the tissue level, cell division is also important in generating daughter cells in the correct number, location and shape to allow formation of tissues and organs of specialized function. For example in the leaf, division in the epidermis generates stomata by subdivision without requiring growth, while internally the apportionment of growth into daughter cells that form pallisade and spongy mesophyll layers generates the photosynthetic tissue.

14.5 Division sustains continuation of growth

The dependence of continued growth on cell formation can be seen from the cessation of growth in lateral roots and shoots when cell division is prevented by irradiation or cytoskeleton inhibition. Lateral root formation begins with localized growth in a few pericycle cells (in Arabidopsis as few as 2–3 longitudinal neighbours) which grow radially and divide periclinally maintaining an isodiametric shape (Sussex *et al.*, 1995). This initial primordial growth can occur if cell division is blocked by colchicine, but growth ceases when cells have become unusually large, having nonetheless formed the shape of an incipient primordium. The blocked primordium can resume growth and development if colchicine is removed (Foard *et al.*, 1965). A similar dependence of lateral root outgrowth upon cell division is seen in the inhibitory effect of upregulated expression of the CDK inhibitor CKI/KRP2 on lateral root formation, which incidentally indicates that CKI protein transcription has a developmental role in addition to a cell cycle regulatory role in plants (Himanen *et al.*, 2002).

In the shoot apex also, leaf primordia can emerge by localized growth even when cell division is blocked by irradiation, as in *Triticum* (Foard, 1971). Furthermore, if division in the developing leaf halts at a smaller cell number, due to suboptimal cytokinin that has been engineered by ectopic expression of cytokinin oxidase, the cells become larger than usual but not sufficiently to compensate for reduced number and a small leaf results (Werner *et al.*, 2003). The eventual cessation of growth for lack of cell division may be caused by physiological inefficiency in abnormally large cells and also, in complex cell division zones such as apical meristems, because further development requires signalling and hormone movement (as will be discussed) between cell layers that have not formed.

14.6 Localized growth

The mechanisms by which growth factors may induce local growth and hence foci of cell division remain mysterious. In the case of lateral root primordia, their initiation occurs opposite protoxylem poles and is negatively regulated by localized expression of the cyclin-dependent kinase inhibitor gene *ICK/KRP2*. This gene reduces lateral root number if ectopically expressed and it is transcribed in interpolar regions of the pericycle where lateral root initiation would disrupt normal radial distribution of laterals (Himanen *et al.*, 2002), but the molecular basis of localized radial expression of ICK/KRP2 and of the spacing of laterals along the longitudinal axis is uncertain. Longitudinal spacing of laterals is set by the proximity to the apical meristem at which a lateral root can be initiated. This distance and, therefore, the spacing between laterals is reduced at higher auxin/cytokinin ratio (Wightman *et al.*, 1980; Hemerly *et al.*, 1993; discussed below).

Also mysterious is the mechanism by which total organ size is perceived and maintained. In the developing leaf it has been noted that cell division is less extensive if cytokinin level is reduced by cytokinin oxidase, which correlates with cytokinin importance in division, but there is also a correlation with reduced overall size of the leaf and therefore it is not impossible that a gradient of cytokinin extending into the enlarging leaf determines the limits of its growth (Werner *et al.*, 2003). Many aspects of hormonal control over growth are still not understood; however, the mechanisms of hormonal control over cell division are perhaps becoming clearer.

14.7 Hormonal impacts at the G1/S phase progression

A major control event of the cell division cycle is the termination of the G1 phase by commitment to enter DNA replication. In microorganisms this commitment, once taken, is sustained even if growth is subsequently slowed or prevented, for example, by low nutrient (e.g. Carter and Jagadish, 1975) or by absence of photosynthesis in plant cells (Donnan and John, 1983). At the molecular level, key events in commitment to the S phase are the attainment of a concentration of G1-CYCLINS (D-CYCLINS) that raises CDK/CYCLIN activity and increases phosphorylation of the CDK inhibitor class of proteins (CKI) that then have increased affinity for the ubiquitination machinery and are consequently proteolyzed. This proteolysis of CKI is a key event since microbial cells where gene knockout is easier reveal that all G1-CYCLINS can be deleted and yet cell viability can be maintained if there is also removal of the CKI protein (Sanchez-Diaz *et al.*, 1998). Consistent with this, elevated expression of plant CKI homologues slows cell division (described in accompanying chapters). Further progression to DNA replication involves CDK/CYCLIN D-catalyzed phosphorylation of pocket proteins (RBR) that unless phosphorylated prevent transcription by E2F class transcription factors that transcribe the genes of DNA precursor synthesis and DNA replication as well as the genes that encode mitotic cyclins (reviewed by Gutierrez *et al.*, 2002, and detailed in accompanying chapters). Establishment of the S phase is therefore accompanied by the appearance of a new class of cyclins – the mitotic cyclins, which are distinctive in supporting higher CDK/CYCLIN activity that is necessary for DNA replication and later for mitosis.

Events at the G1/S phase transition therefore require CDKs and CYCLINS, of which CYCLIN D synthesis can be most clearly linked to stimulation of cell division at this time. In cultured plant cells, cytokinin strongly induces CYCLIN D3 whereas CYCLIN D2 is more responsive to sucrose (Murray *et al.*, 1998; Riou-Khamlichi *et al.*, 2000). Sucrose may be considered in this respect a *de facto* hormone and its effects on shoot/root ratio and flowering (Tran Thanh Van, 1981; Galtier *et al.*, 1993) could derive in part from its effects on cell division through induction of *CYCLIN D2*. Although both auxin and cytokinin have been found to affect the expression of CDKs and CYCLINs, there has been little study of possible hormonal interactions in particular tissues. Competitive effects on gene expression can occur since the rise in CYCLIN D3 transcript that occurs within 4 h of treatment of Arabidopsis callus with cytokinin (kinetin) was prevented if auxin (2,4-D) was present in addition (Murray *et al.*, 1998). Furthermore, post-translational events can affect levels of division catalysts, since in pea root CDK induced by auxin (IAA) is proteolyzed if cytokinin (zeatin riboside) is applied (John *et al.*, 1993). These interactions could have developmental implications, for example as a mechanism of inhibition by root apical meristems of lateral root initiation, as will be discussed.

Induction of CYCLIN D expression is a major component of cytokinin stimulation of the cell cycle since its expression can replace the presence of cytokinin. When leaf tissue of transgenic Arabidopsis expressing *35S::CYCD3* was tested for its ability to form callus on medium with auxin only, or with auxin plus cytokinin, the number of calli able to form without external cytokinin increased from 18% in wild type to between 41% and 50% owing to the elevated *CYCLIN D3* transcript (Riou-Khamlichi *et al.*, 1999). This independence from external cytokinin in some lines that expressed *CYCLIN D3* indicates that a major effect of cytokinin is to induce expression of this cyclin, but the observation that half of the calli still required cytokinin indicates that it has other essential functions.

14.8 Hormonal impacts at the G2/M phase progression

Cytokinin is also required at the initiation of mitosis, which is a control point that operates in all eukaryotes (Norbury and Nurse, 1992; MacNeill and Nurse, 1997) through inhibition of enzyme activity in CDK/mitotic cyclin complexes by phosphorylation of a tyr15 in the CDK supplemented by a weaker additional inhibition caused by phosphorylation of the adjacent thr14. The inhibited state is established in the G2 phase by the tyrosine kinase WEE1, of which plant homologues have been identified (Sun *et al.*, 1999; Sorrell *et al.*, 2002). Mitosis is then initiated by the progressive removal of the inhibitory phosphates by regulated activity of the protein tyrosine phosphatase CDC25 (Millar *et al.*, 1991; Norbury and Nurse, 1992; Kovelman and Russell, 1996), which can be referred to as a dual-specificity phosphatase in higher organisms in recognition of its dephosphorylation of tyr and thr. While the same dephosphorylation occurs in plant CDK (Zhang *et al.*, 2005) the identity of the enzyme that carries out CDC25 phosphatase function in plants is unresolved, as discussed below. This is a major question as the G2/M control point universally provides a means to block mitosis if DNA is damaged or incompletely replicated (Rhind and Russell, 2001; Preuss and Britt, 2003), but it also operates more frequently in plants than in other kingdoms as a point of arrest under physiological or developmental influence (Van't Hof, 1974; Zetterberg and Larsson, 1985). Progression at G2/M can be delayed by water or oxidative stress (Schuppler *et al.*, 1998; Reicheld *et al.*, 1999) or accelerated by hormone stimulus in excised rice stems (Sauter and Kende, 1992) and is also under developmental control since pericycle cells arrested in the G2 phase are the site of the localized growth that initiates lateral root primordia (Beeckman *et al.*, 2001).

A role of cytokinin in stimulating the G2/M phase progression is indicated by G2 arrest when cytokinin is removed from tissue culture (e.g. Fosket, 1977; Tao and Verbelen, 1996; Valente *et al.*, 1998) and by converging evidence from two independent cultured cell lines. A line of *Nicotiana plumbaginifolia* cells repeatedly cultured with cytokinin (kinetin) was found to have lost the capacity to synthesize sufficient cytokinin to be autonomous and in the absence of external cytokinin could only progress through the G1 and S phases to an arrest point in the late G2 phase, which indicated that cytokinin was necessary for entry into prophase (Zhang *et al.*, 1996). Consistent with this, freshly isolated tobacco cells without cytokinin also execute DNA replication but cannot enter mitosis (Valente *et al.*, 1998). The *Nicotiana tabacum* BY2 cell line is autonomous for cytokinin and increases its level through several orders of magnitude at the late G2 phase (Redig *et al.*, 1996) coincident with the time when the *plumbaginifolia* line arrests if cytokinin is not available. Furthermore, the *tabacum* line, like *plumbaginifolia*, is unable to enter mitosis if cytokinin synthesis is inhibited, but mitosis is released by added cytokinin (Laureys *et al.*, 1998).

The block to mitosis without cytokinin in *plumbaginifolia* was found to involve accumulation of CDK in inactive form phosphorylated at Y15, which could be activated by the catalytic portion of CDC25 (Zhang *et al.*, 1996). Furthermore, cells of alfalfa in the late G2 phase accumulate CDK containing tyr-phosphate, which can be activated *in vitro* by CDC25 from Drosophila (Mészáros *et al.*, 2000). From this evidence it was uncertain whether the primary cytokinin-responsive event may have been at some early point in the G2 phase leading to CDK activation later as a part of mitosis. To test this, the CDK activator CDC25 was investigated for possible ability to substitute for cytokinin in stimulating mitosis. The fission yeast *CDC25* gene, which is of well-established specificity for CDK (Millar *et al.*, 1991) was brought under the control of the steroid inducible promoter, and its expression was induced in cells arrested in the G2 phase by lack of cytokinin. Remarkably, in view of the multiple actions of cytokinin (e.g. Hoth *et al.*, 2003), *CDC25* expression was able to fulfil the cell cycle functions of cytokinin, allowing mitosis and proliferation (Zhang *et al.*, 2005). Both normal and induced mitosis involved an increase in CDC25 tyrosine phosphatase, coincident with a maximum rate of activation of CDK, accompanied by a decline in tyrosine phosphate content in the CDK and a decline in capacity of the plant CDK to be activated *in vitro* by CDC25. In tobacco cells also, a cytokinin block to mitosis (Laureys *et al.*, 1998) was circumvented if *CDC25* was expressed (Orchard *et al.*, 2005).

In cell division, therefore, a key outcome of cytokinin presence was the tyrosine dephosphorylation of CDK in the late G2 phase. During evolution, the tyrosine and adjacent threonine that are negatively regulated by WEE1 phosphorylation have been retained in plant cell cycle CDKs (e.g. Joubès *et al.*, 2000; and accompanying chapters), and furthermore, in the plant, developmental consequences of regulatory CDK phosphorylation are consistent with this being a biochemical signal of cytokinin presence, as discussed later.

The identity of the CDC25-like tyrosine phosphatase(s) in plants remains a fascinating question. The primitive unicellular green plant *Ostreococcus tauri* has a gene with clear homology to the highly conserved CDC25 genes of yeasts, insects, and animals and it complements conditional mutation in *cdc25ts* of fission yeast (Khadaroo *et al.*, 2004). However, the green lineage diverged very early in evolution from *O. tauri* (Bhattacharya and Medlin, 1998) and there has apparently been an accompanying divergence of structure in the mitotic phosphatase(s), or else the adoption of mitotic function in other phosphatase(s), since no known higher plant phosphatase has the highly conserved regions characteristic of CDC25 in other kingdoms. Identification of higher plant mitotic phosphatase may therefore come to depend on tests of their function, which can be ambiguous since complementation of the *cdc25*(ts) mutation in fission yeast is rather easily satisfied, for example, by a human protein tyrosine phosphatase that is unrelated to the cell cycle (Gould *et al.*, 1990). Furthermore, it is not certain which of the several mitotic phosphatases in yeast may be the most closely related to the plant enzyme since in fission yeast CDC25 works in tandem with PYP3 phosphatase. Mitosis is delayed if PYP3 is absent and can be sustained entirely by PYP3 alone if *PYP3* expression is raised (Millar *et al.*, 1992), and furthermore, STP1 can also substitute for CDC25 although it has no mitotic phenotype (Mondesert *et al.*, 1994). Higher plants may therefore have evolved a distinct different phosphatase enzyme with mitotic function, or may use a variant of PYP3 or even STP1. In animal cells there are up to three structurally related CDC25 enzymes (e.g. Mailand *et al.*, 2000) and we cannot eliminate the possibility that plants could contain multiple, perhaps structurally unrelated, tyr-thr-phosphatases.

A possible *CDC25* homologue has been named in *Arabidopsis* and indeed both activates CDK *in vitro* (Landrieu *et al.*, 2004) and advances mitosis when overexpressed in yeast (Sorrell *et al.*, 2005), but it cannot complement absence of CDC25 activity (Landrieu *et al.*, 2004) and is not preferentially expressed in tissues containing dividing cells (Sorrell *et al.*, 2005) although the opposing tyrosine kinase *WEE1* is so expressed (Sorrell *et al.*, 2002). The putative CDC25 has no clear sequence similarity to the CDC25 of other kingdoms, although it is predicted to fold similarly, and it is also unusually small, corresponding to only the catalytic region of other CDC25 molecules. Its failure to complement CDC25(ts) may derive from the absence of regulatory regions, although this absence has not prevented other phosphatases from complementing. The enzyme that has CDC25 function at plant mitosis is therefore far more difficult to identify than in other kingdoms and it may be necessary to gain information from the activity of putative CDC25 enzymes through the cell cycle, under normal progression and with checkpoint arrest, to test for functional relationship to division. It should be noted also that function of the same or related tyr-phosphatase may be required at S phase initiation, as in mammalian cells where the S phase–specific CDK2 is under control of a discrete phosphatase (e.g. Mailand *et al.*, 2000). The multiple CDKs found in plants such as alfalfa (Mészáros *et al.*, 2000) could reflect specific S phase function in some.

It is also uncertain how cytokinin, which is clearly necessary at mitosis, can at that time increase mitotic phosphatase relative to WEE1 kinase activity. It is encouraging that alternative phosphorylation patterns have been found to regulate activity in CDC25 phosphatases (e.g. Kovelman and Russell, 1996; Bulavin *et al.*, 2003) and in the opposing WEE1 kinases (Aligue *et al.*, 1997; Rhind and Russell, 2001) and that the activity changes include upregulation of tyr-phosphatase and downregulation of tyr-kinase, either of which could induce mitotic initiation. It may be significant that cytokinin signal transduction proceeds by transfer of protein phosphate groups (Haberer and Kieber, 2002; Kakimoto, 2003).

Operation of hormonal control over mitotic initiation is consistent with the additional importance of mitosis in plants as leading to the formation of structural elements such as cross walls and as committing continued proliferation rather than switching to endoreduplication that frequently provides a means for cell enlargement in the plant kingdom (Larkins *et al.*, 2001; Boudolf *et al.*, 2004). Control of mitosis through regulation of CDK is consistent with the central importance of CDK in plant cell division, indicated by slower division when CDK is mutated (Hemerly *et al.*, 1995) or is inhibited (e.g. Cleary *et al.*, 2002) and faster mitotic progression when active CDK is microinjected (Hush *et al.*, 1996). Significance of the mitotic control point in the intact plant is seen when CDK is activated by local expression of yeast *CDC25*, which results in local zones of division (Wyrzykowska *et al.*, 2002), and similarly constitutive expression increases lateral root and shoot initiation (McKibbin *et al.*, 1998; Suchomelova *et al.*, 2004). Lateral root primordia

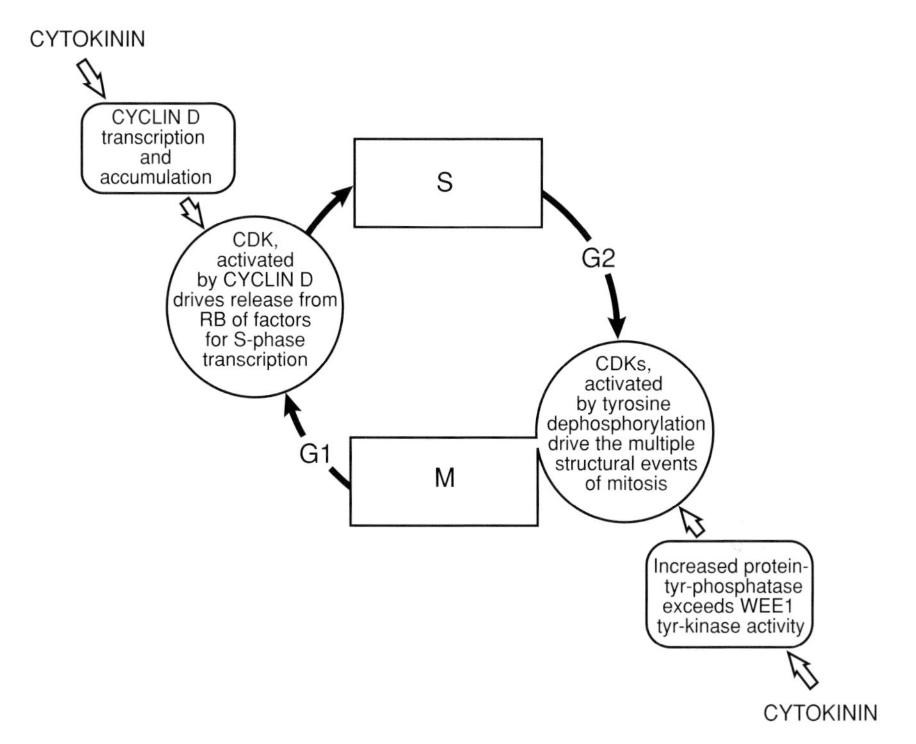

Figure 14.1 Cytokinin control of cell cycle progression. In the G1 phase, cytokinin stimulates CYCLIN D transcription and the increased CDK/CYCLIN D activity commits the cell to subsequent DNA synthesis, by proteolysis of CDK inhibitor proteins, phosphorylation of RB-related protein and activation of transcription by E2F factors. Resulting expression of deoxyribonucleotide-synthesizing enzymes, DNA-replicating proteins and mitotic cyclins drives DNA polymerization. In the G2 phase, cytokinin stimulates protein phosphatase activity, which has the CDC25-like function of progressively removing inhibitory phosphate placed on CDK by the tyrosine kinase WEE1. Rising CDK/CYCLIN B activity is integral with prophase, driving disassembly of the preprophase band, condensation of chromosomes, assembly of the mitotic spindle and nuclear envelope breakdown, which together establish the metaphase of mitosis.

are also indicated to come under post-translational control since the initials begin division in the pericycle while not initially expressing more cell cycle proteins than their neighbours (Hemerly *et al.*, 1993). Cytokinin therefore makes a dual contribution to the cell cycle (John and Zhang, 2001) in which progression at G1/S is advanced by induction of CYCLIN D and mitosis is initiated by activation of CDK tyr-phosphatase activity (Figure 14.1).

The arrest of cells at either the G1 or G2 phase when cytokinin is limiting may depend on slight differences in endogenous levels of CYCLIN D, or of activated phosphatase and CDK/CYCLIN B that determine which transition first becomes limiting as cytokinin concentration falls. Certainly, in plant tissues cells can arrest at either point (e.g. Van't Hof, 1974) and cells in different experimental systems have different arrest tendencies.

14.9 Roots and shoots provide each other with hormones essential for division

The cell division hormones auxin and cytokinin are more abundant at opposite poles of the plant. Higher extractable levels of cytokinin are found in roots (Chen *et al.*, 1985) but auxins are more abundant in shoot tips and young leaves (Srivastava, 2002). These locations are likely sites of synthesis and this is supported for cytokinin by evidence that in vegetative tissue the biosynthetic gene *ISOPENTENYLTRANSFERASE4* (*IPT4*) is preferentially expressed at the root tip (Miyawaki *et al.*, 2004).

Several aspects of meristem activity in the plant can now be understood as derived from the requirement for hormones from these widely separate sources. The higher concentrations of auxin in the root, and cytokinin in the shoot, imply gradients of declining concentration extending from the major sites of synthesis. A more recently appreciated refinement is that both shoot and root apical meristems amplify the concentration of the limiting hormone that arrives from the opposite pole of the plant and it then locates cell division activity within the apex. Furthermore, the initiation of lateral meristems may be located in response to the gradients of auxin and cytokinin extending through the plant from shoot and root, as will be discussed.

The broad trends in hormone concentration through the plant are shown in Plate 14.2, which also indicates the locally raised concentrations within meristems. Cytokinin in the shoot apex is amplified by local synthesis, while auxin in the root is concentrated through local recirculation. These concentration mechanisms for individual hormones that operate at the limit of their long-distance signalling prevent suboptimal concentrations for division where intensive division is required and they also provide demarcation of the regions of rapid cell division within apical meristems.

14.10 Cytokinin contributions to stem cell and meristem identity at the shoot apex

In the shoot apex, a stable core of indeterminate stem cells in the outer three cell layers is created by the interaction of several genes. The *KNOTTED1-like homeobox* (KNOX1*)* family of shoot homeobox genes is upregulated, perhaps indirectly, by cytokinin (Rupp *et al.*, 1999) and in Arabidopsis a key member of this family has been named *SHOOT MERISTEMLESS* (*STM*) for its phenotype when mutated. One function of STM is to specify stem cell proliferation in the central region of the shoot apex in concert with the *WUSCHEL* gene (*WUS*), a homeodomain gene of another family. *WUS* maintains a stable number of stem cells through being sensitive to inhibition by the CLV3 peptide that diffuses from cells that have been induced to stem cell identity, thus creating a negative feedback loop that is stronger if the stem cell area is overlarge (Lenhard and Laux, 2003). In addition, *WUS* enhances the effects of cytokinins, and may thereby stimulate stem cell proliferation, by repressing factors that decrease response to cytokinin (Leibfried *et al.*, 2005). The type-A response regulator proteins ARR5, ARR6, ARR7 and ARR15 are regulated by their phosphorylation and act to moderate overresponse to cytokinin. Expression of *WUS* leads to tenfold lower expression of these regulators, therefore enhancing response to cytokinin and enhanced stem cell proliferation. Furthermore, the key role of cytokinin in stabilizing stem cell proliferation is indicated by the blocking of stem apical meristem growth by over-expression of ARR7 in a constitutively active form (resembling the phosphorylated form that is generated by signal transduction from high cytokinin) which prevents response to cytokinin (Leibfried *et al.*, 2005).

STM has another major function, which is to induce expression of the cytokinin-synthesizing enzyme IPT7 (Yanai *et al.*, 2005). Overexpression of KNOX1 transcription factors (including STM) in Arabidopsis consistently raised expression of *IPT7* and increased the extractable cytokinin levels 30-fold and the intensity of expression from the cytokinin-inducible promoter *ARR5*. This induction of cytokinin synthesis through IPT7 is a major function of *KNOX1* since *stm* mutation, which normally renders the plant nonviable, could be rescued in 60% of transgenics if they received *AtIPT7* attached to the *STM* promoter, implying that cytokinin can largely replace *STM* (Yanai *et al.*, 2005). The zone of raised cytokinin expression that can be induced by *KNOX1* expression is normally more extensive than the central zone of *KNOX1* expression. This is consistent with the requirement for cytokinin to support cell proliferation in the leaf (e.g. Werner *et al.*, 2003) and this cytokinin function is also indicated by mitotic activity (reported by the induction of *CYCLINB:GUS* expression) when *STM* is ectopically expressed in the leaf (Lenhard *et al.*, 2002). The absence of *KNOX1* gene expression from the meristem flanks reflects the requirement that it be strictly downregulated in leaf primordial initials to allow their switch out of proliferation to a determinate cell identity (Long *et al.*, 1996).

In spite of these key functions of cytokinin, auxin remains essential to the shoot apical meristem as can be seen from the block to apex development that follows when auxin transport to the apex is limited by mutation in PIN auxin efflux regulator proteins (e.g. Gälweiler *et al.*, 1998; Blilou *et al.*, 2005), consistent with essential contributions being made by both hormones to cell cycle progression.

14.11 Auxin contributions to stem cell and meristem activity at the root apex

In the root apex, stem cells occupy a region designated the quiescent centre and have the complex function of producing root cap initials in the direction of root growth as well as producing initials for all other tissues behind the extending root tip. A point of similarity with the shoot apex may be that the quiescent centre is under control of a homeobox gene *QUIESCENT CENTRE HOMOBOX* (*QHB*) that resembles *WUS* and, when overexpressed in rice, causes expansion of the quiescent centre in the manner of *WUS*, inducing stem cell identity in the shoot (Kamiya *et al.*, 2003). However, *QHB* is not alone in having the capacity to specify stem cell identity since *PLETHORA* (*PLT*) genes, which encode AP2-domain transcription factors, are auxin inducible and can specify root identity when ectopically expressed (Aida *et al.*, 2004). In the root tip, auxin (the PLT inducer) is tightly focused on the stem cell region by its recirculation through auxin efflux regulators of the PIN-FORMED (PIN) protein family that are asymmetrically located at the periphery of the cell (Gälweiler *et al.*, 1998). Auxin can be tracked arriving at the root tip down the central vascular tissue and, instead of being lost through the central root cap, is recirculated up the epidermis until routed centripetally back to the central stele (Blilou *et al.*, 2005) to reinforce the local high auxin signal in the root tip meristem. Again, the dominant hormone-regulating cell division contributes to meristem organization, since auxin induces *PLT* expression and, therefore, stem cell identity in cells of the quiescent centre, and PLT reinforces this pattern through involvement in *PIN* expression, especially of PIN3 and PIN7, which return auxin from the tip through the epidermis and then direct it centripetally to reinforce the reflux loop. The consequent focusing of auxin concentration at the core of the meristem induces *PLT* expression there and specifies the size of the quiescent centre stem cell region (Blilou *et al.*, 2005). In spite of the importance of auxin, cytokinin is also essential for cell proliferation in the root apical meristem, as illustrated by inadequate division that results from the *woodenleg* mutation in the cytokinin receptor gene *AHK4*, which causes insufficient division in vascular precursors and a deficiency of phloem (Mähönen *et al.*, 2000; Nishimura *et al.*, 2004).

14.12 Hormones and the balance of cell proliferation between root and shoot

The dependence of meristems on auxin from the shoot and cytokinin from the root to sustain division provides a mechanism for balancing shoot and root growth. In general terms, a large shoot will tend to dissipate the cytokinin that it receives in a larger volume of shoot tissue, which will tend to slow the further proliferation of cells in the shoot, while the larger shoot will also have a more extensive capacity to produce auxin, which will tend to stimulate root growth and also act to restore plant proportions. Conversely, a more vigorous root can encourage increased shoot growth through cytokinin supply, and furthermore, this cytokinin signal is enhanced under favourable growth conditions since nitrogen nutrient elevates expression of *IPT3* and presumably cytokinin synthesis in the Arabidopsis root (Miyawaki *et al.*, 2004). The cytokinin signal to shoot growth is sufficiently substantial to be exploited in agriculture since rice cultivars found by QTL to have genes for reduced expression of cytokinin oxidase are better able to provide a strong cytokinin signal to shoot growth and increased grain yield (Ashikari *et al.*, 2005). The most extensive increase in growth in response to hormonal signals occurs through initiation of new points of growth through outgrowth of laterals (Plate 14.2), examined in the next section.

14.13 Auxin/cytokinin ratio and initiation of cell proliferation in lateral meristems

There is evidence that lateral meristem initiation responds to the ratio of hormone concentration in the root/shoot axis (Plate 14.2) in a way that prevents excessive proximity to the tip and spaces laterals in a manner that provides effective interception of water and nutrient in the soil and access to sunlight for photosynthesis.

Hormonal influence over lateral branching has been indicated by classical demonstrations that the inhibition of lateral shoot outgrowth by an active shoot apical meristem (termed apical dominance) could be continued if the apex was removed but replaced by auxin (IAA) diffusing from the lanoline paste applied to the cut tip (Thimann and Skoog, 1933); an effect opposed by additional presence of cytokinin (Wickson and Thimann, 1958) and confirmed in many systems (reviewed by Cline, 1991). Similarly in the root, inhibition of lateral root formation close to the apex could be mimicked by cytokinin applied to excised pea root segments, and this could be reversed by the additional presence of auxin (Torrey, 1962). A long-held expectation, based on such experiments, has been that the ratio of concentration of auxin and cytokinin is as significant as their absolute concentration (Skoog and Miller, 1957; Cary *et al.*, 2002).

Early experimental systems that involved surgery or excision of tissue might not extrapolate to the intact plant, but this concern is removed by use of transgenes and mutants that alter hormone levels (or the perception of hormones) within the intact plant. An early demonstration of the importance of the ratio of cytokinin to auxin concentrations (Romano *et al.*, 1993; Klee and Romano, 1994) was the detection that reduced auxin level in tobacco was equivalent to increased cytokinin level in stimulating lateral shoot outgrowth. The *iaaL* gene, which leads to the formation of inactive auxin-lysine conjugates, lowered effective auxin concentration. The outcome was a raised cytokinin phenotype, of increased lateral shoot outgrowth, presumably caused by raised cytokinin/auxin ratio although cytokinin concentration was not increased.

More recently, a battery of mutations and transgenes has become available to alter internal levels of auxin and cytokinin, and the resulting phenotypes bear out the classical findings that raised cytokinin is stimulatory to shoot development at the expense of the root, whereas raised auxin is stimulatory to root development but can retard cell division in the shoot. Taking only a few illustrative examples, effects of high auxin/cytokinin ratio have been studied using mutations in the AHK cytokinin receptor family that make the cytokinin signal less effective (Mähönen *et al.*, 2000; Nishimura *et al.*, 2004). Stronger effects come from actual reduction of cytokinin levels by expression of the cytokinin-degrading enzyme cytokinin oxidase (*CKX*), which has been thoroughly studied in tobacco and Arabidopsis (Werner *et al.*, 2001, 2003). Strong effects are also obtained if signal transduction from cytokinin is extensively inhibited by expression of the signal transducer ARR7 in a form constitutively activated to attenuate the cytokinin signal (with glutamate substituted for asp 85), which blocks shoot apical development (Leibfried *et al.*, 2005). Furthermore, a high auxin/cytokinin ration can be obtained by increasing auxin synthesis through the regulatory mutation *superroot* (*sur*) in *Arabidopsis* (Boerjan *et al.*, 1995). Phenotypic effects of a high auxin/cytokinin ratio in the root included increased root growth and extent of lateral roots and, with auxin at above normal concentration, a greater density of lateral roots and lateral initiation very close to the apical meristem. While conversely, in the shoot, strong negative effects were observed, including stunting, 20-fold reduction in leaf cell production and, at 4-fold above normal auxin levels, a shoot so reduced that no flowers were produced.

The converse situation of a high cytokinin/auxin ratio has similarly been studied by several means. Mutations that decrease perception of auxin, by reducing the capacity of AUX/IAA negative transcriptional regulator to respond to auxin by becoming ubiquitinated and proteolyzed, have been isolated as a series of alleles with decreasing auxin responsiveness (Rouse *et al.*, 1998). These show increasing stunting of the root system down to less than 20% of wild type and increasing predominance of growth in the shoot, including outgrowth of lateral shoots. These effects obtained through a modified transcription regulator underline the contribution of gene expression in hormonal effects. A further increase in the cytokinin/auxin ratio has been obtained by increasing cytokinin concentration through the *amp1* mutation in *Arabidopsis* (Chaudhury *et al.*, 1993) or by introduction of the bacterial *IPT* cytokinin synthesis gene into many plant species (e.g. Medford *et al.*, 1998). And more recently, the endogenous plant IPT genes have been used and the induction of cytokinin synthesis has been shown to be a major mechanism by which homeobox genes direct shoot meristems (Yanai *et al.*, 2005). High cytokinin/auxin ratios cause reduced growth in the root and in the shoot induce extensive initiation and outgrowth of lateral shoots, and can also initiate ectopic foci of division within the meristem causing supernumerary cotyledons and leaves.

14.14 Possible mechanisms for cell cycle response to hormone concentration and ratio

It is appropriate to consider mechanisms that might allow cell cycle gene control to be influenced not just by single hormones but by hormone ratios, and in addition how a hormone that might be essential for cell proliferation can also have the negative effect of providing an inhibitory signal in apical dominance, such as the effect of auxin imposing dormancy in lateral buds and cytokinin opposing lateral root initiation. Possible mechanisms include gene control by complexes of transcription factors that have a different affinity for DNA than do single factors and can contain proteins that have positive or negative effects on transcription. Investigation of these effects is inevitably painstaking and slow because regions as small as 7 base pairs located within 10 kb either upstream or downstream of the open reading frame can act as enhancers (cis regulatory elements) through binding transcription factors (e.g. the reviews by Blackwood and Kadonaga, 1998; Davidson *et al.*, 2002). Regulatory regions can therefore be too small to be recognized by sequence (Kreiman, 2004). Because each gene requires detailed empirical study, almost everything about the control of expression from cell cycle and meristem organization genes remains to be discovered.

Nevertheless, enough information is available to give some examples of the possible nature of control. The KNOX1 class of transcription factors that in the shoot apical meristem promote indeterminacy (persisting stem cell proliferation) itself comes under transcription control, illustrated by *STM* transcription being under repression by the DORNROSCHEN (DRN) protein (Kirch *et al.*, 2003). In addition, the binding to a unique DNA sequence by a KNOX1 protein and BELL1 protein shows a considerable change in affinity when they complex together (Smith *et al.*, 2002). Furthermore, the mechanism that allows continued division in the flanks of the shoot apex, in islands that form leaf primordia, involves regulation of KNOX1 through its binding of the homeobox transcription protein BELLRINGER (BLR). BLR capacity to bind to KNOX1 is detected in yeast 2-hybrid assay and the *blr* mutation increases leaf primordium number, which suggests that BLR function includes responding to inhibitory signals from pre-existing organs (Byrne *et al.*, 2003). The KNOX1 family of transcription factors therefore come under multiple regulatory influences through the binding of other transcription factors.

Thus, a general mechanism can be postulated by which a tissue may be primed to activate a class of genes under influence of a homoeobox gene that defines tissue identity, such as *STM* and *WUS* do in the shoot apex, and *PLT* and others do in the root. These genes may be expressed only when a hormone activates the transcription factor, by altering its configuration, or by inducing further transcription factor(s) that can complex with and modulate its activity, or by inactivating an existing negative transcription factor. Such inactivation of a negative transcriptional regulator underlies the activation of transcription by auxin and cytokinin. Cytokinin is believed to cancel the negative effects of type A response regulator (ARR) on cytokinin-inducible genes by inducing structural change in the protein (Haberer and Kieber, 2002; Kakimoto, 2003), whereas auxin eliminates the AUX/IAA transcriptional protein that inhibits the auxin response transcription factor (ARF) by inducing proteolysis of AUX/IAA protein through an auxin-induced increase in affinity for the ubiquitination complex (Dharmasiri *et al.*, 2005; Kepinski and Leyser, 2005). Transcription released by removal of negative regulators through cytokinin action is likely to be of particular relevance to cell division in the shoot, while auxin removal of transcription restraint is likely to be more important in the root, although both organs need both hormones for cell division and development as shown above.

Additional transcriptional regulation may provide more complex modulation that allows a hormone that is essential for division to provide a negative signal in certain developmental contexts such as restraining initiation of a lateral meristem close to an active terminal meristem. In the root apex, auxin induces the homeobox factor PLT but a positive signal from cytokinin, operating through other factors, is also required for cell division. However, cytokinin can give a negative signal that prevents initiation of lateral roots too close to the tip and this could derive from expression in lateral precursor cells of a transcription subunit that has a negative effect in response to high cytokinin. Since lateral roots initiate from pericycle cells that are arrested in the G2 phase, mitotic cyclin genes are candidates for repression by high concentrations of cytokinin.

In the shoot, the apical meristem has a dual requirement for cytokinin and auxin, jointly modulating homeobox and other gene actions, as described earlier, but in the region behind the apical meristem auxin exerts a negative effect on lateral bud outgrowth, which imposes apical dominance. A component of apical dominance is auxin inhibition of transcription of the cytokinin biosynthetic genes *IPT1* and *IPT2* (Tanaka *et al.*, 2006), which inhibits cytokinin biosynthesis and thereby maintains dormancy by removing cytokinin from the region of the lateral bud. An absence of cytokinin at the lateral bud is the key outcome and may be explained by the presence of auxin-activated negative transcriptional regulators in the tissue below the shoot apex. An additional reinforcement of this mechanism could come from the observed capacity of auxin to stimulate the breakdown of cytokinin (e.g. Nordstrom *et al.*, 2004). Apical dominance over lateral shoots may therefore be simply caused by an absence of cytokinin, without requiring novel hormone signals.

14.15 Cell cycle control in the spacing of lateral organs

Auxin and cytokinin contribute to the physiological efficiency of plant architecture through their influence over branching in root and shoot. In the root, apical dominance defers initiation of lateral roots to a regulated distance from the established lead apex and the consequent regular spacing allows efficient extraction of nutrients from a large volume of soil. This default location of laterals is fine-tuned by nutritional signals. Low phosphate induces more lateral roots per length of primary root, which more efficiently mines the relatively immobile phosphate compounds (Williamson *et al.*, 2001), and conversely laterals are repressed under high nitrate availability when adequate nutrient can be derived without additional root tissue. Furthermore, this repression is reversed by photosynthate (Zhang *et al.*, 1999), which may signal potential for greater growth if nutrient supply to the shoot is kept high. While it is thought that the low phosphate signal is not primarily mediated by the auxin/cytokinin ratio (Williamson *et al.*, 2001), the repression of laterals by high nitrate may be exerted through auxin (Bhalerao *et al.*, 2002; reviewed by Walch-Liu *et al.*, 2006).

Such physiological responsiveness indicates that cell proliferation in lateral organs is sensitive to control and that mechanisms that influence cell cycle progression are part of the mechanism. A clear cell cycle basis (Zhang *et al.*, 2005) is apparent when increased lateral root and shoot initiation follows expression of yeast *CDC25* (McKibbin *et al.*, 1998; Suchomelova *et al.*, 2002), which is consistent with the stimulatory effect of low concentrations of cytokinin on lateral roots (Wightman *et al.*, 1980; Biddington and Dearman, 1982) and the more easily detected effects of high cytokinin on lateral shoots. This stimulation of lateral organ outgrowth by cytokinin therefore correlates with the cytokinin stimulation of tyr-dephosphorylation in CDK discussed earlier. It is also in accord with the effect of modified *AtCDKA;1* that lacks sites for inhibitory phosphorylation (because it contains the substitutions Ala-14 and Phe-15). This (A14 F15) transgene, when expressed in Arabidopsis (Hemerly *et al.*, 1995), induces increase in lateral shoot initiation in the manner of cytokinin, which we interpret as a further instance of high cytokinin being signalled by tyr-dephosphorylated CDK.

Therefore, hormonal control of cell division makes multiple contributions to the regulation of cell proliferation in meristems and through control over new lateral points of growth, it also strongly influences plant architecture.

References

Aida, M., Beis, D., Hiedstra, R., *et al.* (2004) The PLETHORA genes mediate patterning of the Arabidopsis root stem cell niche. *Cell* **119**, 109–120.

Aligue, R., Wu, L. and Russell, P. (1997) Regulation of *Schizosaccharomyces pombe* Wee1 tyrosine kinase. *J Biol Chem* **272**, 13320–13325.

Ashikari, M., Sakakibara, H., Lin, S., *et al.* (2005) Cytokinin oxidase regulates rice grain production. *Science* **309**, 741–745.

Beeckman, T., Burssens, S. and Inzé, D. (2001) The peri-cell-cycle in *Arabidopsis*. *J Exp Bot* **52**, 403–411.

Bergsagel,Kuehl, W.M., Zhan, F., Sawyer, J., Barlogie, B. and Shaughnessy, J. (2005) Cyclin D dysregulation: an early and unifying pathogenic event in multiple myeloma. *Blood* **106**, 296–303.

Bhalerao, R.P., Eklöf, J., Ljung, K., Marchant, A., Bennett, M. and Sandberg, G. (2002) Shoot-derived auxin is essential for early lateral root emergence in *Arabidopsis* seedlings. *Plant J* **29**, 325–332.

Bhattacharya, D. and Medlin, L. (1998) Algal phylogeny and the origin of land plants. *Plant Physiol* **116**, 9–15.

Biddington, N.L. and Dearman, A.S. (1982) the involvement of the root apex and cytokinins in the control of lateral root emergence in lettuce seedlings. *Plant Growth Regul* **1**, 183–193.

Blackwood, E.M. and Kadonaga, J.T. (1998) Going the distance: a current view of enhancer action. *Science* **281**, 60–63.

Blilou, I., Xu, J., Wildwater, M., *et al.* (2005) The PIN auxin efflux facilitator network controls growth and patterning in *Arabidopsis* roots. *Nature* **433**, 39–44.

Boerjan, W., Cervera, M-T., Delarue, M., *et al.* (1995) *superroot*, a recessive mutation in Arabidopsis, confers auxin overproduction. *Plant Cell* **7**, 1405–1419

Boudolf, V., Vlieghe, K., Beemster, G.T.S., *et al.* (2004) The plant-specific cyclin dependent kinase CDKB1;1 and transcription factor E2Fa-DPa control the balance of mitotically dividing and endoreduplicating cells in Arabidopsis. *Plant Cell* **16**, 2683–2692.

Bulavin, D.V., Higashimoto, Y., Demidenko, Z.N., *et al.* (2003) Dual phosphorylation controls Cdc25 phosphatases and mitotic entry. *Nat Cell Biol* **5**, 545–551.

Burssens, S., de Almeida Engler, J., Beeckman, T., *et al.* (2000) Developmental expression of the *Arabidopsis thaliana CycA2;1* gene. *Planta* **211**, 623–631.

Byrne, M.E., Groover, A.T., Fontana, J.R. and Martienssen, R.A. (2003) Phyllotactic pattern and stem cell fate are determined by the *Arabidopsis* homeobox gene *BELLRINGER*. *Development* **130**, 3941–3950.

Canovas, J.L., Cuadrado, A., Escalera, M. and Navarrete, M.H. (1990) The probability of cells to enter S increases with their size while S length decreases with cell enlargement in *Allium cepa*. *Exp Cell Res* **191**, 163–170.

Carter, B.L.A. and Jagadish, M.N. (1975) Control of cell division in the yeast *Saccharomyces cerevisiae* cultured at different growth rates. *Exp Cell Res* **112**, 373–383.

Cary, A.J., Che, P. and Howell, S.H. (2002) Developmental events and shoot apical meristem gene expression patterns during shoot development in *Arabidopsis thaliana*. *Plant J* **32**, 867–877.

Chaudhury, A.M., Letham, D.S., Craig, S. and Dennis, E.S. (1993) amp1–a mutant with high cytokinin levels and altered embryonic pattern, faster vegetative growth, constitutive photomorphogenesis and precocious flowering. *Plant J* **4**, 907–916.

Chen, C., Ertl, J.R., Leisner, S.M. and Chang, C. (1985) Localisation of cytokinin biosynthetic site in pea plants and carrot roots. *Plant Physiol* **78**, 510–513.

Cleary, A.L., Fowke, L.C., Wang, H. and John, P.C.L. (2002) The effect of ICK1, a plant cyclin-dependent kinase inhibitor, on mitosis in living plant cells. *Plant Cell Rep* **20**, 814–820.

Cline, M.G. (1991) Apical dominance. *Bot Rev* **57**, 318–358.

Cockroft, C.E., den Boer, B.G., Healy, J.M.S. and Murray, J.A.H. (2000). Cyclin D control of growth rate in plants. *Nature* **405**, 575–579.

Daga, R.R. and Jiminez, J. (1999) Translational control of the cdc25 cell cycle phosphatase: a molecular mechanism coupling mitosis to cell growth. *J Cell Sci* **112**, 3137–3146.

Davidson, E., Rast, J., Oliveri, P., *et al.* (2002) A genomic regulatory network for development. *Science* **295**, 1669–1678.

Dewitte, W., Riou-Khamlichi, C., Scofield, S., *et al.* (2003) Altered cell cycle distribution, hyperplasia, and inhibited differentiation in Arabidopsis caused by the D-type cyclin CYCD3. *Plant Cell* **15**, 79–92.

Dharmasiri, N., Dharmasiri, S. and Estelle, M. (2005) The F-box protein TIR1 is an auxin receptor. *Nature* **435**, 441–445.

Dolznig, H., Grebien, F., Sauer, T. and Müllner, EW. (2004) Evidence for a size-sensing mechanism in animal cells. *Nat Cell Biol* **6**, 899–905.

Donnan, L. and John, P.C.L. (1983) Cell cycle control by timer and size in *Chlamydomonas*. *Nature* **340**, 630–633.

Fantes, P.A., Grant, W.D., Pritchard, R.H., Sudbery, P.E. and Wheals, A.E. (1975) The regulation of cell size and the control of mitosis. *J Theor Biol* **50**, 213–244.

Fantes, P. and Nurse, P. (1981) Division timing: controls, models and mechanisms. In: *The Cell Cycle* (ed. John, P.C.L.) Cambridge University Press, Cambridge, pp. 11–33.

Foard, D.E. (1971) The initial protrusion of a leaf primordium can form without concurrent periclinal cell division. *Can J Bot* **49**, 694–702.

Foard, D.E., Haber, A.N. and Fishman, T.N. (1965) Initiation of lateral root primordia without completion of mitosis and without cytokinesis in uniseriate pericycle. *Am J Bot* **52**, 580–590.

Fosket, D.E. (1977) The regulation of the plant cell cycle by cytokinin. In: *Mechanisms and Control of Cell Division* (eds Rost, T.L. and Gifford, E.M.). Dowden Hutchinson and Ross Inc, Strousberg, Pennsylvania, pp. 62–91.

Frei, C. and Edgar, B.A. (2004) Drosophila cyclin D/Cdk4 requires Hif-1 prolyl hydroxylase to drive cell growth. *Dev Cell* **6**, 241–651.

Galtier, N., Foyer, C.H., Huber, J., Voelker, T.A. and Huber, C. (1993) Effects of elevated sucrose-phosphate synthase activity on photosynthesis, assimilate partitioning, and growth in tomato (*Lycopersicon esculentum* var UC82B) *Plant Physiol* **101**, 535–543.

Gälweiler, L., Guan, C., Muller, A., *et al.* (1998) Regulation of polar auxin transport by AtPIN1 in *Arabidopsis* vascular tissue. *Science* **282**, 2226–2230.

Gregory, T.R. (2001) Coincidence, coevolution, or causation? DNA content, cell size, and the C-value enigma. *Biol Rev* **76**, 65–101.

Gorst, J.R., Sek, F.J. and John, P.C.L. (1991) Level of p34^{cdc2}-like protein in dividing, differentiating and dedifferentiating cells of carrot. *Planta* **185**, 304–310.

Gould, K.L., Moreno, S., Tonks, N.K. and Nurse, P. (1990) Complementation of the mitotic activator p80cdc25, by a human protein-tyrosine phosphatase. *Science* **250**, 1573–1576.

Gutierrez, C., Ramirez-Parra, E., Castellano, M. and del Pozo, J.C. (2002) G1 to S transition: more than a cell cycle engine switch. *Curr Opin Plant Biol* **5**, 480–486.

Haberer, G. and Kieber, J.J. (2002) Cytokinins: new insights into a classic phytohormone. *Plant Physiol* **128**, 354–362.

Hemerly, A., de Almeida Engler, J., Bergounioux, C., *et al.* (1995) Dominant negative mutants of Cdc2 kinase uncouple cell division from iterative plant development. *EMBO J* **14**, 3925–3936.

Hemerly, A.S., Ferriera P.C.G., de Almeida Engler, J., Van Montagu, M., Engler, G. and Inzé, D. (1993) *cdc2a* expression in *Arabidopsis* is linked with competence for cell division. *Plant Cell* **5**, 1711–1723.

Himanen, K., Boucheron, E., Vanneste, S., de Almeida Engler, J., Inzé., D. and Beeckman, T. (2002) Auxin-mediated cell cycle activation during early lateral root initiation. *Plant Cell* **14**, 2339–2351.

Hoth, S., Ikeda, Y., Morgante, M., *et al.* (2003) Monitoring genome-wide changes in gene expression in response to endogenous cytokinin reveals targets in *Arabidopsis thaliana*. *FEBS Lett* **554**, 373–380.

Hush, J.M., Wu, L., John, P.C.L., Hepler, L.H. and Hepler, P.K. (1996) Plant mitosis promoting factor disassembles the microtubule preprophase band and accelerates prophase progression in *Tradescantia*. *Cell Biol Int Rep* **20**, 275–287.

John, P.C.L. and Zhang, K. (2001) Cytokinin control of cell proliferation in plant development. In: *The Plant Cell Cycle and Its Interfaces* (ed. Francis, D.). CRC Press, Boca Raton, pp. 190–211.

John, P.C.L., Zhang, K., Dong, C., Diederich, L. and Wightman, F. (1993) $p34^{cdc2}$ related proteins in control of cell cycle progression, the switch between division and differentiation in tissue development and stimulation of division by auxin and cytokinin. *Aust J Plant Physiol* **20**, 503–526.

Joubès, J., Chevalier, C., Dudits, D., *et al.* (2000) CDK-related protein kinases in plants. *Plant Mol Biol* **43**, 607–620

Kakimoto, T. (2003) Perception and signal transduction of cytokinins. *Annu Rev Plant Biol* **54**, 605–627.

Kamiya, N., Nagasaki, H., Morikami, A., Sato, Y. and Matsuoka, M. (2003) Isolation and characterization of a rice WUSCHEL-type homeobox gene that is specifically expressed in the central cells of a quiescent center in the root apical meristem. *Plant J* **35**, 429–441.

Kepinski, S. and Leyser, O. (2005) The Arabidopsis F-box protein TIR1 is an auxin receptor. *Nature* **435**, 446–451.

Khadaroo, B., Robbens, S., Ferraz, C., *et al.* (2004) The first green lineage cdc25 dual-specificity phosphatase. *Cell Cycle* **3**, 513–518.

Kirch, T., Simon, R., Grunewald, M. and Werr, W. (2003) The *DORNROSCHEN/ENHANCER OF SHOOT REGENERATION1* gene of Arabidopsis acts in the control of meristem cell fate and lateral organ development. *Plant Cell* **15**, 694–705.

Klee, H. and Romano, C. (1994) The roles of phytohormones in development as studied in transgenic plants. *Crit Rev Plant Sci* **13**, 311–324.

Kovelman, R. and Russell, P. (1996) Stockpiling of Cdc25 during a DNA replication checkpoint arrest in *Schizosaccharomyces pombe*. *Mol Cell Biol* **16**, 86–93.

Kozar, K.,Ciemerych, M.A., Rebel, V.I., *et al.* (2004) Mouse development and cell proliferation in the absence of D-cyclins. *Cell* **118**, 477–491.

Kreiman, G. (2004) Identification of sparsely distributed clusters of cis-regulatory elements in sets of co-expressed genes. *Nucleic Acids Res* **32**, 2889–2900.

Landrieu, I., da Costa, M., De Veylder, L., *et al.* (2004) A small CDC25 dual-specificity tyrosine-phosphatase isoform in *Arabidopsis thaliana*. *Proc Natl Acad Sci USA* **101**, 13380–13385.

Larkins, B.A., Dilkes, B.P., Dante, R.A., Coelho, C.M., Woo, Y. and Liu, Y. (2001) Investigating the hows and whys of DNA endoreduplication. *J Exp Bot* **52**, 183–192.

Laureys, F., Dewitte, W., Van Montagu, M., Inzé, D. and Van Onckelen, H. (1998) Zeatin is indispensable for the G2-M transition in tobacco BY-2 cells. *FEBS Lett* **426**, 29–32.

Leibfried, A., To, J.P., Busch, W., *et al.* (2005) WUSCHEL controls meristem function by direct regulation of cytokinin-inducible response regulators. *Nature* **438**, 1172–1175.

Lenhard, M., Jürgens, G. and Laux, T. (2002) The *WUSCHEL* and *SHOOTMERISTEMLESS* genes fulfil complementary roles in *Arabidopsis* shoot meristem regulation. *Development* **129**, 3195–3206.

Lenhard, M. and Laux, T. (2003) Stem cell homeostasis in the Arabidopsis shoot meristem is regulated by intercellular movement of *CLAVATA3* and its sequestration by *CLAVATA1*. *Development* **130**, 3163–3173.

Lincoln, C., Britton, J.H. and Estelle, M. (1990) Growth and development of the *axr1* mutants of *Arabidopsis*. *Plant Cell* **2**, 1071–1080.

Long, J.A., Moan, E.I., Medford, J.I. and Barton, M.K. (1996) A member of the KNOTTED class of homeodomain proteins encoded by the STM gene of *Arabidopsis*. *Nature* **379**, 66–69.

McKibbin, R.S., Halford, N.G. and Francis, D. (1998) Expression of fission yeast *cdc25* alters the frequency of lateral root formation in transgenic tobacco. *Plant Mol Biol* **36**, 601–612.

MacNeill, S.A. and Nurse, P. (1997) Cell cycle control in fission yeast. In: *The Molecular and Cellular Biology of the Yeast Saccharomyces: Life Cycle and Cell Biology* (eds Pringle, J.R., Broach, J. and Jones, E.W.). Cold Spring Harbor Press, Cold Spring Harbor, New York, pp. 697–763.

Mähönen, A.P., Bonke, M., Kauppinen, L., Riikonen, M., Benfey, P.N. and Helariutta, Y. (2000) A novel two-component hybrid molecule regulates vascular morphogenesis of the Arabidopsis root. *Genes Dev* **14**, 2938–2943.

Mailand, N., Falck, J., Lukas, C., *et al.* (2000) Rapid destruction of human Cdc25A in response to DNA damage. *Science* **288**, 1425–1429.

Martinez, M.C., Jorgensen, J.-E., Lawton, M.A., Lamb, C.J. and Doerner, P.W. (1992) Spatial pattern of *cdc2* expression in relation to meristem activity and cell proliferation during plant development. *Proc Natl Acad Sci USA* **89**, 7360–7364.

Medford, J.I., Horgan, R., El-Sawi, Z. and Klee, H.J. (1989) Alterations of endogenous cytokinins in transgenic plants using a chimeric isopentyl transferase gene. *Plant Cell* **1**, 403–413.

Mészáros, T., Miskolczi, P., Ayaydin, F., *et al.* (2000)Multiple cyclin-dependent kinase complexes and phosphatases control G(2)/M progression in alfalfa cells. *Plant Mol Biol* **43**, 595–605.

Millar, J.B.A., Lenaers, G. and Russell, P. (1992) *Pyp3* PTPase acts as a mitotic inducer in fission yeast. *EMBO J* **11**, 4933–4941.

Millar, J.B.A., McGowan, C.H., Lenaers, G., Jones, R. and Russell, P. (1991) $p80^{cdc25}$ mitotic inducer is the tyrosine phosphatase that activates $p34^{cdc2}$ kinase in fission yeast. *EMBO J* **10**, 4301–4309.

Mitchison, J.M. (2003) Growth during the cell cycle. *Int Rev Cytol* **226**, 165–258.

Miyawaki, K., Matsumoto-Kitano, M. and Kakimoto, T. (2004) Expression of cytokinin biosynthetic isopentenyltransferase genes in Arabidopsis: tissue specificity and regulation by auxin, cytokinin, and nitrate. *Plant J* **37**, 128–138.

Mondesert, O., Moreno, S. and Russell, P. (1994) Low molecular weight protein-tyrosine phosphatases are highly conserved between fission yeast and man. *J Biol Chem* **269**, 27996–27999.

Murray, J.A.H., Freeman, D., Greenwood, J., *et al.* (1998) Plant D cyclins and retinoblastoma (rB) plant homologues. In: *Plant Cell Division* (eds Francis, D., Dudits, D. and Inzé, D.). Portland Press, London, pp. 99–127.

Nakayama, K., Nakayama, K., Ishida, N., *et al.* (1996) Mice lacking p27*Kip1* display increased body size, multiple organ hyperplasia, retinal dysplasia, and pituitary tumors. *Cell* **85**, 707–720.

Neufeld, T.P., de la Cruz, A.F., Johnston, L.A. and Edgar, B.A. (1998) Coordination of growth and cell division in the Drosophila wing. *Cell* **7**, 1183–1193.

Nishimura, C., Ohashi, Y., Sato, S., Kato, T., Tabata, S. and Ueguchi, C. (2004) Histidine kinase homologs that act as cytokinin receptors possess overlapping functions in the regulation of shoot and root growth in Arabidopsis. *Plant Cell* **16**, 1365–1377.

Norbury, C. and Nurse, P. (1992) Animal cell cycles and their control. *Annu Rev Biochem* **61**, 441–470.

Nordstrom, A., Tarkowski, P.Y., Tarkowska, D., *et al.* (2004) Auxin regulation of cytokinin biosynthesis in *Arabidopsis thaliana*: a factor of potential importance for auxin-cytokinin-regulated development. *Proc Natl Acad Sci USA* **101**, 8039–8044.

Ohtsubo, M. and Roberts, J.M. (1993) Cyclin-dependent regulation of Gl in mammalian cells. *Science* **259**, 1908–1912.

Orchard, C.B., Siciliano, I., Sorrell, D.A., *et al.* (2005) Tobacco BY-2 cells expressing fission yeast cdc25 bypass a G2/M block on the cell cycle. *Plant J* **44**, 290–299.

Polymenis, M. and Schmidt, E.V. (1997) Coupling of cell division to cell growth by translational control of the G1 cyclin CLN3 in yeast. *Genes Dev* **11**, 2522–2531.

Preuss, S.B. and Britt, A.B. (2003) A DNA-damage-induced cell cycle checkpoint in Arabidopsis. *Genetics* **164**, 323–234.

Quelle, D.E., Ashmun, R.A., Shurtleff, S.A., *et al.* (1993) Overexpression of mouse D-type cyclins accelerates Gl phase in rodent fibroblasts. *Genes Dev* **7**, 1559–1571.

Redig, P., Shaul, O., Inzé, D., Van Montagu, M. and Van Onckelen, H. (1996) Levels of endogenous cytokinins, indole-3-acetic acid and abscisic acid during the cell cycle of synchronized tobacco BY-2 cells. *FEBS Lett* **391**, 175–180.

Reicheld, J.-P., Vernoux, T., Lardon, F., Van Montagu, M. and Inzé, D. (1999) Specific checkpoints regulate plant cell cycle progression in response to oxidative stress. *Plant J* **17**, 647–656.

Resnitzky, D., Gossen, M., Bujard, H. and Reed, S.I. (1994) Acceleration of the G1/S phase transition by expression of cyclins D1 and E with an inducible system. *Mol Cell Biol* **14**, 1669–1679.

Rhind, N. and Russell, P. (2001) Roles of the mitotic inhibitors Wee1 and Mik1 in the G2 DNA damage and replication checkpoints. *Mol Cell Biol* **21**, 1499–1508.

Riou-Khamlichi, C., Huntley, R., Jacqmard, A. and Murray, J.A.H. (1999) Cytokinin activation of Arabidopsis cell division through a D-type cyclin. *Science* **283**, 1541–1544.

Riou-Khamlichi, C., Menges, M., Healy, J.M.S. and Murray, J.A.H. (2000) Sugar Control of the Plant Cell Cycle: Differential Regulation of Arabidopsis D-Type Cyclin Gene Expression. *Mol Cell Biol* **20**, 4513–4521.

Romano, C.P., Cooper, M.L. and Klee, H.J. (1993). Uncoupling auxin and ethylene effects in transgenic tobacco and Arabidopsis plants. *Plant Cell* **5**, 181–189.

Rouse, D., Mackay, P., Stirnberg, P., Estelle, M. and Leyser, O. (1998) Changes in auxin response from mutations in *AUX/IAA* gene. *Science* **279**, 1371–1373.

Rupp, H.M., Frank, M., Werner, T., Strnad, M. and Schmülling, T. (1999) Increased steady state mRNA levels of the *STM* and *KNAT1* homeobox genes in cytokinin overproducing *Arabidopsis thaliana* indicate a role for cytokinins in the shoot apical meristem. *Plant J* **18**, 557–563.

Sanchez-Diaz, A., Gonzalez, I., Arellano, M. and Moreno, S. (1998) The Cdk inhibitors p25rum1 and p40SIC1 are functional homologues that play similar roles in the regulation of the cell cycle in fission and budding yeast. *J Cell Sci* **111**, 843–851.

Sauter, M. and Kende, H. (1992) Gibberellin induced growth and regulation of the cell division cycle in deepwater rice. *Planta* **188**, 362–368.

Schuppler, U., He, P.-H., John, P.C.L. and Munns, R. (1998) Effect of water stress on cell division and Cdc2-like cell cycle kinase activity in wheat leaves. *Plant Physiol* **117**, 667–678.

Sherr, C.J. (1994) G1 phase progression: cycling on cue. *Cell* **79**, 551–555.

Skoog, F. and Miller, C.O. (1957) Chemical regulation of growth and organ formation in plant tissue cultured in vitro. *Symp Soc Exp Biol* **11**, 118–131.

Smith, H.M., Boschke, I. and Hake, S. (2002) Selective interaction of plant homeodomain proteins mediates high DNA-binding affinity. *Proc Natl Acad Sci USA* **99**, 9579–9584.

Sorrell, D.A., Chrimes, D., Dickinson, J.R., Rogers, H.J. and Francis, D. (2005) The Arabidopsis *CDC25* induces short cell length when overexpressed in fission yeast: evidence for cell cycle function. *New Phytol* **165**, 425–428.

Sorrell, D.A., Marchbank, A., McMahon., Dickinson, R., Rogers, H.J. and Francis, D. (2002) A *WEE1* homologue from *Arabidopsis thaliana. Planta* **215**, 518–522.

Srivastava, L.M. (2002) *Plant Growth and Development*. Academic Press, San Diego, pp. 303–379.

Suchomelova, P., Velgova, D., Masek, T., *et al.* (2004) Expression of the fission yeast cell cycle regulator *cdc25* induces de novo shoot formation in tobacco: evidence of a cytokinin-like effect by this mitotic activator. *Plant Physiol Biochem* **42**, 49–55.

Sun, Y., Dilkes, B.P., Zhang, C., *et al.* (1999) Characterisation of maize (*Zea mays* L.) Wee1 and its activity in developing endosperm. *Proc Natl Acad Sci USA* **96**, 4180–4185.

Sussex, I.M., Godoy, J.A., Kerk, N.M., *et al.* (1995) Cellular and molecular events in a newly organising lateral root meristem. *Philos Trans R Soc Ser B* **350**, 39–44.

Tanaka, M., Takei, K., Kojima, M., Sakakibara, H. and Mori, H. (2006) Auxin controls local cytokinin biosynthesis in the nodal stem in apical dominance. *Plant J* **45**, 1028–1036.

Tao, W. and Verbelen, J.-P. (1996) Switching on and off cell division and cell expansion in cultured mesophyll protoplasts of tobacco. *Plant Sci* **116**, 107–115.

Theologis, A., Huynh, T.V. and Davies, R.W. (1985) Rapid induction of specific mRNA sequences in pea epicotyl tissue. *J Mol Biol* **183**, 53–68.

Thimann, K.V. and Skoog, F. (1933) Studies on the growth hormones of plants. III The inhibition action of growth substance on bud development. *Proc Natl Acad Sci USA* **19**, 714–716.

Torrey, J.G. (1962) Auxin and purine interactions in lateral root initiation in isolated pea root segments. *Physiol Plant* **15**, 177–185.

Tran Thanh Van, K.M. (1981) Control of morhogenesis in *in vitro* cultures. *Annu Rev Plant Physiol* **32**, 291–311.

Van't Hof, J. (1974) Control of the cell cycle in higher plants. In *Cell Cycle Controls* (eds Padilla, G.M. Cameron, I.L. and Zimmerman, A.). Academic Press, New York, pp. 77–85.

Valente P., Tao, W. and Verbelen, J.-P. (1998) Auxins and cytokinins control DNA replication and deduplication in single cells of tobacco. *Plant Sci* **134**, 207–215.

Walch-Liu, P., Ivanaov, I., Filleur, S., Gan, Y., Remans, T. and Forde, B.G. (2006) Nitrogen regulation of root branching. *Ann Bot* **97**, 875–881.

Werner, T., Motyka, V., Laucou, V., De, Smets R., Van Onckelen, H. and Schmülling, T. (2003) Cytokinin-deficient transgenic Arabidopsis plants show multiple developmental alterations indicating opposite functions of cytokinins in the regulation of shoot and root meristem activity. *Plant Cell* **15**, 2532–2550.

Werner, T., Motyka, V., Strnad, M. and Schmülling, T. (2001) Regulation of plant growth by cytokinin. *Proc Natl Acad Sci USA* **98**, 10478–10492.

Wickson, M. and Thiman, K.V. (1958) The antagonism of auxin and kinetin in apical dominance. *Physiol Plant* **11**, 62–74.

Wightman, F., Schneider, E.A. and Thimann, K.V. (1980) Hormonal factors controlling the initiation and development of lateral roots. II. Effects of exogenous growth factors on lateral root formation in pea roots. *Physiol Plant* **49**, 304–314.

Williamson, L.C., Ribrioux, S.P.C.P., Fitter, A.H. and Leyser, H.M.O. (2001) Phosphate availability regulates root system architecture in Arabidopsis. *Plant Physiol* **126**, 875–882.

Wyrzykowska, J., Pien, S., Shen, W.H. and Fleming, A.J. (2002) Manipulation of leaf shape by modulation of cell division. *Development* **129**, 957–964.

Yanai, O., Shani, E., Dolezal, K., *et al.* (2005) Arabidopsis KNOXI proteins activate cytokinin biosynthesis. *Curr Biol* **15**, 1566–1571.

Zetterberg, A. and Killander, D. (1965) Quantitative cytochemical studies on interphase growth. II Derivation of synthesis curves from the distribution of DNA, RNA and mass values of individual mouse fibroblast *in vitro*. *Exp Cell Res* **39**, 22–32.

Zetterberg, A. and Larsson, O. (1985) Kinetic analysis of regulatory events in G1 leading to proliferation or quiescence of Swiss 3T3 cells. *Proc Natl Acad Sci USA* **82**, 5365–5369.

Zhang, K., Diederich, L. and John, P.C.L. (2005) The cytokinin requirement for cell division in cultured *Nicotiana plumbaginifolia* cells can be satisfied by yeast Cdc25 protein tyrosine phosphatase: implications for mechanisms of cytokinin response and plant development. *Plant Physiol* **137**, 308–316.

Zhang, H., Jennings, A., Barlow, P.W. and Forde, B.G. (1999) Dual pathways for regulation of root branching by nitrate. *Proc Natl Acad Sci USA* **96**, 6529–6534.

Zhang, K., Letham, D.S. and John, P.C.L. (1996) Cytokinin controls the cell cycle at mitosis by stimulating the tyrosine dephosphorylation and activation of $p34^{cdc2}$-like H1 histone kinase. *Planta* **200**, 2–12.

Zhiponova, MK., Pettkó-Szandtner, A., Stelkovics, E., *et al.* (2006) Mitosis-specific promoter of the alfalfa cyclin-dependent kinase gene (Medsa;CDKB2;1) is activated by wounding and ethylene in a non-cell division-dependent manner. *Plant Physiol* **140**, 693–703.

15 Cell cycle and environmental stresses

Christine Granier, Sarah Jane Cookson, Francois Tardieu
and Bertrand Muller

15.1 Introduction

Improved agricultural practices and selective breeding for high-yielding cultivars have greatly increased the productivity of crops over the past century. However, there are numerous environmental constraints that hamper crop performance in the field, for example, high soil salinity, water deficit, hypoxia, periods of high or low temperature, high metal concentration in the soil and the presence of air pollutants such as ozone. The risk of the occurrence of such stresses is likely to increase in most regions of the world according to the predicted global warming scenarios as well as the expected increase in atmospheric CO_2 concentration. Moreover, such changes might also alter the geographical zones of crop cultivation and require the modification of currently cultivated crops and agricultural practices. An understanding of how plants respond to the abiotic environment is essential in order to effectively design selective breeding strategies and construct crops best adapted to such a changing climate.

Plants have developed different mechanisms to cope with the highly variable environment they have to face. Their ability to alter their physiology, morphology and/or phenology in response to environmental changes is called phenotypic plasticity and allows plants to tolerate, avoid or escape the stressing condition (Grime *et al.*, 1986). Plant plasticity has been investigated at developmental and physiological levels for many years, and more recently, these responses have been studied at the molecular level. Besides biochemical modifications, adjustments of phenology and development have been shown to be 'early' responses to environmental changes. Such changes allow plants to modify their architecture and adapt to stress conditions (Wery, 2005). The developmental plasticity of plants in response to contrasted environmental conditions can be illustrated by the extreme example of the heterophylly of amphibious plants, which can grow in water as well as in the air with completely different forms. More subtly, the rate, duration and patterns of plant organ growth are influenced by environmental factors. Plants exposed to soil water deficit, soil salinity, ozone or low nitrogen generally reduce their leaf or root growth rate. The role of cellular processes underlying developmental plasticity in response to environmental stress has been investigated in many species. Both cell expansion and cell division are affected by environmental stresses, and even if the 'cellular theory' is under debate in the literature, the sum of their effects is reflected in the effects on organ size. Furthermore, environmental stresses also affect cell cycle regulation by altering the endoreduplication process and changing cell ploidy.

In planta, the analyses of the changes in the rate and duration of cell division in response to environmental conditions require accurate quantification of cell cycle parameters. Two types of techniques, cytological and kinematic, are used for this purpose and these have been extensively described and discussed in the literature (Baskin, 2000; Tardieu and Granier, 2000). In the cytological approaches, cells are labelled in a particular stage of the cell cycle and their fate is followed over time. Cells may be pulse-labelled, and the extent of labelling is quantified in each cell cycle phase over time or cells may be labelled continuously to determine how fast they accumulate in a specific phase of the cell cycle. These approaches are known to suffer from different drawbacks, described in Baskin (2000), and they have sometimes led to contradictory conclusions, casting doubt on the appropriateness of such studies. In the second type of approach, temporal or spatial changes in cell number, relative tissue expansion rate and cell size are followed in different zones of an organ, allowing calculation of cell division rate with a spatial resolution (Ben Haj Salah and Tardieu, 1995; Sacks *et al.*, 1997; Beemster and Baskin, 1998; Granier and Tardieu, 1998a). Compared with chemical methods, these methods have the benefit of being less intrusive and to allow a detailed spatial analysis. They can therefore be used for analyzing the effects of environmental conditions on the spatial and temporal patterns of cell division rate in plant organs.

This chapter reviews what is known about the circumstances in which cell cycle (cell division and endoreduplication) is affected by environmental stresses in different plant organs, how this plasticity at the cellular level mirrors that of organ plasticity and what are the molecular bases of these changes. Environmental stresses considered here will include temperature, water deficit, light, salt, oxidative, mineral, N and CO_2. Transduction pathways of systemic stress signals, for example, hormones biosynthesis, will not be discussed here.

15.2 Environmental stresses affect spatial and temporal patterns of cell division rate in plant organs

15.2.1 Spatial and temporal patterns of cell division rate in plant organs are a useful framework for analyzing the effects of environmental stresses on cell division

In roots and monocotyledonous leaves, the zone of cell division is restricted to the root tip and the leaf base, respectively. When the organ is growing steadily, the spatial distribution of cell division rate in this zone is stable under constant environmental conditions (Ben Haj Salah and Tardieu, 1995; Muller *et al.*, 1998; Baskin, 2000; Muller *et al.*, 2001). The expansion zone overlaps with that of cell division. The length of the expansion zone and the spatial distribution of the relative expansion rate in this zone are also stable with time under stable environmental conditions (when the relevant variables for the quantification of the processes are used as described

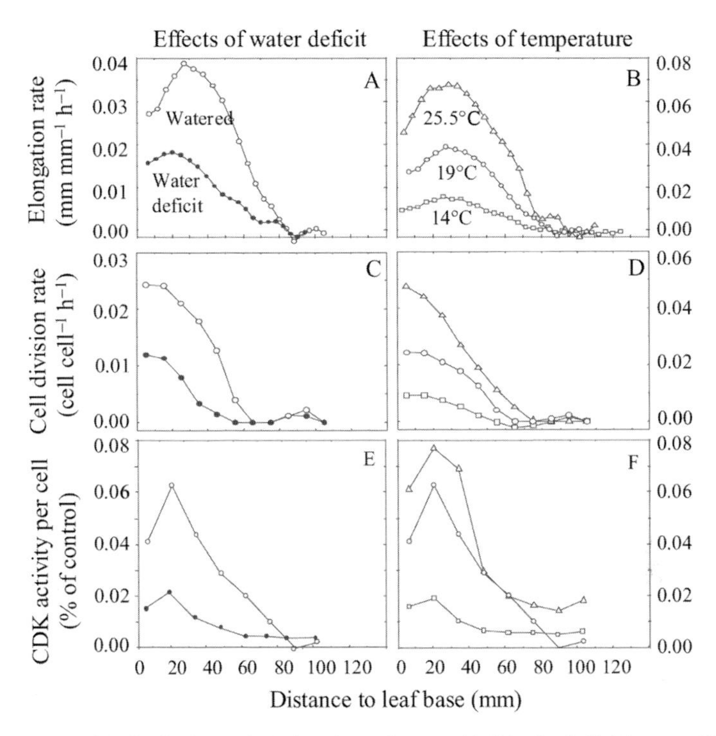

Figure 15.1 Spatial distributions of relative elongation rate (A, B) of cell division rate (C, D) and of CDK activity (E, F) in the sixth maize leaf of plants subjected to different soil water contents or temperature. (○), watered plants at 19°C; (•), plants in water deficit at 19°C; (□), watered plants at 14°C; (Δ), watered plants at 25.5°C. (Figure is redrawn from data in Granier *et al.*, 2000a.)

by Tardieu *et al.*, 2000). The spatial distribution of both cell division and tissue expansion in roots and monocot leaves have been extensively used to study the effects of several environmental stresses on each of these processes independently (Figure 15.1; Granier *et al.*, 2000a; Tardieu *et al.*, 2000).

In dicotyledonous leaves, cell division occurs throughout the leaf at the beginning of their development but as the leaf expands cell division is progressively restricted to the leaf base (Granier and Tardieu, 1998a). As a result, the area of the zone of cell division constantly changes during leaf development. In contrast to that of root and monocotyledonous leaves, cell division is not restricted to a specific zone of the organ but rather to a specific period of development. During a first phase of leaf development, both the cell division rate and the relative leaf expansion rate are maximal, whereas the absolute leaf expansion rate is low. Later on in leaf development, the rate of cell division declines whereas absolute leaf expansion rate increases to a maximum. In the final stage, cell division stops and expansion declines until it ceases (Granier and Tardieu, 1998a). These temporal patterns of cell division and expansion have been used to identify the effect of environmental stresses on cell division alone, cell division with expansion and expansion alone (Lecoeur *et al.*, 1995; Granier and Tardieu, 1999a,b).

15.2.2 Effects of water deficit

15.2.2.1 In roots and monocotyledonous leaves

Spatial distribution of the cell division rate and its response to water deficit have been assessed by kinematic analyses in maize roots (Sacks *et al.*, 1997) and leaves (Figure 15.1; Granier *et al.*, 2000a; Tardieu *et al.*, 2000). Figure 15.1C shows that the cell division rate is reduced by water deficit treatment in all parts of the cell division zone, with a slight reduction in size of this zone. Cytological approaches confirm a shortening of the zone of cell division in sunflower roots (Robertson *et al.*, 1990a), as well as in desert cactaceae roots (Dubrovsky *et al.*, 1998). A sudden widespread decrease in the rate at which the new cells are being produced is suggested by the rapid decrease in mitotic activity that occurs after imposition of water stress to roots (Yee and Rost, 1982; Robertson *et al.*, 1990b; Bitonti *et al.*, 1991; Bracale *et al.*, 1997). A similar response was found in soybean hypocotyls (Edelman and Loy, 1987) and wheat seedlings (Schuppler *et al.*, 1998).

15.2.2.2 In dicotyledonous leaves

A reduction in cell number by soil water deficit has been reported in leaves of many dicotyledonous crop species (Yeggapan *et al.*, 1982, on sunflower; Randall and Sinclair, 1988, on *Phaseolus vulgaris*; Lecoeur *et al.*, 1995, on pea). The effect of water deficit on cell division can be analyzed in more detail by taking advantage of the temporal and spatial patterns of the cell division rate in leaves. This has been done using two approaches. First, periods of soil water deficit treatment with identical intensity and duration were imposed at different stages of leaf development for a given leaf and its subsequent development studied in terms of cell division and cell expansion (Granier and Tardieu, 1999a). Second, a unique period of soil water deficit was imposed when different leaves of a plant were at different stages of their development, and cell division and expansion were followed in all leaves simultaneously (Yeggapan *et al.*, 1982; Lecoeur *et al.*, 1995; Alves and Setter, 2004). The results from these two approaches converge towards similar conclusions: the sensitivity of cell division to water deficit depends on the stage at which the deficit is imposed and is higher during the early phase of leaf development, when the cell division rate is maximal. Accordingly, leaf cell number is more reduced by water deficit when the water deficit treatment is imposed early on in leaf development and there is a short period at the end of leaf development during which cell number is not affected by water deficit (Granier and Tardieu, 1999a).

15.2.2.3 Recovery from water stress

While much is known about stress effects on cell division resulting from stable and static water deficit treatments, much less is known about the ability of cell division rates to recover after rewatering. Recently, Alves and Setter (2004) showed in Cassava that leaves in an early stage of development (with high cell division rates) resumed cell division after rewatering to the extent that final cell number per leaf was the same as the controls. There is also some indication that cell cycle can recover following osmotic adjustment in meristems (Yee and Rost, 1982).

15.2.3 Salt stress, low nitrogen and low phosphorus produce similar effects to those due to water deficit

A smaller cell division zone under saline conditions has been found in the roots of cotton (Kurth *et al.*, 1986) and *Arabidopsis thaliana* (West *et al.*, 2004). In young *A. thaliana* seedlings, a decrease in the size of both shoot and root meristematic zones was observed in response to NaCl application (Burssens *et al.*, 2000). In the cases of low N (MacAdam *et al.*, 1989) or low P (Assuero *et al.*, 2004), a reduction in the leaf meristem size was also observed. Primary root elongation of *A. thaliana* was stopped in response to decreasing P concentrations and this was associated with a reduction in the cell division rate and the exhaustion of the primary root meristem (Sánchez-Calderón *et al.*, 2005).

15.2.4 Effects of light and CO_2

A reduction of incident light causes a reduction in final cell numbers of the leaves of dicotyledonous plants (Wilson, 1966, in xanthium; Dengler, 1980, in sunflower; Cookson and Granier, 2006, in *A. thaliana*). Similar reductions in cell number were found when incident light interception was reduced by 40% by either reducing the light intensity using neutral filters or covering equivalent amounts of the photosynthetic leaf area (Granier and Tardieu, 1999b, in sunflower). These two treatments had identical effects on the temporal pattern of cell division in the leaf as they both reduced the relative cell division rate without affecting the duration of cell division in the leaf. The reduction in absorbed light was accompanied by a reduction in photosynthesis and leaf sugar content, suggesting that the reduction of cell number by this treatment was due to a reduction in carbohydrate availability.

Similarly, the elongation rate decreased in the primary roots of shaded maize plants, and this was accompanied by (i) an early drop in soluble sugar levels in the meristem and (ii) an early reduction of the rate of cell production by the meristem (Muller *et al.*, 1998). This reduction was not due to a lengthening of the cell cycle but rather to a reduction of size of the meristem (Muller *et al.*, 1998).

Elevated CO_2 promotes growth of leaves in different species (Kinsman *et al.*, 1997 on *Dactylis glomerata*; Taylor *et al.*, 2003, on poplar). In different European populations of *Dactylis glomerata*, the increase in leaf and root expansion caused by CO_2 enrichment is associated with an increase in the proportion of cycling cells in the meristem (Kinsman *et al.*, 1997).

15.2.5 Effects of temperature

The cell division rate is strongly dependent on organ temperature. There are several examples in the literature showing that the duration of the whole cell cycle decreases (i.e. the cell division rate increases) as temperature increases. This was the subject of early studies using onion roots as model system (Lopez-Saez *et al.*, 1966; Cuadrado *et al.*, 1989) and more recent studies in *Dactylis glomerata* roots and shoots (Creber *et al.*, 1993; Kinsman *et al.*, 1996), in the maize leaf (Ben Haj Salah and Tardieu, 1995) and in the sunflower leaf (Granier and Tardieu, 1998b). In the maize leaf, the cell division rate was reduced by low temperature in each zone of the leaf, as shown in Figure 15.1D.

However, it has been shown in dicotyledonous leaves that despite changes in the cell division rate, the final cell number in an organ is unaffected in a wide range of temperature. In fact, cell division rate is increased by increasing temperature whereas the duration of cell division is decreased and both variables are affected in a compensatory fashion (Granier and Tardieu, 1998b). Such a total compensation occurred for a range of physiological temperatures and probably does not apply in the case of stressing temperature below or above a threshold, which certainly depends on the species.

In sunflower leaves, the reciprocals of the duration of the phase with cell division and cell division rates are positively related to leaf temperature by a common relationship in a range from 14°C to 28°C with an x-intercept of 5°C (Granier and Tardieu, 1998b). Linear relationships between rates of processes involved in plant development and temperature have been found in other plant species (Gallagher, 1979, on wheat; Ong, 1983, on pearl millet; Ben Haj Salah and Tardieu, 1995, on maize). The x-intercept of these relationships depends on species (e.g. 3°C *for A. thaliana*, Granier *et al.*, 2002; 5°C for sunflower, Granier and Tardieu, 1998b; and 10°C for maize, Ben Haj Salah and Tardieu, 1995). This property is the basis for using the 'thermal time' concept. This concept states that when a process is linearly related to temperature above a threshold, temperature can be integrated over time in order to generate a new Time × Temperature variable. Thermal time has been used for a long time for predicting crop maturity (Reaumur, 1735). The linear relationship between temperature and the cell division rate can thus be integrated over time so cell division rates or cell numbers become temperature-independent functions of thermal time (Granier and Tardieu, 1998b). It follows that the duration of the epidermal cell cycle, which decreased threefold from 14°C to 26°C in clock time, was 22°Cd (degree-days) regardless of leaf temperature if expressed in thermal time (Granier and Tardieu, 1998b, in sunflower leaves). In maize leaves, the time for a cell to travel through the division zone and the elongation zone was 30°Cd, regardless of leaf temperature (Ben Haj Salah and Tardieu, 1995). In conclusion, the concept of thermal-time, as adapted to cell cycle studies, can be useful when the cell division rate or cell number are collected under conditions of fluctuating temperature such as in the greenhouse or in the field.

15.3 Coupling and uncoupling of cell division and tissue expansion in response to environmental conditions

15.3.1 Under several circumstances, cell division and tissue expansion are coupled

15.3.1.1 Coupling of cell division and tissue expansion in response to environmental stresses are often interpreted by a causal relationship

There are many examples in the literature demonstrating that when an organ is challenged by environmental stimuli, cell division and tissue expansion are affected

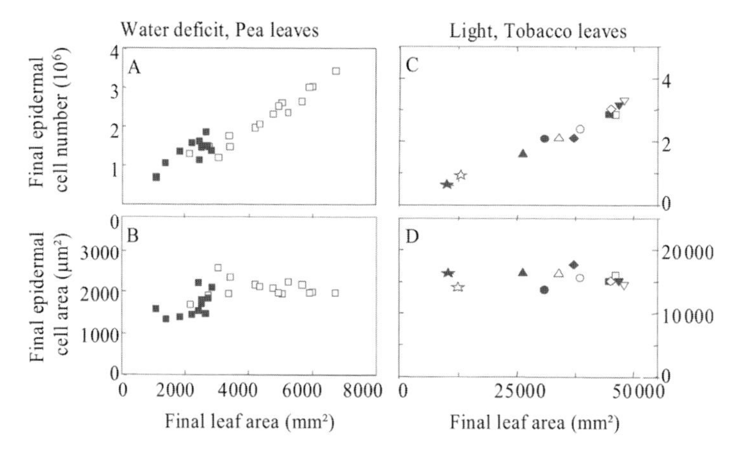

Figure 15.2 Relationship between final area of leaves in pea and final epidermal cell number per leaf (A) or final epidermal cell area (B). Plants subjected to a moderate water deficit in a greenhouse (■) are compared with well-irrigated plants grown under the same other micro-meteorological conditions (□). Relationship between final area of leaf 6 and final epidermal cell number (C) or final epidermal cell area (D) for tobacco plants grown under different light interception. Variability in light interception was obtained either by shading plants or by covering part of the photosynthetic leaf area (closed symbols) or by natural variability of incident light (open symbols). (From Granier *et al.*, 2000b.)

to the same extent, suggesting that both processes are coupled. This is true for environmental stresses decreasing organ size such as water deficit or reduction in incident light (see Figure 15.2) but also for environmental conditions favouring organ size such as high CO_2 concentrations (Masle, 2000). In those examples, there is a strong correlation between final cell number and organ size as shown in Figures 15.2A and 15.2C (see also Cookson *et al.*, 2005) and a lack of correlation between individual cell size and final organ size (see Figures 15.2B and 15.2D; Dale, 1992; Tardieu and Granier, 2000).

In the maize leaf, changes in temperature or changes in soil water status cause similar changes in both the cell division rate and the relative leaf expansion rate in all zones of the growing region (Figures 15.1A–15.1D; Granier *et al.*, 2000a; Tardieu *et al.*, 2000). In the sunflower leaf, both the cell division rate and the relative expansion rate are reduced by temperature to the same extent (Granier and Tardieu, 1998b). This is in agreement with reports of the effect of temperature on roots of sunflower and maize (Silk, 1992) and consistent with the observation that a range of temperature from 14°C to 30°C has no effect on the spatial distribution of cell length in the maize leaf in spite of a fivefold difference in the elongation rate (Ben Haj Salah and Tardieu, 1995). Similarly, when maize plants are subjected to low incident light, both cell division and tissue expansion are affected in the roots (Muller *et al.*, 1998).

The strong correlation between organ size and cell number, shown in Figure 15.2, has often been interpreted as though changes in cell division were responsible for changes in organ dimensions (Granier *et al.*, 2000b). The hypothesized role of cell division in driving tissue expansion and controlling final organ dimensions has been used to model organ growth as a function of cell production rate, this is called 'classical cell theory', in which the expansion of an organ results from the product of the number of cells present and the expansion rate of each individual cell (Arkebauer and Norman, 1995; Lecoeur *et al.*, 1996; Chavarria-Krauser and Schurr, 2004). This model was used to account for the effect of water deficit on pea leaves' expansion (Lecoeur *et al.*, 1996). In this model, 'early' water deficits imposed during the phase with cell division in the leaf cause a reduction in the cell division rate, as a consequence the cell number is reduced at the end of this phase and this in turn causes a reduction in the leaf expansion rate later on in leaf development. Re-watering during this later phase of leaf development causes the re-establishment of the rate of individual cell expansion, but as there are fewer cells present in the organ, the overall leaf expansion is reduced. This model is in agreement with different observations, indicating that when cell number is reduced, by early water stresses or early reductions in incident light, the subsequent leaf expansion rate is reduced even when favourable growth conditions are restored (Lecoeur *et al.*, 1995, on pea; Granier and Tardieu, 1999a,b, on sunflower). The robustness of this model was further tested using other treatments known to affect cell number (such as a reduction in light interception) in different plant species (Granier *et al.*, 2000b). In agreement with the initial hypothesis, reductions in cell numbers were systematically accompanied by similar reductions in the maximum absolute leaf expansion rate.

Results showing that altered cell division in transgenic plants cause changes in organ dimension support the hypothesis that changes in cell division are responsible for changes in organ size. For instance, the overexpression of CYC1A (now *CYCB1*) in *A. thaliana* leads to roots being bigger and more responsive to auxin (Doerner *et al.*, 1996). It has been proposed that such cyclins could mediate environmentally driven changes in root growth (Doerner *et al.*, 1996). The overexpression of *CKS1* in *A. thaliana* led to an overall reduced root extension which was associated with a reduction in the size of the meristem (De Veylder *et al.*, 2001). In tomato, the locus fw2.2 is responsible for fruit size and is linked to a negative repressor of cell division (Frary *et al.*, 2000).

15.3.1.2 An alternative hypothesis is proposed to account for the coupling of cell division and tissue expansion in response to environmental stresses

Despite the evidence described above supporting the coupling of cell division and expansion, the expansion of an organ must obey certain physical laws (Lockhart, 1965). These laws state that cell expansion depends on the balanced water flux into expanding cells and on the loosening of cell walls to facilitate this flux (Lockhart, 1965). These processes strongly depend upon the physical characteristics of the cells in question, such as cell wall mechanical properties (cell walls tend to stiffen, thereby limiting expansion during water deficit; Chazen and Neumann, 1994) and hydraulic gradients between the xylem and the expanding cells (Tang and Boyer, 2002). These physical laws imply that tissue expansion can be analyzed independently of cell division. On several occasions, the biophysicist Paul Green (e.g. Green, 1976) shed light on the driving role exerted by physical forces on growth and morphogenesis with little, if any, role for cell division. Furthermore, Granier and Tardieu (1999a) showed that the effect of water deficit on sunflower leaf expansion was essentially the same whether it was imposed before or after the completion of the final cell number.

To reconcile these results with those described in the previous section one could suggest that environmental stresses affect cell division and tissue expansion in an independent but very similar fashion, involving, for instance, common hormonal signalling pathways (Achard *et al.*, 2006). As a result of such co-responses, changes in cell numbers and organ size would be correlated and cell size would remain largely unaffected (as commonly observed). To validate such an interpretation, a simple model based on this hypothesis was proposed (Granier and Tardieu, 1999a). In this model, tissue expansion did not depend on cell division. Both were affected independently but similarly by environmental stresses (Granier and Tardieu, 1999a,b). With the assumption that abiotic stresses (either water deficit or reduction in light interception) reduced uniformly relative expansion rate and relative cell division rate during the period of stress, without affecting durations of either process, it was possible to simulate the kinetic changes in leaf area and cell number per leaf of sunflower plants subjected to different drought periods (Granier and Tardieu, 1999a). This model also accounted for the permanent reduction in the absolute leaf expansion rate even when stress is imposed solely during the first phase of leaf development (Granier and Tardieu, 1999a).

A series of results showing that altered cell division by chemicals or in transgenic plants does not cause changes in organ shape and dimension support the hypothesis that changes in cell division are not responsible for changes in organ size (Haber and Foard, 1963; Hemerly *et al.*, 1995).

15.3.2 Uncoupling of cell division and tissue expansion is revealed by the analysis of cell size in response to environmental stresses

The detailed examination of the profile of cell length in a meristem can be used to determine whether cell division and tissue expansion are affected by similar factors when plants are exposed to environmental stresses. Local cell size differences result from the balance between cell division and local tissue expansion (Green, 1976, Ivanov *et al.*, 2002). Thus, if cell division decreases in a meristem and tissue expansion continues (at least locally), cell length will increase. Such an increase in cell length in the meristem of organs exposed to stresses has been observed in maize leaves exposed to water deficit (Tardieu *et al.*, 2000), in *A. thaliana* leaves exposed to low incident light (Cookson and Granier, 2006) or in maize leaves exposed to low P (Assuero *et al.*, 2004) or low N (MacAdam *et al.*, 1989). It has also been observed in maize roots exposed to water deficit (Sacks *et al.*, 1997), in pea roots when grown in a soil with high mechanical impedance (Croser *et al.*, 1999) or when *A. thaliana* roots were exposed to high salt concentrations (West *et al.*, 2004). In many of these cases, both cell division rate and the extent of the meristematic zone were reduced by stresses and when measured, these reductions were associated with the maintenance of tissue expansion (e.g. Sacks *et al.*, 1997). These results suggest that in these cases cell division was relatively more affected than tissue expansion and that both processes can be uncoupled to some extent.

15.3.3 Cessation of cell division could be a cause of cessation of elongation in roots in response to environmental stimuli

There are converging reports in the literature suggesting that in roots alterations of cell division are the primary causes of cessation of root elongation under some circumstances. Cessation of elongation in (mostly lateral) roots is a widely reported phenomenon observed in plants subjected to a deficit of carbon supply in annual monocotyledonous (Varney and McCully, 1991; Muller *et al.*, 1998), dicotyledonous (Aguirrezabal and Tardieu, 1996) as well as in perennial plants (Pagès, 1995). There is also accumulating evidence that cessation of elongation can be triggered by environmental stimuli such as drought (Dubrovsky and Gómez-Lomelí, 2003), low P (Sanchez-Calderon *et al.*, 2005), high N/C (Malamy and Ryan, 2001), thereby allowing plasticity of the root architecture in response to environmental stimuli. Work published to date suggest that growth cessation is associated with meristematic cells being longer than in normal elongating roots (e.g. Dubrovsky and Gómez-Lomelí, 2003; Sanchez-Calderon *et al.*, 2005). A straightforward interpretation is that cell production by the meristem was strongly decreased (or even abolished) by a developmental and/or environmental cue while tissue expansion continued for some time, thus resulting in larger cells than in normal meristems. In these cases, cell division would be the primary target of these cues and cessation of organ elongation would be due to cessation of cell production by the meristem. Similar responses have been observed under low P when the primary root shifts from an indeterminate to a determinate growth state, which is associated with increases in cell length near the apex (Sanchez-Calderon *et al.*, 2005). In a desert cactus exposed to severe water stress, roots terminated their growth earlier and the rate of growth was significantly decreased as a result of decreased cell production (Dubrovsky and Gómez-Lomelí, 2003). In maize, a large proportion of lateral roots rapidly ceased elongating and the proportion of which was increased in plants with poor C status due to shading (Muller *et al.*, 1998). Interestingly, this cessation of elongation was preceded by a shortening of the meristem, a disappearance of mitoses at the root tip and exaggerated elongation of apical cells (Muller *et al.*, unpublished data).

15.4 Environmental stresses cause a blockage at the G1–S and G2–M transitions

To determine whether different environmental conditions affect specific cell cycle phase transitions, the relative abundance of cells in the different phases of cell cycle have been measured. If progress through the G1/S transition was preferentially slowed, it would be expected to cause an increase in the proportion of cells in the G1 phase. Conversely, preferential mitotic delay would increase the G2 cell population.

The discovery of cell cycle checkpoints stemmed from work on plant cells in the 1960s by Van't Hof, who showed that when cultured pea root tips were deprived of carbohydrate, meristematic cells stopped dividing and arrested in either G1 or G2. Recently, the analysis of the effects of high temperature or low H_2O_2 concentration on cell cycle progression in tobacco BY-2 synchronous cells confirmed these results (Jang *et al.*, 2005) and revealed that the arrest either at G1–S or at G2–M depends on the cell cycle stage at which the stress was applied. Genes and proteins involved in the regulation of these transitions in response to environmental stresses were identified in the late 1990s.

15.4.1 The plant cell cycle can be regulated at multiple points but it appears that major controls operate at the G1–S and G2–M transitions in response to environmental stresses

15.4.1.1 The G1–S checkpoint

Arrest at the G1 phase is a component of leaf development and occurs as cells cease proliferation and switch to differentiation. In dicotyledonous leaves, the proportion of cells which arrest at the G1 phase increases over time and the eldest cells are arrested first. This results in a tip-to-base gradient of the proportion of cells in G1 which have arrested their cycle (Granier and Tardieu, 1998a). When sunflower plants are subjected to soil water deficit, the proportion of nuclei in G1 is increased in all leaf zones, suggesting an increase in the duration of this phase and/or a blockage at the G1/S transition. In contrast, the duration of the S–G2–M phase is not affected (Granier and Tardieu, 1999a). In monocotyledonous leaves where there is a spatial gradient of cell cycle arrest, the proportion of G1-phase cells increases from 50% at the base to 100% in the cell-differentiation zone (Schuppler *et al.*, 1998). In wheat leaves, a control acting to prevent progress from G1 to S phase was detected under water stress in cells at the distal margin of the meristem (Schuppler *et al.*, 1998). In this zone, the proportion of cells in the G1 phase increased from 62% to 90% after 48 h of drought stress (Schuppler *et al.*, 1998). A lengthening of the G1 phase in plant organs has also been reported for other environmental stresses such as oxidative stress (Reichheld *et al.*, 1999), metal toxicity (Powell *et al.*, 1986), a reduction in incident or absorbed light (Granier and Tardieu, 1999b) and sucrose starvation (Van't Hof, 1973). Together, these results suggest that there is an important checkpoint in the regulation of the cell cycle plasticity in response to environmental conditions at the G1–S transition. This suggestion is in agreement with the observed decrease in mitotic index observed in response to drought (Heckenberger *et al.*, 1998; Schuppler *et al.*, 1998). The important point is that in most of the examples cited, once cells pass the gate of G1–S phase, they are then capable of traversing S, G2 and M phases at the same rate as cells grown in the absence of environmental stresses (Francis, 1992).

15.4.1.2 The G2–M checkpoint

Another checkpoint at the G2–M transition is often reported. Exogenous ABA application lengthens the phase of G2 relative to G1 in maize roots (Müller *et al.*, 1994) but not in pea roots (Bracale *et al.*, 1997). The latter study included a water-stress treatment that did lengthen the G2 phase relative to G1, but the stress was very severe (0.5 M mannitol). In wheat leaves of plants subjected to a mild water deficit the proportion of cells in the different phases of cell cycle was not affected in the first 3 mm of the meristem, although the mitotic activity was reduced (Schuppler *et al.*, 1998). This suggests that the durations of both G1 and G2 phases were increased in the basal part of the meristem. The potential for arrest at a G2 control point is also supported by the work of Reichheld *et al.* (1999) on oxidative stress.

15.4.2 Evidence that non-stressing temperatures lengthen cell cycle without necessarily blocking it at specific checkpoints

As mentioned in Section 15.2.5, cell cycle duration decreases with increasing temperature in several species. The cell doubling time in root meristems of maize decreases 21-fold as the temperature increases from 3°C to 25°C, reflecting a steep increase in the cell division rate (Francis and Barlow, 1998). The relationship between the cell division rate and leaf temperature is linear in the range of 14–28°C in sunflower leaves (Granier and Tardieu, 1998a). In maize leaves, the mitotic index is unaffected by changes in temperature from 18°C to 26°C, although the cell division rate increases twofold in this range (Ben Haj Salah and Tardieu, 1995). In onion roots, the duration of the whole cell cycle decreases as temperature increases (Lopez-Saez *et al.*, 1966; Cuadrado *et al.*, 1989). However, the proportion of cells in each phase of the cell cycle remains unchanged from 10°C to 30°C (Gonzales-Fernandez *et al.*, 1971) and the mitotic index remains constant (Lopez-Saez *et al.*, 1966). Similar results have been observed in barley leaves (Harrison *et al.*, 1998). This suggests that increasing temperature shortens the cell cycle duration without affecting the relative durations of each phase (Tardieu and Granier, 2000). Whether all phases of the cell cycle alter proportionately with temperature has been ascertained by comparing data from the root meristem of five species: *Pisum sativum*, sunflower, *Tradescantia paludosa*, *Allium cepa* and *Triticum aestivum*. In most cases, temperature did not affect the proportion of cells in each phase. However, in three of the five species there was a disproportionate lengthening of the G1 phase at low temperatures (Francis and Barlow, 1998).

15.4.3 Environmental stresses affect the CDK activity

Cyclin-dependent kinase (CDK) activity is involved in the progression of cell cycle and its alteration by environmental stresses in plants. CDK activity and final cell number are both decreased in the leaves of wheat plants under water deficit (Schuppler *et al.*, 1998). Direct evidence that CDK activity is quantitatively related to the cell division rate and its plasticity in response to environmental stresses is shown in Figures 15.1E and 15.1F in maize leaves subjected to a range of growing temperatures and soil water contents (Granier *et al.*, 2000a). In wheat leaves, CDK activity in the 3-to 6-mm segment of water-stressed plants declines by nearly 50% within 3 h (Schuppler *et al.*, 1998). This early decline indicates that the activation of the enzyme is directly affected by stress.

15.4.4 Controls of the G1–S and G2–M transitions in response to environmental stresses depend on the activation state of the CDK

While cell division does not occur without expression of *CDK*, it is reported that the plasticity of cell division in response to environmental stresses is under more subtle control than the amount of CDK proteins. The spatial distribution of CDK activity and its reduction by water deficit in maize and wheat leaves does not correlate with the amount of CDKA (Schuppler *et al.*, 1998, Granier *et al.*, 2000a). Similarly, changes in the relative amount of CDKA caused by temperature in maize leaves do not reflect that of CDK activity or cell division rate (Granier *et al.*, 2000a). Furthermore, the relative amount of CDKA and CDKB proteins is not affected by salt stress in *A. thaliana* roots (West *et al.*, 2004). The level of CDK protein is not limiting their activity and studies using transgenic lines overexpressing different cell cycle genes demonstrate that CDK activity is mainly regulated on a post-translational level (Doerner *et al.*, 1996; Cockcroft *et al.*, 2000). However, candidates regulating the CDK activity are numerous and further work is required to understand more about the molecular control of these transitions in response to environmental stresses.

The expression of cyclins could be a limiting factor for the mitotic activity of the CDK. In Arabidopsis, severe salt stress conditions transiently reduced the *CYCA2;1* and *CYCB1;1* expression (Burssens *et al.*, 2000), and a pivotal role of *CYCB1;2* promoter activity in the activation of CDKs under salt stress conditions has been reported (West *et al.*, 2004). However, CDK activity is not regulated solely by cyclin association but also by post-translational modification of the CDK subunit. For example, the activation of CDK requires the phosphorylation of Thr and Tyr residues and the de-phosphorylation of a Tyr15 residue. In wheat leaves subjected to water deficit, inactivation of the CDK by stress within 3 h was associated with an increase in the proportion of Tyr-phosphorylated protein (Schuppler *et al.*, 1998).

The activation of CDK may also depend on the association of other regulatory proteins. For example, CDK could be inactivated by inhibitors such as *ICK1/KRP1* (Wang *et al.*, 1997) or *ICK2/KRP2* which reduce both kinase activity and cell division rate in Arabidopsis leaves (De Veylder *et al.*, 2001). To our knowledge, the effect of environmental stresses on ICK gene expression has not been studied. However, *ICK1/KRP1* was induced by ABA, and along with *ICK1/KRP1* induction, there was a decrease in CDK activity (Wang *et al.*, 1998), suggesting a molecular mechanism by which plant cell division might be inhibited by environmental stresses interacting with ABA.

15.5 Endoreduplication and abiotic stresses

Endoreduplication occurs when a cell undergoes a round of DNA duplication without cell division, thus resulting in a doubling of the quantity of DNA. This process is widespread in plants, the degree to which different plant species show somatic endoploidy has been related to taxonomic position, genome size and life cycle (Barow and Meister, 2003). The study of the effects of abiotic stresses on endoreduplication in plants has primarily been focused on *A. thaliana* and agriculturally important tissue types of crops such as the maize endosperm, potato tubers and tomato fruit. Endoreduplication is generally reduced in response to abiotic stress; however, the magnitude of this response is less than that of cell division (Artlip *et al.*, 1995; Setter and Flannigan, 2001; Cookson *et al.*, 2006). Like the effects of abiotic stresses on cell division, the timing of the application of the abiotic stress also influences its effect (Artlip *et al.*, 1995). Abiotic stresses affect endoreduplication only if they are applied before the commitment to endoreduplication is made; once the process is underway it is remarkably resilient.

15.5.1 *Effects of water deficit*

A detailed study of the effect of water deficit on cell cycle of the maize endosperm was done by Setter and Flannigan (2001). Water deficit treatments were established 7 days after pollination, and 2 days later the extent of endoreduplication was shown to increase under the water deficit treatment in comparison to the control conditions; however, 13 days after pollination it was reduced by the water deficit treatment. This initial increase in endoreduplication was due to the earlier inhibition of mitosis, thus allowing endoreduplication to begin. Similarly, an increase in the degree of water deficit also progressively reduced the extent of endoreduplication in fully expanded leaves of *A. thaliana* (Table 15.1, Cookson *et al.*, 2006).

15.5.2 *Effects of light and elevated CO_2*

It is well known that when seedlings are germinated in the dark, the hypocotyl is considerably longer (approximately four times longer) than that of plants grown under light conditions. This increase in hypocotyl length is not associated with a change in cell number but with an increase in DNA content (Giles and Myers, 1964). Studies of *Pisum sativum* showed that the increase in DNA content was caused by an increase in the extent of endoreduplication (Van Oostveldt and Van Parijs,

Table 15.1 Effects of shading and different intensities of water deficit on the extent of endoreduplication in *Arabidopsis thaliana* leaves (data issued from Cookson *et al.*, 2006)

Light (mol m^{-2} day^{-1})	SWC (g H_2O g^{-1} dry soil)	% 2C	% 4C	% 8C	% 16C	% 32C	EF
12.0	0.40	20.1	25.8	34.6	17.3	2.2	155.7
3.0	0.40	31.3	25.2	28.3	13.6	1.6	129.0
12.0	0.28	20.5	29.3	35.8	13.2	1.2	145.3
12.0	0.18	26.2	28.8	35.3	9.1	0.6	129.1

Note: For each environmental condition, the percentage of cells with nuclei of 2C, 4C, 8C, 16C and 32C is given and the endoreduplication factor (EF, number of endocycles per 100 cells) is given. Shading was imposed with neutral filters reducing incident light from 12 to 3 mol m^{-2} day^{-1}. Both in the optimal light treatment and shading, plants were grown at a soil water content of 0.40 g H_2O g^{-1} dry soil. Two soil water deficit treatments were imposed in which plants were grown at soil water contents of 0.28 and 0.18 g H_2O g^{-1} dry soil. The soil water deficit treatments were imposed by the automated platform described in Granier *et al.* (2006).

1975). A detailed analysis of this phenomenon in *A. thaliana* revealed that the third endocycle is inhibited by light by the red/far red light photoreceptor phytochrome (Gendreau *et al.,* 1998). The additional endocycle presumably increases the growth potential of plants by facilitating the escape of the seedling from the soil. Ultraviolet light has also been shown to increase the ploidy level of the hypocotyl of tomato plants (Cavallini *et al.*, 2001). The effect of light on endoreduplication in the hypocotyl could be considered as a special case of a specialized structure as generally shading (reductions in light intensity) are associated with reductions in the extent of endoreduplication in *A. thaliana* leaves (see Table 15.1 modified from data in Cookson *et al.*, 2006) and potato tubers (Chen and Setter, 2003). Furthermore, darkness has also been shown to reduce the extent of endoreduplication in the leaf blades of *A. thaliana* in comparison with those of light-treated plants (Kozuka *et al.,* 2005). To our knowledge, only one study has been carried out to determine the effect of elevated carbon dioxide concentrations on endoreduplication in plants and it was not shown to have an effect (Chen and Setter, 2003).

15.5.3 Effects of temperature

The effects of sub-optimal temperature on endoreduplication in plants have been studied in maize endosperm, tomato fruit and soybean roots (Engelen-Eigles *et al.*, 2000; 2001; Stepinski, 2003; Bertin, 2005). Generally, the extent of endoreduplication is reduced by stress temperatures above or below the optimal growing temperature of a given species. For example, the extent of endoreduplication decreases in soybean roots in response to temperatures of 10°C (Stepinski, 2003) and in tomato fruits grown at 20°C in comparison with those grown at their optimum temperature of 25°C (Bertin, 2005). In addition, high temperature (35°C) reduces the extent of endoreduplication in maize endosperm (Engelen-Eigles *et al.*, 2000, 2001).

15.5.4 The role of endoreduplication in adaptation to abiotic stresses

The exact biological significance of endoreduplication remains enigmatic and even it is clear that this process is reproducibly affected by environmental stresses (Cookson *et al.*, 2006), its role in the adaptation of plants to stressing conditions is not established. There is some evidence to suggest that the DNA content of a cell is related to its size (e.g. Castellano *et al.*, 2004; Vlieghe *et al.*, 2005). In this context, modification of the extent of endoreduplication in response to abiotc stresses could modulate cell size. However, the correlation between cell size and the level of endoreduplication is not always observed (Beemster *et al.*, 2002; Gendreau *et al.*, 1998). For example, stressing environmental conditions such as reduction in incident light and water deficit increasing epidermal cell size in *A. thaliana* cause a decrease in the extent of endoreduplication (Cookson *et al.*, 2006).

A study of British flora revealed that evolution of increased genome size was associated with increased capacity for growth at low temperatures (Grime and Mowforth, 1982). The apparent advantage of increased DNA content under conditions of low temperature could be associated with increased protein synthesis to some extent aiding the conservation of normal metabolic function when energy supplies are not limiting. However, when submitted to a shading treatment, plants with an increased extent of endoreduplication in their leaves were not able to maintain their leaf development at the same level as wild-type plants or plants with reduced extent of endoreduplication (Cookson *et al.*, 2006): under such treatments an increased DNA content was of disadvantage and may be more of a waste of energy resources than an increased protein synthesis benefit.

15.6 Conclusion

In response to different environmental conditions (temperature, water deficit, light, nitrogen, phosphorus and salt), plants modify their cell cycle in different organs. These modifications result in changes in cell division rate, in the proportion of cells engaged in the cell cycle, in the duration of the period of cell division and also in the quantity and activity of cell cycle controllers at the biochemical and molecular levels. The link between the upstream regulators of the cell cycle machinery and environmental stresses has only been established for a small number of environmental stresses (Figure 15.3) and will require further work to draw a more complete picture how this regulation functions.

What could be the advantage for a plant to delay its cell cycle progression in response to a stressing environmental conditions? A few ideas have been suggested to answer this question but without clear evidence. For example, blockage at the G1 or G2 checkpoint in response to oxidative stress has been discussed in terms of 'time to repair damaged cellular components and to develop anti-oxidant

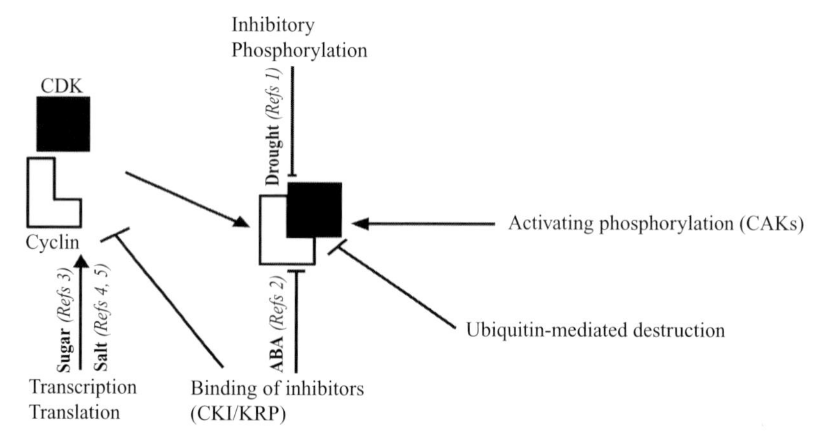

Figure 15.3 Levels of regulation of CDK complexes and evidence in the literature of their alterations by environmental stresses. Arrows indicate processes activating the CDK complexes whereas lines ended by a trait indicate inhibitory processes. Environmental stresses known to interact with those processes are indicated on the lines and the corresponding references are shown in italics and given hereafter. (Ref 1: Schuppler *et al.*, 1998, Ref. 2: Wang *et al.*, 1998, Ref. 3: Riou-Khamlichi *et al.*, 2000, Ref. 4: Burssens *et al.*, 2000, Ref. 5: West *et al.*, 2004.)

defence mechanisms' (Jang *et al.*, 2005; Barnouin *et al.*, 2002). It has also often been suggested that blockage in G1 or G2 phase under different stresses could help in re-entering the cell cycle rapidly after return of favourable conditions (Van't Hof, 1974, 1985; Bergounioux *et al.*, 1988; Francis 1992). If control of cell division and of tissue expansion are considered as partly independent processes, reduction in the cell division rate may also be seen as a mechanism to maintain cell volume. If tissue expansion decreases, cells would have a small non-optimal volume without concomitant reduction in cell division. The ability of transgenic plants altered in cell cycle progression to face environmental stresses has not been examined to our knowledge and will help to answer those questions.

References

Achard, P., Cheng, H., De Grauwe, L., *et al.* (2006) Integration of plant responses to environmentally activated phytohormonal signals. *Science* **311**, 91–94.

Aguirrezabal, L.A.N. and Tardieu, F. (1996) An architectural analysis of the elongation of field-grown sunflower root system. Elements for modelling the effects of temperature and intercepted radiation. *J Exp Bot* **47**, 411–420.

Alves, A.A.C. and Setter, T.L. (2004) The response of cassava leaf area expansion to water deficit. Cell proliferation, cell expansion, and delayed development. *Ann Bot* **94**, 605–613.

Arkebauer, T.J. and Norman, J.M. (1995) From cell growth to leaf growth. I. Coupling cell divison and cell expansion. *Agron J* **87**, 99–105.

Artlip, T.S., Madison, J.T. and Setter, T.L. (1995) Water deficit in developing endosperm of maize: cell division and nuclear DNA endoreduplication. *Plant Cell Environ* **18**, 1034–1040.

Assuero, S.G., Mollier, A. and Pellerin, S. (2004) The decrease in growth of phosphorus deficient maize leaves is related to a lower cell production. *Plant Cell Environ* **27**, 887–895.

Barnouin, K., Dubuisson, M.L., Child, E.S., *et al.* (2002) H_2O_2 induces a transient multi-phase cell cycle arrest in mouse fibroblasts through modulating Cyclin D and $p21^{Cip1}$ expression. *J Biol Chem* **277**, 13761–13770.

Barow, M. and Meister, A. (2003) Endopolyploidy in seed plants is differently correlated to systematics, organ, life strategy and genome size. *Plant Cell Environ* **26**, 571–584.

Baskin, T.I. (2000) On the constancy of cell division rate in root meristem. *Plant Mol Biol* **43**, 545–554.

Beemster, G.T.S. and Baskin, T.I. (1998) Analysis of cell division and elongation underlying the developmental acceleration of root growth in *Arabidopsis thaliana*. *Plant Physiol* **116**, 1515–1526.

Beemster, G.T.S., De Vusser, K., De Tavernier, E., De Bock, K., and Inzé, D. (2002) Variation in growth rate between *Arabidopsis thaliana* ecotypes is correlated with cell division and A-type cyclin dependent kinase activity. *Plant Physiol* **129**, 854–864.

Ben Haj Salah, H. and Tardieu, F. (1995). Temperature affects expansion rate of maize leaves without change in spatial distribution of cell length. Analysis of the coordination between cell division and cell expansion. *Plant Physiol* **109**, 861–870.

Bergounioux, C., Perennes, C., Hemerly, A.S., *et al.* (1988) Relation between protoplast division, cell-cycle stage and nuclear chromatin structure. *Protoplasma* **142**, 127–136.

Bertin, N. (2005) Analysis of the tomato fruit growth response to temperature and plant fruit load in relation to cell division, cell expansion and DNA endoreduplication. *Ann Bot* **95**, 439–447.

Bitonti, M.B., Ferraro, F., Floris, C. and Innocenti A.M. (1991) Response of meristematic cells to osmotic stress in *Triticum durum*. *Biochem Physiol Pflanz* **187**, 453–457.

Bracale, M., Levi, M., Savini, C., Dicorato, W. and Galli, M.G. (1997) Water deficit in pea root tips: effects on the cell cycle and on the production of dehydrin-like proteins. *Ann Bot* **79**, 593–600.

Burssens, S., Himanen, K., Van de Cotte, B., *et al.* (2000) Expression of cell cycle regulatory genes and morphological alterations in response to salt stress in *Arabidopsis thaliana*. *Planta* **211**, 632–640.

Castellano, M.M., Boniotti, M.B., Caro, E., Schnittger, A. and Gutierrez, C. (2004) DNA replication licensing affects cell proliferation or endoreduplication in a cell type-specific manner. *Plant Cell* **16**, 2380–2393.

Cavallini, A., Natali, L., Tonfoni, R., Cionini, G. and Lercari, B. (2001) White and UV light effects on cell nuclei in the aurea genotype of *Lycopersicum esculentrum* L. *Cytobios* **104**, 83–98.

Chavarria-Krauser, A. and Schurr, U. (2004) A cellular growth model for root tips. *J Theor Biol* **230**, 21–32.

Chazen, O. and Neumann, P.M. (1994) Hydraulic signals from the roots and rapid cell wall hardening in growing maize (*Zea mays* L.) leaves are primary responses to polyethylene glycol-induced water deficits. *Plant Physiol* **104**, 1385–1392.

Chen, C.T. and Setter, T.L. (2003) Response of potato tuber cell division and growth to shade and elevated CO_2. *Ann Bot* **91**, 373–381.

Cockcroft, C.E., Den Boer, B.G.W., Healy, J.M.S. and Murray, J.A.H. (2000) Cyclin D control of growth rate in plants. *Nature* **405**, 575–579.

Cookson, S.J. and Granier, C. (2006) A dynamic analysis of the shade-induced plasticity in *Arabidopsis thaliana* rosette leaf development reveals new components of the shade-adaptative response. *Ann Bot* **97**, 443–452.

Cookson, S.J., Radziejwoski, A. and Granier, C. (2006) Cell and leaf size plasticity in Arabidopsis: what is the role of endoreduplication? *Plant Cell Environ* **29**, 1273–1283.

Cookson, S.J., Van Lijsebettens, M. and Granier, C. (2005) Correlation between leaf growth variables suggest intrinsic and early controls of leaf size in *Arabidopsis thaliana*. *Plant Cell Environ* **28**, 1355–1366.

Creber, H.M.C., Davies, M.S. and Francis, D. (1993) Effects of temperature on cell division in root meristems of natural populations of *Dactylis glomerata* of contrasting latitudinal origines. *Environ Exp Bot* **33**, 433–442.

Croser, C., Bengough, A.G. and Pritchard, J. (1999) The effect of mechanical impedance on root growth in pea (*Pisum sativum*). I. Rates of cell flux, mitosis and strain during recovery. *Physiol Plant* **107**, 277–286.

Cuadrado A., Navarrete, M.H. and Canovas, J.L. (1989). Cell size of proliferating plant cells increases with temperature: implications in the control of cell division. *Exp Cell Res* **185**, 277–282.

Dale, J.E. (1992) How do leaves grow? *Bioscience* **42**, 423–432.

De Veylder L., Beeckman, T., Beemster, G.T.S., *et al.* (2001) Functional analysis of cyclin-dependent kinase inhibitors of *Arabidopsis*. *Plant Cell* **13**, 1653–1667.

Dengler, N.G. (1980) Comparative histological basis of sun and shade leaf dimorphiism in *Helianthus annuus*. *Can J Bot* **58**, 717–730.

Doerner, P., Jorgensen, J.E., You, R., Steppuhn, J. and Lamb, C. (1996) Control of root growth and development by cyclin expression. *Nature* **380**, 520–523.

Dubrovsky, J.G. and Gómez-Lomelí, L.F. (2003) Water deficit accelerates determinate developmental program of the primary root and does not affect lateral root initiation in a Sonoran Desert cactus (Pachycereus pringlei, Cactaceae). *Am J Bot* **90**, 823–831.

Dubrovsky, J.G., North, G.B. and Nobel, P.S. (1998) Root growth, developmental changes in the apex and hydraulic conductivity for Opuntia ficus-indica during drought. *New Phytol* **138**, 75–82.

Edelman, L. and Loy, J.B. (1987) Regulation of cell division in the subapical shoot meristem of dwarf watermelon seedlings by gibberellic acid and polyethylene glycol 4000. *J Plant Growth Regul* **5**, 140–161.

Engelen-Eigles, G., Jones, R.J. and Phillips, R.L. (2000) DNA endoreduplication in maize endosperm cells: the effect of exposure to short-term high temperature. *Plant Cell Environ* **23**, 657–663.

Engelen-Eigles, G., Jones, R.J. and Phillips, R.L. (2001) DNA endoreduplication in maize endosperm cells is reduced by high temperature during the mitotic phase. *Crop Sci* **41**, 1114–1121.

Francis, D. (1992) The cell cycle in plant development. *New Phytol* **122**, 1–20.

Francis, D. and Barlow, P.W. (1998) Temperature and the cell cycle. Symp. *Soc Exp Biol* **42**, 181–201.
Frary, A., Nesbitt, T.C., Frary, A., *et al.* (2000) *fw2.2*: A quantitative trait locus key to the evolution of tomato fruit size. *Science* **289**, 85–88.
Gallagher, J.N. (1979) Field studies of cereal leaf growth. I. Initiation and expansion in relation to temperature and ontogeny. *J Exp Bot* **30**, 625–636.
Gendreau, E., Höfte, H., Grandjean, O., Brown, S. and Traas, J. (1998) Phytochrome controls the number of endoreduplication cycles in the *Arabidopsis thaliana* hypocotyl. *Plant J* **13**, 221–230.
Giles, K.W. and Myers, A. (1964) The role of nucleic acids in the growth of the hypocotyl of Lupinus albus under varying light and dark regimes. *Biochim Biophys Acta* **87**, 460–477.
Gonzalez-Fernandez, A., Gimenez-Martýn, G. and de la Torre, C. (1971) The duration of the interphase periods at different temperatures in root tip cells. *Cytobiologie* **3**, 367–371.
Granier, C., Aguirrezabal, L., Chenu, K., *et al.* (2006) PHENOPSIS, an automated platform for reproducible phenotyping of plant responses to soil water deficit in *Arabidopsis thaliana* permitted the identification of an accession with low sensitivity to soil water deficit. *New Phytol* **169**, 623–635.
Granier, C., Inzé, D. and Tardieu, F. (2000a) Spatial distribution of cell division rate can be deduced from that of $p34^{cdc2}$ kinase activity in maize leaves grown at contrasting temperatures and soil water conditions. *Plant Physiol* **124**, 1393–1412.
Granier, C., Massonnet, C., Turc, O., Muller, B., Chenu, K. and Tardieu, F. (2002) Individual leaf development in *Arabidopsis thaliana*: a stable thermal-time-based program. *Ann Bot* **89**, 595–604.
Granier, C. and Tardieu, F. (1998a) Spatial and temporal analyses of expansion and cell cyle in sunflower leaves. A common pattern of development for all zones of a leaf and different leaves of a plant. *Plant Physiol* **116**, 991–1001.
Granier, C. and Tardieu, F. (1998b) Is thermal time adequate for expressing the effects of temperature on sunflower leaf development? *Plant Cell Environ* **21**, 695–703.
Granier, C. and Tardieu, F. (1999a) Water deficit and spatial pattern of leaf development. Variability in responses can be simulated using a simple model of leaf development.*Plant Physiol* **119**, 609–620.
Granier, C. and Tardieu, F. (1999b) Leaf expansion and cell division are affected by reduced intercepted light but not after the decline in cell division rate in the sunflower leaf. *Plant Cell Environ* **22**, 1365–1376.
Granier, C., Turc, O. and Tardieu, F. (2000b) Co-ordination of cell division and tissue expansion in sunflower, tobacco and pea leaves. Dependence or independence of both processes? *J Plant Growth Regul* **19**, 45–54.
Green, P.B. (1976) Growth and cell pattern formation on an axis: critique of concepts, terminology, and mode of study. *Bot Gaz* **137**, 187–202.
Grime, J.P., Crick, J.C. and Rincón, E. (1986) The ecological significance of plasticity. In: *Plasticity in plants*, Symposia of the Society for Experimental Biology, (eds Jennings D.H. and Trewavas A.J.). Scarborough, UK.
Grime, J.P. and Mowforth, M.A. (1982) Variation in genome size – an ecological interpretation. *Nature* **299**, 151–153.
Haber, A.H. and Foard, D.E. (1963) Nonessentiality of concurrent cell divisions for degree of polarization of leaf growth. II. Evidence from untreated plants and from chemically induced changes of the degree of polarization. *Am J Bot* **50**, 937–944.
Harrison, J., Nicot, C. and Ougham, H. (1998) The effect of low temperature on patterns of cell division in developping second leaves of wild-type and slender mutant barley (*Hordeum vulgare* L.). *Plant Cell Environ* **21**, 79–86.
Heckenberger, U., Roggatz, U. and Schurr, U. (1998) Effect of drought stress on the cytological status in *Ricinus communis*. *J Exp Bot* **49**, 181–189.
Hemerly, A., de Almeida Engler, J., Bergounioux, C., *et al.* (1995) Dominant negative mutants of the Cdc2 kinase uncouple cell division from iterative plant development. *EMBO J* **14**, 3925–3936.
Ivanov, V.B., Dobrochaev, A.E. and Baskin, T.I. (2002) What the distribution of cell lengths in the root meristem does and does not reveal about cell division. *J Plant Growth Regul* **21**, 60–67.

Jang S.J., Shin, S.H., Yee, S.T., Hwang, B., Im, K.H. and Ky Park, Y. (2005) Effects of abiotic stresses on cell cycle progression in tobacco BY-2 cells. *Mol Cells* **20**, 136–141.

Kinsman, E.A., Lewis, C., Davies, M.S., *et al.* (1996). Effects of temperature and elevated CO2 on cell division in shoot meristems: Differential responses of two natural populations of *Dactylis glomerata* L. *Plant Cell Environ* **19**, 775–780.

Kinsman, E.A., Lewis, C., Davies, M.S., *et al.* (1997) Elevated CO2 stimulates cells to divide in grass meristems: a differential effect in two natural populations of *Dactylis glomerata*. *Plant Cell Environ* **20**, 1309–1316.

Kozuka, T., Horiguchi, G., Kim, G.-T., Ohgishi, M., Sakai, T. and Tsukaya, H. (2005) The different growth responses of Arabidopsis thaliana leaf blade and petiole during shade avoidance are regulated by photoreceptors and sugar. *Plant Cell Physiol* **46**, 213–223.

Kurth, E., Cramer, G.R., Läuchli, A. and Epstein, E (1986) Effects of NaCl and CaCl2 on cell enlargement and cell production in cotton roots. *Plant Physiol* **82**, 1102–1106.

Lecoeur, J., Wery, J. and Sinclair, T.S. (1996). Model of leaf area expansion in field pea subjected to soil water deficits. *Agron J* **88**, 467–472.

Lecoeur, J., Wery, J., Turc, O. and Tardieu, F. (1995) Expansion of pea leaves subjected to short water deficit : cell number and cell size are sensitive to stress at different periods of leaf development. *J Exp Bot* **46**, 1093–1101.

Lockhart, J.A. (1965). An analysis of irreversible plant cell elongation. *J Theor Biol* **8**, 264–275.

Lopez-Saez, J.F., Gimenez-Martin, G. and Gonzales-Fernandez, A. (1966) Duration of the cell division cycle and its dependence on temperature. *Z Zellforsch* **75**, 591–600.

MacAdam, J.W., Volenec, J.J. and Nelson, C.J. (1989) Effects of nitrogen on mesophyll cell division and epidermal cell elongation in tall fescue leaf blades. *Plant Physiol* **89**, 549–556.

Malamy, J.E. and Ryan, K.S. (2001) Environmental regulation of lateral root initiation in Arabidopsis. *Plant Physiol* **127**, 899–909.

Masle, J. (2000) The effects of elevated CO_2 concentrations on cell division rates, growth patterns, and blade anatomy in young wheat plants are modulated by factors related to leaf position, vernalization, and genotype. *Plant Physiol* **122**, 1399–1415.

Müller, M.L., Barlow, P.W. and Pilet, P.E. (1994) Effect of abscisic acid on the cell cycle in the growing maize root. *Planta* **195**, 10–16.

Muller, B., Reymond, M. and Tardieu, F. (2001) The elongation rate at the base of a maize leaf shows an invariant pattern during both the steady-state elongation and the establishment of the elongation zone. *J Exp Bot* **52**, 1259–1268.

Muller, B., Stosser, M. and Tardieu, F. (1998) Spatial distributions of tissue expansion and cell division rates are related to irradiance and to sugar content in the growing zone of maize roots. *Plant Cell Environ* **21**, 149–158.

Ong, C.K. (1983) Response to temperature in a stand of pearl millet. I. Vegetative development. *J Exp Bot* **34**, 322–336.

Pagès, L. (1995) Growth patterns of the lateral roots of young oak (*Quercus robur*) tree seedlings. Relationship with apical diameter. *New Phytol* **130**, 503–509.

Powell, M.J., Davies, M.S. and Francis, D. (1986) The influence of zinc on the cell cycle in the root meristem of a zinc-tolerant and non-tolerant cultivar of *Festuca rubra* L. *New Phytol* **102**, 419–428.

Randall, H.C. and Sinclair, T.R. (1988) Sensitivity of soybean leaf development to water deficits. *Plant Cell Environ* **11**, 835–839.

Reaumur, R.A. (1735). cited by Durand R (1969) Signification et portée des sommes de température. *Bull Tech Info* **238**, 185–190.

Reichheld, J.-P., Vernoux, T., Lardon, F., Van Montagu, M. and Inzé, D. (1999) Specific checkpoints regulate plant cell cycle progression in response to oxidative stress. *Plant J* **17**, 647–656.

Riou-Khamlichi, C., Menges, M., Healy, J.M.S and Murray, J.A.H. (2000) Sugar control of the plant cell cycle: differential regulation of Arabidopsis D-Type Cyclin gene expression. *Mol Cell Biol* **20**, 4513–4521.

Robertson, J.M., Hubick, K.T., Yeung, E.C. and Reid, D.M. (1990a) Developmental responses to drought and abscisic acid in sunflower roots. 1. Root growth, apical anatomy and osmotic adjustment. *J Exp Bot* **41**, 325–337.

Robertson, J.M., Yeung, E.C., Reid, D.M. and Hubick, K.T. (1990b) Developmental responses to drought and abscisic acid in sunflower roots. 2. Mitotic activity. *J Exp Bot* **41**, 339–350.

Sacks, M.M., Silk, W.K. and Burman, P. (1997) Effect of water stress on cortical cell division rates within the apical meristem of primary roots of maize. *Plant Physiol* **114**, 519–527.

Sanchez-Calderon, L., Lopez-Bucio, J., Chacon-Lopez, A., *et al.* (2005) Phosphate starvation induces a determinate developmental program in the roots of *Arabidopsis thaliana*. *Plant Cell Physiol* **46**, 174–184.

Schuppler, U., He, P.H., John, P.C.L. and Munns, R. (1998) Effect of water stress on cell division and cell-division-cycle 2-like cell cycle kinase activity in wheat leaves. *Plant Physiol* **117**, 667–678.

Setter, T.L. and Flannigan, B.A. (2001) Water deficit inhibits cell division and expression of transcripts involved in cell proliferation and endoreduplication in maize endosperm. *J Exp Bot* **52**, 1401–1408.

Silk, W.K. (1992) Steady form from changing cells. Int. *J Plant Sci* **153**, 49–58.

Stepinski, D. (2003) Effect of chilling on DNA endoreduplication in root cortex cells and root hairs of soybean seedlings. *Biol Plant* **47**, 333–339.

Tang, A.C. and Boyer, J.S. (2002) Growth-induced water potentials and the growth of maize leaves. *J Exp Bot* **53**, 489–503.

Tardieu, F. and Granier, C. (2000) Quantitative analysis of cell division in leaves : methods, developmental patterns and effects of environmental conditions. *Plant Mol Biol* **43**, 555–567.

Tardieu, F., Reymond, M., Hamard, H., Granier C. and Muller, B. (2000) Spatial distributions of expansion rate, cell division rate and cell size in maize leaves: a synthesis of the effects of soil water status, evaporative demand and temperature. *J Exp Bot* **51**, 1505–1514.

Taylor, G., Tricker, P.J., Zhang, F.Z., Alston, V.J., Miglietta, F. and Kuzminsky, E. (2003) Spatial and temporal effects of free-air CO_2 enrichment (POPFACE) on leaf growth, cell expansion, and cell production in a closed canopy of poplar. *Plant Physiol* **131**, 177–185.

Van Oostveldt, P. and Van Parijs, R. (1975) Effect of light on nucleic acid synthesis and polyploidy level in elongating epicotyl cells of *Pisum sativum*. *Planta* **124**, 287–295.

Van't Hof, J. (1973) The regulation of cell division in higher plants. *Brookhaven Symp* **25**, 152–165.

Van't Hof, J. (1974) Control of the cell cycle in higher plants. In: *Cell Cycle Control* (eds Padilla G.M., Cameron I.L. and Zimmerman A.). Academic Press, New York, pp. 77–85.

Van't Hof, J. (1985) Control points within the cell cycle. *Soc. Exp. Biol. Semin. Ser* **26**, 1–13.

Varney, G.T. and McCully, M.E. (1991) The branch roots of Zea. II. Developmental loss of the apical meristem in field-grown roots. *New Phytol* **118**, 535–546.

Vlieghe, K., Boudolf, V., Beemster, G.T.S., *et al.* (2005) The DP-E2F-like gene DEL1 controls the endocycle in *Arabidopsis thaliana*. *Curr Biol* **15**, 59–63.

Wang, H., Fowke, L.C. and Crosby, W.L. (1997) A plant cyclin-dependent kinase inhibitor gene. *Nature* **386**, 45–452.

Wang, H., Qi, Q.G., Schorr, P., Cutler, A.J., Crosby, W.L. and Fowke, L.C. (1998) ICK1, a cyclin-dependent protein kinase inhibitor from *Arabidopsis thaliana* interacts with both cdc2a and cycd3 and its expression is induced by abscisic acid. *Plant J* **15**, 501–510.

Wery, J. (2005) Differential effects of soil water deficit on the basic plant functions and their significance to analyse crop responses to water deficit in indeterminate plants. *Aust J Agric Res* **56**, 1201–1209.

West, G., Inzé, D. and Beemster, G. (2004) Cell cycle modulation in the response of the primary root of Arabidopsis to salt stress. *Plant Physiology* **135**, 1050–1058.

Wilson, G.L. (1966) Studies on the expansion of the leaf surface. V. Cell division and expansion in a developping leaf as influenced by light and upper leaves. *J Exp Bot* **17**, 440–451.

Yee, V.F. and Rost, T.L. (1982) Polyethylene glycol induced water stress in *Vicia faba* seedlings: cell division, DNA synthesis and a possible role for cotyledons. *Cytologia* **47**, 615–624.

Yegappan, T.M, Paton, D.M., Gates, C.T. and Muller, W.J. (1982) Water stress in sunflower (*Helianthus annuus L.*). II. Effects on leaf cells and leaf area. *Ann Bot* **49**, 63–68.

Index